SYMBOL	MEANING
F	F statistic
H_0	Null hypothesis
H_1	Research or alternative hypothesis
	Note: β_0, π_0, μ_0, etc. denote null hypothesis population parameter values that are closest to H_1.
i	Case number
IQR	Sample interquartile range
k	Subscript representing the category (*categorical variables*) or the independent variable (*regression analysis*)
K	*For categorical variables*: Number of categories
	In regression analysis: Number of a and b coefficients in equation
K_c	Number of categories for the column variable in a crosstabulation
K_r	Number of categories for the row variable in a crosstabulation
$\log_e(x)$	Base-e logarithm of x
$\log_{10}(x)$	Base-10 logarithm of x
μ	(*mu*) Population mean
μ_0	Value of μ specified by the null hypothesis
$\mu_1, \mu_2, \ldots$	Means in populations 1, 2, etc.
$\mu_{Y|X}$	*In regression equation*: Conditional population mean of Y given X
$\hat{\mu}_{Y|X_0}$	Estimated conditional mean of Y, given $X = X_0$
$\mu_{Y|X_1, X_2}$	*Multiple regression*: Conditional population mean of Y given X_1 and X_2
Md	Sample median
n	Number of cases in a sample
$n_1, n_2, \ldots$	Number of cases in samples 1, 2, etc.
N	Number of cases in a population
O	Observed frequency in a crosstabulation
O^*	Observed frequency in a crosstabulation, after adjustment by continuity correction for thin cells
$(O - E)^2/E$	Used in chi-square test
p	Population percentage
$\hat{p}$	Sample percentage
π	(*pi*) Population proportion
π_0	Value of π specified by the null hypothesis
$\hat{\pi}$	(*pi-hat*) Sample proportion
$P(A), P(B), \ldots$	Probabilities of events A, B, etc.

MODERN DATA ANALYSIS
A First Course in Applied Statistics

MODERN DATA ANALYSIS
A First Course in Applied Statistics

Lawrence C. Hamilton
University of New Hampshire

Brooks/Cole Publishing Company
Pacific Grove, California

Brooks/Cole Publishing Company
A Division of Wadsworth, Inc.

Printed in the United States of America

10 9 8 7 6 5 4 3 2

Library of Congress Cataloging in Publication Data

Hamilton, Lawrence C.
    Modern data analysis: a first course in applied statistics /
Lawrence C. Hamilton.
        p.   cm.
    Includes bibliographical references.
    ISBN 0-534-12846-7
    1. Statistics.   I. Title.
QA276.12.H355   1990
519.5—dc20   89-27479
                                                   CIP

Sponsoring Editor: John Kimmel
Marketing Representative: John Moroney
Editorial Assistant: Mary Ann Zuzow
Production Coordinator: Joan Marsh
Production Service: Susan L. Reiland
Manuscript Editor: Ellen Z. Curtin
Permissions Editor: Carline Haga
Interior and Cover Design: Roy R. Neuhaus
Front Cover Art: "Teec Nos Pos, 1985," copyright David Em/courtesy Spieckerman
    Associates, San Francisco.
Interior Illustration: Carl Brown
Typesetting: Weimer Typesetting Company, Inc.
Printing and Binding: Arcata Graphics/Fairfield

Preface

Modern Data Analysis is intended for use with a first course in statistics. It starts with the elementary question "What are data?", and ends sixteen chapters later with an introduction to multiple regression. The level of mathematics is kept low, and the emphasis is on applications and real-world data, rather than theory. The book aims to provide students with an experience of statistical analysis as a process of scientific discovery.

Modern statistical methods rely heavily on computers, which are not always practical in an undergraduate statistics course. *Almost all of the problems in this book can be done using only paper, pencil, and a scientific hand calculator.* Computers or statistical calculators could increase the pace of the course and the range of problems assigned, however. They encourage a shift of focus from calculation to interpretation. More subtly, they also make it easier to avoid the implication that statistical analysis is a matter of finding the one "right" (arithmetically correct) answer, and instead to view it as science—a dynamic process of experiment and learning. The book is written in this spirit.

The text contains more material than can reasonably be covered in one semester. Selectivity is required; the computing technology available to students is one major constraint. Four examples illustrate possible plans:

Low-tech: Students have basic scientific calculators. Cover unstarred sections of Chapters 1–4 and 7–12; select topics from other chapters according to time and interest. The focus of the course is on elementary descriptive/exploratory methods (e.g., stem-and-leaf displays, mean–median comparisons) and on concepts of probability and inference. This approach comes closest to a traditional introductory statistics course.

Mid-tech: Students have statistical calculators with linear regression capabilities. Cover unstarred sections of Chapters 1–5 and 7–11 more quickly. Select topics from Chapter 13 (comparison), and emphasize Chapter 14 (regression). Further regression material from Chapters 15 and 16 may be included if time allows. The course leads up to a focus on linear regression.

High-tech: Students have access to interactive computers and a good statistical package.* Chapters 1–5 are much easier with a computer to do the work of calculation (teaching about the computer may eclipse teaching about statistics in this section). The theoretical material of Chapters 7–11 might be treated lightly, in favor of the bivariate methods of Chapters 12–15. Advanced students could delve into multiple regression (Chapter 16). The course emphasizes the routine use of analytical graphics (especially box plots and scatter plots) and leads up to practical experience with regression analysis.

Modern: With or without computers, this book could also serve an unconventional course that emphasizes the interactive use of graphics and nonlinear transformations. Chapter 6 introduces this approach; it incidentally affords an opportunity to review basic mathematical ideas as well. Transformations provide a way to "do something" about distributional problems, so the assumptions of inferential statistics (Chapters 11, 13, and 15) become less ritualistic. Transformations also raise the possibility of curvilinear regression (Chapters 15 and 16). With such tools students are better equipped to attempt research on their own.

These plans are not rigidly built into the book, so any variation or mixture could be tried. One of my goals has been to write a text rich enough that it does not bore the instructor, even after repeated semesters of use. In my own experience I find some version of the "modern" approach to be the most fun to teach, but it is also the hardest, and I do not try it with every class.

---

*A companion text, called *Statistics with STATA*, describes how to perform a wide range of numerical and graphical analyses with the STATA computer program. A student version of this program is available with *Statistics with STATA*.

Modern Data Analysis reflects a belief that most students are more interested in applied statistics than they are in statistical theory. All concepts are introduced in the context of real-world examples, and reinforced by exercises in which students are asked to think out analyses for themselves. I hope to make the concepts more accessible in this way, and also to show that these techniques are actually useful. Students are exposed from the start to the fact that real data are often messy, complex, and ambiguous.

"Introductory statistics" need not be the same as "old-fashioned statistics." Statistical analysis is presented here as an interactive, exploratory process with a large graphical component. This is in keeping with the philosophical changes wrought by recent developments including econometric regression methods, Tukey's exploratory data analysis, robust estimation, criticism and influence analysis, computer graphics, and interactive computing in general. Without going into the technical details of these methods, I lay a foundation for them. Exposure to modern methods should provide all students, including the majority who do not go on into research, with both more sophistication as information consumers and with a strengthened appreciation for the logic of scientific research.

A Note on Rounding Off

Rounding was an unexpected bugaboo in the writing of this book. For most problems different answers are obtained depending on how and when one rounds off. In multistep problems, these small differences tend to accumulate. The higher precision of computer-based calculations often leads to answers notably different from those obtained by students working problems out by hand. I found no consistent way to deal with this problem, and had to settle for making the most seemingly logical choice in each situation. I hope readers will be tolerant of the small discrepancies they will undoubtedly find between their calculations and mine.

ACKNOWLEDGMENTS

When I first thought of writing this book, John Moroney supplied needed encouragement. John Kimmel saw something promising in the fog of my first proposal. Later he did a fine job of guiding the manuscript through a seemingly endless series of reviews and revisions. As the manuscript was revised, it gradually grew larger and more ambitious. Many reviewers contributed useful suggestions, particularly Alan Acock, Louisiana State University; Neil Bennett, Yale University; Susan Borker, Syracuse University; Lawrence Felice, Baylor University; Rudy Freund, Texas A & M University; David Groggel, Miami University; Michael Halliwell, California State University, Long Beach; Donald K. Hotchkiss, Iowa State University; Michael B. Kleiman, University of South Florida; Herman J. Loether, California State University, Dominguez

Hills; David Lund, University of Wisconsin; Daniel Martinez, California State University, Long Beach; Douglas Nychka, North Carolina State University; Fred Pampel, University of Iowa; Robert Parker, University of Iowa; Robert L. Rule, University of Northern Iowa; W. Robert Stephenson, Iowa State University; Bill Stines, North Carolina State University; Daniel J. Troy, Purdue University, Calumet; Joseph Verducci, Ohio State University; Robert Wardrop, University of Wisconsin; and Jeffrey Witmer, Oberlin College. Warren Hamilton read the entire first draft, making stylistic or content suggestions on almost every page. In later drafts I tried to adopt his viewpoint as a "conscience" to critique my writing. Susan Reiland did an impressive job of production editing, and I thank Joan Marsh for all her help.

I am grateful to many people for sharing their data. Among those who went out of their way are Susan Baker, Charlie Blitzer, Bob Flewelling, Robin Gorsky, Jesse LaCrosse, Halûk Ozkaynak, Peggy Plass, Murray Straus, George Thurston, Paul Treacy, Steve Tullar, Sally Ward, and Kirk Williams. Other kinds of assistance came from Tom Ballestero, Gordon Byers, Peter Dodge, Betty Le Compagnon, Stuart Palmer, and Chris Penniman. Dozens of students helped by pointing out fuzziness and errors. I should mention especially John Anderson, Marcia Ghidina, Holley Gimpel, Beth Jacobsen, Amy Oppenheimer, Gloria Straughn, and Stephen Sweet.

The writing of this book required a great deal of patience from those living in the same house with me. While I was immersed in writing, my wife Leslie did all the physical and emotional work of keeping our lives together. Sarah and David looked on, wondering when they would get their father back. To these three people, this book is dedicated.

Lawrence C. Hamilton

Contents

*Denotes optional sections. The material in these sections is not necessarily less important, but it is not required for understanding the nonoptional sections of later chapters. All of Chapters 6, 15, and 16 is marked as optional.

PART III   **BIVARIATE ANALYSIS: RELATIONSHIPS BETWEEN TWO VARIABLES   355**

Chapter 12   ***Two Categorical Variables:***
Crosstabulation and the Chi-Square Test   359

Chapter 13   ***One Categorical and One Measurement Variable:***
Comparisons   397

APPENDICES   607

MODERN DATA ANALYSIS
A First Course in Applied Statistics

UNIVARIATE DESCRIPTIVE ANALYSIS

PART I

A **variable** is an attribute that varies, instead of being always the same. We use variables every day to understand and describe the world around us. People vary in education, weight, health, happiness, ethnicity, and countless other ways, each of which is a variable. Variables can be used to describe almost any attribute of anything.

Statistics provides systematic methods for the study of variables. These methods derive from the same basic ideas, whether the objects of study are people or galaxies or frogs. Similar statistical methods can be found in research journals from fields as diverse as business, social science, medicine, biology, or astronomy. Statistics is a common language for many areas of science. To see how true this is, you might try an experiment: Go to the current periodicals area of your library and page through research journals in any field that interests you. You will likely encounter articles that cannot even be read without advanced training in statistics.

Our news media shower us with statistical information. Daily newspapers have two sections—sports scores and stock market reports—that consist entirely of tiny-print data and statistical summaries. News stories, advertisements, and political campaigns frequently use or misuse statistical information. Their level of sophistication is low, but they still may confuse (sometimes intentionally) a statistically naive consumer. Even people with no interest in science or research should have some background in statistical thinking.

This book aims to provide such a background, which should prove useful whether you stay a consumer of other people's statistical claims or go on to do original research yourself. The book is divided into four parts. Part I, Chapters 1–6, introduces methods for exploring and describing the way in which variables vary. The question of just *how* variables vary is central to all further statistical analysis. We begin by analyzing variables one at a time, within small sets of data.

Part II, Chapters 7–11, looks at **statistical inference.** This is the process of generalizing from the data at hand (the **sample)** to the larger universe **(population)** from which the sample came. Statistical inference requires a marriage of descriptive techniques and mathematical theory. The fit between theory and data becomes a matter of some concern.

Part III, Chapters 12–15, examines how descriptive, exploratory, and inferential methods are applied to understanding relationships between two variables. This is where the interesting questions of **causality** are first addressed.

Part IV consists only of Chapter 16, which extends the methods of previous chapters to the analysis of relationships among three or more variables. It provides a preview of more advanced statistics, for which earlier chapters have built up a foundation.

In Chapter 1 we start with the elementary question, "What do data look like?" The question is not so simple as it first appears. Chapters 2–5 introduce ways for finding out and describing how variables vary. Chapter 6 concludes this first section with a discussion of what to do about data that are found to be "ill behaved."

A First Look at Data

Chapter 1

Data have a story to tell. Statistical analysis is detective work in which we apply our intelligence and our tools to discover parts of that story. The methods of modern statistics are built around an elegant framework of mathematical theory, but mathematical theory is not the focus of this text. Instead we will focus on statistics as a way to learn about the world, through the analysis of quantitative data.

Within each chapter of this book are examples of data stories to discover. There are also many problems that let you do some detective work on your own. First, though, comes the business of learning the basic tools. This chapter presents some sample data sets, discusses their components, and describes simple ways to summarize their information.

Statistics has its own vocabulary, in which some everyday words like *data*, *significant*, and *population* take on technical meanings. Data, for example, are the raw material of statistical analysis. One purpose of this chapter is to introduce some of the basic vocabulary of statistics, especially those terms needed to talk about a set of data.

1.1   A SAMPLE DATA SET

At an early stage in most research, raw data are organized into a rectangular **data set** like that shown in Table 1.1. These data came from a survey of students in introductory statistics courses. Statistics students may not provide the most interesting example we could start out with—later chapters will look into more dramatic topics—but they have the advantage of familiarity.

Each row in Table 1.1 pertains to one of the students. Each column describes something about the students. The 30 students constitute the *cases* for this set of data. Cases are the entities or things that the data describe and are typically used to define the **rows** of a data-set matrix. The **columns** of this matrix are defined by *variables,* or attributes of the cases. Table 1.1 has seven variables: age, gender, major, weight, expected statistics grade, grade point average (GPA), and verbal Scholastic Aptitude Test (VSAT) score. The body of the data set contains numbers indicating the **value** of a given variable, for each case. For instance, 19 is the value of the age variable for the tenth case. The data set shown in Table 1.1 contains 30 cases and 7 variables; the cases are individual students, and the variables are their questionnaire responses concerning age, gender, major, and so on.

> Data sets are typically organized in a row-by-column format, with each row representing one **case** and each column representing one **variable.**

Variables are often represented symbolically using capital letters such as X or Y. The values for specific cases can be represented by adding subscripts to indicate the case number. For example, let X stand for the variable of age. Then X_1 is the value of the age variable for case 1; from Table 1.1 we see that $X_1 = 18$, $X_2 = 26$, and $X_3 = 19$. Similarly, Y could stand for the variable grade point average; then $Y_1 = 2.88$, $Y_2 = 2.40$, and $Y_3 = 3.34$. Symbols make it easier to define statistical concepts. Table 1.2 (page 6) lists some that are used in this chapter.

For some cases (students), the values of certain variables are missing. Such **missing values** are shown by dashes ("—") in Table 1.1. For example, we do not have a college grade point average for student 5. Missing values are common in survey research and many other kinds of data. They complicate our efforts to draw sound conclusions.

TABLE 1.1

Sample Data Set: Questionnaire Responses from 30 Students Beginning Their Statistics Courses

Case	Age in years	Gender[a]	Major[b]	Weight in pounds	Expected grade[c]	GPA[d]	Verbal SAT[e]
1	18	0	7	187	3	2.88	410
2	26	0	3	195	3	2.40	400
3	19	1	2	98	2	3.34	—
4	23	0	3	147	3	2.54	—
5	35	1	2	140	3	—	—
6	36	1	2	140	3	—	—
7	21	0	2	175	3	3.03	490
8	18	1	2	115	2	2.50	450
9	22	0	2	255	4	2.30	650
10	19	1	2	130	2	2.58	490
11	21	1	1	150	4	3.58	510
12	20	1	2	120	1	2.60	370
13	19	0	6	163	2	3.00	450
14	22	1	1	135	2	2.20	—
15	20	1	5	140	3	2.50	480
16	28	0	5	182	4	3.30	—
17	20	1	2	123	3	2.60	560
18	19	1	2	105	3	3.10	600
19	19	1	2	160	3	2.78	420
20	18	1	—	125	4	3.33	570
21	18	1	6	125	4	2.60	460
22	19	0	—	175	3	2.50	550
23	19	1	2	133	3	1.96	520
24	19	1	—	123	3	2.20	310
25	19	1	2	130	—	2.63	550
26	19	1	2	120	2	2.08	530
27	19	1	2	125	3	2.80	—
28	19	1	2	106	3	2.58	475
29	19	1	2	110	3	3.37	470
30	19	1	2	139	4	3.40	560

Missing values shown as "—".
[a]Student's gender: 0 = male, 1 = female.
[b]Student's major: 1 = social science, 2 = health or human services, 3 = economics, 4 = biological sciences, 5 = physical sciences, 6 = undeclared, 7 = other.
[c]Grade students expect to receive in course: 1 = D, 2 = C, 3 = B, 4 = A.
[d]Cumulative college grade point average, on the same 4-point scale as Grade.
[e]Score on the verbal portion of the Scholastic Aptitude Test, which ranges from 200 to 800 points.

These 30 students were randomly chosen from the hundreds taking introductory statistics courses at one university in the spring of 1985. Such a random subset of cases is called a *sample*. The larger set of cases from which a sample is selected is called a *population*. In a later chapter we will look at

TABLE 1.2   ***Summary of Symbols Introduced in Chapter 1***

Symbol	Meaning
X	A variable
X_1	The value of variable X for case 1
Y	A variable
Y_6	The value of variable Y for case 6
n	The number of cases in a sample
f	The frequency or count of cases having a certain value of X
$\hat{\pi}$	Sample proportion or relative frequency: $\hat{\pi} = f/n$
π	Population proportion or relative frequency
$\hat{p}$	Sample percentage: $\hat{p} = 100(\hat{\pi})$
p	Population percentage

exactly what is meant by *random* and examine how to use information from a sample to make guesses about the entire population. Our first task, though, is to understand the sample data at hand.

A **population** is the universe of cases that are of potential research interest. The whole population is usually not available for analysis.

A **sample** is a subset of cases from the population that are available for analysis.

PROBLEM

1. Describe two possible sets of data that might be collected to study a topic that interests you. For each example, specify:

    a. The cases

    b. The variables

    c. The sample

    d. The population

1.2 FREQUENCY DISTRIBUTIONS

If you were asked to give a report about the ages of the students in Table 1.1, you might start by counting the number of students at each age. This is the idea behind a **frequency distribution table** like Table 1.3. The number of cases for which a variable takes on a certain value is called the *frequency* of that value: the frequency of 18 is 4, because 4 students are age 18; the frequency

TABLE 1.3

Frequency Distribution of Student Age Variable from Table 1.1; n = 30 Cases

Variable: Age in years	Frequency: Count	Proportion: f/n	Percentage: $100(\hat{\pi})$
X	f	$\hat{\pi}$	$\hat{p}$
18	4	.133	13.3%
19	14	.467	46.7
20	3	.100	10.0
21	2	.067	6.7
22	2	.067	6.7
23	1	.033	3.3
24	0	.000	0.0
25	0	.000	0.0
26	1	.033	3.3
27	0	.000	0.0
28	1	.033	3.3
29	0	.000	0.0
30	0	.000	0.0
31	0	.000	0.0
32	0	.000	0.0
33	0	.000	0.0
34	0	.000	0.0
35	1	.033	3.3
36	1	.033	3.3
Totals:[a]	n = 30	1.000	100.0%

[a]Due to rounding off, the proportions and percentages do not add up to exactly 1.000 and 100.0%, as they should.

of 19 is 14, because there are 14 19-year-olds. *Frequencies* are shown in the Table 1.3 column labeled *f*. The frequencies add up to the total *number of cases*, called *n*.

Next to the frequencies is a column labeled *proportion*. Proportions (also called *relative frequencies*) are frequencies (*f*) divided by the number of cases (*n*). The age 18 occurs with a frequency of 4, or a proportion of $\hat{\pi} = f/n = 4/30 = .133$. For the value 19, the corresponding proportion is $\hat{\pi} = f/n = 14/30 = .467$. The symbol $\hat{\pi}$ (spoken "pi-hat") indicates that the proportion is calculated from sample data. If the proportion referred to a population, rather than a sample, then we would use the symbol π (the Greek letter *pi*) alone.

Percentages can be found by multiplying a proportion by 100. If the proportion of 18-year-olds in this sample is .133, then the percentage of 18-year-olds is $\hat{p} = 100(\hat{\pi}) = 100(.133) = 13.3\%$. The symbol $\hat{p}$ ("*p*-hat") indicates that this percentage, like the proportion $\hat{\pi}$, is based on sample data. The percentage of 18-year-olds in an entire population would be represented by the letter *p* alone. Table 1.3 shows that this sample contains 46.7% 19-year-olds, 10.0% 20-year-olds, and so on.

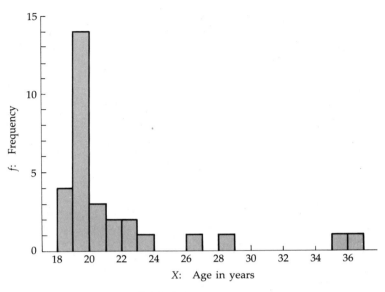

FIGURE 1.1   *Histogram of student age, from Table 1.3*

---

n: the number of cases in a sample.
A **frequency (f)**: the count of cases with a certain value of X.
A **sample proportion** or **relative frequency:** $\hat{\pi} = f/n$.
A **sample percentage:** $\hat{p} = 100(\hat{\pi})$.

Notice that the frequency distribution table includes rows for every age between 18 and 36, even though such values as 24 do not occur in this sample. The reason for including rows for all possible values in the range will soon be apparent.

A frequency distribution can also be represented in graphical form. One simple way to do this is with a **histogram,** like that of Figure 1.1. The horizontal or **X-axis** in the figure is marked off to show the values of the age variable. The vertical axis shows frequency. Bars are drawn over each value of X, to a height representing the frequency of that value. The most frequent value (age 19) appears as a **peak** or high point in this histogram.

Figure 1.1 would look the same if the vertical axis were marked in proportions or percentages instead of frequency. This is shown in Figure 1.2, where all three scales are drawn along the vertical axis.

These histograms show us the **shape of a frequency distribution.** The distribution of student ages is dominated by a high **peak** at X = 19. Around

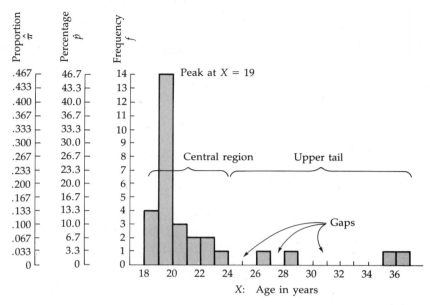

FIGURE 1.2 *Histogram of student age, from Table 1.3, with proportion, percentage, and frequency axes*

the peak is an ill-defined **central region,** where the bars rise to moderate heights indicating moderately frequent values of X. As we leave this central region, we enter the **tails** of the distribution. There is no left-hand tail; frequencies are zero for all ages under 18. On the right, however, a drawn-out tail extends up to ages 35 and 36, each of which occurs with frequency 1. There are several conspicuous **gaps** in this tail, where frequencies fall to zero (at ages 24 and 25, for instance).

Data analysis can show us things about the world, even with so unpromising a variable as the ages of statistics students. We might guess from Figure 1.2 that most of the 30 students are "traditional" students who entered college straight out of high school and are taking statistics in their freshman or sophomore year. The four oldest students are probably back in college after years of doing something else—starting a career, raising children, or being in the Army, perhaps. The distribution's most notable gap comes in the late 20's and early 30's—more likely ages for attending to careers and families than for attending statistics courses.

We will come back to frequency distributions and their shapes in later chapters. They are one of the major concerns of statistical analysis. Once a frequency distribution has been organized into a table or graph, our next step is to seek simple numerical ways to summarize and describe what we see.

2. As a classroom exercise, carry out an informal survey of the age of students in your own class. Construct a frequency distribution table like Table 1.3, and use this as the basis for a histogram. How does the shape of this distribution compare with Figure 1.2? Explain why the two age histograms are similar or different.

3. Prepare a frequency distribution table for the expected grade variable from Table 1.1. Find proportions (based on $n = 29$, because one case has a missing value) and draw a histogram for this distribution, using proportions on the vertical axis. Briefly describe what you see.

1.3 *STATISTICAL SUMMARIES AND TYPES OF VARIABLES*

Even with just 30 cases, Table 1.1 is too complicated for its numbers simply to "speak for themselves." Table 1.3 is an improvement but still has plenty of different numbers. We need to summarize these numbers somehow to describe the students' ages, class standings, test scores, and so on more understandably. This need for summaries grows as we confront larger data sets, like the hundreds of cases in a public opinion poll. More complex questions, such as "Are college students today older than college students were 10 years ago?" or "Do students with higher aptitude test scores expect better statistics grades?" also drive us to seek the simplification of **summary numbers.**

One kind of summary number, percentages, was used in Table 1.3. We could summarize the age distribution by saying that 13.33% of the students are 18, 46.67% are 19, 10.00% are 20, 6.67% are 21, and so on. It would take 10 different percentages to summarize all the ages of just these 30 students. For their grade point averages, it would take at least 22 different percentages. Obviously, percentages will be most useful with variables that take on only a few values, such as gender, major, or expected statistics grade. One way to get around the limitations of percentages with many-valued variables like age, weight, or GPA is to use **grouped data,** constructing a frequency distribution not for every value of age but for groups or **classes,** such as ages 18–19, 20–21, and 22–23. The first class, ages 18–19, has a frequency of 18; 20–21 has a frequency of 5, and so forth. If we want still fewer numbers, we could simply make the **class intervals** wider, even dividing ages into just two classes, such as "under 22" (frequency: $f = 23$; proportion: $\hat{\pi} = f/n = 23/30 = .767$) and "22 and over" ($f = 7$; $\hat{\pi} = f/n = 7/30 = .233$).

A simpler approach is to summarize all 30 ages with a single number, such as an average. To find the average age, we add up all the ages in Table 1.1 ($18 + 26 + 19 + \cdots + 19 = 632$) and divide this sum by the number of cases: $632/30 = 21.1$, so the average age is 21.1 years. Note that averaging cannot be applied to all the variables in Table 1.1. If we calculate the average gender we come up with .73, part way between the codes for male (0) and female (1).[1] The average major is 2.71 on a scale where 1 = social science, 2 = health and human services, 3 = economics, and so on. Clearly the 2.71 is a meaningless number here. On the other hand the average weight, 142.4 lb, and the average GPA, 2.74, are perfectly understandable. Behind these discoveries is a general rule: Averages work only with variables that measure something, where the variable's numerical values indicate "how much" of that something each case has. Such variables are called *measurement variables.* Age, weight, GPA, and VSAT score are examples of measurement variables in Table 1.1. These are the same variables where percentages work least well.

Percentages work best as numerical summaries where the variables take on only a few values. Often such variables are *categorical variables,* whose numerical values just tell us what category a given case fits in. Gender and major are examples of categorical variables in Table 1.1. Whereas measurement variables indicate "how much," categorical variables specify only "what kind." The numbers used as codes, such as 0 for males or 5 for physical science majors, are arbitrary.

Measurement variables tell us "how much" of something each case has.
   Examples: age, weight.
Categorical variables tell us only "what kind" or category a case belongs
   in. Numerical values are arbitrary. *Examples:* gender, major.

The problems of applying percentages to measurement variables, or averages to categorical variables, illustrate the basic fact that *many statistical techniques are appropriate with only certain types of variables and less appropriate or meaningless with other types of variables.* We must always be conscious of what type of variable we face. Categorical and measurement variables are the two principal types.[2]

Some variables inhabit a gray area between measurement and categorical. They are called **ordinal variables;** academic letter grades are a familiar example. The essential characteristic of ordinal variables is that they consist of **ordered categories.** With a categorical variable such as major, the categories (social science, health and human services, etc.) have no inherent order. With ordinal variables like grades, the categories have a definite order: A is higher than B, which exceeds C, and so forth. But simply knowing the order does not tell us much about measurement. We do not know if the difference between an A and a B is the same as the difference between a C and a D, for instance.

Special statistical techniques have been designed for ordinal variables. Many researchers, however, simply analyze ordinal variables with categorical-variable techniques, as if they were categorical, or with measurement-variable techniques, as if they were measurements. For example, we might summarize the expected statistics grades from Table 1.1 by saying that 20.7% of the students expect A's, 55.2% B's, 20.7% C's, 3.4% D's, and none expect F's, using the same percentage-based approach we use with a categorical variable like major or gender. Alternatively, we could report that the average expected grade is 2.93, about a B-minus, following the averaging approach we use with measurement variables like age or weight. In this instance either approach produces a reasonable summary.

PROBLEMS

4. Suppose you planned to conduct a survey regarding drug use by college students. Suggest two questions you might ask about drug use that would produce categorical variables. How might the categories be coded? How could you summarize your findings about these two variables? Suggest two further questions about drug use that would produce measurement variables. How could your findings with these measurement variables be summarized?

5. Distinguish between sample and population in your study of Problem 4.

1.4   AGGREGATE DATA

Each case in Table 1.1 is an individual human being. Many other kinds of cases are possible. Instead of individual people, the cases could be individual frogs, described by such variables as weight, croaking frequency, or maximum jumping distance. In another data set the cases might be the individual moons of Saturn, with variables such as diameter, mass, rotational period, or distance from Saturn. Later we will encounter a data set whose cases are the first 25 flights of the U.S. space shuttle, with variables including launch date, temperature, and booster rocket burn damage. Another example uses drinking water wells as the cases, with such variables as pollution and distance from the nearest road.

Some analyses use a different kind of data set, called **aggregate data,** where many individual cases are combined to form one aggregate case. For example, we can collect data not on students but on entire classes, and report for each the size, percentage of high grades, and other variables. We could aggregate further and construct data sets in which each case was a different university, state, or nation. Table 1.4 shows an aggregate data set where cases are 12 cities, randomly selected from a population consisting of all large U.S. cities.[3] The variables in Table 1.4 include one categorical variable, region, and four measurement variables: median household income; percentage of households with incomes below $5,000; combined rating for six violent television shows; and number of homicide victims per 100,000 population. Unlike the

TABLE 1.4   *Data from a Sample of 12 U.S. Cities, 1980[a]*

City	Region[b]	Median household income[c]	Percentage with low income[d]	Rating 6 violent TV shows	Homicides per 100,000 population[e]
Warren, MI	4	26.5	4.2%	108.3	2.61
Pueblo, CO	2	18.3	11.3	101.5	4.13
Raleigh, NC	1	21.8	9.7	79.8	6.79
Fort Wayne, IN	4	19.6	8.5	98.8	6.97
Tucson, AZ	2	19.4	10.2	70.8	7.32
Anchorage, AK	2	30.7	6.1	—	8.48
Toledo, OH	4	20.2	10.7	110.2	9.53
Portsmouth, VA	1	16.8	15.8	108.0	14.53
Memphis, TN	1	16.9	17.1	119.5	16.93
Hartford, CT	3	14.0	22.5	88.8	19.65
Savannah, GA	1	15.4	18.2	120.3	20.23
Birmingham, AL	1	15.2	23.6	99.4	27.07

[a]Cities were randomly selected from the population consisting of all large (over 100,000 population) U.S. cities.
[b]Census regions: 1 = South, 2 = West, 3 = Northeast, 4 = Midwest.
[c]In thousands of dollars per year.
[d]Percentage of households with annual incomes below $5,000.
[e]Total number of homicide victims per 100,000 people, averaged for 1980–1984.
Source: Data courtesy of Kirk Williams, University of New Hampshire.

anonymous students in Table 1.1, these 12 cities are known to us by name, which makes analysis more lively.

Note that the variables in this aggregate data set are themselves statistical summaries, such as rates or percentages, that must have been derived from earlier individual-level data. The median, for example, is a special kind of average. If we drew up a list of every household in a city, ordered from lowest income to highest, the median income would be that of the household in the middle of our ordered list. We will learn more about medians, and why they are particularly useful with incomes, in Chapter 3.

The categorical variable geographical region could be summarized with percentages: 33% of these cities are in the South, 25% in the Midwest, and so on. The measurement variables could be summarized using averages. The average homicide rate, for example, is 12.02 per 100,000 people per year.[4] If we wanted to construct frequency distributions for these measurement variables, whether tables or histograms, we would have to use grouping.

One way of grouping the homicide rate data is shown in Table 1.5 (page 14). Both the **intervals** and the **true limits** are shown for each class. When we encounter a number such as 9.53 (Toledo's homicide rate), we can assume it was rounded off from its precise value (to one more decimal place). Toledo's homicide rate presumably lies somewhere between 9.525 and 9.535, so in Table 1.5 it falls in the class of rates with rounded values from 5 to 9.99. But because numbers between 4.995 and 5 could have been rounded up to 5, and

TABLE 1.5

Grouped Frequency Distribution of 12 City Homicide Rates (Data from Table 1.4)

True limits X	Class interval X	Frequency f	Sample proportion $\hat{\pi}$	Sample percentage $\hat{p}$
0– 4.995	0– 4.99	2	.167	16.7%
4.995– 9.995	5– 9.99	5	.417	41.7
9.995–14.995	10–14.99	1	.083	8.3
14.995–19.995	15–19.99	2	.167	16.7
19.995–24.995	20–24.99	1	.083	8.3
24.995–29.995	25–29.99	1	.083	8.3
Totals		$n = 12$	1.000	100.0%

numbers between 9.99 and 9.995 could have been rounded down to 9.99, the true limits of the 5–9.99 class are actually 4.995–9.995. Once the data are grouped, frequencies, proportions, and percentages are found in the same manner as with ungrouped data.

A histogram shows this distribution in Figure 1.3. Like the age distribution in Figure 1.1, these homicide rates have a distinct peak and a drawn-out upper (right-hand) tail. The resemblance is not purely coincidental. Many

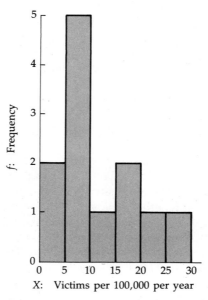

FIGURE 1.3   *Histogram of homicide rates in 12 U.S. cities, based on groupings in Table 1.5*

TABLE 1.6   *Data on a Sample of 25 Countries*

Country	Per cap. GNP 1982 (U.S. $)	Life expect. (Years)	Civil liberties index[a]	Population per doctor	Births per 1,000 pop. 1983
Nicaragua	950	58	3	1,730	44.2
Paraguay	1,670	65	3	1,750	36.0
Venezuela	4,250	68	6	910	35.2
France	11,520	74	6	480	13.7
W. Germany	12,280	73	6	430	10.2
Greece	4,170	73	6	400	15.8
Norway	14,300	75	7	480	12.3
Czechoslovakia	5,540	71	2	350	14.8
Austria	9,830	72	7	440	11.9
Jordan	1,680	61	3	940	44.9
Sri Lanka	320	67	4	7,550	27.0
Brunei	22,260	66	2	1,870	—
Cambodia	—	37	1	29,410	45.5
Indonesia	550	50	2	10,800	30.7
N. Korea	930	66	1	—	30.5
Mongolia	940	64	1	400	33.8
Taiwan	2,670	72	3	840	—
Australia	11,220	74	7	520	16.2
Congo	1,420	48	2	6,000	44.5
Ethiopia	150	41	1	69,620	49.2
Guinea	330	44	3	51,470	46.8
Mauritania	520	44	2	14,340	50.1
Nigeria	940	49	3	9,590	50.4
Togo	350	48	2	18,560	45.4
Zaire	180	48	1	14,700	45.2

[a]Adapted from Kort (1986): 1 = few civil liberties, 7 = many civil liberties.
Sources: Sivard (1985); Kort (1986).

measurement variables that have a lower boundary of zero but no definite upper boundary exhibit such drawn-out upper tails.

We will return to these data in later chapters. Eventually we will use them to investigate the relationship between income and homicide rates. For a preview of how this analysis comes out, you might want to peek at Figure 15.1 in Chapter 15.

Table 1.6 presents a second aggregate data set, in which the cases are 25 countries. Like the cities in Table 1.4, the 25 countries in Table 1.6 are a random subset of a larger population. "All countries" form the population of which these 25 are a random sample, just as "all large U.S. cities" are the reference population for the sample of Table 1.4. Because the cases in Tables 1.1, 1.4, and 1.6 are random samples, there is a scientific basis for using them to draw **inferences** about the larger populations from which they came.

The variables in Table 1.6 include four clear measurement variables: 1982 per capita gross national product (GNP) in U.S. dollars; life expectancy in years; the ratio of population to the number of physicians; and the crude birth rate (births per 1,000 population). There is also one ordinal variable: a "civil liberties index," which ranges from 1 (few civil liberties) to 7 (many civil liberties).

A striking feature of data using nations as cases is the wild variation. For example, per capita GNP among these countries varies from $150 per person, in Ethiopia, to almost 150 times that amount, $22,260 per person, in Brunei. Cambodia's per capita GNP is unknown, but it might be even lower than Ethiopia's. By far the "richest" country here is Brunei, a small oil-exporting sultanate on the island of Borneo. Its per person GNP far exceeds that of the United States ($13,160 per person in 1982), but Brunei's wealth is held by a tiny minority of the population.

Like the 30-student and the 12-city data sets, the 25-country data set of Table 1.6 is relatively small. Many research projects produce large data sets with thousands of cases and hundreds of variables, the analysis of which is impractical without a computer. Most of the sample data sets in this book, however, are small enough to study with a pencil and hand calculator, which also makes it easier for us to see where the numbers come from. Although faster, computer-based analysis runs the risk of insulating the researcher from a firsthand acquaintance with the actual data.

PROBLEMS

6. Construct a grouped frequency distribution for students' grade point averages from Table 1.1. Use intervals of 1.80–1.99, 2.00–2.19, and so on. For each interval, specify:

   a. True limits

   b. Frequency

   c. Sample proportion

   d. Sample percentage

7. Draw a histogram using the grouped frequency distribution from Problem 6. Use proportions as the vertical axis.

8. Construct a grouped frequency distribution for the ratings of six violent TV shows among the cities in Table 1.4. Choose your own intervals and specify the corresponding true limits.

9. Draw a histogram from the TV ratings distribution in Problem 8, using frequency on the vertical axis.

10. Aggregate data on many topics are readily available. The *Statistical Abstract of the United States, World Almanac, Places Rated Almanac,* and sports pages of any newspaper contain countless examples. Consult such a

source to assemble your own aggregate data set. Identify the cases, the variables, and the variables' types in your data set. Construct a histogram for one measurement variable, using grouping if needed, and discuss what this histogram shows about the variable.

1.5 MEASUREMENT ERRORS AND MISSING VALUES

It is convenient to proceed on the assumption that the first student in Table 1.1 really is an 18-year-old male who weighs 187 pounds and has a 2.88 GPA—but none of these things need be true. The student who filled out this questionnaire might have misled us on everything. Probably he is indeed an 18-year-old male. But is 2.88 really his exact GPA? Does he even know what it is? If he does not know, or does not want to tell us, is he more likely to err by over- or underestimating his true GPA? What about his accuracy in reporting his weight or expected statistics grade? Similar questions can be raised about any survey data, for careful research has shown that people often answer inaccurately.

Aggregate data, like those in the states and the nations data sets, do not originate so directly from individual people. Often aggregate data are a product of organizations and bureaucracies. Data quality may vary widely from case to case. How did someone arrive at a value for the percentage of Anchorage households with incomes below $5,000? Which deaths got counted, and which omitted, in calculating Birmingham's homicide rate? Who really knows what the birth rate is in Mongolia? There is much scope for error and subjectivity in these apparently objective numbers. When the cases are nations, problems arise from the huge differences in the information-gathering ability and the honesty of the governments involved. How much faith can we put in estimates that war-wracked Cambodia has 29,410 people per physician? Even the number of people in Cambodia is uncertain. Unlike the figures for Norway or Australia, the values given for such countries as Cambodia may be little more than educated guesses.

Even where the data originate from physical measurements, like rainfall gauges or chemical tests, errors can arise in many ways. Sources of error in data collection are as numerous and complex as the sources of data themselves. It should be presumed that almost any data set will contain errors. Further, the possibilities for error do not end when the data are collected. The original data are often put into some new form, such as a computer file, for further analysis. Data may be recopied several times before reaching print in the form of tables. Careful checking at every stage will reduce the likelihood of errors in the final version, but typographical errors and misplaced decimal points can easily creep in.

Errors in data are a potentially serious problem for any statistical analysis and can undermine the validity of all conclusions that result. Measurement error is a matter not of yes or no but of degree: How much error there might be, and how it might affect the analysis, should be considered in doing or reading about any research. Certain statistical techniques can help in detecting or adjusting for measurement error. These methods do not solve the measurement error problem, however.

A related problem is that of **missing values.** In Table 1.1, for example, we have no value for the verbal SAT score of student 3. Possibly she did not take the test, did not recall her score—or did not feel her test scores are any of our business. We will never be sure why she left this item blank, but any statements we make about VSATs in this sample will have to be based on the students who *did* answer the question.

The missing VSAT values in Table 1.1 would be particularly troublesome if some students left the answer blank to hide embarrassingly low scores. If so, any estimate of the students' average VSAT may be too high, because the lowest VSATs are not in the data. Overestimating an average test score for these 30 students has no serious practical consequences, but in other situations such measurement errors can be damaging.

Values are missing from Tables 1.1, 1.4, and 1.6. In each instance, our analysis must be limited to the known values. We can only speculate how, if at all, our conclusions might change were the data more complete.

PROBLEM

11. Sometimes missing values can be an interesting study in their own right, not just an obstacle to research. Seven of the 30 students in Table 1.1 reported no VSAT score. Do you see any clue in the table why some VSATs are missing? Use averages or percentages to summarize the most obvious way the seven cases, as a group, differ from the remaining 23.

1.6   *UNIVARIATE, BIVARIATE, AND MULTIVARIATE ANALYSIS*

To find the average homicide rate for the 12 cities in Table 1.4, we need only a single column of the data. We could calculate this average equally well even if all the other columns or variables were blacked out. Similarly, we could find the average TV rating or the percentage of Southern cities while blacking out every column except the one we were looking at. Used in this way, averages and percentages are examples of *univariate*, or one-variable, statistics.

> **Univariate** analysis examines variables one at a time, to see how they vary individually.

There are many other univariate statistics besides averages and percentages. All of them give us information about *how one variable varies.* Even if we collect averages or percentages for more than one variable, they are still univariate statistics. For example, the average per capita gross national product in Table 1.6 is $4,540.42. The average life expectancy is 60.3 years. Knowing these two figures tells us something about wealth and something about life expectancy, but nothing about the *relationship between* the two.

Relationships between pairs of variables are the focus of *bivariate* (two-variable) analysis. We cannot do a bivariate analysis by viewing one column of data at a time. Instead, we need to look at two data columns, such as GNP and life expectancy, together. Then we can see what GNP levels go with what life expectancies: that Nicaragua has a GNP of $950 per person and a life expectancy of 58 years; that Paraguay's GNP is $1,670 per person and life expectancy is 65 years; and so forth. (A more detailed analysis of the GNP–life expectancy relationship is found in Chapter 15; see Figure 15.14.)

> **Bivariate** analysis examines the relationship between two variables, or how they vary together.

Problem 11 called for simple bivariate analysis. There is a relationship between missing VSAT scores and age—students with missing scores tend to be older. Even though this relationship is fairly strong, you may not have noticed it simply by scanning the table. And what if the average ages of students with and without VSAT scores were different, yet closer than in Table 1.1? Bivariate techniques give us systematic ways to handle such problems.

Because they deal with how, or if, two variables are related, bivariate questions are often more interesting than the questions that univariate analysis can address. But before we can proceed to bivariate techniques we first need to understand univariate statistical techniques. That is the plan of this book. The first six chapters discuss basic tools for univariate analysis of sample data. Chapters 7–11 introduce concepts needed for drawing inferences from a sample to its population. Bivariate methods are presented in Chapters 12–15, where we will re-examine familiar variables in a new way.

Multivariate analysis involves statistical techniques that can look at three or more columns of data at once. One of its goals is to examine two-variable relationships while statistically controlling for the influences of other variables. This is an important part of **causal** research: if X is a cause of Y, then the effects of X should still be apparent even after we control for other variables.

> **Multivariate** analysis examines interrelationships among three or more variables.

Multivariate analysis is generally done using computers, and it requires much statistical understanding on the part of the researcher. Such analysis nevertheless is commonplace in most fields of research. It constitutes a common language that links many different areas of modern science. This text barely advances to multivariate methods (in Chapter 16), but it provides a solid foundation for such study. Additional coursework and practice with statistics are needed in many fields.

Often students who view their first statistics course with fear and loathing find out, years later, that it taught skills vital to their chosen business or profession. At the very least, this statistics course should deepen your understanding of the rich flow of information around you.

As you develop analytical skills, you will also see how efforts to answer seemingly straightforward questions often turn up unexpected results or raise new sets of questions. The surprise of discovery and the challenge of exploration take data analysis far from the mechanical application of techniques that some people imagine it to be.

PROBLEMS

12. Give your own example of a research question that could be answered by a univariate analysis of the student data in Table 1.1. Do the same for a question in this table that needs bivariate analysis.

13. Give your own examples of questions calling for univariate and bivariate analyses of the cities data in Table 1.4. Do likewise with the nations data of Table 1.6.

SUMMARY

This chapter introduced five central concepts of statistics: sample, population, case, variable, and distribution. We also saw that variables may be of several different types, which affects how they are analyzed. For example, percentages work best with categorical variables, whereas averages better serve measurement variables. Percentages and averages are both statistics commonly used to summarize distributions of single variables. We will eventually encounter bivariate and multivariate statistics that summarize the joint distribution of several variables at once.

Another way to examine variable distributions is to display them graphically. Displays such as histograms allow us to see the shape of the distribution, which is important for subsequent statistical analysis. The next chapter introduces further graphical tools.

PROBLEMS

14. Construct an ungrouped frequency distribution table, like Table 1.3, for each of the following variables:
    a. Gender in Table 1.1
    b. Major in Table 1.1
    c. Region in Table 1.4
    d. Civil liberties in Table 1.6

15. Construct a grouped frequency distribution table and a histogram, like Table 1.5 and Figure 1.3, for each of the following variables:
    a. Weight in Table 1.1
    b. Median household income in Table 1.4
    c. Birth rate in Table 1.6

16. As will be seen in the next chapter, histograms are used with measurement but not with categorical variables. Construct histograms with both frequency and proportion scales along the vertical axis for each of the three variables in Problem 15.

17. Two cases in Table 1.1 have missing values on the college grade point average variable. Examine the values of their other variables and suggest a likely reason why they left GPA blank.

18. It is simple to work back and forth between proportions and percentages. A proportion of .17 means the same thing as 17%, for instance. Though percentages are easier to talk about in English, proportions have certain mathematical advantages. We can use proportions to find out quickly what actual number corresponds to a certain percentage. Suppose you presently earn $12,345, and your employer promises a 17% raise. To see how much money this amounts to, the easiest procedure is to convert 17% to a proportion, .17, and multiply this proportion times your base salary: $.17 (\$12,345) = \$2,098.65$, so you will get a $2,098.65 raise.

    Use this approach to fill out frequencies and proportions in the frequency distribution table at the top of page 22. As of July 1, 1988, there were 680 known AIDS cases in the city of Boston, distributed among risk categories as shown.

Distribution of AIDS by Risk Category

Risk category	Frequency	Proportion	Percentage
Gay/bisexual			63.09%
IV drug user			17.50
Gay & IV user			4.12
Heterosexual contact w/IV user			2.06
Born in Caribbean or Africa			8.53
Other[a]			4.71
Totals:	680		100.00%

[a]Includes children, people infected through transfusions, and cases where risk factors are unknown.
Source: The Boston Health Department, reported in the *Boston Globe,* November 13, 1988.

NOTES

1. The idea of an "average gender," somewhere between males and females, seems like statistical nonsense. It is not quite meaningless, though. For any variable with just two values, coded as 0 and 1, the average will be the proportion of 1's—in this case, the proportion of females (.73, or 73%).

2. Measurement variables may be subdivided into two further types: **interval variables,** which have measurement but no meaningful zero point (like IQ scores or the Fahrenheit thermometer), and **ratio variables,** which do have a meaningful zero point (like height or weight).

3. The population consists of the 168 U.S. cities with over 100,000 people in 1980. Chapter 8 will show how we select samples randomly from such a population.

4. Averages and other statistical summaries can be tricky to calculate and interpret with aggregate data. For example, if we add up all 12 homicide rates in Table 1.4, and divide by 12, we get the average of the city homicide rates. This is not the same as the average homicide rate their *citizens* experience, however. To find the latter, we need to bear in mind that each city's rate is based on its population size, which is six times more for Memphis, Tennessee, than Pueblo, Colorado.

   Such considerations greatly complicate the simple statistical points we are trying to make here, however. We will focus only on the average (and other summaries) of the *cities'* rates, rather than using these rates to obtain estimates for the *people* who live there. Other aggregate-data analyses in this book will also treat the aggregate units themselves as the cases of interest.

Graphing Variable Distributions

Chapter 2

The basic feature of all variables is that they vary: they are not the same for every case. The opposite of a variable is a constant, which *is* the same for every case. A constant can be described easily, because it is only a single value. Variables are less easily described. Averages are one way we try to simplify this description, by summarizing a variable's many values into a single number. But *an average must leave out any information about how a variable actually varies.* For example, an average age of 30 might describe equally well a random collection of people from a city street, or a class in which every person was exactly 30, or an unusual group made up of equal numbers of newborns and 60-year-olds. We need additional tools to help us see and describe such different patterns of variation.

Two kinds of analytical tools are emphasized in this book. One kind, **numerical summaries,** involves calculating special numbers that summarize important features of variable distributions. Averages and percentages are the best-known examples of numerical summaries, but many others exist.

Statisticians often think of numerical summary methods in terms of developing mathematical **models** for the data. In classical statistics, one seeks to understand and describe data largely by calculating the appropriate numerical summaries.

Recent years have seen renewed emphasis on a second kind of analytical tool: methods for displaying data in **graphical** form. A well-constructed graph can show several different features of the data at once. Unexpected or unusual features, which often are hidden within averages, may become strikingly obvious when the data are graphed. *Because the strengths and weaknesses of graphical methods are opposite to those of numerical summary methods, the two work best in combination.*

In Chapter 1 we noted that the type of variable (categorical or measurement) determines what analytical techniques can be used. For example, percentages work best with categorical variables, whereas averages make sense only with measurement variables. For graphical methods, too, it matters what type of variable we have. Measurement-variable distributions can be graphed in a variety of ways including *histograms, frequency polygons, ogives, stem-and-leaf displays,* and *time plots.* Frequency distributions of categorical variables are often displayed in *bar charts* or *pie charts,* but such charts are seldom used for analysis. The information they contain is more easily read from a table. Graphical analysis is generally more important with measurement than with categorical data.

2.1 HISTOGRAMS

Histograms display the frequency distribution of a measurement variable. They show bars with lengths proportional to frequency, proportion, or percentage for each value. If the variable takes on many different values, then grouping may be needed to produce a readable table or graph. Grouping must be done so that the classes cover all possible values within the range of the data, even if some classes have no cases in them. Chapter 1 gave examples of ungrouped (Figure 1.1) and grouped (Figure 1.3) histograms.

Two choices must be made in constructing a histogram: how to group the data and whether frequencies, proportions, or percentages appear on the vertical axis. Grouping is sometimes crucial. If too many classes are used, the graph may appear rough and strung out, with many small peaks and gaps. With too few classes, the graph may appear smoother but hide important features of the data. Choosing the right number of classes and the breaks between them is a matter of judgment.

Table 2.1 contains data on nicotine concentrations in cabin air (smoking sections) for 26 commercial airline flights. These data were collected to study the effects of seating segregation (smoking or nonsmoking) on air quality. (In Chapter 5 we will see how smoking section air compares with samples from nonsmoking and border seats.) Nicotine concentrations are measured in mi-

TABLE 2.1

Nicotine Concentrations in Cabin Air of Smoking Section on 26 Commercial Airline Flights

Flight	Number of passengers in smoking section	Number of cigarettes smoked	Nicotine concentration $\mu g/m^3$
1	13	26	.03
2	—	—	.08
3	25	88	.4
4	20	37	.6
5	21	50	.7
6	22	37	.7
7	—	—	2.1
8	—	—	2.3
9	10	17	3.1
10	—	—	4.5
11	24	20	8.6
12	10	17	8.8
13	10	23	10.2
14	17	32	10.5
15	—	—	11.0
16	—	—	14.9
17	35	123	18.7
18	11	6	22.1
19	7	11	30.2
20	15	19	39.5
21	20	48	42.2
22	22	30	45.0
23	20	17	57.1
24	22	84	59.8
25	23	38	76.7
26	23	31	112.4

Source: Data from Oldaker and Conrad (1987).

crograms per cubic meter of air, or $\mu g/m^3$. Two other variables reported in Table 2.1 are the number of passengers in the smoking section and the estimated number of cigarettes smoked. Values of these variables are missing for six flights.

Table 2.2 (page 26) contains a grouped frequency distribution for nicotine concentration, which is the basis for the histogram in Figure 2.1. Only six classes are used with this variable, otherwise the 26 cases would be spread out too thinly. The histogram shows a distribution sharply peaked at the first class and with a drawn-out upper tail. This tells us that on most flights nicotine concentrations were relatively low, but a few had much higher levels.

Figure 2.2 (page 27) illustrates some features of distributional shape in histograms. A distribution whose left or lower half mirrors the other half is

TABLE 2.2

Grouped Frequency Distribution of Nicotine Concentration on 26 Airline Flights (Raw Data in Table 2.1)

Nicotine concentration[a]	Frequency	Percentage
X	f	$\hat{p}$
0– 19.99	17	65.4%
20.00– 39.99	3	11.5
40.00– 59.99	4	15.4
60.00– 79.99	1	3.8
80.00– 99.99	0	0
100.00–119.99	1	3.8
Totals:	$n = 26$	100%

[a]Values given are class intervals. Since the original data (Table 2.1) have up to two decimal places, the true limits should be set with three decimal places. Thus the true limits for the interval 20.00–39.99 are 19.995 and 39.995.

called **symmetrical:** Both tails are of equal length. A distribution that is not symmetrical is **skewed.** If the upper or right-hand tail is heavier, the distribution is said to be **positively skewed.**[1] Positive skewness is common; every histogram we have viewed so far (student ages, urban homicide rates, and nicotine concentrations) has positive skewness. On the other hand, if the lower tail is heavier the distribution is **negatively skewed.** Skewness and symmetry are issues of increasing importance as we approach more formal analytical techniques.

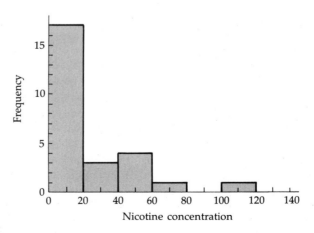

FIGURE 2.1   *Histogram of nicotine concentrations*

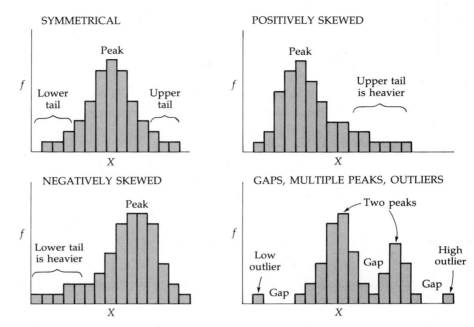

FIGURE 2.2   *Histograms with major features of distributional shape*

In addition to symmetry or skewness, histograms sometimes reveal **multiple peaks, gaps,** or **outliers**—cases that lie far from the bulk of the data (lower right, Figure 2.2). Like other aspects of distributional shape these features have both statistical and real-world implications.

PROBLEM

1. Ignoring cases with missing values, use the data in Table 2.1 to construct grouped frequency distributions and histograms for:
   a. Number of passengers in smoking section (use classes of 0–9, 10–19, etc.)
   b. Number of cigarettes smoked (use classes of 0–19, 20–39, etc.)

2.2 FREQUENCY POLYGONS AND OGIVES

Two close relatives of the histogram are *frequency polygons* and *ogives*. Frequency polygons can be constructed from the same frequency distribution table used for a histogram. Instead of drawing bars to represent the frequency of each X value or class, we simply draw points where the midpoints of the

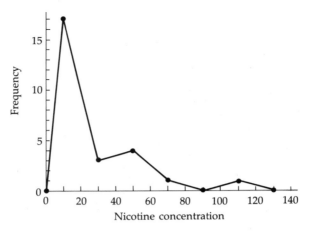

FIGURE 2.3   *Frequency polygon of nicotine concentration*

tops of the histogram's bars would be. Connecting these points by short line segments creates our frequency polygon.

Figure 2.3 shows a frequency polygon for the nicotine concentration distribution of Table 2.2. This polygon conveys the same information as the histogram of Figure 2.1, in a slightly different form. Histograms are easier to read for detail, because we can readily trace the bar heights back to the frequency axis and see what frequency they represent. Frequency polygons, on the other hand, are sometimes better for providing an overall picture of the distributional shape—especially in larger data sets.

> *Frequency polygons* and *histograms* both display the frequency distribution of measurement variables.
>
> In **histograms,** bars are drawn to a height representing the frequency of each X value or class.
> In **frequency polygons,** points at the midpoint of each class interval indicate the frequency. Line segments connect the points.

Either histograms or frequency polygons can also be applied for a second purpose: graphing *cumulative frequency distributions.* An example of a cumulative frequency distribution, using the nicotine concentration data, is shown in Table 2.3. On 17 flights levels were 19.99 $\mu g/m^3$ or less, on 20 flights levels were 39.99 $\mu g/m^3$ or less, and so on. Like ordinary frequencies, cumulative frequencies can be expressed as proportions (by dividing them by n) or as percentages (the proportions times 100). Percentages are shown at the right of Table 2.3.

TABLE 2.3				

Cumulative Grouped Frequency Distribution of Nicotine Concentration on 26 Airline Flights (Raw Data in Table 2.1)

Nicotine concentration	Frequency	Percentage	Cumulative frequency	Cumulative percentage
X	f	$\hat{p}$	$cum(f)$	$cum(\hat{p})$
0– 19.99	17	65.4%	17	65.4%
20.00– 39.99	3	11.5	20	76.9
40.00– 59.99	4	15.4	24	92.3
60.00– 79.99	1	3.8	25	96.2
80.00– 99.99	0	0	25	96.2
100.00–119.99	1	3.8	26	100.0
Totals:	$n = 26$	100%		

> **Cumulative frequency** is defined as the number of cases whose value is less than or equal to a given X value.

Figure 2.4 shows a frequency polygon based on the cumulative frequencies of Table 2.3. Such graphs are called **ogives.** The ogive in Figure 2.4 has cumulative percentages along the vertical axis, but its shape would be no different if cumulative frequencies or proportions were used instead. This ogive climbs steeply at first, because the cumulative percentages increase fastest at low values of X. More than three-fourths of the cases are in the first two classes, 0–19.99 and 20–39.99.

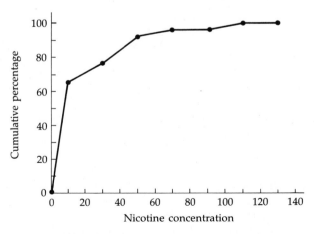

FIGURE 2.4   *Ogive of nicotine concentration*

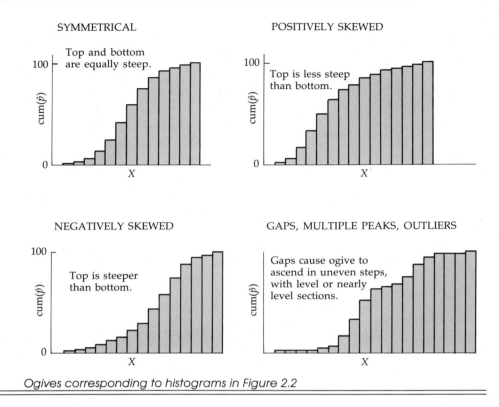

FIGURE 2.5   *Ogives corresponding to histograms in Figure 2.2*

Ogives are often more S-shaped, flattened out at both bottom and top. The details of an ogive's shape reflect the details of the original frequency distribution. Figure 2.5 shows ogive histograms corresponding to the four frequency histograms in Figure 2.2. If the frequency distribution is symmetrical, the corresponding ogive will also be symmetrical, equally steep at the bottom and the top (upper left in Figures 2.2 and 2.5). A positively skewed distribution yields an ogive that first rises steeply then flattens out, approaching the top more slowly (upper right in both figures). We saw this pattern with the nicotine concentration data. The opposite happens with negative skewness (lower left in both figures). Gaps or multiple peaks give an ogive an uneven, stepped appearance. The gaps show up as fairly level sections, interspersed with steeper rises at the peaks (lower right in both figures).

Ogives have applications in medicine, engineering, education, and other fields where it is useful to know what percentage of cases lie below a certain value of X. For example, a testing service might offer teachers ogives showing what percentage of students score at or below each possible level on a standardized test. We could use Figure 2.4 to guess what percentage of airline flights will have nicotine concentrations no higher than 50 $\mu g/m^3$.

It is also possible to construct cumulative frequency distributions and graphs that show how many cases lie at *or above* a certain X value. Graphs of "X or above" cumulative distributions go down from left to right, instead of up as "X or below" distributions like Figures 2.4 and 2.5 do.

PROBLEMS

2. Draw frequency polygons for the following distributions:
    a. Ages of 30 statistics students (Table 1.3)
    b. Homicide rates in 12 cities (Table 1.5)

3. Construct a cumulative frequency distribution for the gross national products of 24 nations (Table 1.6), using class intervals of 0–1,999, 2,000–3,999, 4,000–5,999, etc. Draw a cumulative frequency polygon (ogive), indicating cumulative percentage along the vertical axis. From this ogive, what can we tell about the shape of the frequency distribution?

2.3   *STEM-AND-LEAF DISPLAYS*

New graphing techniques for measurement data are continually being invented. One simple and versatile graph, the *stem-and-leaf display,* is of recent origin.[2] Stem-and-leaf displays are modified histograms where numerical values from the data replace the bars.

Table 2.4 (page 32) lists the 50 U.S. states in order by a variable called "environmental voting." This variable measures how each state's congressional delegation voted on environmental issues during 1984. It reflects the percentage of times that their votes agreed with the recommendations of an environmentalist group, the League of Conservation Voters. Idaho's value of 12, for example, means its representatives agreed with the League on about 12% of the environmental issue votes. The higher this percentage, the more strongly environmentalist a state's representatives are deemed to be.

Congressional voting in 1984 ranged from Idaho's 12% agreement to Vermont's 96% agreement with the League's legislative recommendations. Note that the three lowest states are western and the three highest states are in New England. Listing data in low-to-high order is often a useful step in preliminary analysis.

Looking down the list in Table 2.4, we can see that the first three voting percentages begin with the number 1, the fourth begins with a 2, and the next eight all begin with a 3. This elementary observation is the basis for a *stem-and-leaf display,* shown in Figure 2.6 (page 33). Stem-and-leaf displays are compact versions of the ordered data, with initial digits broken off to form *stems,* shown to the left of the vertical line in Figure 2.6. To the right of this

TABLE 2.4   *Ordered List of 50 States' 1984 Congressional Votes: Percentage of Agreement with Environmentalists*

	State	Environmental voting percentage		State	Environmental voting percentage
1	Idaho	12%	26	S. Dakota	55%
2	Utah	16	27	Illinois	56
3	Alaska	17	28	Montana	56
4	Wyoming	26	29	Missouri	56
5	Alabama	33	30	Ohio	57
6	Mississippi	33	31	Washington	57
7	Virginia	33	32	California	59
8	Nebraska	34	33	N. Dakota	59
9	Arizona	35	34	Maryland	62
10	Arkansas	36	35	Pennsylvania	62
11	Texas	39	36	Hawaii	64
12	Kansas	39	37	Delaware	69
13	Louisiana	40	38	Michigan	69
14	Kentucky	41	39	W. Virginia	69
15	N. Carolina	44	40	Minnesota	70
16	Tennessee	45	41	New York	72
17	New Mexico	47	42	Wisconsin	72
18	Nevada	47	43	New Hampshire	72
19	S. Carolina	47	44	New Jersey	72
20	Colorado	47	45	Iowa	74
21	Georgia	49	46	Maine	79
22	Florida	51	47	Connecticut	79
23	Oklahoma	52	48	Massachusetts	82
24	Oregon	53	49	Rhode Island	86
25	Indiana	54	50	Vermont	96

Source: Data from League of Conservation Voters (1985).

line are the *leaves*, or following digits, for each case in the data. In Figure 2.6 the stems are "tens" and the leaves are "ones." Thus Idaho, with 12% environmental voting, is represented by the 2 leaf (meaning "two ones") to the right of the 1 stem (meaning "one ten"). Utah, at 16%, has the 6 leaf on the same stem, and Alaska (17%) has the 7 leaf. Vermont (96%) is represented by the 6 leaf to the right of the 9 stem.

> **Stem-and-leaf displays** use the initial and following digits of data values. Initial digits determine the **stems,** and following digits become the **leaves.**

Stem-and-leaf displays present an ordered list of the data so compactly and graphically that many features of the distribution become obvious at a

```
1 | 267
2 | 6
3 | 33345699
4 | 014577779
5 | 123456667799
6 | 224999
7 | 02222499
8 | 26
9 | 6
```

FIGURE 2.6

Stem-and-leaf display of 50 states' environmental voting, from Table 2.4. (Stems digits are 10's, leaves digits are 1's. 1 | 2 means 12% agreement with environmentalists.)

glance. For example, Figure 2.6 shows, as we already knew, that the voting percentages range from 12 to 96. We can also see a clear center in the mid-50's, so we might guess the average falls somewhere around 55. Roughly equal numbers of states appear on either side of this center, with most of the states in the 30's through 70's.

Stem-and-leaf displays originated as a paper-and-pencil method. Unlike some other analytical techniques in this book, they can be prepared quickly by hand. Constructing Figure 2.6 from the ordered list in Table 2.4 takes little time or thought. Starting from an unordered list is not much harder. To illustrate, consider the life expectancies in the 25-nation sample from Chapter 1, as shown in Table 2.5. The range, from 37 years (Cambodia) to 75 years (Norway), suggests stems of 3, 4, 5, 6, and 7. Selecting the stems is the first step in constructing a stem-and-leaf display; see Figure 2.7 (page 34), left.

TABLE 2.5   *Life Expectancies in a Sample of 25 Countries*

Country	Life expectancy in years	Country	Life expectancy in years
Nicaragua	58	Indonesia	50
Paraguay	65	North Korea	66
Venezuela	68	Mongolia	64
France	74	Taiwan	72
West Germany	73	Australia	74
Greece	73	Congo	48
Norway	75	Ethiopia	41
Czechoslovakia	71	Guinea	44
Austria	72	Mauritania	44
Jordan	61	Nigeria	49
Sri Lanka	67	Togo	48
Brunei	66	Zaire	48
Cambodia	37		

1. Select an appropriate set of stems.   2. Add leaves from data, unordered.   3. Rearrange leaves, in order.

3		3	7	3	7
4		4	8144988	4	1448889
5		5	80	5	08
6		6	5817664	6	1456678
7		7	43351224	7	12233445

FIGURE 2.7

Steps in constructing a stem-and-leaf display from unordered data: Life expectancy in years from Table 2.5. (Stems digits are 10's, leaves digits are 1's. 3 | 7 means 37 years.)

The next step is to add leaves, or following digits, by simply reading them out of the data. The first life expectancy in Table 2.5 is Nicaragua's 58, so we write an 8 to the right of the 5 stem. Next is Paraguay's 65, so we write a 5 beside the 6 stem. Continuing in this manner produces a stem-and-leaf display where the stems are in order but the leaves are not (Figure 2.7, center). The last step is to rearrange the leaves on each stem to put them in order, too (Figure 2.7, right).

Most striking in the shape of the life expectancy distribution is its two distinct peaks, with a gap between. The first peak represents a group of nations with low life expectancy having values in the 40's. The second peak is made up of countries with higher values, in the 60's and 70's. Only two countries are in between, in the 50's.

The final stem-and-leaf display in Figure 2.7 provides an ordered list of the data values. We can easily read from this display that the lowest value is 37 years, followed by 41, two 44's, and so on. The five highest values are 73's, 74's, and one 75. In later chapters we will see further analytical uses for such ordered lists. Along with their other advantages, stem-and-leaf displays are an efficient way to put a list of numerical values in order.

PROBLEMS

4. The accompanying table has data on 14 major airlines during October and November, 1987.

Airline	Consumer complaints per 100,000 passengers	% flights arriving on time	Baggage complaints per 1,000 passengers
American	4.0	86.1%	7.3
Southwest	1.6	85.2	3.9
Continental	17.2	84.4	7.1
Piedmont	2.2	83.2	6.5
Eastern	12.4	83.0	2.9
United	5.1	80.7	10.3

(continued)

Airline	Consumer complaints per 100,000 passengers	% flights arriving on time	Baggage complaints per 1,000 passengers
Trans World	9.5	79.4	7.4
Pan American	11.8	79.2	4.0
Delta	1.9	77.5	6.0
USAir	3.4	77.3	5.5
Northwest	19.1	76.5	10.6
Alaska	2.6	75.2	7.4
America West	2.5	74.9	7.2
Pacific Southwest	2.5	60.3	4.1

Source: Data from the *Boston Globe,* December 3, 1987.

Construct a stem-and-leaf display for baggage complaints per 1,000 passengers. Use 1's as stems digits and .1's as leaves. As with any stem-and-leaf display, include a label stating what the stems and leaves digits represent, like that in Figures 2.6 and 2.7. Describe the shape of this distribution.

5. After dropping (*not* rounding off) the decimal values in Problem 4, construct and label a stem-and-leaf display for the percentage of flights arriving on time. Why does this display not reveal much about distributional shape?

6. Make a stem-and-leaf display of the consumer complaints per 100,000 airline passengers in Problem 4. Why is this display also unsatisfactory?

2.4   DOUBLE-STEM AND FIVE-STEM VERSIONS

The examples of stem-and-leaf displays given so far have been simple. What happens, though, if the variable's values contain more than two digits, or if they all begin with the same first digit? What if there are too few or too many stems, as happened in Problems 5 and 6? Then the simple approach described above will not work. Fortunately there are ways to adjust the display, as described in the box. Treatment of outliers will be left for a later chapter, when we can define more rigorously what such extreme values are.

Stem-and-leaf displays can be adjusted in several ways.

1. Use **truncation** (cutting off) instead of rounding, if there are more than two digits.
2. **Trim outliers,** listing extremely high or low values separately so they do not distort the scale for the whole display.
3. **Double-stem** and **five-stem** versions spread out the display over a reasonable number of lines.

TABLE 2.6

Overnight Camping in U.S. National Parks (in Thousands), 1960–1986

Year	Back country nights	Tent nights	Recreational vehicle nights	Tent + R.V. nights
1960	—	3,585.5	1,260.2	4,845.7
1961	—	3,586.7	1,473.2	5,059.9
1962	—	4,307.4	1,811.6	6,119.0
1963	—	4,621.2	2,149.5	6,770.7
1964	—	5,043.8	2,413.8	7,457.6
1965	—	5,104.4	2,980.5	8,084.9
1966	—	5,062.1	3,938.3	9,000.4
1967	—	4,715.3	4,594.3	9,309.6
1968	—	4,788.1	4,623.0	9,411.1
1969	—	4,390.1	4,659.1	9,049.2
1970	—	4,684.6	4,339.3	9,023.9
1971	1,095.9	3,482.5	4,451.7	7,934.2
1972	1,494.5	3,651.6	4,731.4	8,383.0
1973	1,953.5	3,801.8	4,883.3	8,685.1
1974	2,172.2	3,791.5	4,620.3	8,411.8
1975	2,346.4	3,735.3	5,081.7	8,817.0
1976	2,608.9	3,870.3	5,397.0	9,267.3
1977	2,569.5	3,973.2	5,345.4	9,318.6
1978	2,589.9	3,775.5	5,353.6	9,129.1
1979	2,397.1	3,424.7	4,446.4	7,871.1
1980	2,395.2	3,934.1	4,378.5	8,312.6
1981	2,329.8	4,221.9	4,663.0	8,884.9
1982	2,424.2	4,154.0	4,596.0	8,750.0
1983	2,579.7	3,601.2	4,232.8	7,834.0
1984	1,978.9	3,747.3	3,943.2	7,690.5
1985	1,680.4	3,586.1	3,759.3	7,345.4
1986	1,644.7	3,360.0	3,788.5	7,148.5

Source: Data from United States National Park Service (1986).

Examine the data in Table 2.6 on camping in U.S. National Parks. The number of tent camping nights ranges from 3,360.0 thousand in 1986 to 5,104.4 thousand in 1965. These values have five digits. To make a stem-and-leaf display, which can accommodate only about two digits, should we round off—round 3,585.5 up to 3,600, for instance? Rounding off makes it harder to reconstruct the ordered list of cases from a stem-and-leaf display, however. For this reason the even simpler method of *truncation* is preferred. Truncation involves cutting off, rather than rounding off: 3,585.5 truncates to 3,500. Similarly, 3,360.0 becomes 3,300, 3,424.7 becomes 3,400, and so on.

Truncation takes care of the extra digits, but a second complication remains. The truncated values range from 3,300 to 5,100, so simply taking their initial digits for stems gives us just three stems—3, 4, and 5—with 16 of the

27 leaves piled up by the 3 stem. It would be more informative if we could stretch out the display a bit. A *double-stem* version is one way to do this: We can create *two* stems for each initial digit. For instance, we can split the values whose first digit is 3 into one stem for 3,000.0 to 3,499.9 and a second stem for 3,500.0 to 3,999.9.

This has been done in the double-stem stem-and-leaf display in Figure 2.8. The first stem (written as 3*) contains values whose first digit (stem) is 3 and whose second digit (leaf) is 0, 1, 2, 3, or 4. The second stem (written as 3.) contains values whose stem is 3 and whose leaves are 5, 6, 7, 8, or 9. Similarly, 4* means values from 4,000.0 to 4,499.9, which have stems of 4 and leaves of 0 to 4.

Double-stem stem-and-leaf displays have two stems for each initial digit. For example:

1*   Initial digit 1, second digits of 0–4
1.   Initial digit 1, second digits of 5–9

The asterisk (*) in a double-stem stem-and-leaf display always marks the stem that holds leaves of 0–4, and the period (.) denotes leaves of 5–9. To help remember these symbols, think of their usual order in a footnote, which typically begins with an asterisk and ends with a period.

Sometimes double-stem versions of stem-and-leaf displays are still not spread out enough, and we need to go to *five-stem* versions. Five-stem versions use five stems for each initial digit, instead of just two as in double-stem versions. Figure 2.9 (page 38) shows how, using the ordered list in Table 2.7 of homicide rates for the 50 U.S. states. All of these homicide rates have tens digits of 0, 1, or 2, so a single-stem version would cram 50 leaves onto just three stems. Even a double-stem version would be too crowded to show much detail.

Five-stem versions still use the asterisk (*) and period (.) symbols, but with a new meaning. The 1* stem in Figure 2.9 holds values whose first digit is 1 (one ten) and whose second digit, after truncation, is 0 or 1. That is, it holds homicide rates from 10.0 to 11.9. The 1. stem in Figure 2.9 holds rates whose second digit is 8 or 9: rates from 18.0 to 19.9. The asterisk and the

3*	344
3.	5556677778899
4*	1233
4.	6677
5*	001

Double-stem stem-and-leaf display of the tent camping data from Table 2.6. (Stems digits are millions, leaves are 100,000's. 3* | 3 means 3,300,000 camping stays.)

FIGURE 2.8

TABLE 2.7   *Ordered List of 1980 Rates of Homicide per 100,000 Population in 50 States*

	State	*Homicide rate*		*State*	*Homicide rate*
1	South Dakota	.7	26	Virginia	8.6
2	North Dakota	1.2	27	Hawaii	8.7
3	Vermont	2.2	28	Kentucky	8.8
4	Iowa	2.2	29	Indiana	8.9
5	New Hampshire	2.5	30	Arkansas	9.2
6	Minnesota	2.6	31	Maryland	9.5
7	Maine	2.8	32	Alaska	9.7
8	Wisconsin	2.9	33	Oklahoma	10.0
9	Idaho	3.1	34	Michigan	10.2
10	Utah	3.8	35	Arizona	10.3
11	Montana	4.0	36	North Carolina	10.6
12	Massachusetts	4.1	37	Illinois	10.6
13	Nebraska	4.4	38	Tennessee	10.8
14	Rhode Island	4.4	39	Missouri	11.1
15	Connecticut	4.7	40	South Carolina	11.4
16	Oregon	5.1	41	New York	12.7
17	Washington	5.5	42	New Mexico	13.1
18	Wyoming	6.2	43	Alabama	13.2
19	Pennsylvania	6.8	44	Georgia	13.8
20	Colorado	6.9	45	Florida	14.5
21	Delaware	6.9	46	Mississippi	14.5
22	Kansas	6.9	47	California	14.5
23	New Jersey	6.9	48	Louisiana	15.7
24	West Virginia	7.1	49	Texas	16.9
25	Ohio	8.1	50	Nevada	20.0

```
0*  | 01
0t  | 22222233
0f  | 4444455
0s  | 6666667
0.  | 88888999
1*  | 00000011
1t  | 2333
1f  | 4445
1s  | 6
1.  |
2*  | 0
```

Five-stem stem-and-leaf display of 50 state homicide rates (Data from Table 2.7) (Stems digits are 10's, leaves digits are 1's. 1t | 2 means 12 victims per 100,000 population.)

FIGURE 2.9

period come first and last for each initial digit, as they do in double-stem versions and in footnotes.

Between the 1* stem (10.0–11.9) and the 1. stem (18.0–19.9) are stems labeled 1t, 1f, and 1s. The "t" stands for leaves of *two* and *three*, so 1t holds values from 12.0–13.9, which truncate to leaves of 2 and 3. Similarly the "f" stands for leaves of *four* and *five*, so 1f holds rates from 14.0–15.9. The "s" stands for leaves of *six* and *seven*, so 1s holds rates from 16.0–17.9.

Five-stem stem-and-leaf displays have five stems for each initial digit. For example:

1* Initial digit 1, second digits of 0–1
1t Initial digit 1, second digits of 2–3
1f Initial digit 1, second digits of 4–5
1s Initial digit 1, second digits of 6–7
1. Initial digit 1, second digits of 8–9

Figure 2.9 shows the highest homicide rate as 2* | 0, meaning a value that truncates to 20. This belongs to Nevada, which stands apart as an outlier. The second highest homicide rate is shown as 1s | 6, or a value that truncates to 16 (Texas). The distribution is positively skewed because most states have low or moderate homicide rates, but a handful are much higher.

Double-stem and five-stem displays are constructed from raw data following the same steps shown in Figure 2.7: First select appropriate stems, then add leaves, and finally put the leaves in order. Selecting the appropriate stems is the only tricky part of this procedure.

Histograms, frequency polygons, and stem-and-leaf displays all provide the same basic information about distributional shape. When grouping is necessary, the stem-and-leaf display retains some detail that the histogram or polygon must leave out. Other advantages of stem-and-leaf displays are their easy construction and ordered listing of the data. Stem-and-leaf displays are impractical for samples with more than a few hundred cases, however. By contrast, since the vertical axis can denote proportions or percentages instead of frequencies, histograms and polygons can represent distributions even with an infinite number of cases.

PROBLEMS

7. Return to the airline delays data in Problem 4. Construct a double-stem stem-and-leaf display for the percentage of flights arriving on time. Are any features of distributional shape now apparent that could not be seen in a single-stem version (Problem 5)?

8. Construct a double stem stem-and-leaf display for consumer complaints per 100,000 passengers in Problem 4. How does this display compare with the single-stem version attempted in Problem 6? What does its shape tell us about the distribution of complaints among these airlines?

9. Construct a five-stem version stem-and-leaf display of the nicotine concentration data in Table 2.1, with stems representing 100's and leaves 10's. Compare this stem-and-leaf display with the histogram in Figure 2.1.

2.5   *GRAPHING NEGATIVE NUMBERS**

The graphical methods just described require no modifications to work with negative numbers. Just remember that the larger a negative number is, the *lower* it is. Otherwise displays with negative numbers follow the same rules as displays with nonnegative numbers.

Table 2.8 shows data from a sample consisting of 33 households in Concord, New Hampshire. During 1980 and 1981 a drought lowered the municipal water supply, and a campaign to conserve water eventually reduced citywide water use by about 15%. Of course not everyone's use declined by 15%; some households saved even more, while others saved little or none. The variables in Table 2.8 include several household details from a survey questionnaire plus actual summer usage measurements from water meter readings.

A crude measure of household water savings is the difference between 1980 (pre-shortage) and 1981 (post-shortage) water use (right-hand column, Table 2.8). Because savings are defined as 1980 use minus 1981 use, a positive value means that less water was used. Household 1 saved 3,600 cubic feet, for instance. Household 11 made negative savings; despite the appeal for conservation, their water use increased by 100 cubic feet. The 33 values are sorted from lowest to highest in Table 2.9 (page 42).

A stem-and-leaf display of the water savings distribution is given in Figure 2.10. The distribution is sharply peaked, meaning most usage did not change greatly. A few households in the tails of the distribution made large positive or negative changes. The majority have positive values, indicating they managed at least some savings.

Water use in three households did not change at all, giving zero values for savings. Since it is arbitrary whether to graph such values on the −0 or the 0 stem, they are divided between the two. Notice the order of the leaves on the −0 stem. Larger negative numbers mean lower values, so the order of leaves on negative stems is opposite that on positive stems: 3, 2, 1 instead of 1, 2, 3.

*Optional material not needed for unstarred sections of later chapters.

TABLE 2.8 ***Characteristics and Summer Water Use of 33 Households During a Water Shortage***

ID	Head: years in school	House. income $1,000s	Head retired	Savings self-estimate	1980 water use[a]	1981 water use	Water savings[b]
1	12	15	No	None	5,700	2,100	3,600
2	12	30	No	None	3,300	2,600	700
3	18	25	No	None	1,200	1,200	0
4	20	53	No	None	4,600	2,300	2,300
5	16	35	No	None	1,800	1,700	100
6	14	10	No	Some	2,300	1,600	700
7	12	30	No	Some	5,900	2,900	3,000
8	16	15	Yes	None	2,500	2,000	500
9	12	20	Yes	Some	800	700	100
10	12	15	No	None	1,700	1,000	700
11	20	50	No	Some	4,300	4,400	−100
12	13	10	Yes	None	5,800	3,000	2,800
13	12	25	No	Some	3,700	2,900	800
14	14	35	Yes	Some	200	100	100
15	12	15	No	None	200	300	−100
16	18	5	Yes	None	1,100	600	500
17	16	25	No	None	2,800	4,000	−1,200
18	12	30	No	Some	4,100	4,200	−100
19	12	20	No	Some	2,100	1,700	400
20	20	20	No	None	2,400	2,400	0
21	13	30	No	None	2,700	2,800	−100
22	20	45	No	None	2,400	1,800	600
23	16	45	No	Some	3,000	3,100	−100
24	12	5	Yes	None	600	600	0
25	14	30	No	None	1,300	1,100	200
26	12	25	No	None	1,800	1,300	500
27	16	10	Yes	Some	2,400	2,300	100
28	20	10	Yes	None	1,700	600	1,100
29	16	100	No	None	6,300	9,200	−2,900
30	10	5	Yes	Some	1,400	1,100	300
31	20	52	No	None	2,700	3,400	−700
32	16	15	No	None	3,700	4,400	−700
33	12	30	No	None	2,100	2,600	−500

[a]Water use is measured from meter readings in cubic feet consumed during the three-month summer period.
[b]Savings are defined as pre-shortage (1980) use minus post-shortage (1981) use, measured in cubic feet of water.
Source: These 33 households are a random sample from a larger data set described in Hamilton (1983).

TABLE 2.9   **Ordered List of Cubic Feet of Water Saved in Drought by 33 Households from Table 2.8**

ID	1980–1981 water savings	ID	1980–1981 water savings
17	−2,900	13	200
4	−1,200	26	300
11	−700	24	400
19	−700	30	500
3	−500	27	500
2	−100	6	500
9	−100	10	600
14	−100	21	700
23	−100	1	700
15	−100	20	700
25	0	7	800
22	0	29	1,100
8	0	12	2,300
32	100	31	2,800
28	100	5	3,000
33	100	18	3,600
16	100		

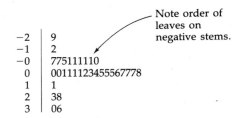

```
−2 │ 9
−1 │ 2
−0 │ 775111110
 0 │ 00111123455567778
 1 │ 1
 2 │ 38
 3 │ 06
```

Note order of leaves on negative stems.

FIGURE 2.10   *Single-stem stem-and-leaf display of 33 households' water savings (data from Table 2.9). Zero values split between −0 and 0 stems. (Stems digits are 1,000's, leaves are 100's. 3 | 6 means water savings of 3,600 cubic feet.)*

PROBLEM

*10. The accompanying table contains data from a study of the validity of survey self-reports. A sample of college students reported test scores on a survey, which were later compared with actual values from college records. Errors are defined as true minus self-reported score; a negative error means that the true score is lower than that self-reported. Construct a double-stem stem-and-leaf display for these data, using stems of $-1*$, $-0.$, $-0*$, $0*$, and $0.$; divide zeroes evenly between the $-0*$ and $+0*$ stems. Describe what this display shows about the self-report errors.

Self-Reported and True Verbal SAT Scores of 16 College Students

Student	True VSAT	Self-reported VSAT	Error: True minus self-report
1	570	590	-20
2	480	480	0
3	560	670	-110
4	340	480	-140
5	460	470	-10
6	620	560	60
7	470	500	-30
8	410	511	-101
9	540	670	-130
10	520	520	0
11	380	480	-100
12	430	500	-70
13	580	590	-10
14	450	450	0
15	490	490	0
16	460	450	10

2.6 TIME PLOTS

Time plots show a variable's changes over time. They require **time series** data, like those in Table 2.10 (page 44). This table reports the annual world-wide whale catch, at 5-year intervals from 1920 to 1985. In such time series the rows (cases) are a sequence of points or intervals in time. Another time series is seen in the National Parks camping data of Table 2.6.

From time plots we can read the story of how a variable has changed. Blue whales are the largest and commercially most valuable whales. As Figure 2.11 shows, they were hunted intensely during the 1930's and early 1940's. World War II brought a sharp drop in whaling activity; after the war whale hunting increased again. But the blue whale catch did not rise much; most of these whales were gone. Despite international protection, by the 1970's the blue whale population was only 6% of its original size.

A second time plot is shown in Figure 2.12 (page 45), based on the tent camping data of Table 2.6. Figure 2.12 shows that tent camping peaked during the 1960's, fluctuating at much lower levels for most of the 1970's and 1980's.

TABLE 2.10　　　　　***World Whale Catch by Species, 1920–1985***

Year	Blue	Fin	Sei	Humpback	Sperm	All whales[a]
1920	2,274	4,946	1,120	545	749	9,634
1925	7,548	9,121	1,093	3,342	1,439	22,543
1930	19,079	14,281	841	1,919	1,126	37,246
1935	16,834	14,078	962	4,088	2,238	38,200
1940	11,559	19,924	541	528	4,091	36,643
1945	1,111	2,653	218	303	1,661	5,946
1950	6,313	22,902	2,471	5,063	8,183	44,932
1955	2,495	32,185	1,940	2,713	15,594	54,927
1960	1,465	31,064	7,035	3,576	20,344	63,484
1965	613	12,351	25,454	452	25,548	64,418
1970	0	5,057	11,195	0	25,842	46,633
1975	0	1,634	4,975	17	21,045	38,892
1980	0	472	102	16	2,091	15,129
1985	0	218	38	8	400	8,350

[a]Total includes other species not listed separately.
Source:　Council on Environmental Quality (1987).

A second smaller peak in the early 1980's preceded another decrease. We can now see that the few years with relatively high levels of camping, which formed the long upper tail of the frequency distribution in Figure 2.8, occurred during the mid-1960's. The three highest years were 1964, 1965, and 1966.

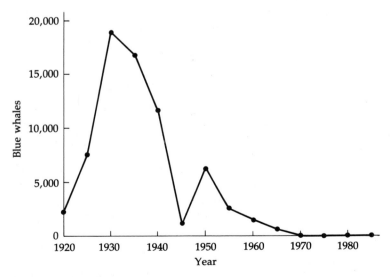

FIGURE 2.11　　　*Time plot of blue whale catch*

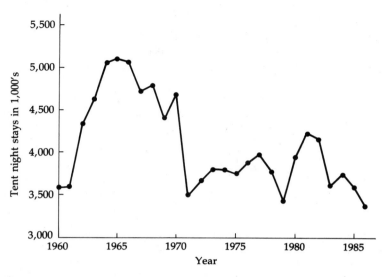

FIGURE 2.12   *Time plot of tent camping in parks*

Such findings raise further questions. Why did tent camping drop off, despite a continually increasing U.S. population during the past two decades? To answer, we would need more information. This brings out another strength of time plots: We can readily add a second or even third variable to the display. For example, perhaps tent camping peaked in the 1960's because so many families were raising "baby-boom" children and unable to afford more expensive vacations. In the 1970's these children had outgrown the prime years for family camping. Also, more Americans could afford the mechanized camping offered by recreational vehicles. Additional time plots for family incomes, recreational vehicles, or other time series could be examined to evaluate these different explanations.

Yearly blue whale catch varied from 0 to 19,079 in Figure 2.11, so the vertical axis was scaled from 0 to 20,000. National Parks tent camping (Figure 2.12) varied only from 3,360.0 to 5,104.4, however, so a vertical scale that began at zero would make a graph with less detail and lots of empty space. Time series variations show up best if we draw the vertical scale not much wider than the range spanned by the actual data. This creates a trap for unwary readers, however. If we had scaled the vertical axis from 0–5,500 in Figure 2.12, the early-1970's fall in tent camping would look moderate. With a restricted scale of 3,000–5,500, the same fall looks more dramatic.

The choice of vertical scale in time plots influences whether any change looks large or small. With a restricted scale, small changes can look like huge peaks and valleys. With a wide enough scale, even large changes can flatten out. Sometimes the vertical axis in a time plot is manipulated deliberately to try to persuade you of something. More often, as in Figure 2.12, the vertical scale is chosen to display as much detail as possible about changes in the variable. In either case, it is essential to *read the scale* and make sure you understand it before trying to interpret a time plot.

PROBLEMS

11. Fin and sei whales, smaller relatives of the blue whale, are the first and second most profitable whales to hunt, after blues. Construct a time plot with separate lines for the fin and sei whale catches from Table 2.10. Use pen and pencil, or two different colors, to distinguish the two lines. Compare your plot with that for the blue whale catch (Figure 2.11). What story do these time plots tell?

12. Return to the National Parks camping data of Table 2.6, and construct a time plot with separate lines for back country nights and recreational vehicle nights. Compare your plot with that for tent camping nights (Figure 2.12), and write a paragraph describing what you see. Does it appear that the decline in tent camping could be due mainly to a switch to recreational vehicles? Can you explain the rise and fall of back-country camping?

2.7   TIME SERIES SMOOTHING*

Changes in the time series shown in Figures 2.11 and 2.12 are mostly gradual. Some variables change more rapidly, however, giving rise to jagged-looking time plots in which it is hard to see any pattern. *Smoothing* is a technique that removes some of the jaggedness from such plots, making them easier to read.

Figure 2.13 is a time plot of the incidence of malignant melanoma (a type of skin cancer) in Connecticut during 1936–1972. Melanoma rates are given as cases per 100,000 people, adjusted for the age composition of the population. Despite year-to-year fluctuations, the overall incidence is obviously rising.[3]

Figure 2.14 is a smoothed plot of the same melanoma data. Smoothing reveals something that was not obvious in the raw data plot: In addition to the general upward **trend,** melanoma rates seem to follow a **cycle** with peaks every 10 years or so. Time series often contain trends and cycles; smoothing helps us to see them.

What could explain the cycles in melanoma incidence? Radiation from the sun is thought to cause melanoma; the general upward trend in Figure 2.14 may partly reflect the growing popularity of tanning. Solar radiation and sunspot activity also follow a cycle with regular peaks. Figure 2.15 (page 48) shows a smoothed time plot of sunspot counts superimposed on the melanoma incidence plot. Peaks in melanoma incidence follow a few years after

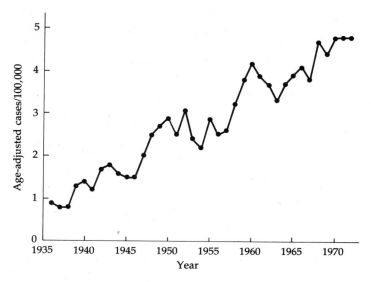

FIGURE 2.13   *Time plot of melanoma incidence, 1936–1972*

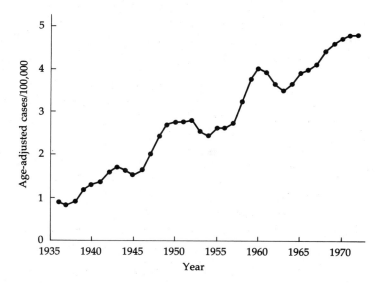

FIGURE 2.14   *Smoothed plot of melanoma incidence*

peaks in sunspot activity. This graph provides visual evidence of a link between solar cycles and cancer rates.

Simple smoothing techniques are needed to interpret the data in Table 2.11, which were assembled by a school board committee trying to forecast enrollment trends in a rural school district. One challenge of such forecasting

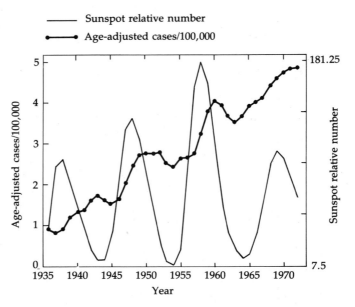

———— Sunspot relative number

•——•——• Age-adjusted cases/100,000

FIGURE 2.15   *Smoothed plots of melanoma and sunspots*

TABLE 2.11   ***Births and School Enrollment in Oyster River School District, 1980–1987***

Year	Number of births	Kindergarten enrollment	Grade 12 enrollment
1980	99	82	150
1981	74	77	147
1982	88	99	132
1983	103	84	127
1984	92	93	123
1985	110	122	145
1986	109	114	143
1987	118	121	138

is distinguishing between short-term fluctuations and longer-term trends. For example, notice that births declined in 1981, 1984, and 1986, but rose in 1982, 1983, 1985, and 1987. In the long run are births going up, down, or staying the same? General trends may be more apparent if we smooth the yearly fluctuations.

Table 2.12 shows a **moving average** smoothing of the birth data. Smoothing creates a new time series, X^*, by transforming values from the raw data time series, X. The endpoint (1980 and 1987) values are copied from the raw

TABLE 2.12   ***Smoothing by Moving Average of Span 3: Births in Oyster River School District (from Table 2.11)***

Year	Raw data	Smoothing transformation	Smooth	Rough
	X_i		X_i^*	$X_i - X_i^*$
1980	99	endpoints copied	99.0	0
1981	74	(99 + 74 + 88)/3	87.0	−13.0
1982	88	(74 + 88 + 103)/3	88.3	−.3
1983	103	(88 + 103 + 92)/3	94.3	8.7
1984	92	(103 + 92 + 110)/3	101.7	−9.7
1985	110	(92 + 110 + 109)/3	103.7	6.3
1986	109	(110 + 109 + 118)/3	112.3	−3.3
1987	118	endpoints copied	118.0	0

data. The smoothed value for 1981 is the average of the number of births in the previous (1980), current (1981), and next (1982) years. For 1982 we average the numbers from 1981, 1982, and 1983, and so on through the series. The number of values included in each average is called the **span,** so Table 2.12 illustrates a *moving average of span 3*.[4] Other spans could also be used.

Smoothing by a **moving average of span 3** involves replacing each value in the original series, X_i, with a new value, X_i^*:

$$X_i^* = \frac{X_{i-1} + X_i + X_{i+1}}{3}$$ [2.1]

In the right-hand column of Table 2.12 is the *rough,* or difference between raw and smoothed data. Here we can see clearly what smoothing has done. Some values were only slightly changed, but smoothing brought up the momentary dips of 1981 and 1984, and flattened out the spike of 1983. In some analyses the rough is more interesting than the smooth, since it highlights any sudden large changes.

Smoothing divides raw data values (X_i) into two parts.

Smooth (X_i^*) values are produced by the smoothing process.
Rough ($X_i - X_i^*$) values are left over when the smooth is subtracted from the raw data.

It follows that

Data = Smooth + Rough

The moving average definition (equation 2.1) could be rewritten as

$$X_i^* = \left(\frac{1}{3}\right)X_{i-1} + \left(\frac{1}{3}\right)X_i + \left(\frac{1}{3}\right)X_{i+1}$$

This version shows that we are giving the previous, current, and next values equal ($\frac{1}{3}$) weight in computing each average. An alternative technique called *hanning* gives the previous and next values lower weights ($\frac{1}{4}$) than the current value ($\frac{1}{2}$). The sunspot and cancer series in Figure 2.15 were smoothed by hanning. Table 2.13 shows results from a hanning smooth of the birth data. The hanned value for 1981, for example, is

$$\left(\frac{1}{4}\right)99 + \left(\frac{1}{2}\right)74 + \left(\frac{1}{4}\right)88 = 83.75$$

Endpoints are copied as before.

> **Smoothing** by **hanning** involves replacing each value in the original series, X_i, with a new value, X_i^*:
>
> $$X_i^* = \left(\frac{1}{4}\right)X_{i-1} + \left(\frac{1}{2}\right)X_i + \left(\frac{1}{4}\right)X_{i+1} \qquad [2.2]$$

Table 2.13 also shows results from a third smoothing technique called the *running median*. A *median* is the value in the middle of an ordered list of data. The median smoothed value for 1981 is 88, because if we put the three surrounding values (99, 74, and 88) in order (74, 88, 99), 88 is in the middle. Similarly the smoothed value for 1985 is the median of 92, 110, and 109, which is 109. Like moving averages, running medians can be found for spans other than 3.

TABLE 2.13

Smoothing by Hanning and by Running Medians of Span 3: Births in Oyster River School District (Data from Table 2.11)

Year	Raw data	Hanning smoothed	Running median span 3 smoothed
1980	99	99.00	99
1981	74	83.75	88
1982	88	88.25	88
1983	103	96.50	92
1984	92	99.25	103
1985	110	105.25	109
1986	109	111.50	110
1987	118	118.00	118

> **Smoothing** by **running medians of span** 3 involves replacing each value
> in the original series, X_i, with a new value, X_i^*:
>
> $$X_i^* = \text{median}\{X_{i-1}, X_i, X_{i+1}\} \qquad\qquad [2.3]$$

Figure 2.16 contains time plots of the raw and smoothed birth data. In
this example all three smoothing techniques produce similar graphs showing
an overall upward trend in number of births since 1981. In general hanning
tends to change raw data the least, so it is appropriate when only mild
smoothing is required. Running medians are the most powerful technique
for smoothing data with wild one-time fluctuations.

Compound smoothers are procedures that smooth the raw data several
times in succession. For example, we could first smooth the data by running
medians of span 5, then smooth the smoothed series itself by hanning. Certain combinations work well together, and computer programs for compound
smoothing are a valuable aid in analyzing time series data.

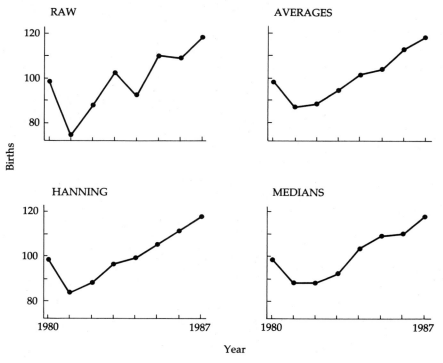

FIGURE 2.16   *Raw and smoothed time plots of births*

PROBLEMS

*13. Apply a running median (span 3) smooth to the kindergarten enrollment data of Table 2.11.

   a. List raw data, smooth, and rough values in a table.

   b. Construct a time plot of the smooth series.

   c. Describe what you see.

*14. Smooth the 12th grade enrollment in Table 2.11 by hanning. Graph the results and describe what you see.

2.8   *THE USES OF GRAPHS*

Statistical graphs have two main uses. First, they help to *present information to others.* This use, **presentation graphics,** is probably the one you are most familiar with. The author of a study already understands his or her conclusions but graphs help convey them to the intended audience, which is usually less knowledgeable than the study's author or less patient in analyzing tables with numbers. Graphs then save the audience the trouble of careful reading, add visual appeal, and present complex information in the clearest way.

Presentation graphics are tools for displaying the data *after* analysis is complete. **Analytical graphics,** in contrast, come earlier in the research process. Analytical graphics, like the examples in this chapter, are displays that inform the researcher about the data. They often reveal unexpected details of variable distributions or show broad patterns that would otherwise go unnoticed.

Unexpected features such as outliers, gaps, or multiple peaks are important because they can cause problems in statistical analysis. If graphs reveal such potential problems, then we can take steps to deal with them. Without graphs, we might remain unaware of statistical problems that distort our conclusions.

Statistical analysis is easier when variables have a symmetrical, bell-shaped frequency distribution (Figure 2.17)—the statistical ideal. Like most ideals, this one is not always a good guide to reality. Of the variable distributions we have seen so far, only a few are at all symmetrical (for example, environmental voting in Figure 2.6 or water savings in Figure 2.10), and even they are not the bell-shaped ideal. Most other examples were skewed, with tails much heavier on one side than the other, and their highest values sometimes stood so far from the rest as to be considered outliers. We have also seen examples with multiple peaks and gaps.

In a relatively well-behaved distribution like environmental voting (Figure 2.6), saying that the average voting percentage is 53.5 summarizes the data reasonably well. This value lies in the obvious central area of the distribution. The distribution of nicotine concentration in airliners (Figure 2.1), by

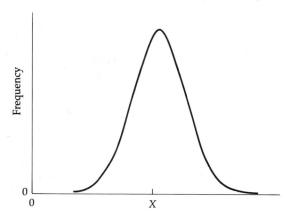

FIGURE 2.17   *Bell-shaped frequency distribution*

contrast, is not well behaved. Relatively low concentrations were measured on most of the flights, but a few highly polluted flights pull up the average concentration (22.9 $\mu g/m^3$), making the average exceed the concentrations measured on 18 of the 26 flights. When averages are untrustworthy, it becomes harder to give simple answers to research questions like "How do nicotine concentrations in smoking and nonsmoking sections compare?"

Whether it is well- or ill-behaved, a distribution's shape can reveal interesting things about the real world. But from a statistical point of view, departures from the bell-shaped ideal create problems. The fact that a widely used summary measure, the average, can be misleading with ill-behaved distributions illustrates these problems. It also illustrates the value of graphical analysis. Without a graph showing the distributional shape, we might not even notice potential problems like skewness or outliers. If we overlook basic distributional features, we risk drawing nonsensical conclusions from our summary measures.

PROBLEM

15. Identify a variable we have studied, besides nicotine concentration (Figure 2.1), that seems to have an ill-behaved distribution: that is, a distribution characterized by extreme outliers, skewness, or major gaps. Explain these distributional features in real-world terms.

SUMMARY

Simple techniques for analytical graphics were introduced in this chapter. Analytical graphics are visual displays used by the researcher to gain insights into the data. They differ in purpose from presentation graphics, which are more common in business and news reports. Presentation graphics are used to present conclusions to an audience, after analysis is complete.

Time plots display changes in measurement variables over time. Because several lines can appear on the same plot, time plots also let us see relationships between variables. Smoothing sometimes helps reveal slow cycles or trends in an erratically fluctuating time series.

Histograms, frequency polygons, ogives, and stem-and-leaf displays convey information about the shape of measurement variable distributions. This information is crucial to analysis because

1. it is part of the story of how the variable varies; and
2. numerical summaries such as averages may be misleading in ill-behaved distributions.

The next two chapters introduce numerical techniques for summarizing measurement-variable distributions. Distributional shape has great impact on how well these techniques work.

PROBLEMS

16. Draw a histogram for the life expectancies data from Table 2.5, using classes of 30–39, 40–49, etc. Compare this histogram with the stem-and-leaf display in Figure 2.7. What details can be seen in both the histogram and the stem-and-leaf display? What details can be seen only in the stem-and-leaf display?

17. Construct a frequency distribution table with cumulative percentages for the student weight variable in Table 1.1. Use class intervals of 80–99, 100–119, etc. Draw a cumulative frequency histogram (ogive) using the cumulative percentages.

18. Construct a single-stem stem-and-leaf display for each of these variables from Table 2.8:

    a.   1980 water use

    b.   1981 water use

    Carefully label the graphs to indicate the meaning of the stems and leaves digits. Comment on distributional shape and on similarities or differences between the two distributions. What do these comparisons tell you about water use?

19. Construct a five-stem stem-and-leaf display for the education of household heads in Table 2.8; include a label. What are the major features of this distribution? Explain what these features tell us.

20. The accompanying table contains time series data on the number of male and female applicants to U.S. medical schools and total applicants to U.S. nursing schools from 1974 to 1986.

Number of Applicants to U.S. Nursing and Medical Schools, 1974–1986

Year	Applicants to nursing school[a]	Female applicants to medical school	Male applicants to medical school
1974	233,511	8,712	33,912
1975	259,810	9,575	32,728
1976	259,943	10,244	31,911
1977	274,785	10,195	30,374
1978	231,425	9,561	27,075
1979	232,730	10,222	25,919
1980	212,813	10,664	25,436
1981	216,957	11,673	25,054
1982	238,834	11,685	24,045
1983	—	11,961	23,239
1984	—	12,476	23,468
1985	175,043	11,562	21,331
1986	—	11,267	20,056

[a]All U.S. RN programs. Male applicants do not exceed 10%. Figures are not available for 1983, 1984, and 1986.
Source: Data from National League for Nursing and Association of American Medical Colleges; reported in the *Boston Globe*, June 1, 1987, p. 42.

a. Construct a time plot with lines for male and female medical school applicants. What trends do you see in these lines?

b. The nursing school series has three missing values, so we cannot plot it as a continuous series. Construct a time plot with the same horizontal axis as your medical school plot, and graph the nursing school series with a dashed line connecting the values for 1982 and 1985. What trend do you see in this plot?

21. Use the data table from Problem 20 to construct a new time series for the *percentage* of all medical school applicants who are female. Construct two time plots for this series. Make them both about the same height, but scale one vertical axis from 0% to 100%, and scale the other vertical axis from 20% to 40%. Describe the impressions made by these two different graphs.

22. Construct and clearly label a stem-and-leaf display for the population per doctor in the 24 countries of Table 1.6. Describe the shape of this distribution, and discuss what this shape tells us about the 24 countries.

23. Construct a double-stem stem-and-leaf display (stems digits 10's) of the city homicide rates in Table 1.4. Compare this distribution with that for the state homicide rates seen in Figure 2.9. Do you see any general similarities?

24. In the 1988 Winter Olympics 28 skiers completed the two races of the women's slalom event. Their combined times are given in the accompanying table. Construct a stem-and-leaf display for these times. Comment on the distributional shape. What does this distribution tell us about the 28 skiers?

Combined Times of Women Slalom Skiers in 1988 Winter Olympics

Place	Skier	Country	Time in seconds
1	Vreni Schneider	Switzerland	96.69
2	Mateja Svet	Yugoslavia	98.37
3	Chrit Kinshoferguetlein	West Germany	98.40
4	Roswitha Steiner	Austria	98.77
5	Blanca Fernandezochoa	Spain	99.44
6	Ida Ladstaetter	Austria	99.59
7	Paoletta Magonisforza	Italy	99.76
8	Dorota Mogoretlalka	France	99.86
9	Mojca Dezman	Yugoslavia	100.21
10	Ulrike Maier	Austria	100.54
11	Beth Madsen	USA	101.18
12	Lenka Kebriova	Czechoslovakia	102.12
13	Lucia Medzihradska	Czechoslovakia	102.18
14	Patricia Chauvet	France	102.79
15	Diann Roffe	USA	102.88
16	Josee Lacasse	Canada	103.14
17	Katarzyna Szafranska	Poland	104.21
18	Michelle Mckendry	Canada	105.79
19	Emiawabata	Japan	109.35
20	Carolina Eiras	Argentina	112.73
21	Kate Rattray	New Zealand	112.94
22	Carolina Birkner	Argentina	117.72
23	Sandra Grau	Andorra	118.44
24	Astrid Steverlynck	Argentina	125.02
25	Thoai Lefousi	Greece	127.01
26	Carolina Photiades	Cyprus	134.92
27	Flammag Smithmini	Guatemala	165.50
28	Shailaja Kumar	India	172.27

Source: Data from the *Boston Globe*, February 28, 1988.

25. The accompanying table contains data on voter turnout in U.S. presidential elections, 1948–1988.

   a. Construct a five-stem stem-and-leaf display of voter turnout.

   b. Comment on the shape of this distribution.

Voter Turnout in U.S. Presidential Elections, 1948–1988

Year	Winner	Loser	Turnout: Percentage of eligible voters
1948	Truman (D)	Dewey	51.1%
1952	Eisenhower (R)	Stevenson	61.6
1956	Eisenhower (R)	Stevenson	59.3
1960	Kennedy (D)	Nixon	63.1
1964	Johnson (D)	Goldwater	61.8
1968	Nixon (R)	Humphrey	60.7
1972	Nixon (R)	McGovern	55.1
1976	Carter (D)	Ford	50.1
1980	Reagan (R)	Carter	52.6
1984	Reagan (R)	Mondale	53.1
1988	Bush (R)	Dukakis	50.0

*26. Smooth the voter turnout data of Problem 25 by moving averages of span 3. Construct a time plot and describe what you see.

NOTES

1. How heavy a tail is depends on how many cases it contains and how far from the distributional center they lie. It corresponds to the literal weight the tail of a histogram would have if the bars were all made of solid blocks.
2. Although graphs resembling stem-and-leaf displays appear in some older writings, the modern form is the invention of statistician John Tukey. His 1977 book *Exploratory Data Analysis* also presented many other innovative graphical and numerical methods for analyzing data.
3. Based on data from Houghton, Munster, and Viola (1978).
4. Another term for *moving average* is *running mean*.

Summarizing Distributions: Measures of Center

Chapter 3

Graphical displays can convey detailed information. Some displays are as complex as the raw data they represent; the information is just rearranged into a visually more accessible form. Sometimes the analysis can stop there, with a few paragraphs describing what the graphs reveal. More often we try to find compact ways—perhaps just a few numbers—to summarize important features of the variables' distributions.

In Chapter 2 we spoke of a distribution's *center*—a vague concept with no single specific meaning. One possible meaning, the *center of gravity*, is the average obtained by adding up the values for each case, then dividing by the number of cases.[1] It could also refer to the *positional center* or median: the value in the middle if we line up the cases from lowest to highest. Alternatively, it could mean the value where the distribution peaks or has the highest frequency—thus, the value that occurs most often is the "center."

Still other definitions of center exist, but these three are the main concern in an introductory statistics course. Each has its own particular merits and disadvantages.

3.1   *THE MODE*

The *mode* is a simple measure of center that can be used with categorical, ordinal, or measurement variables. Graphs provide a general way to define it (see the box). If X is a categorical or ordinal variable, the mode is simply the most frequent value. For the categorical variable student major (Table 1.1, page 5), the mode is "health and human services," which occurs with a frequency of 18. A tally of the expected statistics grades, an ordinal variable in Table 1.1, reveals that B is the mode: 16 out of 29 students replying expected a B.

> The **mode** is the X value corresponding to the peak in a graph of the frequency distribution.

With measurement variables, the definition of mode gets slippery. In an ungrouped frequency distribution the graphical peak still corresponds to the most frequent value of X. For example, in the distribution of student age in Table 1.3 and Figure 1.1 (pages 7 and 8), the mode is 19 years: There are 14 19-year-olds in this sample. The next most frequent age, 18, occurs only four times. When we say that the mode in these data is 19 years, we are saying that 19 is the single most common age.

Many measurement variables require grouping to obtain a reasonable graph. If so, the graphical peak may not coincide with the most frequent single value of X—indeed, there may be no most frequent value. Tent camping nights in the National Parks (Table 2.6, page 36) and international birth rates (Table 1.6, page 15) are examples of measurement-variable distributions where no value occurs more than once. In such distributions the mode refers to a class of X values, rather than to one unique value.

Since the selection of classes for grouping a variable is arbitrary, the mode itself is a vague concept when applied to measurement-variable data. Other measures of center, by contrast, have precise definitions. Nonetheless, the mode is useful for describing distributional shape. Distributions are called **strongly modal** if they peak sharply, like the top left histogram of Figure 3.1. There the modal class is substantially more frequent than the next most frequent classes. Figure 3.1 also illustrates a **weakly modal** distribution (top right histogram), where the peak is only slightly more frequent than other classes.

Distributions with a single peak are called **unimodal.** Some distributions have more than one distinct peak (lower histograms of Figure 3.1). Distributions with two peaks are **bimodal;** more than two peaks makes a distribution **multimodal.** The peaks need not all be of equal height: a distribution can be strongly or weakly bimodal or multimodal.

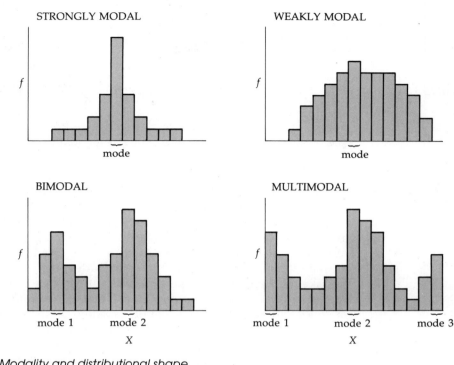

FIGURE 3.1   *Modality and distributional shape*

Figure 3.2, a stem-and-leaf display of life expectancy in 25 nations, is an example of a bimodal distribution, having peaks in both the 40's and the 70's. Bimodality is often a clue that the data contain more than one kind of case. Figure 3.2 suggests that rather than a continuum, where each country is similar to the next highest and next lowest, we can discern two different kinds of country: in one the 40's are the average age at death; in the other the average life span reaches the 70's. A gap separates the two kinds.

To find out what the two kinds might be, it is useful to identify the cases individually. Figure 3.3 (page 62) does this with an **inside-out plot,** which retains the stem structure of the stem-and-leaf display in Figure 3.2. In place of leaves indicating each country's value, the inside-out plot lists each coun-

```
3 | 7
4 | 1448889
5 | 08
6 | 1456678
7 | 12233445
```

FIGURE 3.2   *Stem-and-leaf display of life expectancy in 25 countries, from Table 2.5. (Stems digits are 10's, leaves digits are 1's. 3 | 7 means 37 years.)*

30	Cambodia
40	Ethiopia, Guinea, Mauritania, Togo, Zaire, Congo, Nigeria
50	Indonesia, Nicaragua
60	Jordon, Mongolia, Paraguay, Brunei, N.Korea, Sri Lanka, Venezuela
70	Czech., Austria, Taiwan, W.Germany, Greece, France, Australia, Norway

FIGURE 3.3   *Inside-out plot of countries with life expectancy in the 30's–70's*

try's name. A pattern is obvious. Most of the eight countries with life expectancy below 50 are in Africa. Cambodia, the lone exception, is all the more striking here because most other Asian countries have much greater longevity. At the other end of the scale, six of the eight countries exceeding 70 are in Europe. Thus, the two peaks in the life expectancy distribution largely coincide with two different geographical areas, Africa and Europe. Australia and Taiwan's presence in the longer-lived group suggests, however, that the underlying distinction is not between African and European countries, but between the poorest countries and those better off.

Like multiple peaks, large gaps and outliers in a distribution may signal the presence of more than one kind of case. For example, age distributions for college students typically contain gaps and outliers that reflect the mix of two kinds of student: "traditional" students straight out of high school and older students enrolling after years of doing something else. If cases are of several different kinds, it could be misleading to try to analyze them together or to summarize them with one overall average.

PROBLEMS

1. Examine the displays of distributions listed and describe each in terms of modality: strong or weak, unimodal or bimodal. "Peaks" consisting of a single case should not be counted as modes.
   a. Student age (Figure 1.1)
   b. Urban homicide rate (Figure 1.3)
   c. State homicide rate (Figure 2.9)
   d. Environmental voting (Figure 2.6)
2. Find the mode for these variables:
   a. Tent camping nights (Figure 2.8)
   b. Civil liberties index (Table 1.6)
   c. Water savings (Figure 2.10)

3.2 THE MEDIAN

Another important measure of center is the *median,* mentioned in the section on time series smoothing (Chapter 2); it is defined in the box. A physical example is easy to visualize: if a group of people got in line by height, from shortest to tallest, the median height would belong to the person in the middle of that line. The first step in finding the median is to sort the cases in order, from lowest to highest, for the variable in question.

> The **median** is the value that splits an ordered list in half.

We could find the median life expectancy either from an ordered list of cases, like Table 3.1, or from a stem-and-leaf display, like Figure 3.2. The

TABLE 3.1   *Ordered List of 1982 National Life Expectancy*

	Country	Life expectancy in years
1	Cambodia	37
2	Ethiopia	41
3	Mauritania	44
4	Guinea	44
5	Congo	48
6	Togo	48
7	Zaire	48
8	Nigeria	49
9	Indonesia	50
10	Nicaragua	58
11	Jordan	61
12	Mongolia	64
13	Paraguay	65←Paraguay's value divides this ordered list in
14	Brunei	66 half, so the median life expectancy is 65 years.
15	North Korea	66
16	Sri Lanka	67
17	Venezuela	68
18	Czechoslovakia	71
19	Austria	72
20	Taiwan	72
21	Greece	73
22	West Germany	73
23	Australia	74
24	France	74
25	Norway	75

median is the value in the middle; since we have 25 cases here, the middle case is the 13th lowest or 13th highest—Paraguay, with a life expectancy of 65. Hence, the median life expectancy for the 25 nations is 65 years. Half the countries have life expectancies under 65 years and half over 65 years. (Actually, with 12 below, 12 above, and Paraguay right in the middle, this is only approximately true, but it would approach the strict truth as we included more cases.)

> Given an ordered list of n cases, the **sample median, *Md*,** is the value of the case in position $(n + 1)/2$.

TABLE 3.2

Finding the Median from an Ordered List: Population per Physician in 24 Countries (Data from Table 1.6)

Order[a]	Country	Population per physician	
1	Czechoslovakia	350	Low extreme
2	Greece	400	
3	Mongolia	400	
4	W. Germany	430	
5	Austria	440	
6	France	480	
7	Norway	480	
8	Australia	520	
9	Taiwan	840	
10	Venezuela	910	
11	Jordan	940	
12	**Nicaragua**	**1,730**	
Median position is 12.5.		**Median value = 1,740.**	
13	**Paraguay**	**1,750**	
14	Brunei	1,870	
15	Congo	6,000	
16	Sri Lanka	7,550	
17	Nigeria	9,590	
18	Indonesia	10,800	
19	Mauritania	14,340	
20	Zaire	14,700	
21	Togo	18,560	
22	Cambodia	29,410	
23	Guinea	51,470	
24	Ethiopia	**69,620**	High extreme

[a]Because $n = 24$, the median position is $(n + 1)/2 = (24 + 1)/2 = 12.5$. This median position is the same whether counted from the low extreme (top) or from the high extreme (bottom).

Table 3.1 is an ordered list of 25 cases, so the median is the value of the (25 + 1)/2, or 13th, case. Whenever the number of cases, symbolized by n, is odd, one case falls right in the middle. An even number of cases, however, has none exactly in the middle, so the formula $(n + 1)/2$ will yield a fractional position. For example, Table 3.2, an ordered list of population per physician, covers just 24 nations (in our sample of 25 no data were available for North Korea).

The **low extreme** (lowest value) in Table 3.2 is Czechoslovakia, with only 350 people per physician. The **high extreme** is Ethiopia, with 69,620 people per physician. With 24 countries the median position is (24 + 1)/2, or 12.5—halfway between cases 12 and 13. The 12th case is Nicaragua, with 1,730 people per physician; the 13th case is Paraguay, with 1,750. The median value is therefore 1,740 people per physician, which is halfway between 1,730 and 1,750:

$$\frac{1,730 + 1,750}{2} = 1,740$$

PROBLEMS

3. Most countries put no restrictions on using lead as a gasoline additive. A study in the city of Jos, Nigeria, measured lead concentrations in tree leaves from regions of various traffic densities, with the results shown in the table. Find separate medians for the lead concentrations in leaves near high, medium, and low traffic. What do your findings suggest?

Lead Concentration in Tree Leaves

Tree location	Traffic density	Lead concentration
		$\mu g/cm^2$
1	High	24.9
2	High	21.1
3	High	20.6
4	High	17.4
5	High	17.2
6	High	16.0
7	High	15.4
8	Medium	13.3
9	Medium	12.3
10	Medium	11.1
11	Medium	10.6
12	Low	8.7
13	Low	8.0

Source: Fataki (1987).

4. The 1985–1986 market values of endowments for five colleges and universities are shown.

School	Endowment in millions
University of Washington	$127.4
University of Delaware	232.2
Smith College	272.7
University of Wisconsin/Madison	117.1
Pomona College	199.3

a. What is the median endowment for these schools?

b. Add a sixth school, Harvard University ($3,865.7 million), to the list. What happens to the median?

5. The table presents composite Scholastic Aptitude Test (SAT) scores for 24 suburban communities separated by per pupil expenditure, relative to other communities in this sample. The 14 low expenditure communities and the 10 high expenditure communities are listed in order of increasing average SAT scores.

Scholastic Aptitude Test Scores in 24 School Districts by Expenditure per Pupil

	Low expenditure	Composite SAT score		High expenditure	Composite SAT score
1	Chelsea	808	1	Boston	762
2	Quincy	814	2	Cambridge	831
3	Somerville	834	3	Waltham	860
4	Peabody	836	4	Watertown	877
5	Salem	851	5	Arlington	936
6	Lynn English	860	6	Manchester	960
7	Milton	874	7	Brookline	985
8	Braintree	892	8	Newton	987
9	Burlington	922	9	Winchester	1004
10	Marblehead	927	10	Wellesley	1022
11	Swampscott	956			
12	Needham	975			
13	Canton	979			
14	Belmont	989			

Source: Data from the *Boston Globe*, October 24, 1986.

a. Find the median of the average SAT scores for the 14 low expenditure communities.

b. Find the median of the average SAT scores for the 10 high expenditure communities.

c. Compare both medians. How far apart are they? What do your findings suggest?

3.3 *OTHER ORDER STATISTICS*

The median belongs to a family of summary measures called **order statistics,** which are concerned with position within an ordered list of cases. The median takes the central position, dividing an ordered list in half. The *quartiles* cut an ordered list into quarters, with the *first quartile* dividing the lower one-fourth of the data from the upper three-fourths. The *second quartile* is the same as the median: It divides the lower half of the data from the upper half. The *third quartile* separates the lower three-fourths of the data from the upper one-fourth.

Depth means a value's position relative to the nearest **extreme.** The lowest and highest cases (the extremes), both have a depth of 1. The second lowest and second highest have depths of 2, and so on. The median's depth is the same as its position in an ordered list: $(n + 1)/2$. That is, it will be the $[(n + 1)/2]$th case, counting in from either extreme; see the box. **Truncated median depth** (*tmd*) means the integer (whole number) part of the median depth: drop any decimal part.[2]

The depth of the **first** and **third sample quartiles** (Q_1 and Q_3) is

$$\text{Quartile depth} = \frac{\text{Truncated median depth} + 1}{2}$$

Table 3.3 (page 68) shows how to find the quartiles for life expectancy in our 25 nation sample. With 25 cases ($n = 25$), the median is located in position $(n + 1)/2 = (25 + 1)/2 = 13$. We could also say that the median depth is 13, counting in from either the upper or the lower extreme. The value of the case at depth 13 (Paraguay) is 65, so the median (or second quartile) life expectancy is 65 years.

The median depth (13) is next used to obtain the quartile depths:

$$\text{Quartile depth} = \frac{\text{Truncated median depth} + 1}{2}$$

$$= \frac{13 + 1}{2} = 7$$

The quartile depth is 7, meaning that the quartiles are the 7th cases in, counting from either extreme. The 7th lowest life expectancy in Table 3.3 is Zaire's 48, so the first quartile is 48 years. The 7th highest life expectancy is Austria's 72, so the third quartile is 72 years. (It would not matter if we had listed cases with the same life expectancy, such as Congo and Zaire, or Austria and Taiwan, in a different order; the quartiles would still be the same.) Thus, 48 (first quartile), 65 (median or second quartile), and 72 (third quartile) are the points that divide these 25 cases approximately into quarters (Table 3.3).

When the cases are an odd number, as in Table 3.3, the median depth is a whole number and we need not truncate. An even number of cases, however, results in a fractional median depth that must be truncated to find the quartile depths. For example, with the 24 cases of physician data in Table 3.2, the median depth is $(24 + 1)/2 = 12.5$. Truncation drops the decimal part, so the truncated median depth is 12. The first and third quartile depths are then $(12 + 1)/2 = 6.5$, or 6.5 cases in from either extreme. The 6th lowest and 7th

TABLE 3.3   ***Order Statistics in an Ordered List of Life Expectancy in 25 Countries***

Depth[a]	Country	Life expectancy	Order statistics
1	Cambodia	37	Low extreme = 37; depth = 1
2	Ethiopia	41	
3	Mauritania	44	
4	Guinea	44	
5	Congo	48	
6	Togo	48	
7	Zaire	48	1st quartile = 48; depth = $(tmd + 1)/2 = (13 + 1)/2 = 7$
8	Nigeria	49	
9	Indonesia	50	
10	Nicaragua	58	
11	Jordan	61	
12	Mongolia	64	
13	Paraguay	65	Median = 65; depth = $(n + 1)/2 = (25 + 1)/2 = 13$
12	Brunei	66	
11	North Korea	66	
10	Sri Lanka	67	
9	Venezuela	68	
8	Czechoslovakia	71	
7	Austria	72	3rd quartile = 72; depth = $(tmd + 1)/2 = (13 + 1)/2 = 7$
6	Taiwan	72	
5	Greece	73	
4	West Germany	73	
3	Australia	74	
2	France	74	
1	Norway	75	High extreme = 75; depth 1

[a]Depth means the position relative to the nearest extreme.

lowest cases both have 480 people per doctor, so case 6.5, between them, would also have a value of 480. The 6th highest and 7th highest cases are Mauritania and Indonesia, with respectively 14,340 and 10,800 people per physician. Halfway between them is $(14,340 + 10,800)/2 = 12,570$, so the third quartile is 12,570 people per physician. The list in Table 3.2 is therefore divided into quarters approximately at the points 480 (first quartile), 1,740 (median or second quartile), and 12,570 (third quartile).

Just as medians halve the data and quartiles quarter them, other order statistics can make other divisions. **Deciles,** for example, divide an ordered list into tenths. The first decile divides the lower one-tenth of the distribution from the upper nine-tenths; the second decile divides the lower two-tenths from the upper eight-tenths, and so on. **Percentiles** divide distributions into hundredths. The twenty-fifth percentile is the same as the first quartile; the fiftieth percentile is the same as the median; and the eightieth percentile is the same as the eighth decile. **Quantiles** are similar to percentiles, but they are expressed as proportions: The twenty-fifth percentile is the same as the .25 quantile, and so on.

Percentiles are widely used to indicate where a given value lies with respect to the distribution as a whole. For example, on standardized academic tests like the SAT, performance is reported both as a raw score and as a percentile. If your score is "in the fifty-eighth percentile," you scored higher than about 58% of the people taking the test and lower than about 42%, disregarding the fact that some percentage probably tied your score. Percentiles are also common in medicine. Pediatricians have charts translating children's heights and weights into percentiles. A child whose weight is "in the thirtieth percentile" is heavier than about 30% of children the same age and lighter than about 70% of them.

The *extremes* are also order statistics. They are defined as the values of the first and last cases in an ordered list, such as Cambodia (37) and Norway (75) in Table 3.3. In other words, the extremes are those values at depth 1.

PROBLEMS

6. Calculate the median and quartile depths for samples of sizes
    a. $n = 61$
    b. $n = 250$
    c. $n = 4$

7. Find the first and third quartiles for times of the Olympic slalom skiers in Chapter 2, Problem 24 (page 56). Interpret these values.

8. Find the first, second, and third quartiles of consumer complaints per 100,000 passengers, from Problem 4 in Chapter 2 (pages 34–35).

3.4   *THE MEAN*

The best-known measure of center is the average, more precisely called the *arithmetic mean.* The term is often shortened to just "mean," though other types of mean exist. You already understand how to calculate an arithmetic mean (average) by adding up a set of numbers, then dividing by how many numbers you added up. In Chapters 1 and 2 we referred informally to this statistic several times.

The formal definition of the mean requires an algebraic equation, which is given in the box. The Greek letter Σ (uppercase Greek *sigma*) is called the **summation operator.** It means, "Add up what follows." In Equation [3.1], it instructs us to add up the values of X_i for every value of *i* (case number) from 1 to *n*. X_1 represents the value of variable X for the first case, X_2 represents its value for the second case, and so on to the *n*th or last case. The symbol *n* stands for the number of cases, as with the median. We could rewrite ΣX_i in more cumbersome notation as

$$\Sigma X_i = X_1 + X_2 + X_3 + \cdots + X_n$$

Although the formal definition of the mean in Equation [3.1] may look unfamiliar, remember that it describes a calculation (finding an average) that you already know how to perform.

The **arithmetic mean** of variable X is $\overline{X}$ ("X-bar"). It is defined as

$$\overline{X} = \frac{\Sigma X_i}{n} \qquad\qquad\qquad [3.1]$$

Consider the physician data for 24 countries in Table 3.2. The mean is about 10,149 people per physician:

$$\overline{X} = \frac{\Sigma X_i}{n} = \frac{X_1 + X_2 + X_3 + \cdots + X_{24}}{24}$$

$$= \frac{350 + 400 + 400 + \cdots + 69,620}{24} = 10,149$$

The X values can be added in any order; we need not sort from lowest to highest as we do to find the median.

Unlike the median or mode, the mean is calculated from the actual values of every case in the data. Changing even a single value will therefore change the mean, too. For example, among the 30 students in Table 1.1 the mean age is 21.1 years. If the oldest student, aged 36, dropped out and was replaced by a 96-year-old, the mean age for the 30 students would rise 2 years, to 23.1.

Less drastic changes have smaller effects, but any change alters the mean. By contrast, the median would not change at all if the oldest student were replaced by a 96-year-old. Even if the 14 oldest students were all replaced by 96-year-olds, the median would still be 19.

A statistic that is not much affected by extremely high or low values in the data (such as the imaginary 96-year-olds) is **resistant.** The median is a highly resistant measure of center, whereas the mean is not at all resistant. The statistical problem of the mean's low resistance is already familiar: You know how one low grade can pull an average down, or how a high grade can bring it up.

PROBLEMS

9. Among the 14 low expenditure communities in Problem 5 the median composite SAT score is 883. Among the 10 high expenditure communities the median score is 65 points higher, 948. A comparison of medians thus suggests the latter enjoy a substantial advantage. Calculate and compare the means for both groups. How are your results like the results from comparing medians? How are they different?

10. Explain why comparing means in Problem 9 produces different results from comparing medians. Examine the raw data in the table in Problem 5 for clues.

3.5   COMPARING THE MEDIAN AND THE MEAN

The mode, median, and mean are all intended to summarize where the center of a distribution lies. In Chapter 2 we noted that bell-shaped distributions are a sort of statistical ideal, which analysts often hope to find in real data—partly because they make analysis simpler. The concept of *center* in a bell is unambiguous: There is little doubt about where the center lies, regardless of how we define it.

> In a bell-shaped distribution, *the mode, median, and mean are all the same.*

This is illustrated in Figure 3.4 (page 72). The mode is directly under the highest point of the central peak. Since bell-shaped distributions are symmetrical, the median is also right under the central peak. The same point is the center of gravity (mean), too.

Real frequency distributions, especially in small samples, seldom form perfect bells. They may come close enough, however, that the choice of median versus mean is not very important. In Chapter 2 we identified the envi-

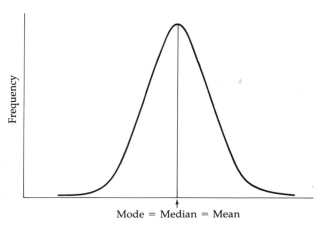

Mode = Median = Mean

FIGURE 3.4　　　*Mode, median, and mean in a bell-shaped distribution*

ronmental voting index for the 50 U.S. states as a variable with a reasonably well-behaved distribution (Figure 2.6, page 33). This variable has a median of 54.5% and a mean of 53.5%. Because the data values range from 12 to 96, a difference of 1 point between median and mean is relatively small. Both statistics are close together, and both lie squarely in the central peak obvious in the stem-and-leaf display of this variable. This distribution is not perfectly bell-shaped, but it is close enough that the three definitions of center converge.

Population per physician among 24 nations (Table 3.2) presents a more complicated picture. We found earlier that the median is 1,740 people per physician, but the mean is nearly six times higher: 10,149. Here, the two definitions of center yield widely divergent results. To understand why, and which "center" summary to believe, we turn to a graphical display.

Figure 3.5 is a five-stem stem-and-leaf display of this variable, with the median and the mean located. Nicaragua's value of 1,730 is shown by the first 1 (meaning "one thousand") to the right of the 0* stem (meaning "zero ten-thousands with leaves of 0 and 1"). Two countries, Ethiopia and Guinea, have values so high they must be considered outliers. We "trim" and list them separately at the bottom of the display, rather than distort the entire scale to accommodate them.

The population per physician data have a severely skewed distribution. Most cases are on the first stem, with fewer than 2,000 people per doctor. There is not much of a lower tail, but the upper tail straggles out to some large numbers. No wonder the median and the mean disagree on where the center lies: This distribution has no clear center, so any location we pick is debatable.

The mean far exceeds the median and most of the countries in Figure 3.5. Looking at this distribution, we can see why: The mean is pulled up by two

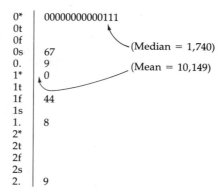

```
0*  | 00000000000111
0t  |
0f  |
0s  | 67
0.  | 9
1*  | 0
1t  |
1f  | 44
1s  |
1.  | 8
2*  |
2t  |
2f  |
2s  |
2.  | 9
```

(Median = 1,740)

(Mean = 10,149)

High outliers: 51,470, 69,620 (Guinea, Ethiopia)

FIGURE 3.5

Five-stem stem-and-leaf display of population per physician in 24 nations, from Table 3.2. (Stems digits 10,000's, leaves digits 1,000's. 2. | 9 means 29,000 people per physician.)

extremely high values (Ethiopia and Guinea). The pull they exert illustrates well the mean's lack of resistance.

The fact that the mean is calculated from the precise value of every case is a theoretical advantage. Yet it is sometimes a practical disadvantage, being so sensitive to the values of a small fraction of the data. The median is theoretically harder to work with than the mean, but it has the practical advantage of being relatively *in*sensitive to the values of any small fraction of the data. Such resistance is especially valuable if some data contain large errors in measurement.

PROBLEM

11. Invent a small data set that illustrates the median's resistance as compared with the mean's.

3.6   MEAN, MEDIAN, AND DISTRIBUTIONAL SHAPE

If a distribution is not symmetrical, it is said to be **skewed**. Distributions with heavier right-hand or upper tails, as in Figure 3.6 (page 74), are called **positively skewed**. In positively skewed distributions, *the mean will be higher than the median.* This is because the weight of cases in the upper tail pulls the mean upward but has little effect on the median.

The distribution in Figure 3.5 for population per physician is positively skewed. Its upper tail (which includes the two outliers Ethiopia and Guinea) is much longer and heavier than the virtually nonexistent lower tail. That upper tail pulls up the mean, making its position relative to the median typical of a positively skewed distribution (Figure 3.6).

The opposite situation, **negative skewness,** occurs when the left-hand or lower tail is heavier. As shown in Figure 3.7, negative skewness results in *a mean that is lower than the median.* The long lower tail pulls down the mean, but again it has little effect on the median. The life expectancy distribution shown in Figure 3.2 also illustrates negative skewness. Its median is 65 years, but the mean is about 60, pulled down by the lower tail of countries with dismal life expectancies.

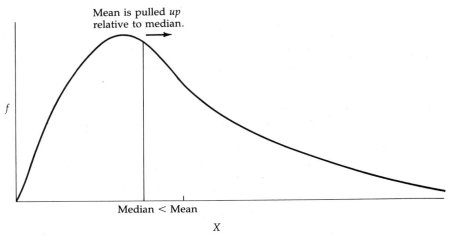

FIGURE 3.6   *Relative positions of median and mean in a positively skewed distribution*

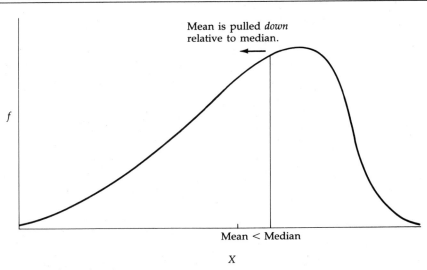

FIGURE 3.7   *Relative positions of median and mean in a negatively skewed distribution*

For these reasons, the median is often preferred to the mean as a measure of center when a distribution is severely skewed. You may have noticed that news reports about trends in family incomes typically refer to the "median family income" rather than the mean or "average." The median is preferred because the family incomes have a positively skewed distribution: Many families have low to moderate incomes, while a minority of incomes range into millions of dollars. If millionaires' incomes are averaged in with those of everyone else, the resulting statistical summary will seem unrealistically high.

Comparing the mean and the median for any measurement variable is a fast way to learn something about the distribution's shape. If the mean and median are similar, then the distribution is approximately symmetrical. If they are different, then the distribution is skewed; the more different, the greater the skewness must be. The direction of the mean and median's difference tells us whether skewness is positive or negative; see the box. A large difference between mean and median warns that skewness may present a problem, in that the distribution's center cannot be unambiguously defined.

Comparing mean and median indicates distributional shape:

Mean = Median: symmetrical
Mean > Median: positively skewed
Mean < Median: negatively skewed

The greater the difference between mean and median, the greater the skewness.

Although mean–median comparisons give clues to distributional shape, they are no substitute for graphs. A distribution can be perfectly symmetrical, with mean equalling median, yet still be ill behaved and far from the bell-shaped ideal. Figure 3.8 (page 76) shows some examples. In each the mean and the median are identical, but note that neither is the same as the mode. Consequently the concept of "center" is still ambiguous.

The variety of possible shapes in Figure 3.8—U-shaped, rectangular, bimodal, and multimodal—illustrates how much we do not know even if we know that a distribution is symmetrical. The best way to use the numerical summaries described in this chapter is *together with graphs* that allow us to see the shape of the entire distribution.

PROBLEMS

12. Pairs of means and medians are given for variables in the statistics students' data set of Table 1.1. What do they tell you about their respective distributions?

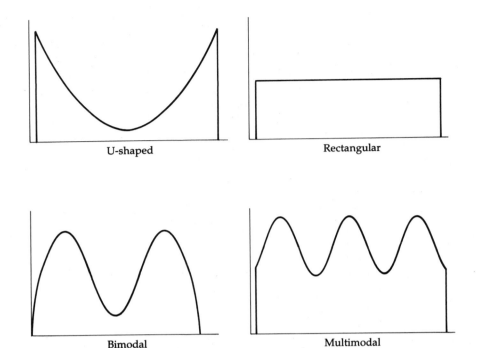

U-shaped   Rectangular

Bimodal   Multimodal

FIGURE 3.8   *Some symmetrical but not bell-shaped distributions*

    a. College grade point average: $\bar{X} = 2.74$, $Md = 2.60$

    b. Verbal SAT score: $\bar{X} = 490.2$, $Md = 490$

    c. Age: $\bar{X} = 21.07$, $Md = 19$

    d. Expected statistics grade: $\bar{X} = 2.93$, $Md = 3.00$

    e. Weight: $\bar{X} = 142.4$, $Md = 134$

13. What can you guess about the relative position of the median and the mean in the following distributions? Explain your reasoning.

    a. Number of children, in a random sample of American families

    b. Number of times arrested, for a random sample of college students

    c. Height in inches, for a random sample of college students

3.7   CHOOSING A MEASURE OF CENTER

Knowing about the shape of distributions in your data is important for two reasons. First, distributional shape often provides descriptive information. The severe positive skewness of population per physician (Figure 3.5), for instance, tells us that most countries in this sample have relatively low ratios of people to doctors, but in a few the ratios are much worse (higher). The bimodal distribution of life expectancy in these countries (Figure 3.2) suggests we might reasonably think in terms of two kinds of countries. The skewness and gaps in the distribution of student age (Figure 1.1) correspond to facts about higher education's place in the life cycle.

Another importance of distributional shape is its effect on statistical summaries such as the mean, median, and mode. With an ideal bell-shaped distribution, all three measures of center are the same. The mean is usually the preferred summary statistic, because of certain mathematical advantages. In a distribution without a bell shape, however, the three measures may not agree. Differences between mean, median, and mode can alert us to distributional problems. Once we are aware of the problems, though, we face a difficult choice: Which number best summarizes where the center of the distribution lies?

Table 3.4 summarizes which measure of center to choose, based on distributional shape. In brief, the advice is: Use modes if the distribution is strongly bimodal or multimodal; use the median if it is unimodal but severely skewed; and use the mean otherwise. When to classify a distribution as *strongly* bimodal or *severely* skewed is a matter of judgment. Consider whether

TABLE 3.4	**Choosing the Best Summary Measure of Center, Based on Distributional Shape**		
		Modality (Peaks)	
Symmetry or skewness	*Approximately unimodal*[a]	*Strongly bimodal or multimodal*[b]	
Approximately symmetrical[c]	Mean	Modes	
Severely skewed[d]	Median	Modes	

[a]Many otherwise well-behaved distributions have minor secondary peaks or gaps; use your judgment in deciding when the bimodality or multimodality is so strong that the median or mean is no longer a good summary for your data.

[b]If you can identify and isolate which different kinds of cases account for the multiple peaks, then you may end up with several smaller 1-peak distributions. Each could be summarized by its own median or mean.

[c]Again, you must judge whether the skewness is so severe that the mean is not a good summary of your data. Perfect symmetry is rare in sample data, but the mean still works even with moderate skewness.

[d]Some techniques for making skewed distributions more symmetrical are presented in Chapter 6. They let us use the mean even with severely skewed data, by first reducing the skewness.

the departures from unimodality or symmetry are so great that the mean seems unreasonable as a summary of the distribution's center.

In later chapters we will see that distributional shape becomes even more important as we move to more elaborate analytical methods. It is therefore wise to investigate the distributions of your variables as soon as possible in any research.

PROBLEM

14. Discuss whether median or mean is the best measure of center for each variable described in Problem 12. Explain your reasoning.

3.8   MEANS AND MEDIANS FROM GROUPED DATA*

It is sometimes more convenient to calculate the mean from a frequency distribution table rather than directly from the raw data. This especially helps with larger samples having hundreds or thousands of cases—too many to add up easily. The frequency distribution works as a shortcut because multiplication can be substituted for repeatedly adding the same numbers.

Table 3.5 illustrates how this is done. The cases here are 276 statistics students. The variable X is the statistics grade they expected at the start of the course. Note that no one expected less than a D; a majority thought they would earn a B. We will treat grade as a measurement variable here, with A = 4, B = 3, C = 2, and D = 1.

There are two ways we could find the mean of the 276 grades. We could add up every grade, then divide by 276. But a frequency distribution like

TABLE 3.5 | Calculating the Mean from an Ungrouped Frequency Distribution of Expected Grade

Expected grade[a]	Frequency	f times grade
X	f	fX
1	2	2
2	52	104
3	172	516
4	50	200
Sums:	$n = 276$	$\Sigma X_i = 822$

Mean: $\overline{X} = \dfrac{\Sigma X_i}{n} = \dfrac{822}{276} = 2.98$

[a] A = 4, B = 3, C = 2, D = 1.

Table 3.6 offers a faster alternative. Fifty students expect A's, so instead of adding the number four 50 times, we can just multiply four by 50 to get the same result. Similarly, 172 students expect B's; instead of adding the number three 172 times, we can multiply three by 172. Multiply all the variable values (X) by their frequencies (f), add up their products, and we get the same quantity (ΣX_i) as if we had added up each of the 276 grades. These 276 students expect a total of 822 grade points. Their mean expected grade is therefore 822/276, or $\overline{X} = 2.98$.

> To obtain the sum of X values, ΣX_i, from an ungrouped frequency distribution, multiply each X value by its frequency, f, and find the sum of the fX products.

The technique just illustrated has several advantages besides simplifying some calculations. One is that it provides a way to find the mean directly from a frequency distribution, without seeing the raw data. We can thus reanalyze any data we find tabulated in a published report.

TABLE 3.6   *Estimating the Mean from a Grouped Frequency Distribution*[a]

True limits	Class interval	Frequency	Interval midpoint	f times midpoint
X	X	f	X	fX
0–1.95	0–1.9	2	.975	1.95
1.95–3.95	2–3.9	8	2.95	23.60
3.95–5.95	4–5.9	7	4.95	34.65
5.95–7.95	6–7.9	7	6.95	48.65
7.95–9.95	8–9.9	8	8.95	71.60
9.95–11.95	10–11.9	8	10.95	87.60
11.95–13.95	12–13.9	4	12.95	51.80
13.95–15.95	14–15.9	4	14.95	59.80
15.95–17.95	16–17.9	1	16.95	16.95
17.95–19.95	18–19.9	0	18.95	0
19.95–21.95	20–21.9	1	20.95	20.95
Sums:		$n = 50$		$\Sigma X_i = 417.55$

Estimated mean based on grouped data (calculations above):

$$\overline{X} = \frac{\Sigma X_i}{n} = \frac{417.55}{50} = 8.35$$

Actual sample mean based on raw data (calculations not shown):

$$\overline{X} = 8.19$$

[a]Data are homicide rates in 50 U.S. states, from Table 2.9.

Table 3.5 shows an ungrouped frequency distribution, but published frequency distributions are often grouped. If so, *it is impossible to reconstruct the original data exactly,* since the process of grouping always throws away information. Table 3.6, for example, contains a grouped frequency distribution for homicide rates in the 50 U.S. states, based on Table 2.7 (page 38). Because of the grouping we can no longer tell the difference between such homicide rates as 2.0 and 3.9. Both fall in the same class, so both rates appear to be the same, and both differ equally from a rate of 4.0. Similar distortions occur with any grouping scheme.

We therefore cannot use Table 3.6 to find the actual mean homicide rate for the 50 states. We could *approximate* the mean homicide rate, however, by applying the method of Table 3.5 to the midpoints of each interval. For instance, the midpoint of the class interval 2–3.9 (and also of the true limits, 1.95–3.95) is 2.95. We can multiply this midpoint by the frequency, then add up the resulting fX values to obtain an approximate ΣX_i as shown in Table 3.6.

> To approximate the sum of X values, ΣX_i, from a grouped frequency distribution, multiply the midpoint of each class interval of X by its frequency, f, and find the sum of the fX products.

The calculations in Table 3.6 lead to an estimated mean state homicide rate of 8.35. The actual mean state homicide rate, which can be calculated by adding up all 50 homicide rates in the raw data (Table 2.7) and dividing by 50, is 8.19. The difference between the estimated mean (8.35) and the actual mean (8.19) reflects the errors introduced by grouping. Obviously it would be best for an analyst to work with the actual mean based on the raw data. But if a table of grouped frequency distributions is the only information available, this approximation technique lets us reasonably estimate the actual mean.

Published frequency tables often use an open-ended top class, such as "65 years old and older" in age, or "$50,000 or more" in income. It is difficult to estimate the mean from such tables, even using the approximation method. It is still possible to use such tables to find a median, however. The median may be the better statistic to calculate anyway, since the need for an open-ended top category suggests that the distribution is positively skewed.

Table 3.7 shows a distribution of incomes for U.S. households with married couples. There is an open-ended top class of "$50,000 or more," so a mean cannot be calculated. Notice also that the intervals for the other classes are of uneven width.

There are 50,351,000 households represented in Table 3.7. The median is the income of the household in position $(n + 1)/2 = (50,351,000 + 1)/2 = 25,175,500.5$, if the households were ordered from lowest to highest in-

TABLE 3.7	**Grouped Income Distribution for U.S. Married Couple Households, as of March 1985**		

Income	Frequency	Cumulative frequency
Dollars	f	cum(f)
0–4,999	1,146,000	1,146,000
5,000–9,999	3,352,000	4,498,000
10,000–14,999	4,847,000	9,345,000
15,000–19,999	5,233,000	14,578,000
20,000–24,999	5,478,000	20,056,000
25,000–34,999	10,333,000	30,389,000
35,000–49,999	10,572,000	40,961,000
50,000 and over	9,390,000	50,351,000
	$n = 50,351,000$	

Source: United States Bureau of the Census (1987).

come. The cumulative frequencies on the right of Table 3.7 indicate that the households at ranks 25,175,500 and 25,175,501 both have incomes in the $25,000–$34,999 class. The median must be somewhere in this range.

To estimate a median (*Md*) from grouped data (as in Table 3.7), use the equation

$$Md = X_L + w\left(\frac{n/2 - \text{cum}(f)}{f}\right)$$ [3.2]

where

X_L is the lower true limit of the interval containing the median;
w is the width of this interval;
n is the sample size;
f is the frequency in this interval; and
cum(f) is the cumulative frequency up to but not including this interval.

The median income in Table 3.7 is in the $25,000–$34,999 interval. The lower true limit of this interval is $X_L = \$24,999.5$; its width is $w = X_U - X_L = 34,999.5 - 24,999.5 = 10,000$. From Table 3.7 we see that the frequency in this interval is $f = 10,333,000$, and the cumulative frequency up to but not including this interval is cum(f) $= 20,056,000$. Using Equation [3.2] we can therefore estimate the median income as

$$Md = X_L + w\left(\frac{n/2 - \text{cum}(f)}{f}\right)$$

$$= 24,999.5 + 10,000\left(\frac{50,351,000/2 - 20,056,000}{10,333,000}\right)$$

$$= 24,999.5 + 4,954.5 = 29,954$$

These calculations lead to a value of $29,954 as the median income in households headed by married couples. Since it is a median, it is not affected by whatever incomes there are in the "$50,000 or more" class. We do not have to know how high these incomes go.

Applications of Equation [3.2] need not be restricted to grouped data. If more than one case shares the median value, according to the simple definition of median given earlier, then Equation [3.2] provides a more precise location for the point that divides the ordered list in half.

PROBLEMS

*15. Many American physicians are graduates of foreign medical schools. More than half such graduates are U.S. citizens who left the country for medical education, then returned for postgraduate training and practice. A 1985 survey by the American Medical Association obtained the data displayed in the accompanying table, showing the percentage of physicians in residency programs (postgraduate specialty training at teaching hospitals) who are graduates of foreign medical schools.

Use the grouped data method to estimate the mean percentage of foreign medical graduates among these 4,452 programs. Be careful: The class intervals are not all the same.

Percentage of Residents in U.S. Teaching Hospitals Who Are Graduates of Foreign Medical Schools (FMG)

FMG percentage of residents	Number of teaching hospitals
X	f
0	2,041
1–10	645
11–20	466
21–25	206
26–30	84
31–40	222
41–50	214
51–99	389
100	185

Source: Crowley and Etzel (1986).

*16. Use the grouped data method of Equation [3.2] to estimate the median percentage of foreign medical graduates from the table in Problem 15. How does this median compare with the mean you estimated in Problem 15? Explain any differences you see.

*17. The accompanying table is a grouped frequency distribution for income in households headed by women (husband absent), similar to that for married couple households in Table 3.7. Apply Equation [3.2] to estimate the median income for female headed households, and compare this value with that obtained from Table 3.7. Could the difference be due to the smaller proportion of very rich people among female household heads?

Grouped Income Distribution for U.S. Households Headed by Females, as of March 1985

Income	Frequency
Dollars	f
0–4,999	1,744,000
5,000–9,999	2,152,000
10,000–14,999	1,635,000
15,000–19,999	1,336,000
20,000–24,999	973,000
25,000–34,999	1,215,000
35,000–49,999	729,000
50,000 and over	347,000

Source: United States Bureau of the Census (1987).

3.9 WEIGHTED MEANS*

A technique similar to the frequency table method can be used to calculate means for *individual cases based on aggregate data.* Table 3.8 (page 84) shows a small aggregate data set: per capita income and populations for the states of Connecticut, Vermont, and Maine in 1986. Since their per capita incomes are 19.6, 13.348, and 12.79 thousand dollars, their mean per capita income is

$$\overline{X} = \frac{\Sigma X_i}{n}$$
$$= \frac{19.6 + 13.348 + 12.79}{3}$$
$$= \frac{45.738}{3} = 15.246$$

or $15,246 per person.

This mean is appropriate if we are interested in *states* as the units of analysis. That is how we have treated state-level data so far, and because it is simple it will remain our usual approach with aggregate data. But what if we were interested not in states, but in people? Far more people live in Connect-

	An Unweighted Mean: The Mean per Capita
TABLE 3.8	Income of Three New England States (1986)

State	Per capita income ($1,000s)	Estimated population
	X	f
Connecticut	19.600	3,169,380
Vermont	13.348	538,650
Maine	12.790	1,167,774
Sums:	ΣX = 45.738	n = 4,875,804

Mean per capita income for 3 New England states:

$$\overline{X} = \frac{\Sigma X_i}{n}$$

$$= \frac{\text{Total per capita incomes}}{\text{Number of states}}$$

$$= \frac{45,738}{3} = 15.246 \text{ thousand dollars (\$15,246)}$$

icut than in Vermont. Connecticut's per capita income of $19,600 is based on more than three million people, whereas Vermont's $13,348 reflects only about half a million people. If people are our interest then we should *weight* each state's value by its population before calculating the mean.

The resulting *weighted mean* calculations are shown in Table 3.9. The population of each state is treated much as frequencies were in Tables 3.5 and 3.6: We multiply each state's value by its population. Summing the columns of Table 3.9 reveals that some $84 billion (or 84,245,577.66 thousand) is spread among a total population of over 4.8 million (4,875,804) people. Dividing the $84 billion by the 4.8 million people yields a per capita income of $17,278 (bottom of Table 3.9).[3]

The people-based figure of $17,278 (Table 3.9) is notably higher than the state-based figure of $15,246 (Table 3.8). This is because when all three states have equal weight, the resulting mean is closer to the values of the two low income states (Vermont and Maine) that make up two-thirds of the cases (states). On the other hand, when populations are taken into consideration, the resulting mean is much closer to the value of the one high income state (Connecticut) that contributes nearly two-thirds of the cases (people). If we use a mean to summarize the affluence of the 4.8 million people in these three states, it is misleading to ignore the fact that many more people live in "rich" Connecticut than in "poor" Vermont and Maine.

A **weighted mean** is required if we wish to estimate an individual-level mean based on aggregate-level data.

TABLE 3.9   *A Weighted Mean: Per Capita Income in Three New England States (1986)*

State	Per capita income ($1,000s)	Estimated population	Total income ($1,000s)
	X	f	fX
Connecticut	19.600	3,169,380	62,119,848.00
Vermont	13.348	538,650	7,189,900.20
Maine	12.790	1,167,774	14,935,829.46
Sums:		$n = 4,875,804$	$\Sigma X_i = 84,245,577.66$

Per capita income for 4,875,804 people:

$$\overline{X} = \frac{\Sigma X_i}{n}$$

$$= \frac{\text{Total dollars}}{\text{Number of people}}$$

$$= \frac{84,245,577.66}{4,875,804} = 17.278 \text{ thousand dollars } (\$17,278)$$

Weighted means have other applications besides that just described. In Table 3.9, state populations provided a set of "weights" to be multiplied by each state's X value before they were added up. Other numbers could also provide the weights. For example, we might wish to calculate an overall grade point average for a student who has attended several different schools. This can be done as a weighted mean: Multiply the GPA earned at each school by the number of credit hours from that school, add these products together, and divide by the total number of credit hours. Because such weights can be viewed as frequencies, weighted mean procedures closely resemble frequency table methods (Tables 3.5 and 3.6).

PROBLEM

*18. The life expectancies and populations for four of the countries in Table 3.1 are shown here.

Country	Life expectancy	1982 population
Mauritania	44	1,561,000
France	74	54,432,000
Austria	72	7,574,000
Zaire	48	30,336,000

a. What is the unweighted mean life expectancy for these four countries?

b. Calculate the weighted mean life expectancy for their over 93 million inhabitants.

c. Explain any differences between the means found in parts a and b.

SUMMARY

The mean, median, and mode are alternative definitions for the center of a measurement-variable distribution. The mean is the "center of gravity"; the median is the "positional center"; and the mode is the most common value or class. The mode can also apply to categorical-variable distributions, and either the median or the mode can be used with ordinal variables.

In bell-shaped distributions, the three definitions of center all point to the same location. The mean is preferred because of its theoretical advantages, but since the three measures of center are all the same we need not choose among them.

Difficulties arise in less well-behaved distributions where the three measures of center disagree. With such distributions there is simply no unambiguous center. Since the median, mode, and mean may all be in different locations, we have to choose among them. Our conclusions may be different depending on which one we look at.

Bimodality suggests that we have mixed together two different kinds of cases. If so, to identify the two kinds and examine them separately may be better than trying to summarize both with a single number. Skewness and outliers particularly affect the mean, pulling it in the direction of the heavier tail. The median is comparatively resistant to these effects. If distributional problems are only mild, however, the mean is still the preferred measure of center. It is used much more often than the median or the mode. We will see in later chapters that many advanced statistical methods are based on the mean.

PROBLEMS

19. News reports about trends in house prices commonly use the median, rather than the mean, as their measure of center. The choice of center is important because with house prices the median and the mean usually differ. For example, in November 1987, the *median* price of a new house in America was estimated as $119,000, while the *mean* price of a new house was estimated as $140,300.

    a. What can we tell about the price distribution of new houses when the mean is so much higher than the median?

    b. Why would the median be preferable to the mean in this situation?

    c. Suggest your own example of a variable with similar distributional problems, where medians probably summarize better than means.

20. Find the medians of:

    a. Baggage complaints per 1,000 passengers (Problem 4 in Chapter 2).

    b. Errors in SAT scores (Problem 10 in Chapter 2).

The accompanying table records accidental oil spills by tankers at sea, from 1973 to 1985. Problems 21–25 refer to this table.

Accidental Oil Spills by Tankers at Sea

Year	Number of tankers afloat	Number of spills	Oil lost, in million metric tons
1973	3,750	36	84.5
1974	3,928	48	67.1
1975	4,140	45	188.0
1976	4,237	29	204.2
1977	4,229	49	213.1
1978	4,137	35	260.5
1979	3,945	65	723.5
1980	3,898	32	135.6
1981	3,937	33	45.3
1982	3,950	9	1.7
1983	3,582	17	387.8
1984	3,424	15	24.2
1985	3,285	8	15.0

21. Calculate the median and the mean number of oil spills for these 13 years. What do these two statistics tell us about the distribution?

22. Construct a single-stem stem-and-leaf display for the number of oil spills in the table. What is the apparent mode of this distribution? Compare this mode with the mean and the median (Problem 21). Is the mode consistent with your earlier conclusion about distributional shape?

23. Find the median and the mean amount of oil lost for the 13 years in the table. What can we infer about distributional shape? Which statistic is most likely to be the better summary?

24. Construct a single-stem stem-and-leaf display (stems digits 100's) for the oil loss variable in the table. Describe the shape of this distribution. What does this shape imply about the median and mean found in Problem 23? What does the shape tell us about oil spills?

25. Calculate and compare separate medians for the amount of oil lost during the 1970's and 1980's.

26. Find the third quartile for the average SAT scores among low expenditure communities in the table in Problem 5. Compare this third quartile with the *median* value among high expenditure communities. What does this comparison tell us?

The table at the top of page 88 contains 1985 data on the ten largest state lotteries. These data were assembled for a study of comparative efficiency; they vary in their administrative costs and in their payoffs to the state or to players. Problems 27–30 refer to this table.

The Ten Largest State Lotteries in 1985

State	Gross sales in millions	Percentage returned to state	Percentage paid in prizes	Number of employees
Connecticut	344.5	43%	52%	105
Illinois	1,207.6	43	49	185
Maryland	681.1	39	54	109
Massachusetts	1,235.3	32	59	430
Michigan	891.2	40	48	194
New Jersey	936.1	42	49	250
New York	1,299.1	46	43	210
Ohio	855.6	40	50	288
Pennsylvania	1,336.5	47	45	210
Washington	150.0	35	47	153

Source: Data are from the *Boston Globe*, October 6, 1986, p. 1.

27. Construct double-stem stem-and-leaf displays for these two variables from the table, and for each one find the median and the mean. Comment on distributional shape.

   a. Percentage of gross sales returned to the state

   b. Percentage of gross sales paid out in prizes

28. Find the mean and median for gross sales in millions. Construct a five-stem stem-and-leaf display (use seven stems) and use it to explain the discrepancy between mean and median.

29. Find the median, quartiles, and extremes of the number of lottery employees. Do you expect the mean to be higher or lower than the median in this distribution? Explain your reasoning.

30. How much money, in millions of dollars, did each lottery bring in to its state government? Recall the rule for finding raw numbers from percentages (convert to a proportion, then multiply), described in Chapter 1.

31. Construct a stem-and-leaf display of recreational vehicle camping from Table 2.6. Use this display to illustrate the locations of:

   a. Low and high extremes

   b. First, second, and third quartiles

*32. What is the median number of children in U.S. families? The table is a frequency distribution for the number of children under 18 living with a population of over 62 million families (1985).

   a. Find the individual-data median number of children.

   b. Find the grouped-data median.

Distribution of Children in U.S. Families (1985)

Number of children	Number of families
X	f
0	31,603,824
1	13,105,554
2	11,663,316
3	4,514,832
4 or more	1,881,180

Source: United States Bureau of the Census (1987).

NOTES

1. The average is a center of gravity in this sense: Build an ungrouped histogram for a distribution out of blocks on a seesaw, and the average is the point at which this seesaw balances.
2. Using median location to define the sample quartiles, as done in this book, makes calculation simple. There are about six other possible ways to define the sample quartiles (and similar ideas called *hinges, fourths,* or *quarters*). Any of these definitions may lead to slightly different values for a given set of data. Some definitions are more trouble to work out by hand but have advantages when computers are used.
3. We have noted that income distributions tend to be positively skewed, so using means with such distributions lets millionaires "pull up the average," making the overall population look richer. That very likely is the case with these data.

Summarizing Distributions: Measures of Spread

Chapter 4

Summary statistics like the median and mean describe the location of a distribution's center. Another kind of summary statistic describes how much **variation** or **spread** there is around this center. Measures of spread are used less often in everyday life than measures of center or "average." They are nonetheless indispensable statistical tools, because they tell us about *how much a variable varies.*

Like center, *spread* is a vague concept. Just as there are several ways to define the center of a measurement-variable distribution (Chapter 3), there are various ways to define a distribution's spread. Some employ order statistics, and so belong to the same mathematical family as the median. Other measures of spread are related to the arithmetic mean.

4.1   MEASURES OF SPREAD BASED ON ORDER STATISTICS

Chapter 3 described how to obtain order statistics, such as the median, quartiles, and extremes, from an ordered list of the data. Order statistics can also be used to derive measures of spread, as the data in Table 4.1 illustrate.

TABLE 4.1   *Ordered List of the Weights of Male Marathon Runners*

	Runner	Weight in pounds	
1	Sandoval	**115** Low extreme	
2	Tabb	116	
3	Rodgers	128	
4	Shorter	132	
5	Callison	135	Range = High extreme − Low extreme
6	Atkins	138	= 156 − 115
7	Wells	140	= 41
8	Thomas	150	
9	Kardong	150	
10	Lodwick	**156** High extreme	

Table 4.1 lists the weights of ten top male marathon runners (qualifiers for the 1980 U.S. Olympic team), in order from lightest to heaviest. The lightest weighed 115 pounds, the heaviest 156, so the *low extreme* is 115 and the *high extreme* is 156. The **range** is defined as the difference between them:

$$\text{Range} = \text{High extreme} - \text{Low extreme} \qquad [4.1]$$
$$= 156 - 115 = 41$$

The range is 41 pounds.[1]

The range is the simplest and best-known measure of spread, often used informally to describe how much something varies. As a statistical summary, though, it has a serious drawback: The range depends entirely on the extreme values, which may be the two most unusual cases in the distribution. The range tells nothing about variation among the cases between either extreme. Adding one 250-pound football player to Table 4.1 would more than triple the range, but 10 of the 11 cases would still be thin marathon runners. Range is one of the least resistant of all summary statistics.

Another measure of spread that is far more resistant is called the *interquartile range*. Recall that the first quartile separates the lowest one-fourth of the data from the upper three-fourths, and that the third quartile separates the lower three-fourths from the upper one-fourth. *The interquartile range is the distance needed to span the middle 50% of the cases.* What happens in the tails of the distribution does not matter; the interquartile range depends on only the variation in the middle.

The **interquartile range,** or **IQR,** is the difference between the first and third quartiles:

$$\text{IQR} = Q_3 - Q_1 \qquad [4.2]$$

where Q_3 is the third quartile and Q_1 is the first quartile.

To find the interquartile range we first must find the quartiles, as shown in Table 4.2. There are 10 runners ($n = 10$), so the median depth is $(n + 1)/2 = (10 + 1)/2 = 5.5$. Thus, the median is halfway between the fifth and sixth runners: halfway between 135 and 138, or $(135 + 138)/2 = 136.5$ pounds. Quartile depth is found from truncated median depth, abbreviated *tmd*. (Recall that truncating is dropping fractions, not rounding off.) The median depth here is 5.5; truncation cuts this to 5. The quartile depths are

$$\text{Quartile depth} = \frac{tmd + 1}{2}$$

$$= \frac{5 + 1}{2} = 3$$

The third lightest runner weighs 128 pounds, so the first quartile is 128. The third heaviest runner weighs 150 pounds, so the third quartile is 150. Following Equation [4.2],

$$\text{IQR} = Q_3 - Q_1$$

$$= 150 - 128 = 22$$

The interquartile range is 22 pounds. Like the extremes, median, quartiles, and range, the IQR is measured in the same units as the variable.

In theory, the first quartile, median, and third quartile divide the data into four parts, each containing 25% of the cases. But ten cases do not actually divide into four equal parts: $10/4 = 2.5$ cases in each part, and there is no such thing as half a case. We can imagine, however, that Bill Rodgers (at the first quartile, depth 3) belongs half in the first part and half in the second.

TABLE 4.2

Finding the Quartiles and Interquartile Range for the Runners' Weights from Table 4.1

	Depth	Runner	Pounds
	1	Sandoval	115
	2	Tabb	116
Q_1 depth: $\dfrac{tmd + 1}{2} = \dfrac{5 + 1}{2} = 3$	3	Rodgers	**128** $\leftarrow Q_1$
	4	Shorter	132
	5	Callison	135
Median depth: $\dfrac{n + 1}{2} = \dfrac{10 + 1}{2} = 5.5$	5.5		$\leftarrow$ Median: 136.5
	5	Atkins	138
	4	Wells	140
Q_3 depth: $\dfrac{tmd + 1}{2} = \dfrac{5 + 1}{2} = 3$	3	Thomas	**150** $\leftarrow Q_3$
	2	Kardong	150
	1	Lodwick	156

Then the first part does have 2.5 cases: Sandoval, Tabb, and half of Bill Rodgers. The second part likewise has 2.5 cases: Shorter, Callison, and Rodgers' other half. Similarly, Randy Thomas, on the third quartile, could be imagined to belong half in the third part (with Atkins and Wells) and half in the fourth part (with Kardong and Lodwick).

Table 4.3 shows the same variable, weight, for a different set of cases: the male statistics students from Table 1.1. Again the data are listed from lightest to heaviest to make it easier to find the necessary order statistics.

These order statistics are located in Table 4.4. There are eight students ($n = 8$), so the median depth is $(n + 1)/2 = (8 + 1)/2 = 4.5$. The fourth student weighs 175, the fifth weighs 182; halfway between them is the median

TABLE 4.3

Ordered List of Weights of 8 Male College Students (from Table 1.1)

	Case	Pounds
1	4	147
2	13	163
3	7	175
4	22	175
5	16	182
6	1	187
7	2	195
8	9	255

TABLE 4.4

Finding the Quartiles and Interquartile Range for the Student Weights from Table 4.3

	Depth	Case	Pounds	
	1	4	**147**	← Low extreme
	2	13	163	
Q_1 depth: $\dfrac{tmd + 1}{2} = \dfrac{4 + 1}{2} = 2.5$				← Q_1: 169
	3	7	175	
	4	22	175	
Median depth: $\dfrac{n + 1}{2} = \dfrac{8 + 1}{2} = 4.5$				← Median: 178.5
	4	16	182	
	3	1	187	
Q_3 depth: $\dfrac{tmd + 1}{2} = \dfrac{4 + 1}{2} = 2.5$				← Q_3: 191
	2	2	195	
	1	9	**255**	← High extreme

(175 + 182)/2 = 178.5 pounds. The quartile depths are $(tmd + 1)/2$; since the median depth is 4.5, the truncated median depth is simply 4. The quartile depths are then $(tmd + 1)/2 = (4 + 1)/2 = 2.5$. This places the first quartile between the second and third lightest students: between 163 and 175 pounds, or (163 + 175)/2 = 169 pounds. The third quartile is correspondingly between the second and third heaviest students, who weigh 195 and 187, respectively: $Q_3 = (195 + 187)/2 = 191$ pounds. These quartiles and the median divide the eight cases into four equal parts of two cases (25%) each.

The IQR for the students' weights is

$$IQR = Q_3 - Q_1$$
$$= 191 - 169 = 22$$

By coincidence, the students and the runners have the same IQR. That is, *the middle 50% of the cases are equally spread out in both distributions.* We will discuss this finding later. First, we will look at mean-based statistics that also measure variation or spread.

PROBLEMS

1. Deforestation, the large-scale cutting of dense forests, is a global problem affecting the world's atmosphere and climate. The data below are estimates of annual deforestation rates in nine South American countries for 1981–1985. Deforestation is measured in thousands of hectares per year; a hectare is a metric unit of area equal to 2.471 acres. (Data from Council on Environmental Quality, 1986.)

Country	Deforestation rate
Bolivia	87
Brazil	1,480
Colombia	820
Ecuador	340
Guyana	3
Paraguay	190
Peru	270
Suriname	3
Venezuela	125

   Order these countries from low to high, and find the following statistics.

   a. Low and high extremes

   b. Median, first quartile, and third quartile

   c. Range and interquartile range

2. In what units is the IQR for the deforestation data measured? Describe the meaning of this IQR.

3. Problem 5 in Chapter 3 (page 66) provides ordered lists of average SAT scores for high and low expenditure school districts.

   a. Find the interquartile range for the average SAT scores in the 10 high expenditure communities.

   b. Find the IQR for the average SAT scores in the 14 low expenditure communities.

   c. Compare the IQRs for both groups of communities. What do we learn from the IQRs' similar size?

4.2   DEVIATIONS FROM THE MEAN

An alternative to summaries based on order statistics (such as the IQR) is the family of summaries based on the arithmetic mean. From a standpoint of mathematical theory, mean-based measures have almost irresistible attractions. Their theoretical properties are relatively easy to work out, so they have been studied thoroughly and are well understood. It is known that certain mean-based measures are, under ideal conditions, statistically better than competing measures. In contrast, order-based statistics like the median and IQR, despite being easy to calculate, have more difficult theoretical properties.

Mean-based measures of spread will be illustrated with the weight data sets described earlier. For a start, we need to find the means themselves. The mean weight of the ten marathon runners in Table 4.1 is

$$\overline{X} = \frac{\Sigma X_i}{n}$$

$$= \frac{X_1 + X_2 + X_3 + \cdots X_{10}}{10}$$

$$= \frac{115 + 116 + 128 + \cdots + 156}{10} = 136$$

Whereas the median and other order statistics are defined by their positions in an ordered list, the mean is found by summing up values for every case in the distribution. This pattern is followed by statistics that are based upon the mean.

To measure spread or variation around the mean, we begin by finding each case's **deviation score,** which is its distance from the mean: $X_i - \overline{X}$. For example, the heaviest runner, the tenth in Table 4.1, weighs 156 pounds. Since the mean is 136 pounds, his deviation score is 20 pounds:

$$X_{10} - \overline{X} = 156 - 136 = 20$$

The lightest runner, listed first, weighs 115 pounds. His deviation score is -21 pounds:

$$X_1 - \overline{X} = 115 - 136 = -21$$

Deviation scores for all ten runners are shown in the second column of Table 4.5. Note that above-average weights ($X_i > \overline{X}$) get positive deviation scores, and below-average weights ($X_i < \overline{X}$) get negative scores. Deviation scores tell us two things: how far from the mean, and in what direction?

Deviation scores indicate how individual cases vary from the mean. To describe how much variation the entire sample has, we need to combine individual deviations into one summary number. But if we just add deviation scores together, positive and negative scores cancel each other out, and their sum is zero. This can be seen by adding up the deviation scores column in Table 4.5:

$$\Sigma(X_i - \overline{X}) = (-21) + (-20) + (-8) + (-4) + (-1) + 2 + 4$$
$$+ 14 + 20 = 0$$

In fact, the sum of deviations from the mean *always* equals zero:

$$\Sigma(X_i - \overline{X}) = (X_1 - \overline{X}) + (X_2 - \overline{X}) + (X_3 - \overline{X}) + \cdots + (X_n - \overline{X})$$
$$= 0 \qquad\qquad [4.3]$$

for any set of data. This is known as the **zero-sum property of the mean.** Because of this property, we cannot just add deviation scores to obtain an overall measure of variation.

TABLE 4.5   ***Deviations from the Mean ($\overline{X} = 136$) of Runners' Weights from Table 4.1***

Weight	Deviation score	Absolute deviation	Squared deviation
X_i	$X_i - \overline{X}$	$|X_i - \overline{X}|$	$(X_i - \overline{X})^2$
115	-21	21	441
116	-20	20	400
128	-8	8	64
132	-4	4	16
135	-1	1	1
138	2	2	4
140	4	4	16
150	14	14	196
150	14	14	196
156	20	20	400
Sums: $\Sigma X_i = 1{,}360$	$\Sigma(X_i - \overline{X}) = 0$	$\Sigma|X_i - \overline{X}| = 108$	$\Sigma(X_i - \overline{X})^2 = 1{,}734$[a]

[a]Total sum of squares, or TSS_X, is in pounds squared.

We can avoid this problem by getting rid of the negative deviation scores. One simple way to do so is to take the **absolute values** of the deviation scores and add them up, as in Table 4.5, column 3. Taking absolute values changes negative scores to positive, thus preventing them from cancelling out positive scores. The sum of absolute deviations in Table 4.5 is 108, meaning that this group of runners differs from the mean weight by a total of 108 pounds.

A second way to get rid of the negative deviations is to square each deviation. **Squared deviations** are shown in the last column of Table 4.5. Multiplying each deviation by itself eliminates negative values because a negative times a negative yields a positive number, as does a positive times a positive. When these squared deviations are added up, the result is a sum of squared deviations, called the *total sum of squares* or TSS_X.

The **total sum of squares** for variable X, TSS_X, is defined as

$$TSS_X = \Sigma(X_i - \overline{X})^2$$
$$= (X_1 - \overline{X})^2 + (X_2 - \overline{X})^2 + (X_3 - \overline{X})^2 + \cdots + (X_n - \overline{X})^2 \qquad [4.4]$$

where $\overline{X}$ is the sample mean of X.

Applying Equation [4.4] to the runners' data,

$$TSS_X = \Sigma(X_i - \overline{X})^2$$
$$= (115 - 136)^2 + (116 - 136)^2 + (128 - 136)^2 + \cdots + (156 - 136)^2$$
$$= 441 + 400 + 64 + \cdots + 400 = 1,734$$

The deviation scores are in pounds, so after squaring they must be in pounds squared. The total sum of squares, TSS_X, is 1,734 pounds squared.

Both the sum of absolute deviations and the sum of squared deviations will equal zero if the variable is really a constant. Where every case is the same, all deviations are zero. If there is any variation at all, then the sums of both absolute and squared deviations will be positive numbers. *The more the variable varies, the higher these positive numbers will be.* Either sum could therefore be used as a measure of overall variation. There are mathematical advantages to working with squared deviations, however.

PROBLEMS

4. Return to the deforestation data of Problem 1. Calculate the mean deforestation rate and each country's deviation from this mean. Explain the meaning of the deviations for Brazil and for Paraguay.

5. Square each deviation in Problem 4 and add them together to obtain the TSS_X. In what units are the deviation scores measured? The total sum of squares?

4.3 VARIANCE AND STANDARD DEVIATION

The TSS_X reflects the total amount of variation around the mean, but it obviously will be greater in large samples, where more deviations are added up. To adjust for sample size, we might simply divide the TSS_X by the number of cases. The resulting statistic is the mean squared deviation:

$$\text{Mean squared deviation} = \frac{\text{TSS}_X}{n}$$

$$= \frac{\Sigma(X_i - \overline{X})^2}{n} \qquad [4.5]$$

Although the mean squared deviation provides a good measure of spread, with sample data most statisticians prefer a slightly different statistic called the *sample variance*. The difference between the mean squared deviation, Equation [4.5], and the sample variance, Equation [4.6], is that the latter involves division by $n - 1$ instead of by just the sample size, n.[2] Like the mean squared deviation, the variance measures how much variation there is around the mean. The more variation, the higher the variance. If X is a constant, the variance will be zero.

The **sample variance**, s_X^2, is defined as

$$s_X^2 = \frac{\text{TSS}_X}{n-1}$$

$$= \frac{\Sigma(X_i - \overline{X})^2}{n-1} \qquad [4.6]$$

It is a general measure of how much X varies.

The variance for the runners' weights can be found by dividing the total sum of squares by $n - 1$. Since there are $n = 10$ runners and the total sum of squares, calculated in Table 4.5, is $\text{TSS}_X = 1{,}734$, the variance is

$$s_X^2 = \frac{\text{TSS}_X}{n-1} = \frac{1{,}734}{10-1} = 192.7$$

The total sum of squares is in pounds squared, so the variance must also be in these awkward units: It is 192.7 pounds squared.

It is hard to comprehend "192.7 pounds squared" in real-world terms. To return to more understandable units, we can use the *square root* of the variance. The square root of the variance is called the *standard deviation*. Since the variance is measured in squared units (such as pounds squared), its square root, the standard deviation, is measured in natural units (such as pounds).

The **sample standard deviation,** s_X, is defined as

$$s_X = \sqrt{s_X^2}$$

$$= \sqrt{\frac{\Sigma(X_i - \overline{X})^2}{n - 1}} \qquad \text{[4.7a]}$$

The standard deviation measures a distribution's variation or spread.

The standard deviation for the runners' data, based on a variance of $s_X^2 = 192.67$, is

$$s_X = \sqrt{s_X^2} = \sqrt{192.7} = 13.88$$

We are now back in the variable's natural units: The standard deviation is 13.88 pounds. The standard deviation tells us something about how much the ten weights spread out around the mean. Such information has many uses in statistical analysis.

PROBLEMS

6. For the deforestation data of Problem 1, the total sum of squares is approximately $\text{TSS}_X = 1,887,380$. Use this quantity to calculate the variance and standard deviation. What are the units of measurement for each statistic?

7. The accompanying table contains data from an observational study of preschool children at play. The researchers counted the number of episodes per child of rough and tumble play and of more aggressive fighting and chasing. Find the mean and standard deviation of the number of rough and tumble episodes for the ten girls in this sample.

Observations on a Sample of 26 Preschoolers at Play

Child	Gender	Age in months	Terms in school	Rough, tumble episodes	Fighting, chasing episodes
1	Boy	54	1	40	4
2	Boy	54	1	5	1
3	Boy	54	1	5	0
4	Boy	53	1	42	4
5	Boy	51	1	19	3

(continued)

Child	Gender	Age in months	Terms in school	Rough, tumble episodes	Fighting, chasing episodes
6	Boy	48	1	3	1
7	Boy	55	0	25	4
8	Boy	53	0	22	2
9	Boy	51	0	61	13
10	Boy	51	0	1	0
11	Boy	51	0	19	20
12	Boy	50	0	18	1
13	Boy	49	0	25	2
14	Boy	48	0	36	7
15	Boy	48	0	19	3
16	Boy	47	0	4	1
17	Girl	56	2	13	3
18	Girl	52	2	25	4
19	Girl	51	1	42	9
20	Girl	51	1	6	3
21	Girl	48	2	21	8
22	Girl	50	0	3	0
23	Girl	50	0	4	0
24	Girl	50	0	8	1
25	Girl	47	0	9	4
26	Girl	47	0	10	2

Source: Data from Smith and Lewis (1985).

8. Find the mean and standard deviation of rough and tumble episodes for the 16 boys in Problem 7. Comparing boys with girls, interpret the differences in

   a. The mean number of episodes

   b. The standard deviation of the number of episodes

4.4   SAMPLE AND POPULATION

In Chapter 1 we encountered the concepts of *sample* and *population*. A population is the universe of cases that is of ultimate analytical interest, such as all airline flights or all college males. A sample is a subset of these cases, such as the 26 airline flights in Table 2.1, or the 8 college males in Table 4.3. The sample comprises the data at hand for statistical analysis.

Numerical summaries that describe sample data are called **sample statistics.** The sample mean, variance, and standard deviation are examples of sample statistics. In theory, corresponding summaries such as the *population* mean, variance, and standard deviation exist. Numerical summaries for entire populations are called **population parameters.** Population parameters (such as the mean nicotine concentration on all airline flights, or the mean weight of all college males) are usually unknown. One task of statistical analysis is to **estimate** unknown population parameters on the basis of known sample statistics.

Population parameters are represented by symbols different from the corresponding sample statistics. For example, the symbol for the population mean is not $\overline{X}$, the sample mean, but the Greek letter μ (*mu*). The symbol σ_x (lowercase Greek *sigma*) is used for the population standard deviation, and the population variance is σ_x^2. Table 4.6 summarizes the symbols that represent the sample and population mean, variance, and standard deviation.

The sample mean $\overline{X}$ has been defined as

$$\overline{X} = \frac{\Sigma X_i}{n} \tag{4.8}$$

where n is the number of cases in the sample. The summation involves adding up the values of X for all n cases. In a parallel fashion the population mean, μ, could be defined as

$$\mu = \frac{\Sigma X_i}{N} \tag{4.9}$$

where the capital N represents the number of cases in the entire population ($N \geq n$). This definition will not work for a population of infinite size (such as the population consisting of "all possible rolls of these dice"), which needs a more refined definition employing probability and calculus. For simpler situations with a finite population, however, the equations for sample mean, Equation [4.8], and population mean, Equation [4.9], correspond closely.

The equations for sample and population variances likewise correspond, as do equations for sample and population standard deviations. The sample variance was defined in Equation [4.6] as

$$s_X^2 = \frac{\Sigma(X_i - \overline{X})^2}{n - 1}$$

TABLE 4.6   *Symbols for Sample and Population*

Numerical summary	Sample statistic[a]	Population parameter[b]
Mean	$\overline{X}$	μ
Variance	s_X^2	σ_X^2
Standard deviation	s_X	σ_X

[a]Sample statistics are usually *known,* since they can be calculated from a relatively small sample of data at hand.
[b]Population parameters are often *unknown,* since they refer to a large universe of cases "out there." Known sample statistics may be used to *estimate the values* of unknown population parameters.

where n is the sample size. The variance for a finite population is

$$\sigma_X^2 = \frac{\Sigma(X_i - \mu)^2}{N} \qquad [4.10]$$

This equation for the population variance differs from that for the sample variance in three details:

1. The numerator is the sum of squared deviations around the population mean, μ, rather than around the sample mean, $\overline{X}$.
2. The denominator involves division by N, the population size, instead of n, the sample size.
3. This denominator is N itself, whereas the sample statistic's denominator is $(n - 1)$.[3]

The population standard deviation σ_X is defined as the square root of the population variance:

$$\sigma_X = \sqrt{\sigma_X^2}$$

$$= \sqrt{\frac{\Sigma(X_i - \mu)^2}{N}} \qquad [4.11]$$

It differs from the sample standard deviation in the same respects that the population and sample variances differ.

When we discuss numerical summaries such as means, medians, or standard deviations in this book, we usually refer to sample statistics. If a population parameter is being discussed, it will be identified as such.

PROBLEM

9. Magazines often survey their subscribers, partly to learn about incomes, since potential advertisers often desire relatively affluent or "upscale" readership. Suppose we want to conduct such a survey, to gain information about the mean and standard deviation of income of subscribers to *Rock and Ice* magazine. There are 10,000 subscribers, so we can't afford to contact them all. For such a survey, specify:

   a. The sample
   b. The population
   c. The sample statistics
   d. The population parameters

4.5   CALCULATING THE STANDARD DEVIATION

Equation [4.7a] is the *definitional formula* for the sample standard deviation. Rearranging the terms creates a *computational formula,* which is easier for hand calculation. Although it speeds hand calculation, the computational formula is harder to grasp intuitively. It is no longer obvious that we are working with a sum of squared deviations from the mean, for instance.

A **computational formula** for the **sample standard deviation** is

$$s_X = \sqrt{\frac{\Sigma X_i^2 - [(\Sigma X_i)^2/n]}{n-1}}$$ [4.7b]

This equation produces results identical to those from the **definitional formula,** Equation [4.7a].

Because most statistical calculations are now done with computers or hand calculators, computational formulas are less essential in an introductory course than they once were. It is more important to understand the definitional formulas, where we find the meaning of the statistics.

If you use a scientific calculator to find the standard deviation, you often face a choice. The calculator may have keys for either of two standard deviations:

1. One choice is the "sample standard deviation," or perhaps the "$n-1$" standard deviation. This corresponds to Equation [4.7a] or [4.7b] and is the one to use for any standard deviation problem in this book.
2. The second choice is the "population standard deviation," or the "n" standard deviation. This is based on the equation

$$s_X = \sqrt{\frac{\Sigma(X_i - \overline{X})^2}{n}}$$ [4.12]

Equation [4.12] is identical to the usual standard deviation, Equation [4.7a], except for the denominator n instead of $(n-1)$.

When calculated from sample data, the "n" standard deviation is not really a population standard deviation. The calculator's manual may call it that because of the resemblance between Equations [4.12] and [4.11].

Some calculator manuals use the symbol σ to refer to either of these standard deviations, or they define the variance as the square of Equation [4.12]. To avoid such confusion, *look at the equations, not the symbols or names* the manual uses to describe these statistics. Standard deviations and variances with $n-1$ in the denominator are the usual sample statistics, corresponding to the definitional formulas of Equations [4.7a] or [4.6]. These are the ones we will use in this text. Standard deviations and variances with n in the denominator resemble the population definitions, Equations [4.10] and [4.11], but are usually avoided for sample data (see Note 3).

Most computer programs designed specifically for statistical analysis automatically use the $n-1$ standard deviation, Equation [4.7a]. Many general-purpose spreadsheet and database programs, on the other hand, automatically use the n standard deviation, Equation [4.12], and their manuals may fail to say which formula is used. To resolve any confusion about whether a machine-calculated standard deviation divides by n or $n - 1$, try the following simple test:

Enter as data the numbers 1, 2, and 3. If the standard deviation comes out as 1, the preferred $n - 1$ definition is being used. If the standard deviation comes out as a fractional value (about .816) then the n definition is in use.

Are you stuck with a program that computes only the n standard deviation? You can convert it to the $n - 1$ version by multiplying the n standard deviation by $\sqrt{\dfrac{n}{n - 1}}$.

The larger the sample size n, the less difference it makes whether we divide by n or $n - 1$. In small samples, though, the numerical difference between results obtained from Equations [4.7a] and [4.12] may be substantial.

PROBLEMS

10. Return to the data of Problem 5 in Chapter 3, on the average SAT scores in high and low expenditure communities.
    a. Find the mean and standard deviation of average SAT score for the 10 high expenditure communities.
    b. Find the mean and standard deviation of average SAT score for the 14 low expenditure communities.
    c. Compare both groups in terms of their means and standard deviations.
11. The table at the top of page 106 contains data from an experiment in which 6 normal and 12 neurologically impaired (aphasic) subjects were videotaped during conversation. Researchers determined what proportion of the conversation time was spent in speech, and what proportion in gestures related to the speech. Proportions are reported as the relative duration of verbalization and gestures, respectively.
    Find the sample means and standard deviations for the relative duration of gestures for the 6 normal subjects. Calculate similar statistics for the 12 impaired subjects. Compare both groups in terms of
    a. The mean duration of gestures
    b. The standard deviation of duration of gestures

Relative Duration[a] of Verbalization and Gestures in Conversation

Subject	Sex	Age	Neurological status	Verbalization	Gestures
1	Male	63	Normal	.68	.12
2	Female	65	Normal	.92	.07
3	Female	64	Normal	.84	.36
4	Male	57	Normal	.88	.14
5	Male	69	Normal	.95	.20
6	Male	71	Normal	.95	.06
7	Male	55	Impaired	.46	.22
8	Male	66	Impaired	.84	.48
9	Male	81	Impaired	.61	.45
10	Male	48	Impaired	.78	.44
11	Male	36	Impaired	.81	.38
12	Female	47	Impaired	.61	.15
13	Female	71	Impaired	.80	.29
14	Male	68	Impaired	.73	.40
15	Male	39	Impaired	.78	.69
16	Male	49	Impaired	.58	.32
17	Female	50	Impaired	.75	.52
18	Male	51	Impaired	.45	.23

[a]As a proportion of conversation time.
Source: Data from Feyereisen (1982).

4.6   THE PSEUDO-STANDARD DEVIATION

In Chapter 3 we saw that mean–median comparison tells us whether a distribution is skewed and, if so, in which direction. Another informative comparison is between standard deviation and interquartile range divided by 1.35. The quantity IQR/1.35 is known as the *pseudo-standard deviation.* Why we use the constant 1.35 is explained in Chapter 9.

The **pseudo-standard deviation** or **PSD** is:

$$PSD = \frac{IQR}{1.35} \qquad\qquad [4.13]$$

where IQR is the interquartile range.

For a particular kind of bell-shaped distribution, a **normal distribution,** the standard deviation and pseudo-standard deviation are equal. Comparing the standard deviation with the PSD therefore easily checks whether a distribution is approximately normal. Such checks are valuable because many statistical techniques work best with normally distributed data. The term *normal*

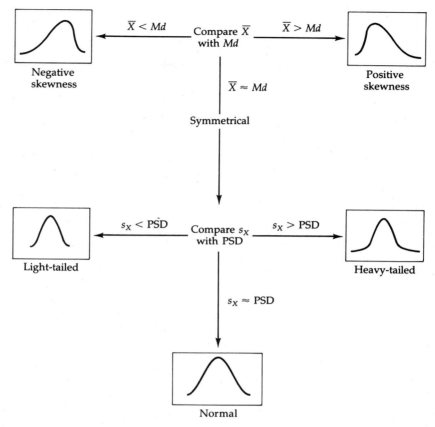

FIGURE 4.1 *Flow chart for inferring distributional shape from comparisons of order- and mean-based statistics*

here does not mean usual or typical; it refers to a distribution defined by a specific mathematical equation (given in Chapter 9).

Compare standard deviation with pseudo-standard deviation only if a mean–median comparison has already established that the distribution is approximately symmetrical. If it is not symmetrical it cannot be normal, and the s_X –PSD comparison may be misleading.

In a symmetrical distribution, we read s_X–PSD comparisons as follows:

1. If s_X roughly equals PSD, we infer that *the distribution is approximately normal.*
2. If s_X is less than PSD, the distribution has *lighter than normal* tails.
3. If s_X is greater than PSD, the distribution has *heavier than normal* tails.

Figure 4.1 shows this in flowchart form. The s_X–PSD comparison of spread is analogous to the mean–median comparison of centers, and it works for much the same reason: The median and PSD are resistant, not much influenced by distributional tails. In contrast, the mean and standard deviation are strongly influenced by the tails.

For the runners' weights, we saw that the mean is 136 pounds and the median is 136.5—virtually no difference. A standard deviation–pseudo-standard deviation comparison is therefore appropriate. The interquartile range is 22 pounds, so the pseudo-standard deviation is PSD = IQR/1.35 = 22/1.35 = 16.3 pounds. We earlier found the standard deviation to be 13.88 pounds. Since the standard deviation is less than the PSD (13.88 < 16.3), we conclude that the weight distribution has slightly lighter than normal tails. That is, fewer runners have high or low weights than we would expect if this were a normal distribution. The discrepancy between 13.88 and 16.3 is not large, however.

For the students' weights in Table 4.3, the following summary statistics can be calculated:

Center	Spread
$\overline{X} = 184.87$	$s_X = 31.94$
$Md = 178.5$	$PSD = 16.3$

The mean and median are 184.87 and 178.5, respectively, indicating mild positive skewness. For purposes of illustration, and since the skewness is not great, we will treat this distribution as approximately symmetrical and proceed with an s_X–PSD comparison. The comparison suggests that the distribution has heavier than normal tails (31.94 > 16.3). That is, more students have especially high or low weights than we would expect if this distribution were normal.

The next section will put together these diverse numerical summaries with a graphical display, to examine what they are telling us about the two sets of weight data.

PROBLEMS

12. For each pair of standard deviation and interquartile range given, find the PSD and compare it with s_X. Assuming that the distributions are roughly symmetrical, what does each comparison tell you about distributional shape?

    a. Average SATs in high expenditure communities (Chapter 3, Problem 5): $s_X = 85.8$, IQR = 127.

    b. Average SATs in low expenditure communities: $s_X = 63.7$, IQR = 120.

    c. Savings in a water conservation campaign (Table 2.9 and Figure 2.10): $s_X = 1,201$, IQR = 800.

13. These summary statistics describe deforestation rates in nine South American countries (Problem 1):

$$Md = 190 \quad \overline{X} = 368.7$$
$$IQR = 253 \quad s_X = 485.7$$

Use these statistics to discuss the shape of this distribution. What does this shape mean, in terms of deforestation rates?

4.7   UNDERSTANDING MEASURES OF SPREAD

Figure 4.2 summarizes our calculations so far concerning the weights of the ten runners in Table 4.1 and the eight students in Table 4.3. It features back-to-back stem-and-leaf displays. Comparing the distributions visually helps us better understand what these numerical summaries mean.

The most obvious difference between groups is that the students are heavier. The mean weight of the ten marathon runners is 136; the mean weight of the eight male students is almost 50 pounds more, 184.87. World-class marathon runners train to the point that their bodies contain very little fat, and their exercise does not build up much weight in upper-body muscles. Further, people with heavy body structures are unlikely to run marathons well; the sport favors people who are lightly built. Thus it is not surprising to find a large difference between mean weights. The difference appears as the offset between the centers of the stem-and-leaf displays in Figure 4.2.

A more subtle difference is that there is *more variation* among the students' weights than among the runners' weights. The runners' lower standard de-

STUDENTS		RUNNERS	
	1*	11	
	1t	2333	
4	1f	4555	
776	1s		*(back-to-back*
988	1.		*stem-and-leaf displays)*
	2*		
	2t		
5	2f		

# of cases:	$n = 8$	$n = 10$
Mean:	$\overline{X} = 184.87$	$\overline{X} = 136$
Variance:	$s_X^2 = 1,020.12$	$s_X^2 = 192.67$
Std. dev.:	$s_X = 31.94$	$s_X = 13.88$
Median:	$Md = 178.5$	$Md = 136.5$
1st quartile:	$Q_1 = 169$	$Q_1 = 128$
3rd quartile:	$Q_3 = 191$	$Q_3 = 150$
Interquartile range:	$IQR = 22$	$IQR = 22$
Pseudo-std. deviation:	$PSD = 16.3$	$PSD = 16.3$

FIGURE 4.2   *Comparison of the weight distributions for runners and students from Tables 4.1 and 4.3. (Stems digits 100's, leaves digits 10's. 1f | 4 means 140 pounds.)*

viation (13.88) indicates that they are more alike in terms of weight. The students' higher standard deviation (31.94) shows that they are more diverse. Students come in a variety of weights, heavy and light, whereas top marathon runners come only one way—light. The contrast is again visible in the stem-and-leaf displays.

The stem-and-leaf displays reveal something else that may partly account for the difference in means and standard deviations. The student group contains a high outlier, the fellow weighing 255. His presence pulls up the student mean; he is the reason the mean (184.87) noticeably exceeds the median (178.5). In contrast, the runners' mean and median weights are almost identical (136 vs. 136.5). Their PSD and standard deviation are also fairly close, 16.3 and 13.88. The students' PSD and standard deviation are much farther apart, another consequence of the heavy tail made up of one outlying student.

If we compare the two groups' medians, the students (178.5) still outweigh the runners (136.5)—but not as much as the means would suggest. The larger difference in means is due to the 255-pound student. Comparing the two groups' interquartile ranges, we find that the middle 50% of both groups have the same amount of spread. The large difference in standard deviations entirely reflects that heavy student. Thus, we should modify our conclusions a bit: The students definitely tend to be heavier than the runners, as the medians and the means confirm. Student weights also vary more widely, but the difference stems from one unusual case. Looking only at the middle of the distributions, with the IQR or PSD, we can report that these students are actually no more variable in weight than the runners.

In this example graphical displays, order-statistic summaries, and mean-based summaries all work together as analytical tools. Combining them helps show how each one works and also helps us understand the data better than any single approach can. The next chapter will extend this idea of using graphical and numerical methods in combination.

PROBLEMS

Problems 14–16 call for analysis of the numbers of aggressive fighting and chasing episodes among preschool children, based on the table in Problem 7.

14. Who was more aggressive, boys or girls? Find the median number of aggressive fighting and chasing episodes for each gender, and compare the two. Also find the two IQRs, and compare boys with girls in terms of their variation.

15. Repeat the comparison of Problem 14, using means and standard deviations. Which gender was more aggressive? Which was more variable?

16. Construct back-to-back stem-and-leaf displays (use a five-stem version) to compare the distributions of fighting and chasing episodes for boys and girls. How can you explain the contradictory conclusions reached in Problems 14 and 15? Summarize your conclusions about boys, girls, and aggressive play.

SUMMARY

Measures of spread such as the standard deviation and interquartile range tell us how much a variable varies. They define *variation* differently and have different statistical properties. The interquartile range is based on order statistics: It is the distance between the first and third quartiles. This distance spans the middle 50% of the cases, so the IQR reflects variation in the central part of the distribution, not the tails. Like the median and quartiles, the IQR is resistant to the effects of extreme values.

The standard deviation, in contrast, is calculated from the value of every case in the distribution. Deviations of each case from the mean are squared, then summed, to produce an overall measure of variation called the total sum of squares (TSS_X). The standard deviation is derived from the TSS_X. Like the mean, the standard deviation is well understood theoretically, but it is not resistant to extreme values. Indeed, the standard deviation and variance are *much less resistant* than even the mean itself.

In a normal distribution, the standard deviation will equal a fraction of the IQR called the pseudo-standard deviation, or PSD. Like the IQR, the PSD is unaffected by extreme values in the tails of a distribution. Comparing the standard deviation with the PSD provides a rough test of normality, in distributions that have already been found to be symmetrical. The amount and direction of any differences between standard deviation and PSD inform us about the shape of the distribution, as do comparisons of means and medians.

The standard deviation is a very useful statistic. Its applications will appear in nearly every chapter of this book. The PSD is used primarily to check on whether there might be problems with the standard deviation. If no major problems appear in mean–median or standard deviation–PSD comparisons, then the combination of mean (center) and standard deviation (spread) provides a good two-number summary of a variable's distribution.

Chapters 2–4 introduced a number of basic ideas about statistics and data analysis. Chapter 5 applies these ideas to examples, to show how graphical and numerical-summary methods can be used together to understand measurement-variable distributions.

PROBLEMS

17. Suggest three examples *not* used in this book that illustrate the distinction between sample statistics (mean and standard deviation) and population parameters (again, mean and standard deviation). In at least one example use cases that are not individual people.

18. In your own words, explain the concept of spread in variable distributions. Illustrate this concept with an example.

19. The 1981–1985 average annual deforestation rates, in 1,000's of hectares per year, are listed for seven Central American countries (data from Council on Environmental Quality, 1986).

Country	Deforestation rate
Costa Rica	65
Cuba	2
Guatemala	90
Honduras	90
Mexico	595
Nicaragua	121
Panama	36

    Find the median, quartiles, and interquartile range for these data.

20. Find the mean and standard deviation for the Central American deforestation rates in Problem 19. Compare the mean with the median, and comment on the shape of this distribution. How does the shape of this distribution compare with that for the South American countries in Problem 1?

21. The accompanying table gives an ordered list of the weights of the 22 female statistics students from Table 1.1 Calculate the median and IQR for these weights, and use these statistics to compare the 22 female students with 8 male students (Table 4.3) and also with the 10 marathon runners (Table 4.1) in terms of weight. The summary statistics for weight in the two male samples are:

10 marathon runners	8 male students
$Md = 136.5$	$Md = 178.5$
$IQR = 22.0$	$IQR = 22.0$

Ordered List of the Weights of 22 Female Statistics Students (Table 1.1)

	Case	Pounds		Case	Pounds
1	3	98	12	21	125
2	18	105	13	10	130
3	28	106	14	25	130
4	29	110	15	23	133
5	8	115	16	14	135
6	12	120	17	30	139
7	26	120	18	6	140
8	17	123	19	15	140
9	24	123	20	5	140
10	27	125	21	11	150
11	20	125	22	19	160

22. Find the mean and standard deviation for weights of the 22 female students in Problem 21. Compare these statistics with the corresponding order statistics (median and PSD) for female students in Problem 21, and comment on distributional shape.

23. Much theoretical work has been done on the problem of measurement errors. In theory, if measurement errors are random they should *increase the standard deviation of a measured variable,* relative to the standard deviation of its true values. If measurement errors are not random, then this may not happen. Non-random errors would arise, for instance, if errors tend to be greater for certain true values of the variable.

    Problem 10 in Chapter 2 (page 42) gives both true and self-reported values for college students' SAT scores. The self-reports obviously contain errors. If these errors are random, the standard deviation of the self-reports should be higher than the standard deviation of the true values. Is it? What can you conclude?

NOTES

1. An alternative definition of the range is the distance between the *true limits* of the upper and lower extremes. For the runners' weights this quantity is $156.5 - 114.5 = 42$ pounds, instead of $156 - 115 = 41$ pounds.

2. If we know the mean and the first $n - 1$ deviation scores, we can predict what the nth deviation score will be. A sum of n deviation scores is thus said to have $n - 1$ **degrees of freedom:** only $n - 1$ of the deviations are free to vary, before the nth one, given the mean, is fixed. We divide the sum of squared deviations (TSS_x) by its degrees of freedom ($n - 1$) to find the sample variance. Degrees of freedom become important in inferential statistics, from Chapter 11 on.

3. If sample variances were calculated with n in their denominator, instead of $n - 1$, they would in the long run tend to be *systematically lower* than the true population variance defined in Equation [4.10]. This is an example of a statistical problem called **bias.** Sample variances calculated with the $n - 1$ denominator (Equation [4.6]) are better for estimating σ_x^2, because in the long run they tend to be neither too high nor too low. The $n - 1$ versions are **unbiased.**

Comparing Variable Distributions

Chapter 5

Students encountering a new statistical method tend to concentrate first on the details of calculation. The immediate challenge is getting the right answer. Even while grappling with this problem, though, they are wondering what the new method is good for. Textbooks and teachers promise that each method eventually proves useful, but reassurances are cold comfort when a student has just managed to calculate a standard deviation and hasn't the faintest idea what to make of it. Demonstrations of usefulness are needed.

The last three chapters introduced a variety of ways to analyze variable distributions. Now we are ready for examples of how they help us understand real data. We will draw comparisons between two or more distributions, because comparisons make it easier to interpret each numerical summary. Comparisons are also more revealing than most single-distribution analyses.

5.1   CONSTRUCTING BOX PLOTS

Graphical techniques (Chapter 2) and numerical summaries (Chapters 3 and 4) should be teamed in a statistical analysis. One valuable method that is a hybrid of both approaches, with some of the advantages of each, is the **box plot:** a graphical display based on the order-statistic summaries of median and quartiles.[1]

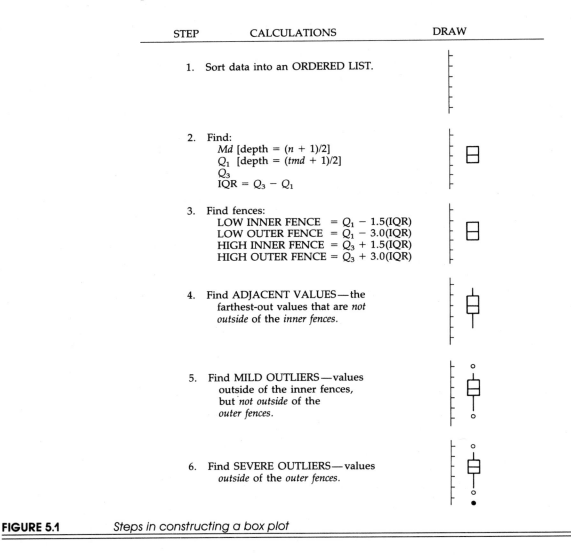

STEP	CALCULATIONS	DRAW
1.	Sort data into an ORDERED LIST.	
2.	Find: Md [depth $= (n + 1)/2$] Q_1 [depth $= (tmd + 1)/2$] Q_3 IQR $= Q_3 - Q_1$	
3.	Find fences: LOW INNER FENCE $= Q_1 - 1.5(\text{IQR})$ LOW OUTER FENCE $= Q_1 - 3.0(\text{IQR})$ HIGH INNER FENCE $= Q_3 + 1.5(\text{IQR})$ HIGH OUTER FENCE $= Q_3 + 3.0(\text{IQR})$	
4.	Find ADJACENT VALUES—the farthest-out values that are *not outside* of the *inner fences.*	
5.	Find MILD OUTLIERS—values outside of the inner fences, but *not outside* of the outer fences.	
6.	Find SEVERE OUTLIERS—values *outside* of the *outer fences.*	

FIGURE 5.1   *Steps in constructing a box plot*

The steps for constructing a box plot are outlined in Figure 5.1. At each step, certain values are calculated or taken from the data, then drawn on the

TABLE 5.1   ***Annual Alcohol Sales per Person 1975–1977***

	State	Alcohol sales[a]
1	Utah	1.69
2	Idaho	2.55
3	Oregon	2.79
4	New Mexico	2.90
5	Washington	2.97
6	Montana	3.09
7	Arizona	3.15
8	Colorado	3.23
9	Hawaii	3.26
10	Wyoming	3.37
11	California	3.38
12	Alaska	3.96
13	Nevada	6.88

[a]Annual sales in gallons/person aged $\geq$ 15 years.
Source: Linsky, Colby, and Straus (1985).

box plot. We will follow each step using for our data set the yearly alcohol sales per person (aged 15 years and older) in each of 13 Western states (Table 5.1).

Step 1: Make an ordered list of the data, as in Table 5.1. Your list might come from a stem-and-leaf display (Figure 5.2). Draw a ruler-like scale so that we can locate any value within the range of these data.

Step 2: Find the median, quartiles, and interquartile range. The median is the value of the case at the $(n + 1)/2$ position or depth in an ordered list. For our 13 cases,

$$\text{Median depth} = \frac{n + 1}{2} = \frac{13 + 1}{2} = 7$$

```
1.  | 6  ◄────── Utah (1.69 gallons)
2*  |
2.  | 5799
3*  | 012233
3.  | 9
4*  |
4.  |
5*  |
5.  |
6*  |
6.  | 8  ◄────── Nevada (6.88 gallons)
```

Double-stem stem-and-leaf display of annual per capita state alcohol sales, from Table 5.1. (Stems digits 1's, leaves digits 0.1's. 3* | 0 means 3.0 gallons per person per year.)

FIGURE 5.2

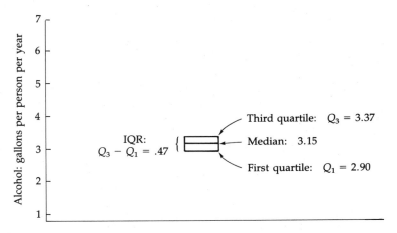

FIGURE 5.3   *Drawing a box plot from Table 5.1: Steps 1 and 2*

The 7th highest or 7th lowest state in Table 5.1 is Arizona, so the median is Arizona's 3.15 gallons per capita. Next,

$$\text{Quartile depth} = \frac{\text{Truncated median depth} + 1}{2}$$

$$= \frac{7 + 1}{2} = 4$$

The 4th lowest state is New Mexico, so the first quartile is $Q_1 = 2.90$. The 4th highest state, Wyoming, gives us the third quartile: $Q_3 = 3.37$. The interquartile range is the distance between these two quartiles:

$$\text{IQR} = Q_3 - Q_1 = 3.37 - 2.90 = .47$$

On our box plot we draw lines at the first quartile, median, and third quartile, then connect the quartile lines to form a box that encloses the median. The box's height equals the interquartile range (Figure 5.3).

Step 3:   Find two pairs of imaginary points called *fences.*

The **inner fences** are located one and one-half IQRs beyond the first and third quartile:

Low inner fence = $Q_1 - 1.5(\text{IQR})$	[5.1]
High inner fence = $Q_3 + 1.5(\text{IQR})$	[5.2]

Since Q_1 is at 2.90 and the IQR is .47, the low inner fence is

$$\text{Low inner fence} = Q_1 - 1.5 \, (\text{IQR})$$

$$= 2.90 - 1.5(.47)$$

$$= 2.90 - .705 = 2.195$$

The third quartile is at 3.37, so

High inner fence = Q_3 + 1.5(IQR)

$= 3.37 + 1.5(.47)$

$= 3.37 + .705 = 4.075$

The *outer fences* are still farther out. The low outer fence is

Low outer fence = Q_1 − 3(IQR)

$= 2.9 − 3(.47)$

$= 2.9 − 1.41 = 1.49$

The high outer fence is

High outer fence = Q_3 + 3(IQR)

$= 3.37 + 3(.47)$

$= 3.37 + 1.41 = 4.78$

The inner and outer fences are *not drawn* onto the box plot; we merely note their location and proceed to Step 4.

The **outer fences** are located three IQRs out from the first and third quartiles:

Low outer fence = Q_1 − 3(IQR)   [5.3]

High outer fence = Q_3 + 3(IQR)   [5.4]

Step 4: Use the ordered data list to find the *adjacent values*. The low inner fence is at 2.195; the lowest case that is not lower than 2.195 is Idaho, at 2.55. The low adjacent value is therefore Idaho's 2.55. The high adjacent value is found similarly: The high inner fence is at 4.075, and the highest case no higher than this fence is Alaska, at 3.96—the high adjacent value. On the box plot, we now draw lines (sometimes called **whiskers)** from the quartiles to the adjacent values. The plot at this point looks like Figure 5.4 (page 120).

The **adjacent values** are the most extreme cases that are *not outside the inner fences.*

Steps 5 and 6: Plot any *outliers*. Any value more than 1.5(IQR) beyond the first or third quartile is considered an outlier. We now have a specific definition of this previously vague concept. Each mild outlier is drawn separately as a light circle (Step 5) and each severe outlier as a dark circle (Step 6).

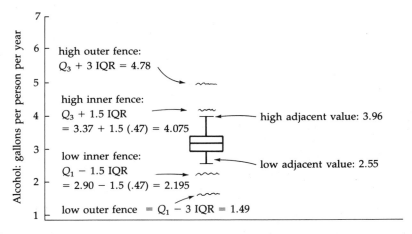

FIGURE 5.4   *Drawing a box plot from Table 5.1: Steps 3 and 4*

Outliers are defined as values beyond the inner fences: x is a

Low outlier if $x < Q_1 - 1.5(\text{IQR})$

High outlier if $x > Q_3 + 1.5(\text{IQR})$

Values beyond the inner fences, but not beyond the outer fences, are considered to be **mild outliers.** Any value beyond the outer fence is a **severe outlier.** We consider x to be a

Low severe outlier if $x <\ \ Q_1 - 3(\text{IQR})$
Low mild outlier if $Q_1 - 3(\text{IQR}) \leq x < Q_1 - 1.5(\text{IQR})$
High mild outlier if $Q_3 + 1.5(\text{IQR}) < x \leq Q_3 + 3(\text{IQR})$
High severe outlier if $Q_3 + 3(\text{IQR}) < x$

Since they are beyond the inner fences, both Utah and Nevada are outliers. Utah (1.69) is only a mild outlier, below the low inner fence (2.195) but not below the low outer fence (1.49). Utah is therefore shown as a light circle. Nevada's 6.88, well above the high outer fence (4.78), is a severe outlier graphed as a dark circle. Figure 5.5 shows the box plot with outliers added (Steps 5 and 6). If there were other outliers, each would be plotted separately as a light or dark circle.

In its final form (Figure 5.6), the box plot is drawn without notations. The six steps and numerous new terms just introduced may make box plots seem like a difficult technique to use. They are indeed cumbersome to construct by

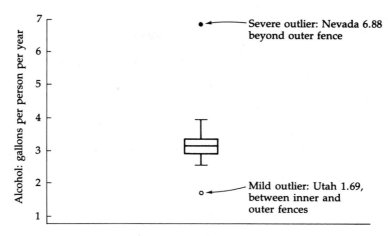

FIGURE 5.5   *Drawing a box plot from Table 5.1: Steps 5 and 6*

hand, yet box plots are easily understood. The following sections provide real examples of how to read them.[2]

Notice that the box plot employs a measure of spread, the IQR, as a yardstick to measure distances. For instance, outliers are defined as values that are more than a certain number of IQRs out from the quartiles. Measuring distance in a distribution is one of the most common uses for measures of spread such as the IQR and standard deviation.

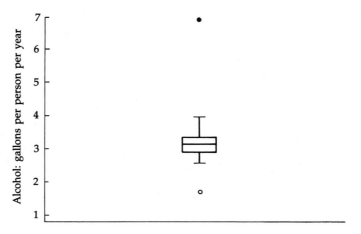

FIGURE 5.6   *Completed box plot, per capita alcohol sales in 13 states*

PROBLEMS

1. Problem 21 in Chapter 3 (page 87) gave the estimated annual oil loss by spills from tankers at sea, over the years 1973–1985. A list (in millions of metric tons), re-ordered from the smallest to the largest annual loss, is shown.

Year	Loss
1982	1.7
1985	15.0
1984	24.2
1981	45.3
1974	67.1
1973	84.5
1980	135.6
1975	188.0
1976	204.2
1977	213.1
1978	260.5
1983	387.8
1979	723.5

   a. Find the median, quartiles, and interquartile range for these data.

   b. Identify the inner fences, 1.5(IQR) beyond each quartile. What are the adjacent values (last values not outside these fences)?

   c. Identify the outer fences, 3(IQR) beyond each quartile. Are any years mild outliers, beyond the inner fence but not beyond the outer fence?

   d. Are any years severe outliers, beyond the outer fence?

2. Use your answers to Problem 1 to construct a box plot for annual oil losses. Comment on what this plot shows.

5.2 *READING BOX PLOTS*

Box plots provide information about four aspects of a distribution: center, spread, symmetry, and outliers. The distributional center is indicated by the line at the median, within each box. Spread is shown by the box's height, which is the interquartile range. Symmetry can be examined at several levels in the distribution. Is the median about halfway between the quartiles? Are the high and low adjacent values equally far from the quartiles? Are any high outliers balanced by corresponding low outliers? A distribution may be symmetrical in its central region, within the box, while its tails are very skewed—or vice versa. Outliers, whether balanced or not, are obvious in a box plot.

These distributional details are important not only from a statistical standpoint. Outlying cases, for example, are by definition different from the

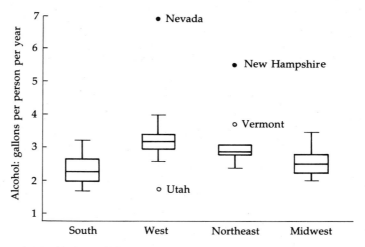

rest of the data. We should ask ourselves *why* they differ. What is there about Utah and Nevada that so strongly affects alcohol sales? (The answer lies in Nevada's success as a gambling attraction and in Utah's predominantly Mormon population.) Asking such questions may shed light on the data.

Obviousness is one of the greatest advantages of box plots. Although outliers and skewness are often present in measurement data, many analysts forget about them. Unsought can mean unnoticed, and the analyst may have no clue that outliers or skewness are affecting his or her main conclusions. With box plots, however, it is almost impossible *not* to notice outliers and skewness. Box plots provide a safeguard against the human tendency to see only what we expect to see.

Box plots are especially valuable for comparing two or more distributions. Figure 5.6 was a single plot of the distribution of per capita alcohol sales in 13 Western states. Figure 5.7 shows that plot side by side with three box plots representing the alcohol sales in Southern, Northeastern, and Midwestern states. Summary statistics are given in Table 5.2 (page 124).

Figure 5.7, backed up by the numerical summaries of Table 5.2, supports the following observations about how the distributions of alcohol sales differ across the four regions:

1. *Center*   Median alcohol sales are highest in the West (3.15 gal/person) and lowest in the South (2.225). Comparing means produces the same regional ordering as the medians. Thus, the conclusion that state per capita alcohol sales are highest in the West, followed by the Northeast, Midwest, and South, in that order, *does not depend on whether the median or the mean is our measure of center.* Such unanimity makes our conclusions more convincing.

TABLE 5.2	*State Rates of Annual Alcohol Sales in Gallons per Person (Aged $\geq$ 15), 1975–1977, by Geographical Region*

South (n = 16 cases)		
First quartile: Q_1 = 1.935		
Median: Md = 2.225	Mean: $\overline{X}$ = 2.325	
Third quartile: Q_3 = 2.625		
IQR = $Q_3 - Q_1$ = .690	Standard	
PSD = IQR/1.35 = .511	deviation: s_X = .475	

West (n = 13 cases)	
First quartile: Q_1 = 2.900	
Median: Md = 3.150	Mean: $\overline{X}$ = 3.325
Third quartile: Q_3 = 3.370	
IQR = $Q_3 - Q_1$ = .470	Standard
PSD = IQR/1.35 = .348	deviation: s_X = 1.191

Northeast (n = 9 cases)	
First quartile: Q_1 = 2.750	
Median: Md = 2.840	Mean: $\overline{X}$ = 3.182
Third quartile: Q_3 = 3.050	
IQR = $Q_3 - Q_1$ = .300	Standard
PSD = IQR/1.35 = .222	deviation: s_X = .939

Midwest (n = 12 cases)	
First quartile: Q_1 = 2.200	
Median: Md = 2.465	Mean: $\overline{X}$ = 2.520
Third quartile: Q_3 = 2.765	
IQR = $Q_3 - Q_1$ = .565	Standard
PSD = IQR/1.35 = .419	deviation: s_X = .425

When they are upheld by several different methods of analysis, the choice of which statistic to rely on matters less.

2. *Spread*   Judging by the interquartile ranges, or by the heights of the boxes in Figure 5.7, the spread or variation is greatest in the South (IQR = .69) and least in the Northeast (IQR = .30). In other words, the Southern states are *least alike,* and the Northeastern states are *most alike,* in per capita alcohol sales. Interquartile ranges measure variation only in the middle 50% of these distributions, however. The standard deviations, which measure variation based on every case in a distribution, tell a different story: They show Western states to be most variable (s_X = 1.191), and Midwestern states the least so (s_X = .425). Unlike our conclusions about center, *our interpretation of spread differs according to which summary statistic we consider.*

3. *Symmetry*   All of the distributions show slight positive skewness—a few high states pull up each region's mean. The mean–median discrepancy is greatest in the Northeast, as might be expected from Figure 5.7.

4. *Outliers*   We saw earlier the high (Nevada) and low (Utah) outliers among the Western states. Two outliers, both high, occur in the Northeast as well:

Vermont (3.66 gal/person) and New Hampshire (5.9). As in the West, the Northeastern outliers inflate the standard deviation. Unlike the West, the Northeastern outliers do not balance each other out, so they have a larger effect on the mean.

New Hampshire has the second highest per capita sales, after Nevada, among the 50 states. The state government in New Hampshire derives considerable income by selling discount liquor to residents of neighboring states. This illustrates a key point: *There is often some special reason why outliers stand apart.* Nevada and New Hampshire's high per capita alcohol sales may be largely unrelated to how much alcohol their residents actually consume. This would be important if, for example, we were studying whether state alcohol sales are connected to rates of alcohol-related illness. For such a study, our figures for Nevada and New Hampshire might be seriously misleading.

The sections that follow give further examples of how box plots are read. In each example, combined graphical and numerical-summary techniques will be used to compare measurement-variable distributions.

PROBLEMS

3. Figure 5.8 shows box plots for the number of episodes of rough and tumble play in the preschool data of Problem 7 in Chapter 4 (page 100). Compare the boys with the girls in terms of
   a. Center (medians)
   b. Spread (interquartile ranges)
   c. Skewness

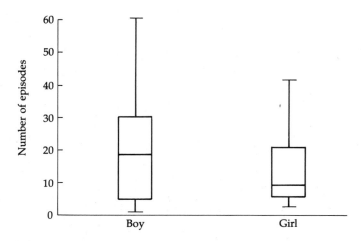

FIGURE 5.8   *Episodes of rough and tumble play among 26 children*

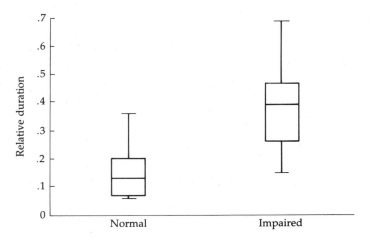

FIGURE 5.9   *Relative duration of gestures by neurological status*

4. Figure 5.9 has box plots for the neurological impairment data from Chapter 4, Problem 11 (page 105). The variable shown is the proportion of conversation time accompanied by gestures, during experimental conversations with normal and impaired subjects.

   a. Compare the two distributions in terms of centers (medians). Which group tends to do more gesturing during conversation?

   b. Are there other notable differences between the two distributions? Explain.

5.3 SPEED OF PERSONAL COMPUTERS*

The earliest personal computers appeared in about 1977. Expensive and awkward to use, they at first appealed mainly to hobbyists. Their computational power was quite limited, far below that of the large mainframe computers that serious business and science employed. Many people regarded the small computers as little more than toys.

Within ten years a revolution had occurred. Personal computers became cheaper, more powerful, and indispensable for many jobs. Basic models boasted 40 times more memory than their 1977 predecessors, at less than one-fifth the price. Some desktop computers had more processing power than room-sized mainframes of a decade before. Performance improvements and price reductions were announced almost continuously.

Table 5.3 shows data from speed tests on 40 models of high-performance personal computers. The computers were built around one of two central processing unit (CPU) chips, called the Intel 80286 and 80386. The 80386 is

TABLE 5.3   ___*Benchmark Tests of 40 Makes of Personal Computer Based on Intel CPU Chips*___

Make	Chip type	Calculation test Seconds	Memory test Seconds	System speed MHz
PC Designs GV286	80286	18.82	.71	12.0
IndTech	80286	22.23	.74	10.0
Standard	80286	23.16	.84	12.0
ARC Turbo 12	80286	23.56	.86	12.5
Compaq Port.III	80286	23.80	.90	12.0
FiveStar	80286	27.98	1.00	10.0
QSP	80286	28.00	.90	8.0
PC Designs ET	80286	28.00	1.00	8.0
Multitech	80286	28.00	1.00	8.0
CompuAdd	80286	28.00	.90	8.0
Wang 280	80286	28.01	1.02	10.0
Mitsubishi	80286	28.12	.93	15.8
Maxum	80286	29.00	.90	8.0
Kamerman	80286	35.00	1.30	8.0
Victor	80286	35.00	1.30	8.0
ARC 286 Turbo	80286	35.00	1.30	8.0
IBM AT	80286	35.60	1.32	8.0
Compaq Port. II	80286	35.70	1.30	8.0
Compaq 386/20	80386	10.50	.40	20.0
PC's Limited	80386	13.40	.44	16.0
PC Designs GV386	80386	13.62	.57	16.0
NCR	80386	14.28	.50	16.0
Acer	80386	14.90	.61	16.0
Televideo	80386	15.16	.62	16.0
CCC	80386	15.50	.77	16.0
Compaq 386/16	80386	15.50	.78	16.0
IBM PS/2 80	80386	15.60	.63	16.0
Tandy	80386	17.10	.94	16.0
ALR 386–2	80386	17.10	.77	16.0
Wang 380	80386	17.30	—	16.0
Kaypro	80386	17.30	.85	16.0
Laser	80386	17.30	.85	16.0
Corvus	80386	17.30	.71	16.0
ARC 386i	80386	17.30	—	16.0
Osicom	80386	17.30	—	16.0
Micro 1	80386	17.35	—	16.0
CAE SAR	80386	17.36	—	16.0
Discount	80386	17.40	.72	16.0
CCI	80386	17.40	.69	14.2
Dynamics	80386	22.14	.66	18.0

Source: Data from various issues of *PC Magazine*, 1987.

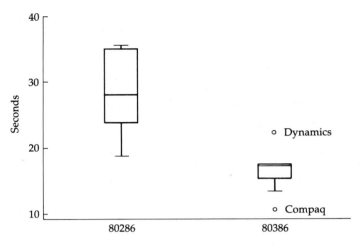

FIGURE 5.10   *Calculation test times of 40 personal computers*

the more advanced chip; these tests were done during the transition period from the older 80286-based systems to a new generation of 80386 computers. The results are from two speed tests, called benchmarks, checking mathematical calculation and memory speed. Both test results are given in seconds; the fewer seconds taken to complete the test, the faster the computer is.

Box plots of calculation times are given for each type of chip in Figure 5.10. The cases in Table 5.3 are in order by calculation speed, so you can easily verify the calculations for this box plot and for the summary statistics (Table 5.4). The following observations can be made:

TABLE 5.4

Calculation Speed Test by CPU Chip Type for 40 Makes of Personal Computer (Data from Table 5.3)

80286 Chip (n = 18 cases)

First quartile:	$Q_1 = 23.8$	
Median:	$Md = 28$	Mean: $\overline{X} = 28.50$
Third quartile:	$Q_3 = 35.0$	
IQR $= Q_3 - Q_1 = 11.2$		Standard
PSD $=$ IQR/1.35 $= 8.30$		deviation: $s_X = 5.08$

80386 Chip (n = 22 cases)

First quartile:	$Q_1 = 15.16$	
Median:	$Md = 17.20$	Mean: $\overline{X} = 16.28$
Third quartile:	$Q_3 = 17.30$	
IQR $= Q_3 - Q_1 = 2.14$		Standard
PSD $=$ IQR/1.35 $= 1.56$		deviation: $s_X = 2.23$

1. *Center* Computers built with the more advanced 80386 chip are faster. The median time on the calculation test is 28.0 seconds for the 80286 computers, and only 17.2 seconds for the 80386. A similar difference in speed is apparent in mean comparisons.
2. *Spread* The 80386 systems are more alike than the 80286 systems are. The 80286 IQR is more than five times that for the 80386 computers (11.2 sec vs. 2.14 sec), as the more compact box in Figure 5.10 indicates. The difference in spread is less pronounced if we compare standard deviations, but even so there is more than twice as much variation among the 80286 machines (5.08 vs. 2.23).
3. *Symmetry* There is evidence of mild negative skewness among the 80386 systems, where the mean is less than the median (16.28 vs. 17.20). The mean and median for 80286 systems are nearly the same (28.32 vs. 28.00).
4. *Outliers* The 80386 group has two mild outliers: the Dynamics (22.14 sec, high) and the Compaq 386/20 (10.50 sec, low). The Dynamics is the only 80386 machine that performed as slowly as the fastest of the 80286 group. The Compaq 386/20, at the opposite extreme, was well ahead of the pack in calculation speed.

The two outliers in the 80386 distribution are roughly balanced, so their effect on the mean is small. They noticeably inflate the standard deviation, however. The standard deviation for 80386 computers is higher than the pseudo-standard deviation, as can be seen in Table 5.4. With two outliers, the distribution has heavier than normal tails, whereas the 80286 distribution is light tailed.

PROBLEMS

The newer, faster, 80386-based personal computers replaced 80286 machines in many scientific and business applications. Only a few years earlier, the 80286 computers had themselves replaced an older generation of machines based on the Intel 8088 chip. The accompanying table gives benchmark data on 11 models of 8088-type computers, comparable to the data on 80286 and 80386 machines in Table 5.3. Problems 5 and 6 use these data.

Benchmark Tests of 11 Makes of Personal Computer Based on Intel Chips

Make	Chip type	Calculation test Seconds	Memory test Seconds	System speed MHz
NEC Multispeed	8088	51.3	1.9	9.54
Wang Laptop	8088	54.2	1.9	8.00
Toshiba T1100	8088	76.9	3.7	7.16
Datavue Spark	8088	84.6	2.2	9.54

(continued)

Benchmark Tests (continued)

Make	Chip type	Calculation test Seconds	Memory test Seconds	System speed MHz
Data General 1	8088	108.5	4.0	8.00
Sharp PC 7100	8088	109.9	5.7	7.37
GridLite Portable	8088	113.8	5.5	4.77
Zenith Z-183	8088	114.2	4.0	8.00
IBM Convertible	8088	150.5	5.7	4.77
IBM PC XT	8088	159.2	5.9	4.77
Bondwell Model 8	8088	164.3	6.0	4.77

Source: Data from various issues of *PC Magazine,* 1987.

*5. Using a vertical axis from 0 to 170 seconds, construct a box plot for the calculation test speeds of the 11 8088-type computers in this table. Sketch in the box plots for 80286 and 80386 computers (copied from Figure 5.10) on the same graph, and write a paragraph comparing the calculation speeds of these three generations of computers.

*6. Use the data in Tables 5.3 and here to compare 8088, 80286 and 80386 computers in terms of memory speed. Construct box plots and describe what they show. (You will have to ignore the cases with missing values.)

5.4   *TRUE AND SELF-REPORTED TEST SCORES**

Our next example involves a type of data that is typically well behaved: standardized test scores. Problem 10 of Chapter 2 had a small sample of VSAT scores matching students' self-reports with true scores from university records. It was a subset of a larger data set from a study of patterns in survey response errors.

Figure 5.11 shows three distributions of VSAT scores from this larger study. On the left is the distribution of self-reports by 353 students; the median self-reported score is 500 (Table 5.5). The middle distribution in Figure 5.11 shows their true test scores, with a median of 470—thirty points lower. Whether intentional or not, some exaggeration mars the self-reports.

Many students surveyed did not fill in the VSAT blank, but their true scores were obtained from university records. The right-hand box plot in Figure 5.11 shows the distribution of these 112 true scores. The median is 420, fifty points lower than the median true score of students who self-reported. This suggests that students with poor scores are understandably less likely to recall them or to feel like divulging them on a survey.

The distributions in Figure 5.11 illustrate two problems that confront survey researchers. The first is that people tend to answer so as to present themselves in a favorable light. The first and second box plots in Figure 5.11 show a systematic exaggeration of SAT scores. Other researchers have found exaggerations in survey reports of such things as whether people own library

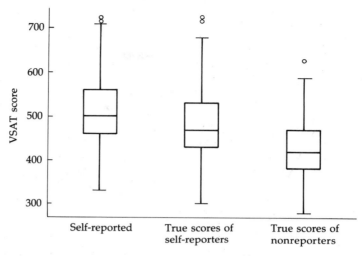

FIGURE 5.11   *Self-reported and true VSAT scores for 465 students*

cards or voted in the last election. When it comes to more sensitive issues such as sexual behavior or criminal activities, distortion is almost certain.

The second problem is nonresponse or missing values. If people are more likely to leave questions blank when the true answer would be embarrassing,

TABLE 5.5   ***Self-Reported and True VSAT Scores in a Sample of 465 College Students***

Self-reported VSAT (n = 353 cases)	
First quartile: $Q_1 = 460$	
Median: $Md = 500$	Mean: $\overline{X} = 511.3$
Third quartile: $Q_3 = 560$	
IQR $= Q_3 - Q_1 = 100$	Standard
PSD $=$ IQR$/1.35 = 74.1$	deviation: $s_x = 75.6$

True VSAT of self-reporters (n = 353 cases)	
First quartile: $Q_1 = 430$	
Median: $Md = 470$	Mean: $\overline{X} = 481.1$
Third quartile: $Q_3 = 530$	
IQR $= Q_3 - Q_1 = 100$	Standard
PSD $=$ IQR$/1.35 = 74.1$	deviation: $s_x = 75.8$

True VSAT of nonreporters (n = 112 cases)	
First quartile: $Q_1 = 380$	
Median: $Md = 420$	Mean: $\overline{X} = 428.1$
Third quartile: $Q_3 = 470$	
IQR $= Q_3 - Q_1 = 90$	Standard
PSD $=$ IQR$/1.35 = 66.7$	deviation: $s_x = 68.8$

then conclusions based on the nonmissing cases will likely be misleading. Going only by the self-reports (Figure 5.11), we might conclude that the median VSAT score was much higher than it really is. The problem is more serious when it interferes with the Census and similar surveys that are the basis for congressional redistricting, funding allocations, and other policy decisions.

The different medians are the only notable differences among the three distributions in Figure 5.11. All three have about the same spread and at most a few mild outliers. They are also nearly symmetrical, so our conclusions would be the same whether based on medians or means. Finally, the three pseudo-standard deviations are within a few points of the corresponding standard deviations: The distributions are approximately normal.

Standardized tests like the SAT are often *designed* to produce bell-shaped, approximately normal frequency distributions. This makes their analysis much simpler. For example, *almost the whole story of these three verbal SAT distributions can be summed up by comparing their means.* Students tended to report higher scores than they actually got. If their score was low, they were less likely to report anything at all. Not much else is going on in Figure 5.11 or Table 5.5.

PROBLEMS

*7. The accompanying table gives gross sales, in millions of dollars, for the ten largest state lotteries, 1983–1985. Construct parallel box plots for sales in each of these three years, and describe what they show about changes in lottery sales.

Gross Sales in Millions of Dollars for the 10 Largest State Lotteries, 1983–1985

State	Sales		
	1983	1984	1985
Connecticut	188.4	254.4	344.5
Illinois	494.5	888.0	1,207.6
Maryland	462.8	536.8	681.1
Massachusetts	617.0	961.2	1,235.3
Michigan	560.9	594.1	891.2
New Jersey	699.0	856.2	936.1
New York	651.4	900.0	1,299.1
Ohio	399.0	604.3	855.6
Pennsylvania	893.6	1,263.2	1,336.5
Washington	200.1	165.8	150.0

Source: Data from the *Boston Globe*, October 6, 1986, p. 1.

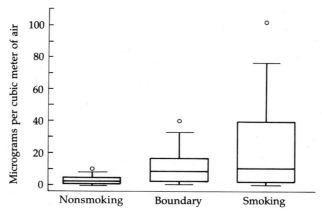

FIGURE 5.12   Nicotine concentration by airline seating

*8. Recall the study of nicotine concentrations in air over the smoking section seats on 26 airline flights (Table 2.1, page 25). Figure 5.12 shows a box plot of these data, together with plots of nicotine concentrations from nonsmoking and boundary sections. Summary statistics are shown in the table.

	Nonsmoking	Boundary	Smoking
$\overline{X}$	3.38	10.63	22.39
s_X	3.53	9.69	28.41
Md	2.1	8.6	14.35
IQR	3.6	14.35	37.4

Source: Oldaker and Conrad (1987).

The mean nicotine concentration of course is highest in smoking sections and lowest in nonsmoking sections. In what other ways do distributions of nicotine concentrations for the three sections differ? In what ways are they similar? From the standpoint of an airline passenger who does not want to breathe nicotine, what might these data imply?

*9. The accompanying table contains data on occupational death rates for selected occupations. Construct parallel box plots for blue-and white-collar jobs, and compare both distributions in terms of center, spread, symmetry, and outliers.

Occupational Death Rates[a] in Selected Blue- and White-Collar Jobs

Blue-collar jobs	Rate	White-collar jobs	Rate
Loom operator	12.5	Advertising agent	3.6
Butchers	13.8	Editor	3.6
Shipfitters	14.2	Ticket agent	3.7

(continued)

Occupational Death Rates (continued)

Blue-collar jobs	Rate	White-collar jobs	Rate
Farm machine op.	14.2	Inspector	3.9
Forge operator	14.2	School administrator	4.2
Plasterer	14.2	Chemist	4.2
Tailor	15.0	Architect	4.3
Sawyer	15.4	Funeral director	4.5
Millwright	15.5	Union official	4.6
Engraver	16.6	Sales manager	4.7
Baker	16.9	Insurance sales	4.9
Police	17.5	Computer operator	5.0
Crane operator	19.3	Restaurant manager	5.1
Grading	20.9	Veterinarian	5.2
Oiler	22.5	Assessor	5.2
Flight attendant	23.0	Superintendent	5.8
Molder	26.6	Surveyor	6.1
Roofer	31.9	Athlete	6.5
Sheriff	32.4	Pharmacist	6.5
Surveyor's apprentice	33.3	Manager	6.6
Miller	33.3	Realtor	6.6
Construction	33.5	Coach	6.6
Taxicab driver	34.0	Technician	6.7
Boilerman	35.0	Weigher	7.1
Crafts	37.5	Public administration	7.2
Miner	37.5	Engineer	7.3
Driller	38.8	Office manager	7.4
Bulldozer operator	39.3	Construction inspector	7.6
Truck driver	39.6	Physicist	7.6
Garbage collector	40.0	Dispatcher	8.3
Firefighter	48.8	Agricultural scientist	9.0
Cable installer	50.7	Geologist	9.5
Metal worker	72.0	Sales	12.3
Asbestos worker	78.7	Office worker	14.5
Logger	129.0	Pilot	97.0

[a] Deaths per 100,000 workers per year.
Source: Data from the *Boston Globe*, January 8, 1989.

5.5 *SEASONAL PATTERNS IN WATER USE**

Table 5.6 shows a time series of average daily water consumption, in millions of gallons, for the entire city of Concord, New Hampshire. The first case is the month of January 1979, during which the water usage averaged 4.517 million gallons per day. In February 1979, daily consumption averaged 4.678 million gallons, and so forth. A time plot appears in Figure 5.13 (page 136).

Table 5.6 and Figure 5.13 cover 29 consecutive months, from January 1979 to May 1981. This particular period is of research interest because the city

TABLE 5.6   *Daily Water Use[a] for City of Concord, 1979–1981*

	Year	Month	Water use
1	1979	Jan.	4.517
2	1979	Feb.	4.678
3	1979	Mar.	4.563
4	1979	Apr.	4.651
5	1979	May	4.785
6	1979	June	5.461
7	1979	July	5.201
8	1979	Aug.	4.829
9	1979	Sep.	4.473
10	1979	Oct.	4.377
11	1979	Nov.	3.971
12	1979	Dec.	3.842
13	1980	Jan.	4.110
14	1980	Feb.	4.175
15	1980	Mar.	4.034
16	1980	Apr.	4.062
17	1980	May	4.491
18	1980	June	4.823
19	1980	July	4.861
20	1980	Aug.	4.560
21	1980	Sep.	4.258
22	1980	Oct.	3.987
23	1980	Nov.	3.778
24	1980	Dec.	3.894
25	1981	Jan.	4.031
26	1981	Feb.	3.552
27	1981	Mar.	3.477
28	1981	Apr.	3.775
29	1981	May	3.922

[a]Average daily usage in millions of gallons.

was experiencing a serious water shortage, and officials were appealing to Concord citizens to reduce their use of water. (The survey data in Table 2.8, page 41, were also collected as part of this study.) Figure 5.13 indicates that overall water use declined during this period, so the conservation appeals could be considered a success.

In northern climates like Concord's, household water use follows a seasonal cycle. Winter is a time of relatively low use, whereas in summer people wash cars more, water lawns, fill pools, and so on. Summer peaks are visible in the time plot of Figure 5.13. Box plots provide another way to view regular cycles.

Figure 5.14 shows 12 box plots, one for each month of the year, based on water use data from 1970 to 1981. The January box plot, for instance, is

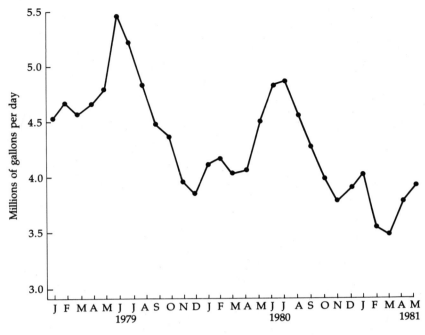

FIGURE 5.13   *Time plot of average daily water use in Concord*

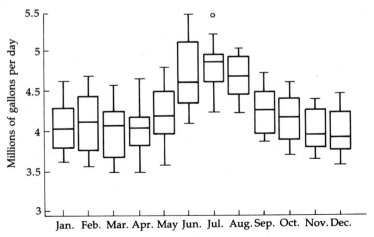

FIGURE 5.14   *Box plots of water use by month, 1970–1981*

constructed from 12 cases: the January water consumption in each of these 12 years. The median for these 12 Januaries is 4.03 million gallons per day, which is the center of the first box plot in Figure 5.14. Summary statistics are given in Table 5.7.

The seasonal rise and fall of water use is clearly visible in Figure 5.14. Median use is lowest in the winter and early spring, rises to a peak in July, and declines during the fall. June is the month with the highest variation, judging by the height of its box (interquartile range). This reflects Concord's

TABLE 5.7

Average Daily Water Use in Concord, 1970–1981, by Month (in Millions of Gallons)

January (n = 12 cases)

First quartile:	$Q_1 = 3.79$	
Median:	$Md = 4.03$	$\overline{X} = 4.07$
Third quartile:	$Q_3 = 4.30$	
	IQR $= Q_3 - Q_1 = .51$	Standard
	PSD $=$ IQR$/1.35 = .38$	deviation: $s_x = .32$

February (n = 12 cases)

First quartile:	$Q_1 = 3.77$	
Median:	$Md = 4.12$	Mean: $\overline{X} = 4.11$
Third quartile:	$Q_3 = 4.43$	
	IQR $= Q_3 - Q_1 = .66$	Standard
	PSD $=$ IQR$/1.35 = .49$	deviation: $s_x = .40$

March (n = 12 cases)

First quartile:	$Q_1 = 3.67$	
Median:	$Md = 4.06$	Mean: $\overline{X} = 4.01$
Third quartile:	$Q_3 = 4.25$	
	IQR $= Q_3 - Q_1 = .58$	Standard
	PSD $=$ IQR$/1.35 = .43$	deviation: $s_x = .34$

April (n = 12 cases)

First quartile:	$Q_1 = 3.82$	
Median:	$Md = 4.03$	Mean: $\overline{X} = 4.03$
Third quartile:	$Q_3 = 4.19$	
	IQR $= Q_3 - Q_1 = .37$	Standard
	PSD $=$ IQR$/1.35 = .27$	deviation: $s_x = .31$

May (n = 12 cases)

First quartile:	$Q_1 = 3.95$	
Median:	$Md = 4.19$	Mean: $\overline{X} = 4.22$
Third quartile:	$Q_3 = 4.49$	
	IQR $= Q_3 - Q_1 = .54$	Standard
	PSD $=$ IQR$/1.35 = .40$	deviation: $s_x = .38$

(continued)

TABLE 5.7   (*continued*)

June (*n* = 11 cases)

First quartile: Q_1 = 4.33
Median: Md = 4.60   Mean: $\overline{X}$ = 4.71
Third quartile: Q_3 = 5.11
        IQR = $Q_3 - Q_1$ =   .78   Standard
        PSD = IQR/1.35 =   .58   deviation: s_X =   .45

July (*n* = 11 cases)

First quartile: Q_1 = 4.60
Median: Md = 4.86   Mean: $\overline{X}$ = 4.82
Third quartile: Q_3 = 4.93
        IQR = $Q_3 - Q_1$ =   .33   Standard
        PSD = IQR/1.35 =   .24   deviation: s_X =   .35

August (*n* = 11 cases)

First quartile: Q_1 = 4.44
Median: Md = 4.68   Mean: $\overline{X}$ = 4.66
Third quartile: Q_3 = 4.93
        IQR = $Q_3 - Q_1$ =   .49   Standard
        PSD = IQR/1.35 =   .36   deviation: s_X =   .27

September (*n* = 11 cases)

First quartile: Q_1 = 3.94
Median: Md = 4.26   Mean: $\overline{X}$ = 4.23
Third quartile: Q_3 = 4.47
        IQR = $Q_3 - Q_1$ =   .53   Standard
        PSD = IQR/1.35 =   .39   deviation: s_X =   .28

October (*n* = 11 cases)

First quartile: Q_1 = 3.87
Median: Md = 4.15   Mean: $\overline{X}$ = 4.12
Third quartile: Q_3 = 4.38
        IQR = $Q_3 - Q_1$ =   .51   Standard
        PSD = IQR/1.35 =   .38   deviation: s_X =   .29

November (*n* = 11 cases)

First quartile: Q_1 = 3.76
Median: Md = 3.93   Mean: $\overline{X}$ = 3.96
Third quartile: Q_3 = 4.23
        IQR = $Q_3 - Q_1$ =   .47   Standard
        PSD = IQR/1.35 =   .35   deviation: s_X =   .25

December (*n* = 11 cases)

First quartile: Q_1 = 3.74
Median: Md = 3.89   Mean: $\overline{X}$ = 3.95
Third quartile: Q_3 = 4.22
        IQR = $Q_3 - Q_1$ =   .48   Standard
        PSD = IQR/1.35 =   .36   deviation: s_X =   .29

variable climate, where June may or may not bring the first of the warm summer weather and its accompanying rise in outdoor water use.

Many natural and social phenomena have a pattern of cycles. Ice cream sales and robberies, like water use, peak in the summer. Other variables are higher in the cold months or holiday seasons. There may be cycles within cycles: fatal car accidents are more common in the summer, when people do more driving (seasonal cycle), and on Fridays (weekly cycle), peaking Friday at night (daily cycle). Cycles are so common in time series data that they are one of the first things an analyst should look for. Parallel box plots provide a simple visual way to examine cyclical patterns. They may reveal cyclical patterns in spread or variation, as well as more obvious changes in centers.

PROBLEM

*10. Suggest your own example of a time series you believe might show cyclical patterns. Sketch what you think the time plot and box plots would look like, if such cycles are indeed present.

5.6 NOTES ON COMPARING DISTRIBUTIONS

In this chapter we drew conclusions such as "80386 computers tend to be faster" or "self-reported test scores tend to be higher than true scores." These conclusions follow common sense and are readily apparent from the data. We use similar comparison-based ideas in everyday life. It is worth looking more closely, though, at what such conclusions actually mean.

If we claim that "nicotine concentrations tend to be higher in smoking section seats," for example (Figure 5.12), we do not mean that smoking sections *always* have worse air. Some had better air than in some nonsmoking sections. The three distributions in Figure 5.12 overlap considerably.

The statement that "nicotine concentrations tend to be higher" actually refers to *differences between the centers (medians or means) of the two distributions.* The nonsmoking sections' distributions have lower centers. This does not necessarily imply anything about other aspects of the two distributions. Many statistical and everyday generalizations are of this sort: They amount to the simple claim that the center of one distribution is higher or lower than the center of another.

A claim like this *cannot be disproved by citing a counterexample.* People often misunderstand this point. For example, someone might object to the statement about nicotine by saying, "I sat in a nonsmoking seat and the air was awful!" This may be true; in the distribution of Figure 5.12, we see some nonsmoking seats had nicotine concentrations as high as the smoking section median. But the original claim was not that all nonsmoking seats were better, only that the *average* concentration was lower there.

Statements making statistical comparisons are often based only on distributional centers. Such comparisons are easy to make and easy to explain. They provide an accurate description of the differences between distributions provided that two things are true:

1. *The location of the distributional "centers" must be unambiguous.* Specifically, the median and the mean should agree fairly well. If not, group A may be higher than group B in terms of medians, but lower regarding means (for example, Chapter 4, Problems 14–16). The *amount* of difference between groups could also vary widely depending on which "center" you choose. If the median and the mean are similar, your choice is immaterial, as both lead to the same conclusions. We again emphasize the importance of detecting outliers and skewness, which can cause the median and mean to differ. Skewed distributions have no unambiguous "center."
2. *Apart from any differences in center, the distributions must be similar.* That is, the differences in center should be more or less the "whole story" of how the distributions differ. If one of the distributions being compared is shaped differently, contains outliers, or is less spread out, then differences in center are not the whole story. For example, the speeds of the older 80286 computers are not only slower but also much more varied than the speeds of the newer 80386 machines (Figure 5.10). A comparison of these two distributions should note this fact, as well as the obvious difference in center.

We will see that distributional problems like skewness and different variability also complicate more advanced statistical analyses. It is important to catch any such problems at an early stage in research. Ideally, all your distributions will look like the VSAT scores in Figure 5.11, and these difficulties will not arise. To assume so before checking the data, however, is unwise.

PROBLEM

11. Box plots of life expectancies in 142 countries are grouped approximately by continent (see figure, based on data from Sivard, 1985). Summary statistics, including low and high extremes for each continent, are given in the accompanying table. A horizontal line indicates the 142-country median, 62 years.

    Describe how life expectancy varies by continent. Discuss differences in centers, spread, symmetry, and outliers. What do these distributional features tell us about life in the countries of these four regions?

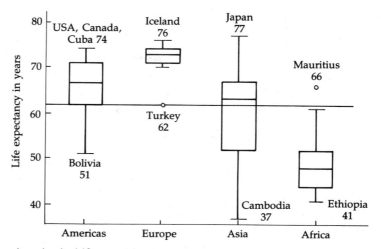

Life expectancies in 142 countries by continent

Life Expectancies in 142 Countries, by Continent

Americas (n = 26 cases)

Low extreme:	51 (Bolivia)	
First quartile:	$Q_1 = 62$	
Median:	$Md = 66.5$	Mean: $\overline{X} = 65.6$
Third quartile:	$Q_3 = 71$	
High extreme:	74 (Canada/Cuba/U.S.)	
IQR = $Q_3 - Q_1 = 9$		Standard
PSD = IQR/1.35 = 6.7		deviation: $s_X = 6.4$

Europe (n = 29 cases)

Low extreme:	62 (Turkey)	
First quartile:	$Q_1 = 71$	
Median:	$Md = 73$	Mean: $\overline{X} = 72.4$
Third quartile:	$Q_3 = 74$	
High extreme:	76 (Iceland)	
IQR = $Q_3 - Q_1 = 3$		Standard
PSD = IQR/1.35 = 2.2		deviation: $s_X = 2.6$

Asia (n = 42 cases)

Low extreme:	37 (Cambodia)	
First quartile:	$Q_1 = 52$	
Median:	$Md = 63$	Mean: $\overline{X} = 60.2$
Third quartile:	$Q_3 = 67$	
High extreme:	77 (Japan)	
IQR = $Q_3 - Q_1 = 15$		Standard
PSD = IQR/1.35 = 11.1		deviation: $s_X = 10.3$

(continued)

Life Expectancies in 142 Countries, by Continent

	Africa ($n = 45$ cases)	
Low extreme:	41 (Ethiopia)	
First quartile:	$Q_1 = 44$	
Median:	$Md = 48$	Mean: $\overline{X} = 49.1$
Third quartile:	$Q_3 = 52$	
High extreme:	66 (Mauritius)	
IQR $= Q_3 - Q_1 = 8$		Standard
PSD $=$ IQR/1.35 $= 5.9$		deviation: $s_X = 5.7$

SUMMARY

Box plots display order-statistic summaries in graphical form. By combining the strengths of graphical and numerical summary methods, box plots convey detailed information as an easily understood picture. They are especially helpful for comparing two or more distributions.

The classical approach in such comparisons is to focus on differences among means—the only important ways in which some data distributions differ. Often, however, more complicated patterns emerge when the distributions are carefully examined. Box plots help check whether mean-based analysis is sufficient. They also alert us to other interesting or unexpected features of the data, thus keeping the analysis realistic.

Some of the comparisons in this chapter illustrate a simple kind of **bivariate,** or two-variable, analysis. The states' alcohol sales example could be viewed as *an analysis of the relationship between two variables:* a categorical variable, region, and a measurement variable, per capita alcohol sold. Other examples include the relationships between children's gender (categorical) and the number of rough and tumble play episodes (measurement); between computer chip type (categorical) and speed (measurement); and between airline seating section (categorical) and nicotine concentration (measurement).

Parallel box plots help us explore this type of problem involving one categorical and one measurement variable. More formal methods for such problems are introduced in Chapter 13. Chapters 11, 13, 15, and 16 also give further reasons to make box plots your first analytical step.

The next chapter tackles the problem of ill-behaved distributions. We have now seen a variety of ways to detect them, but they still complicate many analyses. In Chapter 6, we will look at some things we can do to minimize these problems.

PROBLEMS

12. Draw sketches to show how you might use box plots to investigate the following research questions, assuming that the data were available. First specify exactly the cases and variables in your data.

a. Do patients who receive Therapy A after hand surgery recover faster than patients who receive Therapy B, or than patients who receive no therapy at all?

b. Are elementary school students taught under the experimental New-fangled Curriculum learning to read any better (or any worse) than other students under the Old-Fashioned Curriculum?

c. Does a city police department tend to give out more parking tickets on any particular day?

d. Compare the following (somewhat overlapping) groups in terms of their household income: owners of at least one American-made automobile; owners of at least one Japanese-made automobile; owners of at least one European-made automobile; people who own no car.

e. Invent a research question of your own.

13. Table 2.9 in Chapter 2 (page 42) contains an ordered list of savings (reductions) in household water use by 33 households during a conservation campaign. Use this list to construct a box plot, and describe its main features.

14. The standard deviation of water savings (Problem 13) is $s_x = 1,201$ cubic feet. Use your work for Problem 13 to calculate the pseudo-standard deviation. Compare the PSD and s_x. How does this comparison confirm what can be seen in your box plot for Problem 13?

The accompanying table contains data on the unemployment rates for graduates of 44 British universities, between 1976 and 1981. Other variables are average entrance exam scores (on the British A-level exam), an index of the per student cost of academic staff, and the ratio of male to female full-time undergraduates. Ten of these universities were formerly Colleges of Advanced Technology (CATs), emphasizing technical and scientific training. Former CATs and other schools are listed separately in the table. Problems 15–19 refer to these data.

Unemployment of Graduates and Other Data on 44 British Universities

University	Graduate unemployment[a]	Entrance exam scores[b]	Academic staff cost index	Male/female ratio
Former colleges of advanced technology (CAT)				
Bath	8.8	10.5	102.6	1.87
Brunel	8.3	7.1	102.5	3.48
Loughborough	10.4	9.1	111.2	2.38
Aston	6.8	7.2	104.5	2.54
Bradford	13.0	7.9	104.5	1.94
Strathclyde	10.1	7.7	104.4	1.89
City	8.5	8.4	106.3	3.64
Heriot-Watt	10.6	8.7	102.0	3.20

(continued)

Unemployment of Graduates (continued)

University	Graduate unemployment[a]	Entrance exam scores[b]	Academic staff cost index	Male/female ratio
Former colleges of advanced technology (CAT)				
Surrey	11.7	8.2	103.2	1.54
Salford	9.1	6.8	102.7	2.86
Other British universities				
Warwick	18.4	9.0	97.4	1.15
Kent	19.9	7.9	96.3	1.40
East Anglia	23.6	7.3	95.8	1.23
London	14.4	9.7	99.5	1.45
Ulster	28.8	9.7	98.5	.77
Sussex	23.4	9.4	97.6	1.15
Hull	16.8	7.9	96.6	1.05
Aberdeen	11.4	8.5	97.1	1.12
Cambridge	9.1	13.6	98.5	2.42
Dundee	9.2	8.2	97.9	1.89
Bristol	14.0	12.1	98.7	1.46
Sheffield	12.9	10.0	102.7	1.47
Edinburgh	15.1	10.7	98.3	1.11
Leicester	16.5	9.2	97.8	1.20
Durham	12.9	11.8	97.7	1.21
St. Andrews	16.9	10.5	96.0	.87
Stirling	21.0	7.8	96.8	1.16
Nottingham	15.3	10.3	101.5	1.47
Leeds	13.6	9.7	102.0	1.40
Wales	16.1	7.8	99.0	1.38
Southampton	11.5	10.8	100.2	1.64
Liverpool	10.7	9.5	100.5	1.74
Oxford	10.4	13.1	98.0	2.12
Lancaster	20.4	8.1	98.7	1.28
Birmingham	9.8	10.3	100.6	1.50
Exeter	13.5	9.6	98.2	1.07
Newcastle	11.4	8.6	102.0	1.55
York	18.4	10.2	96.2	1.13
Keele	21.0	7.0	97.4	1.19
Essex	18.1	6.3	96.7	1.63
Reading	16.4	9.8	101.8	1.18
Glasgow	8.7	9.2	98.7	1.25
Queens	11.5	9.7	99.5	1.58
Manchester	13.0	10.4	101.6	1.66

[a]Estimated average 1976–1981 unemployment rate of graduates.
[b]Mean A-level score of entrants in Fall 1978.
Source: Data from Taylor (1984).

15. Construct parallel box plots comparing the unemployment rates of graduates from the CATs and other universities. Discuss these distributions in terms of center, spread, and symmetry. Identify the one outlier individually on your plot. Explain why graduates of that particular university might have such an unusual unemployment rate.

16. Use box plots to compare the A-level examination scores at the two types of university, and describe the differences you see. Are the results surprising? What do they suggest about the students attracted to CATs and other universities?

17. Construct box plots and compare the academic cost per student across the two university types. What economic realities could explain the difference in median costs?

18. Compare box plots of the male/female ratios of CATs and other universities. Do your findings here have any possible relevance to your findings in Problem 15?

19. Summarize your main findings from Problems 15–18 in a brief report about the former Colleges of Advanced Technology and other British universities. What are the main points of difference?

20. U.S. homicide rates and other violent crimes show marked regional differences. The accompanying table lets you explore such differences. The cases are a sample of 36 U.S. cities, classified as either South or other. The cities are listed in order of increasing homicide rates. Use the data in the table to construct parallel box plots of homicide rates in Southern and non-Southern cities.

Homicides by Region in 36 U.S. Cities, 1980–1984

	City	Region	Homicides[a]
1	Alexandria, VA	South	6.78
2	Greensboro, NC	South	7.58
3	Hollywood, FL	South	7.58
4	Tulsa, OK	South	8.64
5	Pasadena, TX	South	9.42
6	Amarillo, TX	South	11.26
7	Louisville, KY	South	12.13
8	Jacksonville, FL	South	13.57
9	Waco, TX	South	14.22
10	Orlando, FL	South	14.65
11	Jackson, MS	South	19.12
12	Tampa, FL	South	19.30
13	Fort Worth, TX	South	23.57
14	Richmond, VA	South	26.00
15	Birmingham, AL	South	27.07

(continued)

Homicides by Region (continued)

	City	Region	Homicides[a]
1	Sterling Hts., MI	Other	.55
2	Cedar Rapids, IA	Other	1.63
3	Ann Arbor, MI	Other	2.04
4	Fullerton, CA	Other	2.35
5	Boise, ID	Other	2.54
6	Madison, WI	Other	2.70
7	Eugene, OR	Other	2.84
8	Concord, CA	Other	3.10
9	Stamford, CT	Other	4.10
10	Pueblo, CO	Other	4.13
11	Allentown, PA	Other	4.24
12	Albany, NY	Other	4.33
13	Erie, PA	Other	4.70
14	Davenport, IA	Other	4.84
15	Lansing, MI	Other	5.37
16	Salt Lake, UT	Other	6.01
17	Topeka, KS	Other	6.25
18	Tucson, AZ	Other	7.32
19	Kansas City, MO	Other	19.73
20	Los Angeles, CA	Other	21.66
21	Las Vegas, NV	Other	42.29

[a]Average per 100,000 population.

21. Use your box plots from Problem 20 to compare the distributions of homicide rates among cities from each region. Discuss center, spread, symmetry, and outliers.

22. What findings would you expect if we compared the two regions in Problem 20 using the mean homicide rates, instead of the medians?

23. Acid rain is a growing problem in many parts of the world, even far downwind from the industrial and urban areas where the heaviest pollution originates. The accompanying table contains data from studies of summer (1983) storms at mountain locations in Colorado and Pennsylvania. Acidity of precipitation is measured as pH (*power of Hydrogen*), a scale on which 7 is neutral and lower values mean greater acidity. (Pure water has a pH of 7; vinegar is about 2.8–3.0.)

    Construct box plots, and use them to compare the acidity of precipitation from summer storms on Niwot Ridge and Allegheny Mountain.

Acidity of Summer Rainstorms at Niwot Ridge, Colorado, and Allegheny Mountain, Pennsylvania

Storm	Location	Acidity (pH)	Storm	Location	Acidity (pH)
1	Niwot	2.97	21	Niwot	4.24
2	Niwot	3.37	22	Niwot	4.28
3	Niwot	3.50	23	Niwot	4.30
4	Niwot	3.58	24	Niwot	4.35
5	Niwot	3.60	25	Niwot	4.40
6	Niwot	3.61	26	Niwot	4.47
7	Niwot	3.62	27	Niwot	4.57
8	Niwot	3.68	28	Niwot	4.92
9	Niwot	3.70	29	Allegheny	3.13
10	Niwot	3.88	30	Allegheny	3.19
11	Niwot	3.91	31	Allegheny	3.24
12	Niwot	3.97	32	Allegheny	3.26
13	Niwot	3.98	33	Allegheny	3.39
14	Niwot	4.01	34	Allegheny	3.42
15	Niwot	4.01	35	Allegheny	3.52
16	Niwot	4.02	36	Allegheny	3.54
17	Niwot	4.06	37	Allegheny	3.57
18	Niwot	4.17	38	Allegheny	3.60
19	Niwot	4.23	39	Allegheny	3.71
20	Niwot	4.23	40	Allegheny	3.75

Sources: Reddy et al. (1985); Pierson et al. (1987).

NOTES

1. Like stem-and-leaf displays, box plots are a technique from the recently developed approach called exploratory data analysis, or EDA; see Tukey (1977). Some basic principles of EDA are:
   1. Make few assumptions about the data.
   2. Emphasize graphical displays that highlight unexpected features of the data.
   3. Use resistant numerical summaries, such as those based on order statistics.

   Box plots embody all three principles and are one of the most successful of EDA's many innovations.

   For further information about EDA, consult:

   DAVID C. HOAGLIN, FREDERICK MOSTELLER, and JOHN W. TUKEY, eds. (1983). *Understanding Robust and Exploratory Data Analysis.* New York: Wiley.

DAVID C. HOAGLIN, FREDERICK MOSTELLER, and JOHN W. TU-
KEY, eds. (1985). *Exploring Data Tables, Trends, and Shapes.* New York:
Wiley.

PAUL F. VELLEMAN and DAVID C. HOAGLIN (1981). *Applications, Basics,
and Computing of Exploratory Data Analysis.* Boston: Duxbury.

The first two books are edited volumes that cover a broad range of topics,
often at an introductory level. The third book focuses on a smaller set of
specific techniques, but includes worked-out examples and computer programs. These programs are also available in the MINITAB statistical
package.

2. More than some other univariate graphs, box plots really come into their
own when a computer program takes over the work of construction. Obtaining the plots is then effortless, and they can be used routinely to check
on any analysis with measurement variables.

In Chapter 2 it was noted that stem-and-leaf displays are most readable
when any outliers are trimmed and listed separately. The scale for the
entire display then need not accommodate the most extreme values. Computer programs that do stem-and-leaf displays often automatically trim
outliers. For this purpose an *outlier* is defined as for box plots: any value
more than 1.5(IQR) beyond the first or third quartile.

Coping with Outliers and Skewness*

Chapter 6

We have seen that real data often do not resemble the bell-shaped statistical ideal. Many distributions are not even symmetrical; positive skewness and outliers are common problems. Other distributions have large gaps or multiple peaks. Even if a distribution *is* unimodal and fairly symmetrical, it may have much heavier or lighter tails than a normal distribution.

Many statistical methods work best with normal distributions and are misleading when applied to less well-behaved data. Chapters 2–5 introduced a variety of graphical and numerical-summary techniques for detecting possible distributional problems. This chapter now asks what we can do about such problems.

There are two general approaches. One is to change the data, so they are less ill behaved. This may require nothing more than temporarily dropping an outlier, or it may involve *nonlinear transformations,* which alter the shape of a whole distribution. A second approach, called *robust estimation,* applies special statistical techniques that serve even ill-behaved distributions. Robust estimation is beyond the scope of this book, but nonlinear transformation procedures are comparatively simple and well worth learning.

6.1 DELETING OUTLIERS

An outlier is a value that lies far from the central part of a distribution—more than 1.5(IQR) beyond the first or third quartile. A single outlier can greatly influence the mean and even more so the standard deviation. A good example of outlier distortion appears when we examine marriage rates in the 50 U.S. states by geographical region (Table 6.1).

TABLE 6.1   *1980 Marriages per 1,000 Population in 50 U.S. States, Sorted by Region*

	State	Region	Marriage rate		State	Region	Marriage rate
1	Delaware	South	7.46	26	Alaska	West	13.34
2	N. Carolina	South	7.94	27	Idaho	West	14.23
3	W. Virginia	South	8.92	28	Wyoming	West	14.63
4	Kentucky	South	8.94	29	Nevada	West	142.83
5	Louisiana	South	10.33	30	New Jersey	Northeast	7.58
6	Maryland	South	10.97	31	Pennsylvania	Northeast	7.90
7	Mississippi	South	11.07	32	Rhode Island	Northeast	7.91
8	Florida	South	11.12	33	Massachusetts	Northeast	8.07
9	Virginia	South	11.26	34	New York	Northeast	8.23
10	Arkansas	South	11.60	35	Connecticut	Northeast	8.38
11	Alabama	South	12.59	36	New Hampshire	Northeast	10.05
12	Texas	South	12.77	37	Vermont	Northeast	10.22
13	Tennessee	South	12.89	38	Maine	Northeast	10.71
14	Georgia	South	12.93	39	Wisconsin	Midwest	8.74
15	Oklahoma	South	15.37	40	Nebraska	Midwest	9.07
16	S. Carolina	South	17.27	41	Minnesota	Midwest	9.23
17	Oregon	West	8.74	42	Ohio	Midwest	9.25
18	California	West	8.91	43	N. Dakota	Midwest	9.34
19	Montana	West	10.60	44	Michigan	Midwest	9.38
20	Arizona	West	11.12	45	Iowa	Midwest	9.43
21	Washington	West	11.55	46	Illinois	Midwest	9.61
22	Utah	West	11.61	47	Kansas	Midwest	10.51
23	Colorado	West	12.08	48	Indiana	Midwest	10.54
24	Hawaii	West	12.29	49	Missouri	Midwest	11.11
25	New Mexico	West	12.77	50	S. Dakota	Midwest	12.74

Table 6.2 gives the summary statistics. Judging from the means, marriage rates are lowest in the Northeast ($\overline{X}$ = 8.78) and somewhat higher in the Midwest ($\overline{X}$ = 9.91) and South ($\overline{X}$ = 11.47). The mean marriage rate for the 13 Western states ($\overline{X}$ = 21.90) far exceeds the rest. Also, the West's standard deviation is over 10 times higher than elsewhere. Something unusual seems to be going on out West.

Since it is hard to imagine why Western state marriage rates would differ so, we should immediately suspect a statistical problem. Comparing means with medians in Table 6.2 confirms this suspicion. In the South, Northeast, and Midwest, the medians are about the same as the corresponding means. But the West has a big discrepancy between the median (12.08) and the mean (21.90). Because the mean exceeds the median, we can guess that some state or states with a high marriage rate pull up the Western mean. Checking the ordered data in Table 6.1, we spot Nevada, with 142.83 marriages per 1,000 people in 1980—much higher than any other state in the nation. The next highest, South Carolina, has only 17.27.

TABLE 6.2   *Statistical Summaries for 1980 U.S. Marriage Rates by Region (Data from Table 6.1)*

South (n = 16 cases)

First quartile:	Q_1 = 9.64	
Median:	Md = 11.19	Mean: $\overline{X}$ = 11.47
Third quartile:	Q_3 = 12.83	
IQR = $Q_3 - Q_1$ = 3.19		Standard
PSD = IQR/1.35 = 2.36		deviation: s_X = 2.57

West (n = 13 cases)

First quartile:	Q_1 = 11.12	
Median:	Md = 12.08	Mean: $\overline{X}$ = 21.90
Third quartile:	Q_3 = 13.34	
IQR = $Q_3 - Q_1$ = 2.22		Standard
PSD = IQR/1.35 = 1.64		deviation: s_X = 36.38

Northeast (n = 9 cases)

First quartile:	Q_1 = 7.91	
Median:	Md = 8.23	Mean: $\overline{X}$ = 8.78
Third quartile:	Q_3 = 10.05	
IQR = $Q_3 - Q_1$ = 2.14		Standard
PSD = IQR/1.35 = 1.59		deviation: s_X = 1.19

Midwest (n = 12 cases)

First quartile:	Q_1 = 9.24	
Median:	Md = 9.41	Mean: $\overline{X}$ = 9.91
Third quartile:	Q_3 = 10.52	
IQR = $Q_3 - Q_1$ = 1.28		Standard
PSD = IQR/1.35 = .95		deviation: s_X = 1.13

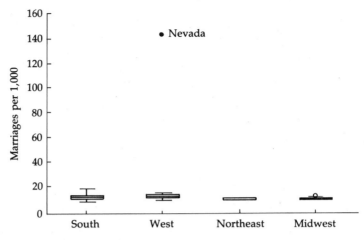

FIGURE 6.1   *Marriage rates in 50 U.S. states by region*

Nevada's high marriage rate, like its high alcohol sales (Chapter 5), does not directly reflect the behavior of Nevadans themselves. Famous for its roadside wedding chapels and loose laws, Nevada draws thousands of visitors for quick marriages every year. Counting these marriages as part of the state's overall marriage rate greatly inflates it. The reason nonresidents have such an impact on alcohol and wedding rates is partly due to Nevada's relatively low population. In California the same number of nonresident weddings would scarcely alter the overall rate.

In the box plots of Figure 6.1 Nevada stands alone as a severe outlier at the top of the graph. All the other states, having relatively tiny variations, look squashed flat. Figure 6.1 dramatizes how different Nevada is, but this graph reveals little else. For example, we cannot tell whether, excluding Nevada, Western marriage rates really are any higher than elsewhere.

This point leads to the simplest method of dealing with outliers: Just set them aside and re-analyze the remaining data without them. This is called **outlier deletion.** Deleting Nevada produces new Western state summary statistics (Table 6.3) and new box plots (Figure 6.2). Our data set now consists of only 49 cases. No summary statistics except the Western ones have changed from Table 6.2.

The change for the West (Table 6.3) is substantial: The mean now almost equals the median—a nearly symmetrical distribution. The Western standard deviation (36.38 with Nevada) falls to only 1.84. Nevada had a large effect on the mean and a much larger effect on the standard deviation—as outliers typically do. Also typically, the outlier had little influence on the median or interquartile range: The median went from 12.08 to 11.84 and the IQR from 2.22 to 2.20—a trivial difference.

TABLE 6.3 ***Statistical Summaries for 1980 Marriage Rates Among 12 Western States (Excluding Nevada)***

West ($n = 12$ cases)	
First quartile: $Q_1 = 10.86$	
Median: $Md = 11.84$	Mean: $\overline{X} = 11.82$
Third quartile: $Q_3 = 13.06$	
IQR $= Q_3 - Q_1 = 2.20$	Standard
PSD $=$ IQR$/1.35 = 1.63$	deviation: $s_x = 1.84$

The statistics in Table 6.3 revise our conclusions about how regional marriage rates compare. The Western mean (11.82) remains highest, but only slightly above the South's (11.47). Northeastern rates are distinctly the lowest. Both the standard deviation and the IQR now confirm that marriage rates vary most in the South. The box plots in Figure 6.2, unlike those in Figure 6.1, clarify many respects in which these regions differ.

Obviously our next question is *why* the regions differ. We seem to see a sunbelt–frostbelt split: Marriages occur more often in the South and West than in the Northeast or Midwest. Economics might be a factor, or different age mixes in their populations. To investigate such possibilities would require bivariate or multivariate analytical methods, which we are not yet ready for—but these methods are also sensitive to the effects of outliers. Outlier deletion therefore remains a useful tool even in more complex analyses.

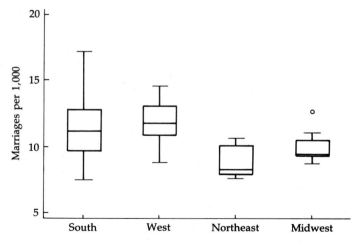

FIGURE 6.2   *Marriage rates in 49 U.S. states, omitting Nevada*

6.2   *WHEN SHOULD OUTLIERS BE DELETED?*

Outlier deletion is certainly justified with the marriage rates example, but many other situations are less clear-cut. It is *not* a good idea just to drop any outlying cases automatically. Outlier deletion amounts to throwing away data, which should be done only after careful thought. In some instances, the outliers are the most important cases in the data.

Graphs such as stem-and-leaf displays and box plots will reveal whether there are any outliers to worry about. Even if outliers exist, though, they may pose no serious problem. Try repeating the same analysis with and without the outliers. Are the results very different? Do they lead to different conclusions? If deleting the outliers makes little difference, then go ahead and keep them in. You have proven that they are not causing serious problems.

If the outliers do make a difference, then *your conclusions depend on the values of a small fraction of the data.* Look at the outlying cases, and try to figure out why they are so unusual. Sometimes the problem results from an error in measurement or in copying the data. If the outlying values are accurate, perhaps they reflect some unusual circumstance that is irrelevant to the rest of the data—like Nevada's marriage rate. On the other hand, sometimes outliers are very relevant indeed. Suppose we were studying the effects of air pollution on death rates in American cities. We might find that Los Angeles is an outlier on our air pollution variable, because of that city's exceptionally high levels of smog. Deleting L.A., however, throws away information about what happens at such a high level of pollution. In this instance the outlier is the most valuable part of the data. It *should* influence our conclusions.

Some outliers, like Nevada, are so far out they appear unrelated to the rest of the distribution, but frequently outliers are just part of the tails. This often arises with positive skewness, when the long upper tail trails off into a scattering of outliers, as in Figures 5.12 (nicotine concentration, page 133) and 3.5 (population per physician, page 73). When outliers are part of overall skewness, it is generally better to *remedy the skewness, rather than remedy the outliers alone.* A versatile method for handling skewness and its associated outliers is introduced later in this chapter. Unlike outlier deletion, this method—*nonlinear transformation*—does not require discarding data.

As you analyze measurement data, ask the following questions:

1. *Do outliers exist?*   Graphs and mean–median comparisons help make any outliers apparent. If there are outliers, proceed to the next question.
2. *Are outliers influencing your conclusions?*   Try the analysis including and excluding the outliers. If the results differ much, keep questioning.
3. *Why are the outlying cases so extreme?*   Use everything you know about these cases to figure out what makes them unusual. Are the reasons irrelevant to your study's goals? Should these particular cases be allowed to influence your conclusions?

4. *Are the outliers just the long tail of a skewed distribution?* If so, then outlier deletion may be inappropriate. Instead, consider nonlinear transformation, which can change the shape of the entire distribution.

If, after all this thinking and analysis, you still cannot decide what to do, report your results *with and without* the outliers. Chances are that the decision is not clear-cut, and different people would choose differently. Give your audience this option by reporting it both ways.

PROBLEMS

*1. The box plots of life expectancy by continent (Chapter 5, Problem 11, page 140) show two outliers, Turkey and Mauritius. If our study calls for comparing the mean life expectancies of European and African countries, should these outliers be deleted? Explain. What would be the statistical consequences of deletion?

*2. In the student weight example (Table 4.3, page 94, and Figure 4.2, page 109), one heavy student appeared to be an outlier. If his weight is deleted, the remaining students are closer to the marathon runners in mean weight. Should we delete the outlier before making this comparison? Explain.

6.3 LOGARITHMS

When the distribution as a whole is skewed, outliers may be just a natural extension of the longer tail. Instead of deleting the outliers as if they were invalid cases, it is better to try to correct the basic skewness itself. This can be done with **nonlinear transformations.** A *transformation* is any mathematical change applied to data. Converting miles to kilometers is a transformation, done by multiplying distance in miles by about 1.6. Another simple transformation is converting hours into minutes, multiplying by 60. Such conversions are examples of **linear transformations,** which *change the scale but not the shape* of a variable's distribution. Nonlinear transformations change *both scale and shape.* Sometimes a simple nonlinear transformation can be found that changes a skewed distribution into a symmetrical one, which is easier to analyze.

One nonlinear transformation that is widely used to correct positive skewness is the *logarithm.* Two kinds of logarithm are popular: *base-10 logarithms,* sometimes called *common* logarithms, and *base-e logarithms,* also known as *natural* logarithms. This book mainly uses base-10 logarithms, which are easier to explain, but both kinds have the same effect on distributional shape.

> The **base-10** or **common logarithm** of a number x, written $\log_{10}(x)$, is *the power to which you would have to raise* 10 *to get* x. That is, $\log_{10}(x) = p$ if $10^p = x$.

The base-10 logarithm of 10 is 1: $\log_{10}(10) = 1$, because 10 to the first power equals 10. Similarly the base-10 log of 100 is 2: $\log_{10}(100) = 2$, because $10^2 = 100$. The base-10 log of 1,000 is 3, because $10^3 = 1,000$. Any number raised to the 0 power is 1, so $\log_{10}(1) = 0$. $\text{Log}_{10}(x)$ will be negative if x is between 0 and 1; as x approaches 0, $\log_{10}(x)$ becomes a larger and larger negative number, approaching negative infinity. The logarithm of 0 itself is undefined: There exists no real number p such that $10^p = 0$. Logarithms of negative numbers are also undefined.

Values of the base-10 logarithm of x:

$\log_{10}(x) > 0$	if $x > 1$
$\log_{10}(x) = 0$	if $x = 1$
$\log_{10}(x) < 0$	if $0 < x < 1$
$\log_{10}(x)$ is undefined	if $x \leq 0$

Any positive number can be transformed into a logarithm. Hand calculators and computer programs often have a function that finds logarithms automatically. Playing with this function—seeing what happens when you take logarithms of various numbers—is a good way to grow comfortable with the effects of this mathematical transformation.

Examples of base-10 logarithms are given in Table 6.4. Simple algebraic relationships among these logarithms may become apparent if you study them for a moment. Logarithms have mathematical properties that prove useful in many areas besides statistics. For example, logarithms transform multiplication into addition—the property that underlies the design of the slide rule. Until the microchip revolution the slide rule was an indispensable aid to hand calculation: You can multiply on it by adding different lengths of two sliding scales, marked in logarithmic units.

Another property of logarithms is that they shrink any number greater than 1. The larger the number, the more drastically logarithms shrink it. This is illustrated in Figure 6.3, for the sequence of numbers 10, 20, 30, . . ., 100. $\text{Log}_{10}(10)$ equals 1; $\log_{10}(20)$ is about 1.30; $\log_{10}(30)$ is about 1.48. Notice that the difference between $\log_{10}(10)$ and $\log_{10}(20)$ is larger than the difference between $\log_{10}(20)$ and $\log_{10}(30)$. The difference between the logarithms of 30 and 40 is larger than that for 40 and 50, and so on.

Similar shrinkage can be seen in Table 6.4. As x goes from 10 to 100, $\log_{10}(x)$ goes only from 1 to 2. Each time x increases by a factor of ten, as from

TABLE 6.4   ***Examples of Base-10 Logarithms (Approximate Calculations)***[a]

x	$Log_{10}(x)$	*Reason*
-10	Undefined	No p exists such that $10^p = -10$
-1	Undefined	No p exists such that $10^p = -1$
0	Undefined	No p exists such that $10^p = 0$
0.1	-1	$10^{-1} = 1/10^1 = 1/10 = 0.1$
0.2	-0.7	$10^{-.7} = 1/10^7 = 1/5 = 0.2$
0.5	-0.3	$10^{-.3} = 1/10^3 = 1/2 = 0.5$
1	0	$10^0 = 1$
2	0.3	$10^{.3} = 2$
5	0.7	$10^{.7} = 5$
10	1	$10^1 = 10$
20	1.3	$10^{1.3} = 20$
50	1.7	$10^{1.7} = 50$
100	2	$10^2 = 100$
200	2.3	$10^{2.3} = 200$
500	2.7	$10^{2.7} = 500$
$1,000$	3	$10^3 = 1,000$
$1,000,000$	6	$10^6 = 1,000,000$

[a]Definition: $\log_{10}x = p$, where $10^p = x$.

10 to 100 or from 100 to 1,000, $\log_{10}(x)$ increases only by 1. The logarithm of one million is 6, only twice as high as the logarithm of one thousand: $\log_{10}(1,000) = 3$. In logarithms, the difference between one million and two million is exactly the same as the difference between 1 and 2, or between 10 and 20: $\log_{10}(1,000,000) = 6$ and $\log_{10}(2,000,000) = 6.3$.

One statistical consequence of such shrinkage is that large data values, such as high outliers or the long upper tail of a positively skewed distribution, are "pulled in" relative to the rest of the data. The more extreme an outlier is, the more strongly taking logarithms pulls it in, as we shall see in the next example.

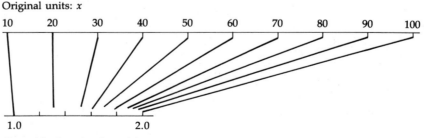

Original units: x

Logarithmic units: $\log_{10}(x)$

FIGURE 6.3   *Base-10 logarithms "shrink" numbers greater than one*

PROBLEMS

*3. Use a calculator or computer to obtain these values to five decimal places.
   a. $\log_{10}(50)$
   b. $\log_{10}(986)$
   c. $\log_{10}(4.6)$
   d. $\log_{10}(.001)$
   e. $\log_{10}(1)$
   f. $\log_{10}(.9)$
   g. $\log_{10}(1.1)$
   h. $\log_{10}(0)$
   i. $\log_{10}(-100)$

*4. Raise 10 to the power of the logarithms you obtained in Problem 3. Are the results as expected? Explain.

6.4   PER CAPITA GROSS NATIONAL PRODUCT

Table 6.5 lists the per capita gross national product (GNP) of 24 countries in U.S. dollars. A box plot of these data (Figure 6.4) shows one high outlier, Brunei. The distribution would be positively skewed even without Brunei, however. If Brunei is deleted the next highest country, Norway, becomes an outlier. Deleting Norway makes Australia, France, and West Germany all outliers! The basic problem here is the distributional shape, not the presence of one outlier. Outlier deletion is therefore inappropriate; we should try a

TABLE 6.5   *Ordered List of 1982 per Capita GNP in 24 Countries*

	Country	GNP in dollars		Country	GNP in dollars
1	Ethiopia	150	13	Paraguay	1,670
2	Zaire	180	14	Jordan	1,680
3	Sri Lanka	320	15	Taiwan	2,670
4	Guinea	330	16	Greece	4,170
5	Togo	350	17	Venezuela	4,250
6	Mauritania	520	18	Czechoslovakia	5,540
7	Indonesia	550	19	Austria	9,830
8	North Korea	930	20	Australia	11,220
9	Nigeria	940	21	France	11,520
10	Mongolia	940	22	West Germany	12,280
11	Nicaragua	950	23	Norway	14,300
12	Congo	1,420	24	Brunei	22,260

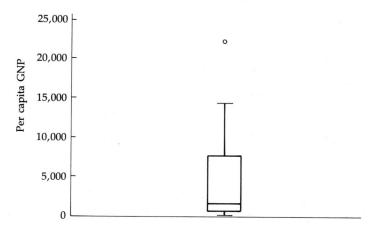

FIGURE 6.4   *Box plot of 1982 per capita GNP in 24 nations*

shape-changing transformation instead. Since the skewness is positive, logarithms are a reasonable choice for this transformation.

Table 6.6 gives logarithms of per capita GNP for each country. The poorest country, Ethiopia, has a per capita GNP of $150, which in logarithms is $\log_{10}(150) = 2.176$. The wealthiest country in this sample is Brunei, with a per capita GNP of $22,260, which in logarithms is $\log_{10}(22,260) = 4.348$.

Glancing down the columns of Table 6.6, we can see how much logarithms pull in the upper tail, thus changing the distribution's overall shape. Compare the box plot of the logarithm of GNP (Figure 6.5) with that for the

TABLE 6.6   *Ordered List of 1982 per Capita GNP in 24 Countries, in Raw and Base-10 Logarithm Forms*

	Country	GNP in dollars	$Log_{10}(GNP)$ in log dollars		Country	GNP in dollars	$Log_{10}(GNP)$ in log dollars
1	Ethiopia	150	2.176	13	Paraguay	1,670	3.223
2	Zaire	180	2.255	14	Jordan	1,680	3.225
3	Sri Lanka	320	2.505	15	Taiwan	2,670	3.427
4	Guinea	330	2.519	16	Greece	4,170	3.620
5	Togo	350	2.544	17	Venezuela	4,250	3.628
6	Mauritania	520	2.716	18	Czechoslovakia	5,540	3.744
7	Indonesia	550	2.740	19	Austria	9,830	3.993
8	North Korea	930	2.968	20	Australia	11,220	4.050
9	Nigeria	940	2.973	21	France	11,520	4.061
10	Mongolia	940	2.973	22	West Germany	12,280	4.089
11	Nicaragua	950	2.978	23	Norway	14,300	4.155
12	Congo	1,420	3.152	24	Brunei	22,260	4.348

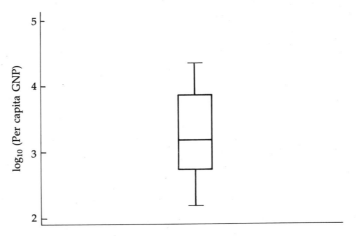

FIGURE 6.5   *Box plot of the logarithm of per capita GNP (Table 6.6)*

raw data (Figure 6.4). The skewness is nearly gone in the logarithmic version, and there are no outliers.

Figure 6.6 contains two double-stem stem-and-leaf displays. The left-hand display shows the raw distribution of per capita GNP. Its severe positive skewness is obvious, and the mean ($4,540) is about 3 times higher than the median ($1,545). The mean–median disagreement highlights the fact that such a skewed distribution lacks a clear "center."

To the right in Figure 6.6 is a stem-and-leaf display of the logarithms of per capita GNP. Although the logarithmic distribution is not perfectly sym-

RAW DATA		LOGARITHMIC DATA	
0*	00000000000111244	2*	12
0.	59	2.	555779999
1*	1124	3*	1224
1.		3.	6679
2*	2	4*	00013

Median = 1,545   Median = 3.188
Mean = 4,540   Mean = 3.253

Stems digit: $10,000/person   1 log-dollar/person
Leaves digit: $ 1,000/person   .1 log-dollar/person

2* │ 2 means: $22,000/person   2.2 log-dollars/person

Per capita 1982 GNP of 24 countries, raw data (in dollars) and logarithms (in log dollars)

FIGURE 6.6

metrical either, the long upper tail of the raw distribution has been eliminated. The median (3.188) and the mean (3.253) are in much closer agreement.

This is not to say that we have made the distribution *normal*. The log distribution in Figures 6.5 and 6.6 is less skewed than the raw distribution, and it contains no outliers, so in these respects it is closer to the ideal normal distribution. But even in log form it does not look at all bell-shaped. Obtaining a truly normal distribution often is impossible even with nonlinear transformations like the logarithm. Still, the relative symmetry and lack of outliers in the transformed distribution should make it simpler to analyze.

The logarithmic distribution is better behaved than the original, but do you wonder if we have merely fiddled with the data? After all, the real differences between these countries are measured in dollars, not log dollars. In some respects, however, real-world differences are actually more like the logarithms. For example, Ethiopia is a devastatingly poor country that has experienced mass starvation. Taiwan, while not rich, has a well-developed industrial economy that provides millions of jobs and a rising standard of living. The difference in per capita GNP between impoverished Ethiopia and prospering Taiwan is $2,520—a vast gulf, in its consequences for health, comfort, and quality of life. Now consider the other end of the scale, where we find a substantially larger difference between Australia and Norway, $3,080, yet fairly similar standards of living—much more alike than in Ethiopia and Taiwan.

Thus, the $2,520 difference between Ethiopia and Taiwan is really far "larger" than the $3,080 difference between Australia and Norway: A dollar is worth more in poverty than in wealth. Logarithmic scales, not the original dollar scales, reflect this reality. In log dollars, the difference between Ethiopia and Taiwan is substantial, 2.176 versus 3.427. The log-dollar difference between Australia and Norway is less than one-tenth as large as that between Ethiopia and Taiwan.

Logarithms have other statistical and theoretical advantages, besides those just described. Because of these advantages, and because measurement variables are so often positively skewed, many scientists work primarily with logarithmic data. Logarithms are the most widely used of the family of nonlinear transformations that are discussed in the following section.[1] Such transformations are versatile tools for coping with skewness and outliers.

PROBLEMS

The table on page 162 contains data on the adult weights and quality of diet for 15 primate species. According to an ecological relationship called the Jarman-Bell Principle, among related species of animals the larger the body weight, the lower the quality of diet. The data in the table were assembled to study this relationship separately by sex, and to examine effects of the species' mating systems on the body weight–diet relationship. Problems 5 and 6 refer to these data.

Adult Weight and Dietary Quality for 15 Primate Species

Species	Male weight in kg	Male dietary quality[a]	Female weight in kg	Female dietary quality
Tarsius bancanus	1.20	580	1.10	580
Cebus capucinus	3.80	410	2.70	410
Alouatta seniculus	8.10	351	6.40	351
Ateles belzebuth	7.40	362	7.60	362
Saimiri oerstedii	.89	460	.74	460
Saimiri sciureus	1.04	460	.67	460
Cercopithecus cephus	4.10	406	2.90	426
Cercopithecus neglectus	7.00	392	4.00	392
Cercopithecus nictitans	6.60	394	4.20	366
Cercopithecus pogonias	4.50	428	3.00	429
Cercopithecus ascanius	9.20	311	6.40	311
Cerocebus albigena	9.00	416	6.40	443
Nasalis larvatus	20.40	210	9.98	210
Symphalangus syndactylus	11.10	326	10.30	326
Pongo pygmaeus	69.00	392	37.00	332

[a]Index of dietary quality, derived from the proportions of animal matter (highest quality), plant reproductive parts (middle quality), and plant structural parts (lowest quality) in diet. Higher numbers indicate better quality of diet.
Source: Gaulin and Sailer (1985).

*5. The authors of the primate study chose to use logarithms of body weight, rather than body weight itself, in their analysis. They had two reasons: Logs improve the statistical properties of this variable, and they fit with theoretical ideas about the relationship between body weight and metabolism. To see the statistical effects, take base-10 logs of the male body weights in the table. Use the raw and log weights to fill out the stem-and-leaf displays provided.

Male weight (kg)		*Log$_{10}$ (Male weight)*	
0		− 0*	
1		0*	
2		0.	
3		1*	
4		1.	
5			
6			

Stems digits 10's.   Stems digits 1's.
6 | 9 means 69 kg   1. | 8 means 1.8 log(kg)

Describe the shapes of the two distributions. What is the effect of taking logarithms?

*6. Take base-10 logs of the female body weights in the table and construct two new stem-and-leaf displays, using the same stems as in Problem 5. Comment on what you see.

6.5   *OTHER NONLINEAR TRANSFORMATIONS*

Though they often work, logarithms are not a universal cure for skewness. If the raw distribution is negatively skewed, logarithms worsen this skewness. With mild positive skewness, logarithms may prove *too powerful* (changing the positively skewed raw distribution into a negatively skewed logarithmic one). With severe positive skewness logarithms may *not be powerful enough* (leaving the distribution still positively skewed).

To deal with these occasions, we can use a graduated family of transformations. Let X^* represent transformed X. Then

$$X^* = X^q \qquad \text{if } q > 0$$

$$X^* = \log(X) \qquad \text{if } q = 0$$

$$X^* = -(X^q) \qquad \text{if } q < 0$$

These are called **power transformations,** because they mostly involve raising X to some power q.[2] Effects of these transformations depend on the value of q chosen:

$q > 1$ *reduces negative skewness;* the higher q is, the stronger this effect.

$q = 1$ leaves the raw data *unchanged.*

$q < 1$ *reduces positive skewness;* the lower q is, the stronger this effect.

A scheme called the **ladder of powers** (Table 6.7, page 164) simplifies the choice of q to round numbers. $q = 2$ and 3 are increasingly strong transformations to reduce negative skewness. $q = \frac{1}{2}$, 0 (logarithms), $-\frac{1}{2}$, and -1 are increasingly strong transformations to reduce positive skewness.

The ladder of powers thus provides a systematic approach to improving the symmetry of skewed distributions. For example, the X^2 transformation

TABLE 6.7			

The Ladder of Powers: Nonlinear Transformations for Changing Distributional Shape[a]

q	Transformation	Name	Effect
3	X^3	Cube	Reduces extreme negative skewness
2	X^2	Square	Reduces negative skewness
1	$X^1 = X$	Raw	No effect
½	$X^{1/2}$	Square root	Reduces mild positive skewness
0	$\log_{10}(X)$	Log	Reduces positive skewness
$-½$	$-(X^{-1/2}) = -1/\sqrt{X}$	Negative reciprocal root	Reduces extreme positive skewness
-1	$-(X^{-1}) = -1/X$	Negative reciprocal	Reduces even more extreme positive skewness

[a]Powers higher than 3 could be used to correct for extreme negative skew, but rarely are. Similarly, it is theoretically possible to use powers lower than -1 (-2, -3, etc.) to correct for extreme positive skew. The minus sign is applied to powers lower than 0 to keep the transformed data in the same order as the raw data. The same cases will be highest, lowest, and so on.

reduces negative skewness, much as the logarithm does for positive skewness. If negative skewness is severe, try the X^3 transformation.

Square roots ($\sqrt{X} = X^{1/2}$) can be used to reduce positive skewness, if logarithms are too strong. They pull less extremely on the upper tail of a distribution. For positive skewness so severe that even logarithms are not enough, try the negative reciprocal root, $-(X^{-1/2})$. This transformation can be performed with a calculator or computer that does exponential functions. Its actual meaning is

$$-(X^{-1/2}) = -\left(\frac{1}{X^{1/2}}\right) = -\frac{1}{\sqrt{X}}$$

The reciprocal root transformation itself, $X^{-1/2} = 1/\sqrt{X}$, reverses the order of the data, so that the formerly high values become low and the formerly low values become high. To restore the original order, we add a minus sign to the transformed data, so they become $-1/\sqrt{X}$. A minus sign is likewise added after other negative-power transformations, such as X^{-1} (reciprocal, or $1/X$) or X^{-2} (reciprocal square, or $1/X^2$).

Confronted with an asymmetrical distribution, we can try moving up or down the ladder of powers, seeing which transformation has the best effect. Box plots and mean–median comparisons help in judging symmetry. More elaborate graphs and tests are available, but for many purposes simple methods suffice to show whether skewness and outliers have been improved.

Not all distributions can be made symmetrical, no matter what transformation we try. Furthermore, applying the wrong transformation can easily make distributional problems worse instead of better. Even when properly used, transformations have an obvious drawback: Your results become harder to understand. The next section suggests one way to make sense of results from transformed data.

PROBLEMS

The accompanying table contains 1980 census data on a sample of 16 U.S. cities. (For variety, these are not the same cities studied earlier.) Problems 7–10 refer to these data.

1980 Population and Density in 16 U.S. Cities

	City	Population	Population aged ≥ 45	Population density[a]
1	Berkeley, CA	103,328	25,934	9,480
2	Aurora, CO	158,588	31,468	2,652
3	Stamford, CT	102,453	37,480	2,689
4	Atlanta, GA	425,022	121,676	3,244
5	Warren, MI	161,134	52,473	4,684
6	Springfield, MO	133,116	41,694	2,051
7	Albuquerque, NM	331,767	89,102	3,481
8	Buffalo, NY	357,870	128,752	8,561
9	Winston–Salem, NC	131,885	43,206	2,173
10	Philadelphia, PA	1,688,210	599,987	12,413
11	Irving, TX	109,943	24,134	1,634
12	Lubbock, TX	173,979	40,543	1,920
13	Pasadena, TX	112,560	24,865	2,962
14	Arlington, TX	152,599	49,101	5,984
15	Chesapeake, VA	114,486	29,477	337
16	Tacoma, WA	158,501	51,334	3,323

[a]People per square mile.

*7. Population density has a positively skewed distribution. Transform population density by taking square roots ($X^{1/2}$), and fill out the stem-and-leaf displays shown at the top of page 166. How did the transformation change the distribution? Is there any evidence that this transformation was too strong or too weak? What would be the next transformation to try?

Population density		Square root (Pop. density)	
0*		0*	
0t		0t	
0f		0f	
0s		0s	
0.		0.	
1*		1*	
1t			

Stems digits 10,000's.
  1t | 2 means 12,000 people
          per square mile

Stems digits 100's.
  1* | 1 means 110.

*8. The distribution of population 45 years old and over is also positively skewed, much more so than population density. To make it symmetrical requires the strongest transformation listed in the ladder of powers table (Table 6.7): the negative reciprocal, $-X^{-1}$. Calculate negative reciprocal values for each value of population ≥ 45 in the table.[3]

*9. Use the negative reciprocals from Problem 8 to fill out the stem-and-leaf displays provided.

Population ≥ 45		Negative reciprocal (pop. ≥ 45)	
0		-4	
1		-3	
2		-2	
3		-1	
4		-0	
5			

Stems digits 100,000's.
  5 | 9 means 590,000 people.

Stems digits .00001's.
  -1 | 9 means $-.000019$.

Describe the effect of this transformation on the distributional shape.

*10. Calculate the mean and median of population $\geq$45, before and after transformation in Problem 8. How does the mean–median difference before transformation compare with the mean–median difference after transformation? What does this tell us about "center" in the two distributions?

6.6 *USING RESULTS FROM TRANSFORMED DATA*

The gross national product example (Figure 6.6) shows how taking logarithms can change a skewed distribution into one whose median and mean are close together, simplifying the choice of what "center" to use. But the positional center (median) is about the same either way: The log of the raw data's median nearly equals the log data's median. If all we needed was a measure of center, we could have just reported the raw-data median in the first place, ignoring the mean (which skewness distorts) and not bothering with logarithms. For simple analytical purposes, resistant order statistics will often serve us better than complicated nonlinear transformations.

On the other hand, our job is not always so simple as just locating the sample's center. The remainder of this book will introduce techniques for answering questions such as, "What does this small sample tell us about the larger population?" or "What is the relationship between these two variables?" These harder questions are generally more approachable through mean-based techniques than through order statistics. Nonlinear transformations give us ways to apply mean-based techniques to ill-behaved data.

Besides making skewed distributions more symmetrical, nonlinear transformations have other properties that show their worth in more complicated analyses. Skip ahead to Figure 15.14 in Chapter 15, for instance, and you will see an analysis of the relationship between GNP and life expectancy that starts by transforming GNP into logarithms. Other examples using nonlinear transformations appear in Figures 11.5 and 15.18. Published research like the primate study in Problem 5 employs transformations routinely. They are indispensable in many fields.

Although we do not usually need nonlinear transformations for the sorts of analyses shown in Chapters 1–5, it is easier to explain them at this simple level. Another idea easier to explain now, which will help later in understanding analytical results from transformed data, is the **inverse transformation.** It amounts to just undoing the original transformation. If we transformed the raw data by taking square roots, then the inverse transformation is to square the transformed data, thereby returning to the original units of X. That is, if

$$X^* = \sqrt{X} \qquad \text{(transformation)}$$

then

$$(X^*)^2 = X \qquad \text{(inverse transformation)}$$

Applying the inverse transformation to the transformed data (X^*) regains the values of the raw data (X). This proves that *no information was lost* in transformation: The transformed and raw data contain the same information in different form. Apart from this insight, applying the inverse transformation to the data is not very interesting, as it just returns us to where we started.

A more important application of inverse transformations is to statistical summaries, rather than to the data themselves. For example, the mean of the transformed GNP distribution in Figure 6.6 is 3.253 log dollars. To find out how much that is in dollars, apply the inverse transformation. We originally transformed by taking base-10 logs. To undo this transformation, we take **antilogarithms** (raise 10 to a power). Thus, if

$$X^* = \log_{10}(X) \qquad \text{(transformation: logs)}$$

then

$$X = 10^{X^*} \qquad \text{(inverse transformation: antilogs)}$$

TABLE 6.8   ***Inverse Transformations for the Ladder of Powers (Table 6.7)***

q	Transformation	Inverse transformation	Name of inverse transformation
3	X^3	$(X^*)^{1/3}$	Cube root
2	X^2	$(X^*)^{1/2} = \sqrt{X^*}$	Square root
1	$X^1 = X$	None needed	Raw data
1/2	$X^{1/2}$	$(X^*)^2$	Square
0	$\log_{10}(X)$	10^{X^*}	Antilog
$-1/2$	$-(X^{-1/2})$	$(-X^*)^{-2} = 1/(X^*)^2$	Reciprocal square
-1	$-(X^{-1})$	$(-X^*)^{-1} = 1/(-X^*)$	Reciprocal negative

The mean of the logarithms is 3.253 *log dollars,* so its antilog is $10^{3.253} = 1{,}791$ *dollars.* (Note that the antilog of the mean logarithm, $1,791, is *not the same* as the raw-data mean, $4,540, although both are measured in dollars.) Inverse transformation returns us to units we understand.

Table 6.8 shows inverse transformations for the ladder of powers. Just as each of the original transformations (except the logarithm) involves raising X to some power q, the corresponding inverse transformation involves raising transformed X, X^*, to the power $1/q$. For example, if $X^* = X^2$, then $X = (X^*)^{1/2}$. Raising to the power ½ is the same as taking a square root. Some of the inverse transformations in Table 6.8 can be simplified algebraically. They are easier to remember, though, if we just think of inverse transformations (except antilogs) as raising X^* to the power $1/q$.

If the original transformation required changing the sign of data values after raising to a power, the inverse transformation will require changing the sign back again, *before* raising to the inverse of that power. If we added a constant to X before transforming, to eliminate nonpositive values, then this constant should be subtracted again *after* the inverse transformation is accomplished.

PROBLEMS

*11. Apply inverse transformations to the following summary statistics. Include the units of measurement with your answer.

  a. Mean log of male primate weight (from Problem 5): .76

  b. Mean log of female primate weight (Problem 6): .61

  c. Median log of female primate weight (Problem 6): .62

  d. Mean square root of population density (Problem 7): 60.81

  e. Median square root of population density (Problem 7): 55.69

f. Mean negative reciprocal of population ≥45 (Problems 8–10): − .000023261

g. Median negative reciprocal of population ≥45 (Problems 8–10): − .000023565

*12. A small, perfectly symmetrical set of artificial data, for which the mean and the median are both 200, is given:

100   200   300

a. Take square roots of these values, and find the mean and the median of the square roots. Is the distribution still symmetrical?

b. Perform the appropriate inverse transformation to return your mean and median from part a to their original units.

c. Explain why the square of the mean square root is not the same as the original mean.

d. Explain why in this example the square of the median square root is the same as the original median.

SUMMARY

In the remainder of this book, we will be working with statistical techniques that require us to make certain **assumptions** about sample data or about the population from which they come. The validity of our analytical conclusions may hinge on whether these assumptions are true. Ill-behaved data sometimes directly contradict our assumptions, calling the whole analysis into question. For this reason, two things become increasingly important: using methods such as graphing that help us *detect* distributional problems, and knowing about such methods as outlier deletion and nonlinear transformations, which allow us to improve the distributional problems we detect.

Detecting and handling distributional problems are harder than just assuming none exist. To simplify explanations, most examples in the chapters ahead avoid severe distributional problems. From time to time we will tackle more complex examples where statistical problems do arise, and we will see what can be done to cope with them.

PROBLEMS

*13. The square root transformation can improve the symmetry of some distributions with mild positive skewness. Thus, if you encounter a distribution with the shape illustrated at the top of page 170, the square root would be a reasonable guess for the transformation you should try first.

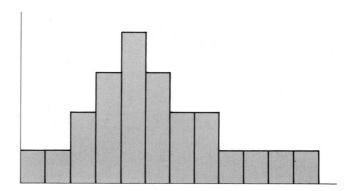

Sketch a histogram for a distribution that might be made symmetrical by each of the following nonlinear transformations from the ladder of powers (Table 6.7).

a. Cube

b. Square

c. Logarithm

d. Negative reciprocal root

e. Negative reciprocal

*14. Sketch a histogram of a distribution for which outlier deletion, rather than nonlinear transformation, would be the best way to improve symmetry.

*15. Statisticians often work with base-*e* or natural logarithms, rather than the base-10 logarithms used in this chapter. The base-*e* logarithm of *x*, denoted $\log_e(x)$, is the power to which you must raise *e* to get *x*. The *e* is a constant, equal to about 2.71828. Scientific calculators usually have keys for either kind of logarithm, which have exactly the same effect on distributional shape. Return to the GNP data of Table 6.6.

a. Find the natural logarithm ($\log_e$) of each GNP value in this table.

b. What is the relationship between the natural logarithms and the base-10 logarithms for each value?

*16. Use a calculator or computer to find the following values.

a. The cube of 25

b. The square of 25

c. The base-10 log of 25

d. The negative reciprocal root of 25

e. The negative reciprocal of 25

*17. Find values of x for which the following calculations are impossible.

   a. $\text{Log}_{10}(x)$

   b. $x^{1/2}$

   c. $-(x^{-1})$

*18. Use inverse transformations to find x, if:

   a. $x^3 = 100$

   b. $x^2 = 100$

   c. $x^{1/2} = 9$

   d. $\text{Log}_{10}(x) = 4$

   e. $-(x^{-1/2}) = -.05$

   f. $-(x^{-1}) = -.0125$

*19. The frequency distribution for an imaginary data set is shown. The variable X ranges from 1 to 9, and there are 420 imaginary cases.

X	f
1	10
2	20
3	40
4	80
5	120
6	80
7	40
8	20
9	10

Construct a histogram for this variable's distribution. How would you describe its shape?

*20. Apply the X^3 transformation to each value of X in Problem 19. Since the resulting X values are not evenly spaced, construct a new grouped frequency distribution. Use classes of 0 to 99, 100 to 199, 200 to 299, etc. for this table. Use the new table as the basis of a new histogram.

*21. Apply the $-(X^{-1})$ transformation to each value of X in Problem 7. Again, construct a new grouped frequency distribution for the resulting X^* values. Class intervals of -1 to $-.91$, $-.9$ to $-.81$, $-.8$ to $-.71$, etc. may be convenient. Use this table as the basis of another histogram.

*22. What generalization can you suggest, from your histograms in Problems 20 and 21, about the effects of the ladder of power transformations on distributions that are already symmetrical?

*23. Invent a small set of data in which the logarithm of the median is *not* close to the median of the logarithms. What is needed for this approximation not to work?

The accompanying table presents data on ten private water wells in the town of Lee, New Hampshire. The variables are the water's chloride concentration, in milligrams per liter, and the well's distance from the nearest road that is salted to melt ice in the winter. Since salts contain chloride, we suspect some relationship between chloride in the water and distance from a salted road. Such a relationship would imply that road salting can affect the health of nearby residents. Before investigating this relationship, we should take a close look at the two particularly ill-behaved distributions, as done in Problems 24–29.

Data on Ten Private Water Wells in Lee, New Hampshire

Well	Chloride concentration in mg/liter	Distance to nearest salted road in feet
1	10	300
2	10	300
3	10	500
4	10	35
5	17	300
6	21	300
7	110	42
8	150	50
9	620	50
10	680	37

Source: Data courtesy of New Hampshire Water Supply and Pollution Control Commission.

*24. Construct a single-stem stem-and-leaf display for the chloride concentration. What possible problems do you see in this distribution? Is it a good idea to delete the outliers before proceeding with further analysis? Why or why not?

*25. Select a nonlinear transformation from the ladder of powers that you think might make the chloride distribution more symmetrical. Explain your choice, try it out, then construct a stem-and-leaf display of the results.

*26. Compare the raw (Problem 24) and transformed (Problem 25) chloride distributions. In what (if any) ways did the transformation improve distributional shape? In what ways did it *not* improve shape?

*27. Construct a stem-and-leaf display (single-stem version) for the variable of distance from the road in the table. Comment on the shape of this distribution.

*28. A prominent feature of the distance distribution (Problem 27) is the large gap between the five wells that are near a road, and the remaining five that are far from one. Separate the chloride concentrations from the table into two groups, based on whether they come from a well that is near or far from a road. Find the median chloride concentration for the near wells, and compare it with that for the far wells. What does comparing medians suggest about the relationship between distance from salted road and chloride concentration?

*29. Just by looking at the data in the table, can you make any guesses about the accuracy with which either variable is measured? For each variable, what do you think the measurement process was like?

NOTES

1. One complication with using logarithms is that they are undefined for negative numbers and zero. Thus if your data contain negative values or zeroes, you cannot take their logs. This problem is commonly overcome by adding a constant to each case before taking logs. The constant should be large enough to *raise the lowest raw-data value to 1*. For example, if the lowest value X takes on is 0, then add 1 to each case before taking logs. Since the log of 1 is 0, this will have the effect of making the lowest value in the *log* data equal to 0. Similarly if the lowest value of X is -31, adding 32 to each case will ensure that logs can be taken, and the lowest value of $\log_{10}(X+32)$ will be 0.

2. Logs are used for $q=0$. Any number to the 0 power equals 1, so X^0 would be useless as a transformation. Apply a negative sign after raising to negative powers, to preserve the order of the raw data.

   You may need to add a constant to each value of X before transforming, to eliminate nonpositive X values and ensure that the transformation is possible (see note 1).

3. You will see some very small numbers when you perform this transformation. Many calculators or computer programs will shift into **scientific notation** to express them. Scientific notation breaks the number into a decimal value times a power of ten. For example, the negative reciprocal of population ≥ 45 in Berkeley is

$$\begin{aligned}
X^* &= -(X^{-1}) \\
&= -(25{,}934^{-1}) \\
&= -(.0000386) = -.0000386
\end{aligned}$$

In scientific notation, the number $-.0000386$ appears as $-3.86\text{E}-05$, meaning "-3.86 times 10 to the -5 power." (Some calculators do not show the "E" with this value, but still use an exponent of 10. In a published report, the value might be given in more formal scientific notation: -3.86×10^{-5}.) To write out a value with an exponent $\text{E}-05$, move the decimal five places to the left. Large numbers are shown by positive exponents. The 1980 population of Philadelphia, for example, is approximately $1.688\text{E}+06$.

INFERENCE IN UNIVARIATE ANALYSIS

PART II

The methods introduced in Chapters 1–6 help us explore and describe sample data. Since our sample is at hand for analysis, we can study it thoroughly. Often the sample is just a subset of some larger population, though, which is the real focus of our research interest. Unlike the sample, the population is not available for analysis and cannot be studied directly. We must form conclusions about the unknown characteristics of the population, based on what we know about the sample. This is the goal of statistical *inference.*

Characteristics of a sample are described by *sample statistics* such as the mean ($\overline{X}$), standard deviation (s_x), and proportion ($\hat{\pi}$). In theory, corresponding numerical summaries could describe the population: the population mean (μ), standard deviation (σ_x), proportion (π), and so forth. Numerical summaries for the population are called *population parameters.* Population param-

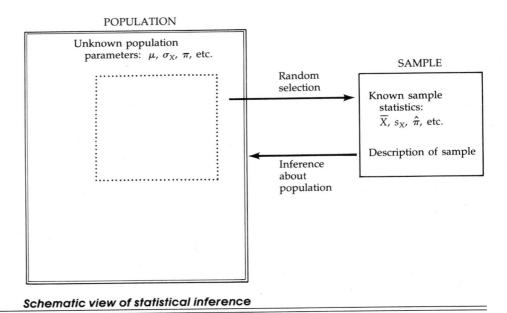

Schematic view of statistical inference

eters are usually unknown. The statistician's task is to draw conclusions or inferences about the values of unknown population parameters, on the basis of known sample statistics.

Two things keep this task from being mere guesswork. The first is **random selection:** the cases that make up the sample are chosen randomly from the population, which helps ensure that the sample is representative of the whole. In some fields such as opinion polling, selecting a representative sample is harder and absorbs more effort than the statistical analysis itself.

Secondly, we are guided by **statistical theory.** Theory describes the long-term behavior of sample statistics: what would happen if we drew many random samples from a population and obtained sample statistics from each. These statistics would vary from sample to sample, but in a theoretically predictable pattern. In recent years, theoretical knowledge has been confirmed and extended by experimental research with computers.

Because we are trying to estimate the unknown values of population parameters, inferential conclusions must be stated in terms of probabilities, not certainties. Chapter 7 introduces basic concepts of probability, letting us work out the likelihood of simple events. Chapter 8, on random sampling, describes a computer experiment to illustrate long-term statistical behavior. Normal distributions are examined in Chapter 9. Chapters 10 and 11 focus all these topics on the two main themes of statistical inference: estimating the values of unknown population parameters, and testing hypotheses about those values.

An Introduction
to Probability

Chapter 7

Sample data are at hand for thorough analysis, so we can describe them exactly. What we learn from a sample helps us draw conclusions about the larger population. But as we cannot analyze the population itself, our conclusions inevitably contain a degree of uncertainty.

This degree of uncertainty can be quantified as a *probability*. Probabilities guide our conclusions regarding the population. To obtain probabilities we employ the techniques of **statistical inference,** which combine mathematical theories of probability with descriptive statistical techniques like those seen in previous chapters. This chapter introduces some basic ideas of probability, including rules by which probabilities are combined. Next we will apply these rules to categorical and measurement data. Last, we will look at normal distributions, which are important theoretical probability distributions.

7.1 BASIC CONCEPTS OF PROBABILITY

We begin by defining some basic terms. Any phenomenon whose outcome is uncertain, such as tossing a coin or contracting a disease, can be viewed as a **statistical experiment.** The set of *all possible outcomes* of an experiment is called its **sample space,** represented by *S*. An **event** is some *specific collection* of outcomes, or a subset of *S*. We will denote different events by capital letters such as *A*, *B*, or *C*.

Suppose the experiment consists of throwing a single six-sided die. Then the sample space *S* contains six possible outcomes: one dot showing, two dots showing, three dots showing, and so on. We represent this symbolically as *S* = {1, 2, 3, 4, 5, 6}. If we define an event *A* to be "three dots showing," *A* includes only one of the six possible outcomes in *S*: *A* = {3}. Defining event *B* as "at least three dots showing" means *B* would include four of the six possible outcomes in *S*: *B* = {3, 4, 5, 6}. That is, the event *B* occurs if any of the bracketed outcomes occurs—if 3, 4, 5, or 6 dots show.

For every event, we can define a second event called its **complement:** the set of all outcomes that are in *S*, but are *not* in that first event. If *A* is "three dots showing," then its complement, written *not A*, consists of all the other outcomes: *not A* = {1, 2, 4, 5, 6}.

If we tossed a die repeatedly, we could construct a frequency distribution table showing how many times each face came up. These frequencies (*f*) can be expressed as proportions or relative frequencies by dividing them by the number of tosses (*n*): $\hat{\pi} = f/n$. If we saw three dots showing on 120 out of 600 tosses, that face's proportion or relative frequency is $\hat{\pi} = f/n = 120/600 = .2$. As more tosses were made, we could expect the proportion of threes to stabilize eventually around $\frac{1}{6} \approx .167$.

> An event's **probability** is the proportion of times that we expect the event to occur, if an experiment were repeated a large number of times.

This idea that relative frequencies eventually stabilize around a certain value underlies the **relative frequency** definition of probability. In the long run we expect that $\frac{1}{6}$ of our tosses will result in three dots showing, so the probability of event *A*, **P(A)**, is $\frac{1}{6} \approx .167$. Table 7.1 summarizes these basic terms and symbols as they apply to this example.

Event *A* occurs with only one of the six possible outcomes, but event *B*, "at least three dots showing," occurs with any of four possible outcomes. Intuitively, to find the probability of event *B* we think of adding up the probabilities of each of *B*'s four individual outcomes:

$$P(B) = \frac{1}{6} + \frac{1}{6} + \frac{1}{6} + \frac{1}{6} = \frac{4}{6} \approx .667$$

TABLE 7.1   *Some Basic Probability Terms*

Term	Definition	Example
Experiment	Phenomenon where outcomes are uncertain	Single throws of a six-sided die
Sample space	Set of all possible outcomes of the experiment	$S = \{1, 2, 3, 4, 5, 6\}$ (1, 2, 3, 4, 5, or 6 dots show)
Event	A collection of outcomes; a subset of S	$A = \{3\}$ (3 dots show) $B = \{3, 4, 5, \text{ or } 6\}$ (3, 4, 5, or 6 dots show)
Complement	All outcomes in S that are not in the particular event	*not A* $= \{1, 2, 4, 5, 6\}$ (Anything but 3 shows) *not B* $= \{1, 2\}$ (Anything but 3, 4, 5, or 6 shows)
Probability	Proportion of experiments in which event is expected to occur, over many repetitions	$P(A) = \frac{1}{6} \approx .167$ $P(B) = \frac{4}{6} \approx .667$

Intuition is correct here, but we will need formal rules to handle more complicated questions of probability and to recognize situations where this intuitive solution is wrong.

An event expected to occur *every time* an experiment is repeated has a probability of 1, meaning certainty. An event expected *never* to occur has a probability of 0, meaning impossibility. All probabilities must be somewhere between 0 and 1, inclusive.

> For any event A, the probability that A will occur is a number between 0 and 1, inclusive:
>
> $$0 \leq P(A) \leq 1 \qquad\qquad [7.1]$$

The probability of at least one of the outcomes in the sample space S occurring is 1 (certainty):

$$P(S) = 1 \qquad\qquad [7.2]$$

Events are said to be **mutually exclusive** if they have no outcomes in common. In other words, if it is impossible both could occur in a single trial of the experiment, then

$$P(A \text{ and } B) = 0$$

In the die-toss example, $A = \{3\}$ and $B = \{3, 4, 5, 6\}$ are not mutually exclusive, since the outcome "three dots showing" belongs to both of them. On the other hand, $A = \{3\}$ and $C = \{1, 2\}$ are mutually exclusive.

For mutually exclusive events, the probability that at least one of them occurs is

$$P(A \text{ or } C) = P(A) + P(C)$$

For example, if the probability of event $A = \{3\}$ is .167, and the probability of mutually exclusive event $C = \{1, 2\}$ is $\frac{2}{6} = .333$, then the probability of A or C is

$$P(A \text{ or } C) = P(A) + P(C)$$
$$= .167 + .333 = .500$$

This property can be generalized to any number of mutually exclusive events:

$$P(A \text{ or } C \text{ or } D \text{ or } E \text{ or } \dots) = P(A) + P(C) + P(D) + P(E) + \cdots \qquad [7.3]$$

The events A and *not* A are mutually exclusive by definition. Consequently,

$$P(A \text{ or } \textit{not } A) = P(A) + P(\textit{not } A)$$

Since we also know from the definition of *not* A that it includes all the events in the sample space, S, that are not in A,

$$P(A) + P(\textit{not } A) = P(S) = 1$$

For any complementary events A and *not* A,

$$P(A) + P(\textit{not } A) = 1$$
$$P(A) = 1 - P(\textit{not } A) \qquad\qquad [7.4]$$
$$P(\textit{not } A) = 1 - P(A)$$

Equation [7.4] states formally a property that you already know; common sense says a .30 probability of rain tomorrow means a .70 probability of no rain. This equation simplifies many probability problems. If $P(\textit{not } A)$ is easier to calculate than $P(A)$, then $P(\textit{not } A)$ and Equation [7.4] let us obtain $P(A)$ indirectly. This and other properties of probability are summarized in Table 7.2.

A conditional probability is the probability of one event if another event occurs. For example, the probability of event A, three dots showing, is $P(A) = \frac{1}{6} = .167$ on a single toss. But what if we know that event B, at least three dots showing, will occur? Then there are only four possible outcomes, one of

TABLE 7.2   ***Basic Properties of Probabilities***

Property	Symbols
If event A will *always* occur, its probability is 1.	$P(A) = 1$
If event A will *never* occur, its probability is 0.	$P(A) = 0$
Probabilities are always between 0 and 1, inclusive.	$0 \leq P(A) \leq 1$
The probability that one of all possible outcomes will occur is 1.	$P(S) = 1$
Events A and B are mutually exclusive if they have no outcomes in common.	$P(A \text{ and } B) = 0$
If A and B are mutually exclusive then $P(A \text{ or } B)$ can be found by addition.	$P(A \text{ or } B) = P(A) + P(B)$
If A, B, C, . . . are all mutually exclusive then $P(A \text{ or } B \text{ or } C . . .)$ can be found by addition.	$P(A \text{ or } B \text{ or } C . . .)$ $= P(A) + P(B) +$ $P(C) + . . .$
Since A and *not A* are mutually exclusive and between them include all possible outcomes, $P(A \text{ or } not A)$ is 1.	$P(A \text{ or } not A)$ $= P(A) + P(not A)$ $= P(S) = 1, \text{ and}$ $P(not A) = 1 - P(A)$

which is A. The probability of $A = \{3\}$ is $\frac{1}{4} = .25$, *given* that $B = \{3, 4, 5, 6\}$ occurs. The "conditional probability of A given B" is written $P(A|B)$.

Event A is **independent** of B if the conditional probability of A given B is the same as the unconditional probability of A. That is, they are independent if

$$P(A|B) = P(A)$$

In the die-toss example, $P(A) = \frac{1}{6} = .167$ and $P(A|B) = \frac{1}{4} = .25$, so the events A and B are not independent.

The probability that two events A and B will both occur is obtained by applying the **multiplication rule:**

$$P(A \text{ and } B) = P(B)P(A|B) \qquad [7.5]$$

where $P(A|B)$ means the probability of A given B.

As defined in the box, the probability of both A and B occurring equals the probability of B times the probability of A if B occurs. For independent events only, Equation [7.5] simplifies to

$$P(A \text{ and } B) = P(B)P(A) \qquad [7.6]$$

Since Equation [7.6] holds only if *A* and *B* are independent, it enables us to *test* for independence in data.

The multiplication rule tells us how to find probabilities for the combined event (*A* and *B*). The probability of (*A* and *B*) is used in the general *addition rule* for finding the probability of (*A* or *B*). The addition rule states that the probability of either *A* or *B* occurring equals the probability of *A* plus the probability of *B*, minus the probability that they *both* occur.

> For any two events *A* and *B*, the probability that either *A* or *B* will occur is given by the **addition rule:**
>
> $$P(A \text{ or } B) = P(A) + P(B) - P(A \text{ and } B) \qquad [7.7]$$

Sometimes we cannot obtain the probability of a combined event such as (*A* and *B*), which we need to solve Equation [7.7]. Even so, we can still use the addition rule to find the **upper bound** of the probability of (*A* or *B*):

$$P(A \text{ or } B) \leq P(A) + P(B) \qquad [7.8]$$

By definition, for mutually exclusive events $P(A \text{ and } B)$ equals 0. Thus, for *mutually exclusive events only* Equation [7.7] simplifies to:

$$P(A \text{ or } B) = P(A) + P(B) \qquad [7.9]$$

This is a two-event version of Equation [7.3]. Since the six possible outcomes of a single throw of a die can all be regarded as mutually exclusive events, it follows that Equation [7.9] should apply. Then

$$P(\text{"one dot" or "two dots"}) = P(\text{"one dot"}) + P(\text{"two dots"})$$

$$= \frac{1}{6} + \frac{1}{6}$$

$$= \frac{2}{6}$$

$$\approx .333$$

There is thus a sound mathematical basis for our intuitive idea that adding up probabilities of individual outcomes will yield the probability of an event. This works, however, only if the events are mutually exclusive, otherwise their probabilities must be found from the more complicated Equation [7.7].

Table 7.3 summarizes all these rules. Although in principle straightforward, in practice they are easily misunderstood or misapplied—especially the multiplication rule. The next section illustrates how to perform probability reasoning in a context more dramatic than rolling dice.

TABLE 7.3   *Some Rules About Probabilities for Any Events A and B*

Rule	Symbols
Definitions:	
The *conditional probability of A* given *B* is the probability of event *A*, if event *B* occurs.	$P(A\|B)$
A is *independent* of *B* if the conditional probability of *A* given *B* is the same as the unconditional probability of *A*.	$P(A\|B) = P(A)$
Multiplication rule:	
The general *multiplication rule* for probabilities	$P(A \text{ and } B) = P(B)P(A\|B)$
For *independent events* only, the multiplication rule is simplified.	$P(A \text{ and } B) = P(B)P(A)$
Addition rule:	
The general *addition rule* for probabilities	$P(A \text{ or } B) = P(A) + P(B) - P(A \text{ and } B)$
The addition rule implies an *upper bound* for the probability of (*A* and *B*).	$P(A \text{ or } B) \leq P(A) + P(B)$
For *mutually exclusive events only*, the addition rule is simplified.	$P(A \text{ or } B) = P(A) + P(B)$

PROBLEMS

1. For an experiment consisting of a single draw from a well-shuffled deck of playing cards, give examples of the following.
   a. An event and its complement
   b. Two mutually exclusive events
   c. Two independent events
   d. Two events that are neither independent nor mutually exclusive

2. A deck of playing cards consists of 52 cards, including an ace, king, queen, jack, and two through ten in each of four suits. If a single card is drawn from a shuffled deck, what is the probability that this card is
   a. A queen: $P(Q)$
   b. Anything but a queen: $P(not\ Q)$
   c. A heart: $P(H)$
   d. The queen of hearts (apply the multiplication rule): $P(Q \text{ and } H)$
   e. Either a queen *or* a heart (apply the addition rule): $P(Q \text{ or } H)$

3. Define events X and Y for the single-card draw experiment such that
   a. $P(X) = 0$
   b. $P(Y) = 1$

7.2 *REASONING WITH PROBABILITY: AN EXAMPLE*

Imagine yourself on a cross-country airplane flight, opening an onboard magazine to an article about problems with the type of jet engine that powers your plane. Disturbingly, the article reports a malfunction rate equivalent to one engine failure per 1,000 flights, but it adds that your plane can land safely even with only one of its two engines operating. The probability of single-engine failure being only 1/1,000 or .001, an expert is quoted cheerfully predicting only "one chance in a million" that both engines would quit on any one flight.

The one in a million probability (P = .000001) sounds reassuring until you look closer. The expert obviously just multiplied the probability that the right engine would fail, $P(R)$ = .001, by the probability that the left engine would fail, $P(L)$ = .001, to estimate the probability that both engines would quit,

$$P(L \text{ and } R) = P(R)P(L)$$
$$= .001(.001)$$
$$= .000001$$

This is an application of the simplified multiplication rule, Equation [7.6], which applies only if the two events are independent.

It is reasonable to suspect that if one engine fails, there is a higher chance that the other will fail too. Flying on one engine highly stresses it; moreover, whatever did in the first engine (age, environmental conditions, faulty maintenance, fuel problems, etc.) could also affect the second. If so, events R and L are *not independent:* The probabilities should have been calculated using Equation [7.5], not Equation [7.6]. The probability of both engines failing is really

$$P(L \text{ and } R) = P(R)P(L|R)$$

To find this probability, we must know the conditional probability of a left-engine failure given a right-engine failure, $P(L|R)$. Such a probability will be hard to come by, without crashing a lot of planes. $P(L|R)$ must thus be considered unknown.

Without $P(L|R)$, the probability of (L and R) is also unknown. We can, however, find the *upper bound* for this probability. The worst case is that the left engine *always* fails if the right one does, so $P(L|R)$ = 1. Then

$$P(L \text{ and } R) = P(R)P(L|R)$$
$$= P(R)(1)$$
$$= P(R)$$

Since the probability of a right-engine failure by itself was reported as $P(R)$ = .001, this is also the worst-case probability, or upper bound, for the probability of double-engine failure (L and R).

We now see that the probability both engines will fail could far exceed "one chance in a million" if left- and right-engine failures are not independent. At worst, the probability could be as high as one chance in a thousand—far less reassuring than the expert's "one chance in a million."

The discrepancy between these two probability estimates arises from different assumptions about the independence of events L and R. The magazine's expert made the most optimistic assumption possible: that L and R are independent, so $P(L|R) = P(L) = .001$. A gloomy passenger, on the other hand, might reason from the pessimistic extreme: that L and R are perfectly dependent, so $P(L|R) = 1$. The result of these two different assumptions is that one final estimate is 1,000 times higher than the other. *Assumptions are crucial*, even when both sets of calculations use the same data.

This example is imaginary, but it illustrates a genuine type of problem where probability reasoning is important: the assessment of technological risks. Other real-world areas that use probability reasoning daily include gambling, insurance, and industrial quality control.

PROBLEMS

Inherently risky systems such as spacecraft and nuclear power plants often count on redundancy, or multiple backup systems, to reduce the probability of disaster. A power plant might have four separate cooling pumps, each one individually capable of maintaining water flow during an emergency. If one pump fails, the next can be turned on to replace it.

Failures in such redundant safety systems are often *not* independent events. They are susceptible to *common mode failures,* where multiple components fail for similar reasons. For example, one possible cause for pump failure is that in an emergency the workers forget to turn it on.[1] If they forget to turn on pump 1, they will likely forget to turn on pumps 2, 3, and 4 as well—so all four pumps could be inoperative for the same reason.

The multiplication rule, Equation [7.5], can be extended to more than two events. For three events A, B, and C it becomes

$$P(A \text{ and } B \text{ and } C) = P(C)P(B|C)P(A|B \text{ and } C)$$

which reduces to the simpler

$$P(A \text{ and } B \text{ and } C) = P(C)P(B)P(A)$$

only if the events are independent.

4. Suppose we accept that any one reactor cooling pump has a .05 probability of failing during an emergency. What is the probability that all four pumps will fail, under the optimistic assumption that the failures are *independent*? Show how your calculations derive from the multiplication rule.

5. What is the probability that all four pumps will fail, if we make the *most pessimistic* assumption about their independence? To support your answer, extend the multiplication rule to four events and state your assumption about conditional probabilities.

6. Comment on the difference between your best-case and worst-case estimates in Problems 4 and 5.

7.3   TREE DIAGRAMS*

One useful aid to thinking about probability is the **tree diagram.** The simple example in Figure 7.1 is set up to address the question, "What is the probability of drawing 2 aces in two consecutive draws from a well-shuffled deck of cards?" There are 4 aces in a 52-card deck, so the probability of an ace on the first draw is 4/52 = .0769. The probability of an ace on the second draw

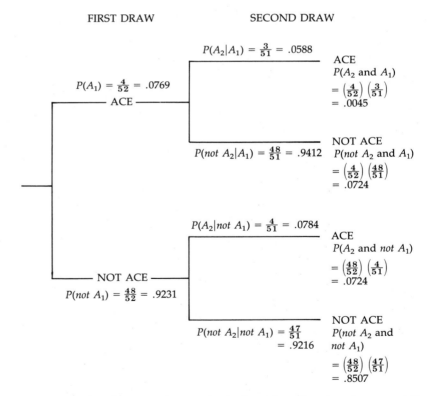

FIGURE 7.1   *Tree diagram for drawing aces from a deck of cards, without replacement. Event A_1 = {ace on first draw}, A_2 = {ace on second draw}.*

depends on whether we drew one on the first draw; the events are not independent, unless we replace each drawn card.

We can represent the event {ace on first draw} as A_1. Its complement, *not A_1* or {not an ace on first draw}, has a probability $P(\text{not } A_1) = 48/52 = .9231$. The events A_1 and *not A_1* are shown as the first two branches of the tree in Figure 7.1.

On the second draw, only 51 cards are left in the deck. Let A_2 represent the event {ace on second draw}, and *not A_2* represent its complement {not an ace on second draw}. If we drew an ace on the first draw (A_1 occurred) then 3 aces and 48 other cards remain in the deck. The conditional probability of A_2 given A_1 must be $P(A_2|A_1) = 3/51$. The conditional probability of drawing other than an ace on the second draw, given that we drew one on the first, is $P(\text{not } A_2|A_1) = 48/51 = .9412$. Similar calculations obtain the conditional probabilities given that we did *not* draw an ace on the first draw, in which case 4 aces and 47 non-aces would remain in the deck. Conditional probabilities are shown beside the appropriate branches in the tree diagram of Figure 7.1.

The multiplication and addition rules can be applied to tree diagrams to find probabilities for any sequence or combination of events. To find the probability of a sequence along one branch of the tree, multiply the associated probabilities. For example, the sequence A_1 followed by A_2, which is the topmost branch in Figure 7.1, has probability $P(A_2 \text{ and } A_1) = P(A_1)P(A_2|A_1) = .0769(.0588) = .0045$. There is a .0045 probability, or less than 1 chance in 200, of drawing 2 aces on two consecutive draws. Likewise, we can see that the probability of drawing an ace on the first draw, but not on the second, is $P(\text{not } A_2 \text{ and } A_1) = P(A_1)P(\text{not } A_2|A_1) = .0769(.9412) = .0724$.

Each sequence of branches in a tree diagram represents a combined event that is mutually exclusive of the others. We can therefore find the probabilities of more than one of these events by addition. Suppose we are interested in finding the probability of drawing at least 1 ace in two draws. Three outcomes produce at least 1 ace: (A_2 and A_1), (*not A_2* and A_1), or (A_2 and *not A_1*), with respective probabilities of .0045, .0724, and .0724 (Figure 7.1). Adding gives us a probability of $.0045 + .0724 + .0724 = .1493$ that we will draw at least 1 ace.

The sum of the probabilities for all branches in a tree diagram must be 1. We could thus have found the probability of "at least one ace" by subtracting the probability of "no aces" [the event (*not A_2* and *not A_1*)] from 1: $1 - .8507 = .1493$, the same figure obtained above.

Purely algebraic reasoning would lead to the same conclusions as tree diagrams, but the diagrams let us think out the problems in visual form. They are especially helpful when we are interested in combinations of events that are not independent.

Up to this point, we have discussed probability in general, without specific reference to its statistical applications. The following sections introduce some examples with such applications.

PROBLEMS

*7. Set up tree diagrams and find probabilities for the following.

   a. Drawing at least one diamond on two consecutive draws from a deck of 52 cards

   b. Drawing first a king, then a queen on two consecutive draws from a deck of 52 cards

   c. Drawing one king and one queen (in any order) on two consecutive draws from a deck of 52 cards

*8. A married couple plans to have three children, and they hope for at least one boy and at least one girl. The probability that any one child will be a boy is .51 (slightly more boys are born than girls). Assume that this probability is the same for all births (the events are independent).

   a. Construct a tree diagram showing possible outcomes and probabilities for the genders of three children.

   b. How many different ways could they end up with at least one boy *and* at least one girl?

   c. What is the probability that they will have either three boys *or* three girls?

7.4   PROBABILITY DISTRIBUTIONS FOR CATEGORICAL VARIABLES

In the 1980 election, Ronald Reagan ran against incumbent President Jimmy Carter. More than 86 million votes were cast, but that represented only about 53% of the voting-age population. The low turnout was not unusual; voter participation is generally much lower in America than in other industrial democracies. Pollsters and political scientists have studied this problem closely in an effort to find out who does not vote and why.

For individual people, the act of voting can be described by a categorical variable. Let *Y* be the variable "voting status for 1980 presidential election," which has as its possible values the two categories "voted" and "did not vote." We can put this discussion in terms of the probability concepts discussed earlier:

Experiment: Select an individual at random from the voting-age population as of November 1980 and ask whether he or she voted.

Events: Let event *V* = {person reports voting}. Its complement is event *not V* = {person does not report voting}.

Probabilities: $P(V)$ is the probability that a randomly selected individual reports having voted. $P(not\ V) = 1 - P(V)$ is the probability that a randomly selected individual does not report having voted.

Note that the action of randomly selecting a case and collecting data on a categorical variable is a kind of statistical *experiment*. Possible experimental outcomes are the values or categories of the variable.[2]

Earlier we defined probability as the proportion of times that an event is expected to occur, if an experiment is repeated a large number of times. This definition implies a straightforward relationship between probability and population proportion, as stated in the box. Hence, if we know population proportions, we know the probabilities under random selection.

The probability of obtaining any event A if one case is selected randomly from a population is the same as the population proportion of cases with event A:

$$P(A) = \pi$$

The Current Population Survey of the U.S. Census Bureau provides estimates of voting population proportions. Their data resemble what ours would be if we performed the random selection experiment repeatedly. The Population Survey data lead to the following estimates: in 1980, about 157.1 million Americans were of voting age, and about 93.1 million people reported having voted. Thus the proportion of the voting-age population who reported they had voted is about 93.1/157.1 = .593, or 59.3%. The probability of any randomly chosen individual reporting that he or she voted is therefore the same as this proportion: $P(V) = .593$. The probability that he or she does not report voting is $P(not\ V) = 1 - P(V) = 1 - .593 = .407$. These calculations are summarized in Table 7.4.

Political observers are often interested in how voting behavior is related to demographic variables such as age, race, sex, income, or occupation. For example, does voter participation have anything to do with occupational

---

TABLE 7.4   *Self-Reported Voter Participation in the 1980 Presidential Election[a]*

Voting-age population in November 1980 (millions):   157.1

Persons reporting they voted (millions):   93.1

Probability of voting: $P(V) = \dfrac{93.1}{157.1} = .593$

Probability of not voting: $P(not\ V) = 1 - P(V)$

$$= 1 - .593$$

$$= .407$$

---

[a]Official vote counts and population estimates suggest that the actual participation rate was closer to 53%; the 59.3% shown above reflects the exaggeration of self-reports.
Source: United States Bureau of the Census (1983).

status? Like voting, occupational status can be viewed as a categorical variable, with categories such as white collar, blue collar, or unemployed. These
categories could help define further events, elaborating on our earlier statistical experiment:

Experiment: Select an individual at random from the U.S. population of voting
age at the time of the 1980 presidential election and ask whether he or
she voted and what his or her occupation is.
Events: Event V = {person reports voting}
Event W = {person reports white-collar occupation}
Event B = {person reports blue-collar occupation}
Event U = {person reports being unemployed}
Event N = {person reports not being in the labor force}
Probabilities: $P(V)$ is the *unconditional* probability that a randomly selected in-
dividual reports having voted; $P(V|W)$ is the *conditional* probability of
voting given a white-collar occupation; etc.

Since our experiment now includes outcomes that are not mutually exclusive,
such as voting and white-collar occupation, conditional probabilities should
be examined.

If voting is not related to occupation, then the outcomes for these two
categorical variables are *independent:* that is, conditional probabilities are the
same as unconditional probabilities. For example, the conditional probability
of voting, given that job status is reported as "unemployed," should be the
same as the unconditional probability of voting:

$$P(V|U) = P(V)$$

Similarly, the probability that a randomly selected blue-collar worker reports
voting would be the same as the probability that any randomly selected
person reports voting:

$$P(V|B) = P(V)$$

The unconditional probability of voting, $P(V)$, has been calculated as .593,
implying that if we calculate the probability of voting for blue-collar workers
only (or for any other occupational status only) it will also be .593. That is,
we would find this to be true *if* voting and occupational status are unrelated
or independent.

Table 7.5 reveals otherwise. Based on estimates from the Current Population Survey, about 52.8 million American white-collar workers were of voting age in November 1980, 37.4 million of whom voted. Consequently, the
conditional probability of voting, given a white-collar occupation, is

$$P(V|W) = \frac{37.4}{52.8}$$
$$= .708$$

Their probability of voting is thus substantially higher (.708) than that among the overall population (.593). Among blue-collar workers the probability of voting is much lower: only 16 million out of 33.4 million reported voting:

$$P(V|B) = \frac{16.0}{33.4}$$
$$= .479$$

Among the unemployed, it is lower still. Those "not in the labor force," who are primarily retired or keeping house, report a relatively higher turnout: $P(V|N) = .569$.

Table 7.5 shows that occupation and voting are not independent, for if they were the conditional probabilities should match the unconditional probabilities. Probability reasoning thus gives us a way to *test whether two variables are related.* Furthermore, the conditional probabilities allow us to *describe* whatever relationship does exist. For instance, from the table we can see that white-collar workers are much more likely to report having voted.

The probabilities in Table 7.5 could also be expressed in terms of **frequencies,** to help analyze a sample of any given size. Suppose we interview a sample of 235 white-collar workers. How many of them would be expected to report voting, if voting report and occupation were independent? The answer can be found simply by multiplying the probability by the sample size: $.593(235) = 139.4$.

TABLE 7.5	**Probabilities of Reported Voting in 1980 Election, by Employment Status.**[a]		
Employment status[b]	*Number voted/ Number eligible*	*Estimated probability of reported voting*	
	Millions	*Conditional*	*Unconditional*
White-collar	37.4/52.8	$P(V|W) = .708$	$P(V) = .593$
Blue-collar	16.0/33.4	$P(V|B) = .479$	$P(V) = .593$
Unemployed	2.8/6.9	$P(V|U) = .406$	$P(V) = .593$
Not in labor force[c]	31.4/55.2	$P(V|N) = .569$	$P(V) = .593$

[a]Conditional probabilities were estimated from Census data on self-reports of voting. As in Table 7.4, the self-reports exaggerate how many people actually voted. Unconditional probabilities, from Table 7.4, are the probabilities we would see *if* (as is not the case) voting and employment status were *independent* of each other.
[b]The four categories are not a full listing for the U.S. population; farm and service workers are excluded, for instance.
[c]Includes retirees and homemakers.

To find how many cases in a sample are expected to fall in a given category, multiply the *probability, P,* that any one case falls in that category by the *size of the sample, n:*

Expected frequency $= Pn$

If voting report and occupation were independent, we would expect about 139 people, out of a sample of 235 white-collar workers, to report having voted in the election. This **expected frequency** of 139 could then be compared with the actual count of how many of those 235 did report having voted. If the actual or **observed frequency** were very different from the expected frequency, that would be evidence that voting report and occupation are not independent. This idea will be developed further in Chapter 12.

PROBLEMS

9. In a study of the geographical knowledge of college students training to teach in elementary and secondary school (Herman, Hawkins, and Berryman, 1985), 90% were elementary-school education majors and 10% were secondary-school social studies education majors. In general, their geographical knowledge was poor. For example, only 44.7% of them could locate England on a map of the world.

   Use these symbols to represent events in the experiment of selecting one student at random from this group:

   E: randomly chosen student is training to teach elementary school
   S: randomly chosen student is training to teach secondary-school social studies
   L: randomly chosen student can locate England on a map of the world

   a. What is the probability that you select an elementary education major? That your choice can correctly locate England?

   b. If geographical knowledge and major are independent, what is the probability that a randomly chosen student is *both* an elementary-education major and able to locate England? A secondary/social studies major and able to locate England?

   c. Do you think geographical knowledge and major actually are independent? Explain.

   d. State your answer to part c in terms of what you would expect to find comparing the unconditional and conditional probabilities, as in the voting example in Table 7.5.

10. The study described in Problem 9 involved 282 students.

   a. How many were elementary education majors? How many could locate England on a map?

b. If major and ability to locate England were independent, how many of the 282 students would we expect to be both elementary majors *and* able to locate England?

c. In the actual study, 99 students were both elementary majors and able to locate England. Compare this with the number expected under independence (part b). What does the difference suggest?

7.5 PROBABILITY DISTRIBUTIONS FOR MEASUREMENT VARIABLES

Our discussion so far has concerned variables that are **discrete,** meaning that their possible values are distinct and finite. All categorical variables are discrete, and so are some measurement variables such as "number of siblings" or "size of state legislature." With discrete variables, we can consider the probabilities of each different value of the variable.

Many measurement variables, on the other hand, theoretically have infinitely many possible values. Such measures as time, distance, weight, or temperature are **continuous** rather than discrete—there is an infinite number of temperatures between 34° and 35°, for instance. If a variable has infinitely many possible values, then the probability of any one exact value (such as temperature = 35.00000...°) becomes vanishingly small. Realistically, we must focus on the probability of a range of values such as $P(X > x)$ or $P(X \leq x)$.

In theory, for a continuous variable X, the probability that X takes on any one specific value x is 0:

$$P(X = x) = 0$$

Even when a measurement variable cannot take on infinitely many values, enough values may exist that it is cumbersome to assign each one a probability. We saw a similar problem in Chapter 1: Percentages do not work well as summary statistics for variables that assume more than a few values. Other methods, such as means and standard deviations, summarize measurement variable distributions better. Likewise, we need other methods to discuss probabilities in such distributions.

One method is graphical. A **probability distribution** is a distribution of the probabilities or population proportions associated with all possible values of a variable. For continuous variables, such distributions may be graphed as smooth curves, and *probability is represented by areas under the curve.* The total area under the curve, like the probability of a sample space, is 1.

Figure 7.2 (page 194) shows an approximation of the probability distribution of cumulative grade point average for a population of college students. Since GPA is continuous, the probability that a student chosen at random has any particular GPA value, such as 1.99999, is very small. It is more useful to consider probabilities for a range of values, such as GPA less than 2.0.

Areas corresponding to $P(\text{GPA} < 2.0) = .147$ and $P(\text{GPA} > 3.2) = .156$ are shaded in Figure 7.2. The sum of the areas or probabilities corresponding to all possible values must be 1:

$$P(\text{GPA} < 2.0) + P(2.0 \leq \text{GPA} \leq 3.2) + P(\text{GPA} > 3.2) = 1$$

We can therefore find the probability (or population proportion) of GPA between 2.0 and 3.2 by subtraction:

$$P(2.0 \leq \text{GPA} \leq 3.2) = 1 - P(\text{GPA} < 2.0) - P(\text{GPA} > 3.2)$$
$$= 1 - .147 - .156 = .697$$

The usual rules of probability (Tables 7.1–7.3) apply to continuous variables. For example, (GPA < 2.0) and (GPA > 3.2) are mutually exclusive, so

$$P[(\text{GPA} < 2.0) \text{ and } (\text{GPA} > 3.2)] = 0$$
$$P[(\text{GPA} < 2.0) \text{ or } (\text{GPA} > 3.2)] = P(\text{GPA} < 2.0) + P(\text{GPA} > 3.2)$$
$$= .147 + .156 = .303$$

Shaded graphs of probability distributions, like Figure 7.2, help us think about probabilities of continuous variables.

Figure 7.2 is based on data for a population of college students. Probability distributions for continuous variables often are derived not from data but from mathematical theory. The next section discusses one such theoretical probability distribution.

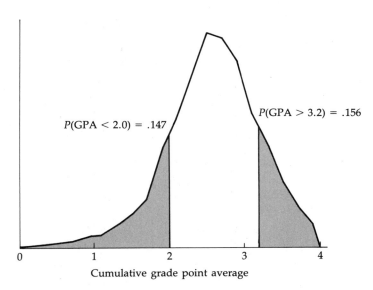

FIGURE 7.2   *Probability distribution of student GPA*

PROBLEMS

11. Some further probabilities for the GPA distribution of Figure 7.2 are

$$P(X < .5) = .004$$
$$P(X < 1.5) = .048$$
$$P(X < 2.5) = .420$$
$$P(X < 3.5) = .936$$

Use this information to find the following probabilities. Draw and shade curves to illustrate your answers.
a. $P(X \geq 2.5)$
b. $P(X \geq 3.5)$
c. $P(2.5 \leq X < 3.5)$
d. $P(.5 \leq X < 2.5)$
e. $P(X < .5 \text{ or } X \geq 3.5)$

These probabilities are based on census estimates of the distribution of household income (in dollars) for the U.S. population in 1985:

$$P(X < 5,000) = .077$$
$$P(X < 10,000) = .201$$
$$P(X < 15,000) = .316$$
$$P(X < 20,000) = .425$$
$$P(X < 25,000) = .525$$
$$P(X < 35,000) = .695$$
$$P(X < 50,000) = .853$$

Problems 12 and 13 refer to these probabilities.

12. For each of the following provide a shaded sketch of the problem, the appropriate symbolic expression [e.g., $P(X \geq 25,000)$], and the numerical answer. What is the probability that a household selected at random from this population has an income
a. Of at least $10,000?
b. Of at least $50,000?
c. Of at least $5,000, but less than $20,000?
d. Either less than $5,000 or at least $50,000?
e. Either less than $10,000, or from $20,000 to $34,999, or at least $50,000?

13. If the probabilities given in Problem 12 are applied to a population of 88,458,000 households, how many households would we expect with incomes
a. Of $50,000 or more?
b. Of at least $35,000, but less than $50,000?

7.6   NORMAL DISTRIBUTIONS

A **normal distribution** is a continuous theoretical probability distribution, defined by a specific mathematical equation. Graphs of these distributions have a characteristic symmetrical bell shape, the **normal curve,** seen in Figure 7.3.

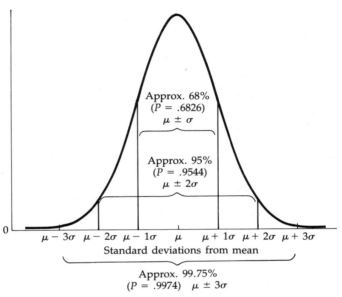

FIGURE 7.3   *Normal probability distribution*

Normal-distribution probabilities have a known relationship to the population mean (μ) and standard deviation (σ). In normal distributions about 68.26% of the area will be within plus or minus 1 standard deviation ($\pm 1\sigma$) of the mean. As shown in Figure 7.3, 95.44% of the area is within $\pm 2\sigma$ of the mean, and 99.74% is within $\pm 3\sigma$ of the mean.

The total area under the normal curve equals 1. Since the curve is symmetrical, the area under either half equals .5. Also, for any number z, the probability of values at least z standard deviations *above* the mean is the same as the probability of values at least z standard deviations *below* the mean. These observations, together with the probabilities shown in Figure 7.3, let us make further deductions:

What is the probability of values more than 1 standard deviation from the mean? The probability of values *less* than 1 standard deviation from the mean, according to Figure 7.3, is .6826. The probability of values *more* than 1 standard deviation from the mean, shaded in Figure 7.4, is 1 − .6826 =

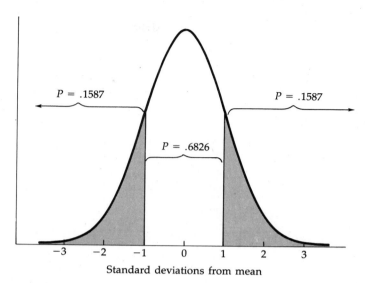

$P = .1587$

$P = .1587$

$P = .6826$

−3 −2 −1 0 1 2 3

Standard deviations from mean

FIGURE 7.4 *Normal distribution showing probabilities of values more than 1 standard deviation from the mean: $P(|X| > \mu + 1\sigma) = 2(.1587) = .3174$*

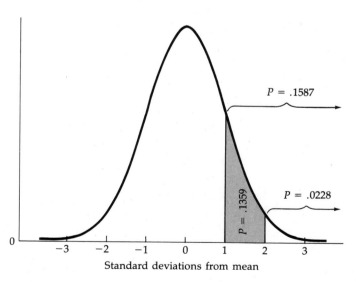

$P = .1587$

$P = .1359$

$P = .0228$

0

−3 −2 −1 0 1 2 3

Standard deviations from mean

FIGURE 7.5 *Normal distribution showing probabilities of values more than 1 but less than 2 standard deviations above the mean: $P(\mu + 1\sigma < X < \mu + 2\sigma)$*

.3174. (We treat the probability of values *exactly equal* to 1 standard deviation from the mean as 0.)

What is the probability of values more than 1 standard deviation above the mean? If .3174 of the area in both tails combined is more than 1 standard deviation from the mean, one tail contains half this proportion: .3174/2 = .1587.

What is the probability of values more than 1 but less than 2 standard deviations above the mean? The area corresponding to this question is shown in Figure 7.5. To find it, we need to know two probabilities: $P[X > (\mu + 1\sigma)] = .1587$, as found above, and similarly, $P[X > (\mu + 2\sigma)] = (1 - .9544)/2 = .0228$. If .1587 is the probability of values above $\mu + 1\sigma$, and .0228 is the probability of values above $\mu + 2\sigma$, the probability of values between these two points must be .1587 − .0228 = .1359.

More detailed normal probabilities are given by tables such as Tables A.1 and A.2 in the Appendix. Their use is described in Chapter 9. By consulting these tables, we can achieve fairly precise probabilities for any range of values in a normal distribution.

PROBLEMS

14. Use Figure 7.3 to find the following probabilities. Sketch shaded curves to illustrate your answers.
    a. $P\{[X < (\mu - 3\sigma)] \text{ or } [X > (\mu + 3\sigma)]\}$
    b. $P[X < (\mu - 2\sigma)]$
    c. $P\{[X < (\mu - 1\sigma)] \text{ or } [X > (\mu + 2\sigma)]\}$
    d. $P\{[X < (\mu - 3\sigma)] \text{ and } [X > (\mu + 3\sigma)]\}$

15. Use Figure 7.3 to find values for z for which:
    a. $P(X > \mu + \dot{z}\sigma) = .5$
    b. $P(X < \mu + z\sigma) = .8413$
    c. $P(X > \mu + z\sigma) = .9772$

16. Is the GPA distribution in Figure 7.2 approximately normal? Apply the reasoning described in Chapter 4 (Figure 4.1, page 107), using these population parameters:
    $$\mu = 2.58 \qquad \sigma_X = .62 \qquad Md = 2.61 \qquad IQR = .78$$

17. Suppose the GPA distribution were approximately normal, with $\mu = 2.58$ and $\sigma = .62$.
    a. What GPA would put a student in the top 2.28% of this distribution?
    b. What proportion of the students have GPAs below 1.34?

SUMMARY

This chapter introduced some basic ideas and rules of probability. Of central importance are the concepts of statistical experiments, outcomes, events, the relative frequency definition of probability, and the rules for deducing some probabilities from others. Tree diagrams visually aid our thinking about possible combinations of events.

Historically, probability was first studied by gamblers, and games of chance still illustrate conveniently many probabilistic concepts. More mainstream analytical activities including insurance, industrial quality control, technological risks, and managerial decision making rely on methods adapted from the gamblers' techniques.

Most statistical applications involve one kind of experiment: selecting cases at random from a population and observing the value(s) of one or more of their variables. The probability that the value of variable X for a single randomly selected case equals x is the same as the proportion of cases in the population for which $X = x$. This correspondence provides the foundation for statistical inference: drawing conclusions about the population on the basis of sample data.

Categorical and measurement variables require different analytical approaches. With categorical variables, or any that take on few values, we can discuss the probability of each value individually. Measurement variables may assume so many values that this is impractical. Instead, with measurement variables we typically deal with inequalities: the probabilities of values within given ranges, rather than values exactly equal to something.

A normal distribution is a specific type of theoretical probability distribution with many statistical applications. Probabilities in normal distributions depend on distance from the mean. We will return to examine normal distributions and their statistical applications in more detail in Chapter 9.

Several examples in this chapter described statistical experiments that involved selecting cases at random from a larger population. The following chapter discusses the concept of random selection, which is crucial to statistical research. The mathematical theories of statistical inference and the practical success of political pollsters both rest on random selection.

PROBLEMS

18. If you toss a coin three times, give the probability of these outcomes.

    a. Three heads

    b. Two heads *or* two tails

    c. Two heads *and* two tails

    d. Tails at least once

19. Give your own examples of each of the following pairs of events. Clearly describe the relevant statistical experiment and the events themselves.

    a. An event and its complement

    b. Two events that are mutually exclusive

    c. Two events that are not mutually exclusive

d. Two events that are independent

e. Two events that are not independent

20. A simple lottery involves spinning a pointer that may end (with equal probability) on any of the numbers 1–32. The pointer is spun four times to select four numbers from 1 to 32 (any number can be chosen more than once). Paying $1 to buy a ticket entitles you to guess what the four numbers will be.

   a. What is the probability that with a single guess you correctly specify all four numbers, in order?

   b. One way to interpret the probability of any event A is that, over the long run, we expect A to occur about once in every $1/P(A)$ trials. Over the long run, how often could you expect to guess the winning lottery numbers correctly?

A typical Las Vegas slot machine or "one-armed bandit" has three dials, each with 20 different symbols. Suppose that hitting the jackpot on such a machine pays $100 and requires a bar symbol on all three dials. The number of bars on each dial is as shown.

Symbols	Dial 1	Dial 2	Dial 3
Bar	1	3	1
Other	19	17	19

Problems 21 and 22 refer to this slot machine. Assume that the machine is fair, and that each of the 20 symbols on each dial is equally likely to be showing at the end of a spin.

*21. Construct a tree diagram for possible outcomes from one play of this slot machine.

   a. What is the probability of three bars showing?

   b. What is the probability of at least two bars?

   c. What is the probability of at least one bar?

   d. Use your answers to parts b and c to find the probability of exactly one bar showing.

   e. Use your answer to part c to find the probability of no bars showing.

22. How often should we expect to hit the jackpot (three bars showing) on this slot machine? If each trial costs $.25, how much money would we spend, over the long run, for each $100 jackpot won?

23. An aircraft manufacturer conducted studies of an accident scenario: a plane loses engine power; slats on leading edges of the wing become damaged; and this occurs during takeoff. The probability of all three events coinciding was put at less than one in a billion. Yet this sequence of events actually happened four times during 1977–1981 (once causing a

crash that killed 273 people), before the manufacturer modified the aircraft (Perrow, 1984). In your answers use the symbols E for loss of engine power, S for slat damage, and T for takeoff, and refer to the appropriate probability rules.

a. The "one chance in a billion" estimate was evidently in error. What assumptions about probability likely contributed to this very low figure?

b. Suggest reasons why the assumptions in part a might be unrealistic.

24. Describe how a common mode failure might make the following conditional probabilities much different from the corresponding unconditional probabilities.

a. The probability that your television will not work, given that your neighbor's television does not work.

b. The probability that a ship's radar does not work, given that the ship's radio does not work.

c. The probability that a space shuttle's fourth redundant on-board computer will give the wrong answer, given that the other three computers on board are wrong.

d. The probability that chemical plant personnel will make the wrong decision about how to relieve a dangerous rise in pressure, given that they made decisions that created the rise.

*25. M. L. Murray and others (1987) studied possible health risks in piles of Canadian uranium mill tailings, which contain radioactive sand and clay from the processing of uranium ore. These wastes remain potentially dangerous for thousands of years, and they could be spread into the environment by a variety of natural processes. For example, a forest fire or drought could send radioactive dust into the air; a flood could wash it into a nearby lake. Over 1,000 years, these events were judged to have the following probabilities of occurring at least once to a given tailings pile:

Event	$P(\geq 1$ occurrence in 1,000 years)
Flood	.049
Drought	.095
Forest fire	.865

Assume for the moment that flood, drought, and fire occur as independent events. Construct a tree diagram based on these probabilities, and use it for your answers.

a. What is the probability that none of these three types of events will disturb the pile in 1,000 years?

b. What is the probability that at least one dust-causing event (drought or fire) will occur?

c. Confirm your answer to part b by applying the multiplication and addition rules directly to the probabilities in the table. Assume that drought and fire are independent but not mutually exclusive events.

d. How likely is it that both airborne (due to fire or drought) and water contamination (due to flood) will occur? Show how the answer can be obtained first by using a tree diagram, then by applying the multiplication and addition rules.

26. The probabilities given in Problem 25 are for occurrences in 1,000 years. Assume that the 1-year probabilities are 1/1,000th of the 1,000-year probabilities. Find the probability that in any one year, a forest fire occurs at the tailings site. Apply the $1/P(A)$ rule given in Problem 20 to estimate how often (once every how many years?) forest fires occur over the long run.

27. Following the reasoning of Problem 26, how often are floods expected to occur at a given uranium tailings site? How often are droughts expected?

28. The probability that none of the three types of events will occur to a given tailings pile in 1,000 years is about $P = .116$ (Problem 25, part a).

a. Find the probability that *at least one* contaminating event will affect a given pile in 1,000 years.

b. Uranium mining areas may contain hundreds of separate tailings piles. Suppose we expand our focus from just one pile to two. Under the unrealistically optimistic assumption that events affecting the two piles are independent, what is the probability that *either* of two tailings piles will experience one or more contaminating events in 1,000 years?

c. Explain why the independence assumption of part b is optimistic.

Problems 29–32 refer to the following statistical experiment: Randomly select one of the 33 households from the water conservation data set of Table 2.8 (page 41). Express probabilities symbolically [e.g., $P(R \text{ or } C)$], and find their numerical values.

29. Use the symbols R to represent the event "household head is retired" and C to represent the event "household head is a college graduate."

a. What is the probability that the selected household is headed by a retiree?

b. What is the probability that the selected household's head is a college graduate (≥ 16 years of education)?

c. What is the probability that the selected household's head is someone who is retired *and* has a college education?

d. What is the probability that the selected household's head is someone who is retired *or* has a college education?

e. Apply the addition rule to the probabilities you obtained in parts a–c to confirm your answer to part d.

30. Use the symbol X to represent the variable household income in thousands, and Y to represent 1980 water use. Find the following probabilities from the data in Table 2.8.

   a. What is the probability that a household randomly selected from Table 2.8 has an income over $40,000 ($X > 40$)?

   b. What is the probability that a randomly selected household used more than 3,000 cubic feet of water in 1980 ($Y > 3,000$)?

   c. What is the probability that the selected household has an income over $40,000 *and* used more than 3,000 cubic feet?

   d. What is the probability that the selected household has an income over $40,000 *or* used more than 3,000 cubic feet?

   e. Show how the addition rule can be used to confirm your answer to part d.

31. What would be the probability of having income over $40,000 and 1980 water use over 3,000 cubic feet if these two events were independent? Compare this probability with the one you obtained from the actual data, in Problem 30, part c, and comment on the difference.

32. Use the variables of Table 2.8 to suggest your own examples of

   a. Two events that are mutually exclusive.

   b. Two events that are not mutually exclusive.

33. Imagine a large lecture course with an enrollment of 140 students, including 80 women and 60 men; 15% are seniors. To facilitate random selection, we put each student's name on a scrap of paper and mix them thoroughly in a fishbowl.

   Apply the appropriate probability rules and show how the following probabilities may be calculated. Use the symbols S = {student is senior}, M = {student is male}, and F = {student is female}.

   a. If gender and class year are independent, what is the probability that a single student drawn at random is both female *and* a senior?

   b. Assuming that gender and class are indeed independent, what is the probability that a student drawn at random is female *or* a senior?

   c. If we twice draw a name, write it down, and mix it back into the fishbowl, we get a list of two names that may or may not be different. What is the probability that *both* students drawn are male?

   d. What is the probability that *five* students drawn with replacement (as in part c) all are male?

34. What is the probability that both students are seniors, if we make two drawings *without* replacing either of the drawn names (so both must be

different)? What is the probability that at least one of the students is a senior? Construct a tree diagram to illustrate your reasoning.

35. In November 1984, an outbreak of intestinal illness struck a Caribbean resort club (Spika et al., 1987). Among one group of 411 tourists visiting then, 300 reported illness. Of the 411, 349 had been drinking bottled water.

   a. What is the probability that a tourist selected at random from this group became ill? Had drunk bottled water?

   b. If drinking bottled water and becoming ill were independent events, what is the probability that a randomly selected tourist was a bottled-water drinker who became ill?

   c. In fact, 261 of the bottled-water drinkers in this group fell ill. What is the *conditional* probability of a bottled-water drinker becoming ill?

   d. Contrast the conditional probability (part c) with the unconditional probability (part a). Was drinking bottled water much protection against illness?

   e. Apply the multiplication rule to find the actual probability that a randomly selected individual is both ill and a bottled-water drinker. Compare this probability with the probability expected under independence (part b), and explain why they are or are not very different.

36. Many of the 411 resort club tourists of Problem 35 had been taking anti-malarial medication (chloroquine). Of 380 taking chloroquine, 282 fell ill. Calculate the conditional probability of illness given chloroquine, and compare this with the unconditional probability of illness. Does this evidence suggest that the chloroquine protected people against intestinal illness?[4]

The probabilities shown are based on census estimates of the years of schooling completed by people 25 years old and over in the U.S. population in 1985 (8 = eighth grade, 12 = high school graduate, 16 = college graduate, etc.).

$$P(X < 5) = .027$$
$$P(X < 8) = .075$$
$$P(X < 9) = .139$$
$$P(X < 12) = .261$$
$$P(X < 13) = .643$$
$$P(X < 16) = .806$$

Problems 37–38 refer to these probabilities.

37. For each of the following, provide a shaded sketch of the problem, the approximate symbolic expression [e.g., $P(X \geq 13)$], and the numerical

answer. What is the probability that a person selected at random from this population

a. Has at least an eighth grade education?

b. Is at least a college graduate?

c. Has exactly 12 years of education?

d. Has at least an eighth grade education, but did not finish high school?

e. Attended college, but did not graduate?

38. If the probabilities given for Problem 37 are applied to a population of 124,905,000 people, how many people would we expect

a. With less than an eighth grade education?

b. With at least some college?

NOTES

1. United States Atomic Energy Commission (1974), Appendix IV, p. 20.

2. Random selection requires choosing cases in such a way that *every case in the population has the same probability of being chosen.* If there are N cases in a population and the experiment consists of choosing one case, then each case in the population should have a probability $1/N$ of being selected. Chapter 8 looks at random selection in more detail.

3. The proportion *reporting* that they voted, .593, is higher than the proportion of people who actually *did* vote in the election (less than .53). This is a common problem in survey research: Even on anonymous questionnaires, many people give inaccurate answers to present themselves in a more favorable light. Obviously this obscures the researcher's view of the reality behind survey responses. Note that in our example here, $P(V)$ refers to "the probability of reporting voting." For convenience this has been shortened to "the probability of voting" in the discussion, but *reported* voting is not identical to *actual* voting.

4. Problems 35 and 36 are an example of *epidemiological* research, in which statistical analysis is used to try to understand the causes of disease in an uncontrolled, real-world setting. This particular disease was spread by ill North American food handlers; factors such as eating raw hamburger or having an ill roommate increased the tourists' chances of becoming ill themselves.

Random Sampling and Sampling Distributions

Chapter **8**

Games often rely on procedures that produce chance outcomes, such as throwing dice, dealing from a shuffled deck of cards, spinning roulette wheels, tossing a coin, and drawing names from a hat or numbered balls from an urn. The purpose of such **random generators** is to generate outcomes that are impossible to predict or control. Not all such procedures actually work: A stacked deck of cards or loaded dice will not generate random outcomes. Sometimes nonrandomness is so subtle it can be detected only by statistical analysis.[1] In principle, however, random generators produce results that appear to be controlled entirely by chance.

The concept of randomness plays a central role in statistics. It is crucial to theories of how to draw inferences about a population by analyzing only a small sample of cases. If a sample consists of cases chosen randomly from the population, then it makes a sound basis for inference. We can expect a random sample to be representative of the population, within the limits of chance variations that are predicted by statistical theory.

This chapter first examines *random numbers,* which are numbers that might have been generated by a random process. We then use random numbers to select random samples. The problem of estimating a population parameter, such as the mean age for a large population, on the basis of a small random sample is pursued with a computer experiment. The experimental results illustrate two important—and often difficult—ideas: *sampling distributions* and *standard errors.*

8.1   *RANDOM NUMBERS*

A single throw of a six-sided die has six possible outcomes, from one to six dots showing. Assuming that the die is fair, any of these six numbers has an equal probability ($\frac{1}{6}$ or about .167) of coming up. A probability distribution in which all possible outcomes have the same probability of occurring is called a **uniform distribution.** Figure 8.1 is a histogram of the uniform distribution that describes single throws of a fair six-sided die. Because the bars all rise to the same height, forming a rectangle, uniform distributions are also known as **rectangular distributions.**

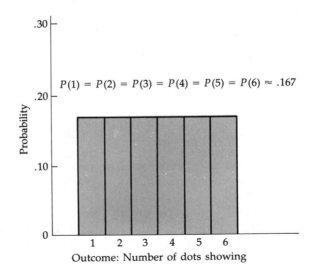

$$P(1) = P(2) = P(3) = P(4) = P(5) = P(6) \approx .167$$

Outcome: Number of dots showing

FIGURE 8.1   *Example of a uniform distribution: Single throws of a fair six-sided die*

Note that Figure 8.1 is a distribution for the *theoretical* probabilities of die throwing. If we actually conduct an experiment with a few hundred die tosses, our resulting *empirical* distribution will be, by chance, less uniform.

Imagine that our die has ten sides instead of six, with the faces numbered 0 through 9 (such odd-looking dice are possible). Any of these ten digits would have an equal probability ($\frac{1}{10}$ or .10) of coming up on any given throw. Imagine throwing tirelessly for hours, each time writing down what number shows, then throwing again. The result would be a **random number table** like Table 8.1 (page 210), which contains 1,800 random digits corresponding to the outcomes from 1,800 throws of a ten-sided die.

The outcomes of successive die throws are completely *independent* of each other; a die has no memory of what it has previously done. Similarly, any of the digits 0–9 has a probability of $\frac{1}{10}$ or .10 of appearing at any given position in a random number table, regardless of what has come before. For example, it is unlikely in a relatively short list of random numbers to see a string of ten 3's in a row, and even less likely that the string would extend to 11. Nevertheless, if in throwing the die we toss ten 3's in a row, the probability of getting another 3 on our next throw is the same as always: $\frac{1}{10}$. Stated in the probability notation of Chapter 7,

$$P(3|\text{ten previous 3's}) = P(3) = \frac{1}{10}$$

That is, the conditional probability of rolling a 3, given ten previous 3's, is the same as the unconditional probability of rolling a 3. This follows the formal definition of independence (Table 7.3).

Statisticians, cryptographers, computer programmers, and other researchers have many uses for random numbers, and whole books of such numbers have been published. The 1,800 digits in Table 8.1 are too few for some forms of research. Later we will discuss a computer experiment employing more than nine million random numbers. Of course, these numbers are not really generated by tossing ten-sided dice. The most convenient way to generate so many random numbers is by a type of mathematical equation called a **pseudorandom number generator.** These equations produce numbers that are not really random, in that they are determined by the equation, but that for all practical purposes *appear* random. Computers programmed with such equations can generate long lists of random numbers, as was done for Table 8.1.[2]

Statistical analysis helps us judge whether a supposedly random process is indeed producing random outcomes. For example, to test a table of random numbers like Table 8.1, we can simply check whether all of the digits from 0 to 9 occur with a probability of $\frac{1}{10}$. Since there are 1,800 digits in this table, we expect each digit to show up about 180 times. Table 8.2 shows their actual frequency distribution. No digit occurs exactly 180 times, but most are close; 9, which occurs 202 times, is the farthest away. Figure 8.2 shows this distri-

TABLE 8.1   *Random Number Table (1,800 Digits)*

24679	24655	75970	85292	24429	22660	41106	66621	62724	15055
11896	05212	02348	95800	16874	51650	73357	78654	15313	88811
98182	94875	94573	71509	12832	29303	29270	93421	87824	52745
85760	30078	95351	79073	18351	69546	60569	03328	53840	07491
08425	65259	19137	99585	95462	18381	60672	85371	85743	51832
70880	29136	73527	63465	68509	58548	72900	55802	45819	06653
78709	46697	97289	18568	48303	99486	61650	15469	98139	57038
08168	67125	78084	77402	40905	06494	31908	75149	85052	89520
12779	95960	25920	39177	14051	42682	10086	17736	89177	95177
61548	27226	04191	53097	13739	56957	77947	18478	05516	93751
80867	95863	18035	23682	95091	98410	49819	07466	26695	16051
07117	12800	93806	57732	55399	33934	87499	93185	19549	95323
27915	84750	49209	38778	95994	06563	37001	29319	90226	12374
84873	27112	62544	05360	78041	12484	55030	62900	79534	76345
09937	27206	86478	10230	15254	80380	14984	65773	99356	98091
59182	94429	66156	22286	24137	17464	63664	18631	46772	57673
52752	43666	65773	18843	24232	26588	27596	89245	44133	50300
31860	71017	84848	98844	98596	67631	91669	16769	48893	82375
46865	73631	78551	09330	40849	70234	25205	23207	18820	76956
55599	85987	82556	26868	12953	23818	30625	02007	11059	27472
93046	85481	04042	96948	56314	40960	43739	40909	61024	10864
55662	99590	33478	44850	18292	97511	76301	05921	06369	69068
45182	15390	30568	01297	55227	96988	93883	24866	30293	99963
34155	20448	49537	41653	19973	80100	44705	90817	37281	79858
10273	68675	97678	62451	32659	63858	35923	86010	73540	14727
53857	33405	40138	54814	39871	13209	13371	43638	82524	48184
09961	98872	10364	51633	06370	97010	85276	48494	69194	75488
27477	06045	39464	90883	70779	86912	41929	66129	72883	24210
33778	43186	77412	26928	98078	90568	44487	94363	67743	65809
23380	14608	96725	42895	02735	75318	36074	05101	44384	71415
71301	87437	57967	54039	64744	99224	39547	81872	03390	92552
90226	08936	91174	33972	42484	47902	35536	70547	08831	80569
48723	09442	13678	62898	49099	65893	01621	52000	89803	29062
50608	71223	56361	26225	44910	80112	79255	49521	29280	38441
64846	56778	08310	92965	38200	04849	29849	69743	82468	84927
67705	64777	27013	50516	79205	78204	79893	90100	41967	84957

bution as a histogram, with a horizontal line indicating the theoretically expected proportion for each digit, $\pi = .10$. Table 8.2 and Figure 8.2 do not strongly challenge our expectation of one-tenth 0's, one-tenth 1's, and so on. The slight variation around the $\pi = .10$ line could easily be due to chance.[3]

TABLE 8.2

Frequency Distribution of the 1,800 Random Digits in Table 8.1

Digit	Frequency f	Proportion $\hat{\pi}$
0	185	.103
1	163	.091
2	170	.094
3	167	.093
4	178	.099
5	183	.102
6	172	.096
7	182	.101
8	198	.110
9	202	.112
Totals:	1,800	1.0

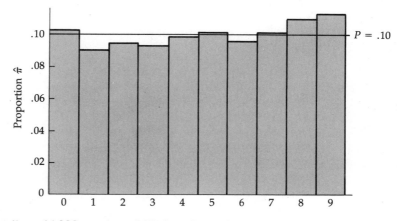

FIGURE 8.2 *Distribution of 1,800 random digits from Table 8.1*

PROBLEMS

1. If we generate five genuinely random digits, what is the probability that
   a. All five of them are 3's?
   b. The sequence is 2, 4, 6, 7, 9?
   c. None of the five is a 0?
   d. All five are odd numbers?

2. Use a random number table from another textbook, the random number function of a statistical calculator or computer program, or a random

device of your choosing to construct your own random number table consisting of 100 digits.

3. Construct a histogram along the lines of Figure 8.2 for the digits in your table from Problem 2. Use proportions as the scale for the vertical axis. Comment on how closely your empirical distribution of digits resembles the uniform probability distribution we would theoretically expect.

8.2   *SIMPLE RANDOM SAMPLES*

A **population** is the total collection of cases that are of interest to the researcher. When it is impractical to study every case in the population, a **sample** or subset of these cases is examined instead. If every case in the population has a known probability of being selected for the sample, we obtain a **probability sample.** In contrast, when the probabilities are not known we have a **nonprobability sample.** Nonprobability samples result from selection criteria such as "looks typical" or "is conveniently available for study."

The most basic kind of probability sample is a *simple random sample.*

> In a **simple random sample,** each case in the population has an equal chance of being selected, and the chance of selecting one case is independent of whether some other case is selected.

If N cases are in the population, and we wish to draw a simple random sample consisting of n different cases, then each case in the population should have an n/N probability of being included. Random numbers often serve as the basis for selection.

Telephone surveys are one area where simple random sampling is readily applied. Suppose we wish to contact a random sample of households with telephones in a community where all telephone numbers begin with 555. We could use random numbers to supply the last four digits of a phone number; applying Table 8.1, we would first dial 555-2467, next (reading across the columns) 555-2465, 555-7597, 555-8529, and so on. This technique, *random digit dialing,* ensures that every phone number beginning with 555 has an equal probability of being dialed. It does not matter whether the phone number is unlisted, unlike choosing our sample from a phone book. If a phone number we dial does not exist or is nonresidential, we just go on to the next number.

This system easily extends to more complicated situations. To dial more than one community, we can choose the first three digits or the area code randomly as well. With two area codes, for example, we could use eight-digit sequences of random numbers. The first seven digits would be the phone number to dial, and we could make up a rule such as dialing into one area

TABLE 8.3

Percentage of White and Black Heads of Households at 26 Boston-Area Public Housing Projects

	Housing project	White heads of households	Black heads of households
1	Commonwealth	60.2%	28.0%
2	Faneuil	62.1	25.3
3	Charleston	94.2	0.9
4	Columbia	1.7	81.7
5	Franklin N.	0.8	81.9
6	Franklin S.	4.7	78.2
7	Orient	79.8	6.6
8	Maverick	71.4	6.1
9	Fairmount	83.4	15.1
10	Bromley	1.7	84.3
11	South Street	52.7	31.3
12	Gallivan	43.8	54.2
13	Archdale	43.8	39.8
14	Beech	44.9	33.8
15	Orchard	1.1	81.9
16	Whittier	2.5	87.8
17	Mission N.	2.5	83.7
18	Mission S.	4.1	41.1
19	Camden	2.8	88.9
20	Lenox Street	3.0	78.0
21	West Broadway	98.3	0.0
22	Old Colony	98.2	0.0
23	McCormack	98.0	0.0
24	South End	13.0	43.0
25	West Newton	19.4	34.1
26	Rutland	30.8	23.1

code if the eighth digit is odd, and into the other area code if the eighth digit is even. Similar rules can determine whether we ask to talk to a male or a female, or to the oldest resident, second oldest, third oldest, or whatever.

Random sampling can also be applied where there are lists that include every case in the population. Lists of taxpayers, registered voters, drivers' licenses, utility customers, and college alumni are possible examples. For illustration we will use the data in Table 8.3, which describe 26 Boston-area public housing projects. Suppose we wish to randomly select four of these 26 projects for detailed study.

Each housing project is designated by a case number from 1 to 26. For our purposes all case numbers must have the same number of digits, so case numbers are written as 01–26 in Table 8.4 (page 214). In a random number table such as Table 8.1, all two-digit combinations should be equally likely. By reading off two-digit sequences from Table 8.1, we can choose projects at

TABLE 8.4   *Choosing a Random Sample Without Replacement*[a]

Case	Housing project	Selection procedure: Sampling without replacement
01	Commonwealth	
02	Faneuil	Two-digit random numbers from Table 8.1,
03	Charleston	reading left to right: **24**, 67, 92, 46, 55, 75,
04	Columbia	97, **08**, 52, 92, **24** (ignore, because already
05	Franklin N.	selected), 42, 92, **26**, 60, 41, **10** (stop—we
06	Franklin S.	now have our $n = 4$ cases)
07	Orient	
08	**Maverick**	
09	Fairmount	
10	**Bromley**	
11	South Street	
12	Gallivan	
13	Archdale	
14	Beech	
15	Orchard	
16	Whittier	
17	Mission N.	
18	Mission S.	
19	Camden	
20	Lenox Street	
21	West Broadway	
22	Old Colony	
23	McCormack	
24	**South End**	
25	West Newton	
26	**Rutland**	

[a]Selecting $n = 4$ housing projects from the population ($N = 26$) in Table 8.3.

random from Table 8.4. From left to right the random numbers begin 24679, 24655, and so on; the first five two-digit combinations are 24, 67, 92, 46, and 55. Only one of these, 24, corresponds to a case number in Table 8.4. Project 24, South End, is therefore selected for our sample. We ignore the other combinations (67, 92, 46, 55), and keep reading until we have three other usable two-digit combinations. The resulting sample consists of projects 8, 10, 24, and 26, as shown in Table 8.4.

The number 24 comes up twice in this procedure. There are two ways to deal with such repetition:

1. Count it twice, so our sample consists of projects 8, 26, and 24 twice. This is called sampling **with replacement**.
2. Ignore 24 on its second occurrence, and go on with selection until four *different* projects are chosen, as done in Table 8.4. This is sampling **without replacement**.

Sampling without replacement, as in Table 8.4, results in a sample for which every case has an equal chance of being selected, but the selection probabilities are not independent. Thus, it is not a simple random sample.

The larger the population is relative to the sample, the less difference it makes whether we sample with or without replacement. In a typical survey the sample is much less than 1% of the population, and this difference is negligible. In such cases sampling without replacement and simple random sampling produce approximately the same results.

Sometimes people say informally that they have chosen some cases "at random." Reporters, for instance, may hold "random interviews" of people on the street. Teachers may "randomly" grab several exams out of a pile. Usually the selection involved here is not really random at all. Interviewers tend to make subjective judgments in deciding whom to question (**judgmental** sampling), preferring people who look approachable, interesting, and perhaps even photogenic. In any event, only those people walking down that street at that time have any chance of being chosen.

Snatching several exams from a pile may seem more random, but it would be better described as **haphazard.** As with judgmental sampling, haphazard sampling is a nonprobability method—we simply do not know what the probabilities of selection are.

Of course, a stack of exams can be shuffled so that grabbing one off the top *is* a random process. Likewise, it is possible to design a system for randomly choosing people on the street. The point is not that random sampling is impossible in these situations, but that *random sampling* is not the same as *careless sampling.* Truly random sampling generally requires of the researcher a great deal of thought and effort.

Random samples give researchers a reasonable basis for drawing conclusions about a population. Most statistical procedures assume simple random sampling, which entails more straightforward calculations than other types of probability sample. Simple random sampling is not always the best possible approach, however. Other types of probability samples may have advantages in precision, feasibility, or cost.

PROBLEMS

4. The state of Florida is covered by four telephone area codes: 305, 407, 813, and 904.

   a. Devise a random digit dialing scheme such that every Florida telephone number has an approximately equal chance of being called. Assume that each area code has about the same number of telephones.

   b. Use Table 8.1 to specify the first six numbers you would dial.

5. How could Table 8.1 simulate rolls from a single six-sided die, if we wanted to play a dice game without the dice? How could the table be used to simulate rolls from a pair of six-sided dice?

6. Begin in the fifth "paragraph" of random numbers in Table 8.1 (52752 43666 . . .), and select a random sample of six students From Table 1.1 (page 5). Do your sampling *without replacement*, and write down each student's gender and grade point average.

7. Repeat the experiment of Problem 6, but select six students by sampling *with replacement*.

8.3 RANDOM SAMPLES FROM COMPUTER DATA FILES*

Sometimes a list of the population of interest is available in a computer file. The computer can then select a random sample directly. Computer programs designed to work with numerical data often have the capability to generate pseudorandom numbers, which can be used to select samples in several ways. For example, consider the following situation. At midsemester, a university day-care center announces openings for children from five additional families. Since subsidized day care is rare, immediately 25 applications are made. A lottery will randomly select which children to admit; the problem is to select a random sample of size $n = 5$ from the population of $N = 25$. One way to do this, other than drawing names out of a hat, is to enter the names of the 25 applicants into a computer file and use the program to generate a random number for each applicant.

Table 8.5 shows the resulting file. The program's pseudorandom generator produces numbers between 0 and 1; such a number has been generated for each family. Next we sort the 25 families in order by random number, as shown in Table 8.6. The first five families in the sorted list can be the ones

TABLE 8.5   *Twenty-Five Families and Computer Generated Random Numbers*

Family	Pseudorandom number *(uniform distribution)*	Family	Pseudorandom number *(uniform distribution)*
A	.0321949	N	.7549810
B	.6710823	O	.7855499
C	.9830985	P	.1429575
D	.6296493	Q	.9293609
E	.2459989	R	.0243085
F	.8984687	S	.9645180
G	.3359111	T	.2912911
H	.0407609	U	.1880395
I	.3131475	V	.7001474
J	.7812470	W	.4811665
K	.9458091	X	.6912491
L	.0863035	Y	.8804231
M	.8953475		

	Families from Table 8.5, Sorted by
TABLE 8.6	**Assigned Random Number**

Rank[a]	Family	Random number
1	R	.0243085
2	A	.0321949
3	H	.0407609
4	L	.0863035
5	P	.1429575
6	U	.1880395
7	E	.2459989
8	T	.2912911
9	I	.3131475
10	G	.3359111
11	W	.4811665
12	D	.6296493
13	B	.6710823
14	X	.6912491
15	V	.7001474
16	N	.7549810
17	J	.7812470
18	O	.7855499
19	Y	.8804231
20	M	.8953475
21	F	.8984687
22	Q	.9293609
23	K	.9458091
24	S	.9645180
25	C	.9830985

[a]Five lowest accepted for day care. Remaining 20 families form waiting list in order shown.

selected for day-care admission: Families R, A, H, L, and P. We have selected a random sample of $n = 5$ from a population of $N = 25$ in such a way that each of the original 25 families had the same probability, $\frac{5}{25}$, of being chosen.

The method has the further advantage of providing a readymade waiting list. If any family selected declines, next in line is Family U, then Family E, and so on. Thus we can not only select random samples of any size this way, but also randomly rank any number of cases. The more cases there are to rank, the greater the computer's advantages over manual selection methods like names in a hat.

Another possibility is randomly selecting not a certain number of cases, but a certain (approximate) fraction of the population. For example, we might want a random sample consisting of about 25% of the cases in Table 8.5. These pseudorandom numbers follow a uniform distribution, which means that all numbers in the interval 0 to 1 are equally likely to occur.[4] Therefore numbers less than .5 will occur about 50% of the time, less than .2 about 20%, and so forth. To select an approximately 25% random sample, we choose cases that have been assigned random numbers less than .25: Families R, A, H, L, P, U, and E. To select an approximately 90% random sample, we could choose those with random numbers less than .9. The actual sampling fractions obtained by these methods will often not be exactly what we aimed for (that is, not exactly 25%, 90%, or whatever), but for many purposes this does not matter.

The usual purpose of random sampling is to draw a sample that is representative of the population. As the day-care example illustrates, random sampling can be applied to other purposes as well. Although it may seem like an unlikely way to allocate day-care slots, this approach has actually been used.

The computer sampling just described is simple to carry out in practice. Some computer programs have built-in random-sampling functions, making the process of drawing random samples even easier. There are several potential pitfalls, however, in using computers for random sampling. These come from the fact that we are not using real random numbers, just pseudorandom numbers generated by an equation. Some equations produce much better pseudorandom numbers than others, with a longer cycle before numbers begin repeating. This is mainly of concern for advanced statistics and other applications where many thousands of numbers are needed.

A more basic problem is that all such equations begin with a *seed value*, the starting number from which the first pseudorandom number is calculated. Each calculated number then becomes the seed value for the number that follows it. If we start with the same initial seed value, we can produce exactly the same sequence of "random" numbers over and over again. In some programs the initial seed is always the same, unless the user specifically resets it. Other programs take the seed value from some aspect of the computer's continually changing internal state, so they are less likely to repeat a seed value. Resetting is better for sophisticated users who know what they want, whereas the more common internal state approach better protects naive users.

PROBLEMS

*8. The data in the accompanying table are the 1985–1986 value of endowments at 14 U.S. colleges and universities. For the purposes of this exercise, consider these 14 schools as a population from which to draw random samples.

Assign five-digit random numbers, beginning in the eighth "paragraph" of Table 8.1 (33778 43186 . . .), to each of the 14 schools. Draw up a new list, with the 14 schools in order by random number.

a. Apply the procedure shown in Tables 8.5 and 8.6 to select a random sample of $n = 3$.

**Market Value of the Endowments of 14 U.S. Colleges
and Universities, 1985–1986**

School	Endowment in millions of dollars
Brown University	315.4
Cornell University	711.7
Dartmouth College	520.6
Duke University	338.7
Emory University	731.8
Georgetown University	174.0
Princeton Theological Seminary	284.2
Rockefeller University	475.7
St. Louis University	120.5
Tulane University	182.1
University of Texas, Austin	244.6
Washington University	972.5
Wellesley College	265.0
Yale University	1,750.7

Source: The Information Please Almanac (1988), p. 823.

    b. If the sample in part a were expanded to size $n = 6$, which schools would be added?

    c. Describe how the approximation method could be used to select an approximately 70% random sample. Which schools are included? What actual percentage of the population does this sample include?

*9. If you are familiar with a computer program that has a random number function, do the following.

    a. Type in a small set of data, like those in this book.

    b. Assign random numbers to each case.

    c. Sort the cases according to their assigned random numbers.

    d. Which cases would you pick if you needed a sample of $n = 5$? How would you instruct the program to conduct an analysis using only those five cases?

    e. How could you instruct the computer to drop randomly about half the cases and analyze only the remaining half?

8.4   SAMPLE AND POPULATION: AN EXAMPLE

Random samples have been used several times in this book. The 30 statistics students in Table 1.1 are a random sample from a larger population of several hundred surveyed one semester. The 25 countries in Table 1.6 were randomly sampled from 142 countries. When original population data sets are too large to illustrate simple statistical calculations, random sampling brings them down to more manageable sizes. In other examples, random or nonrandom sampling was part of the original data collection. All of these sample data sets raise the issue of statistical inference.

The basic problem of statistical inference is uncertainty. When we are interested in a population of cases too large or otherwise inaccessible for direct study, we select a sample to study and base our conclusions about the population on our sample findings. Our conclusions are necessarily uncertain, because the sample is only a small part of the population. Choosing a second or third sample might get other results and still leave the true population parameters in doubt. Statistical theory helps us make the best possible guesses about the population. Theory also helps us assess how much uncertainty our guessing entails.

To make the concepts behind statistical inference more concrete, the remainder of this chapter will focus on a specific example. Each year, the FBI collects data on serious crimes in the United States, as part of its Uniform Crime Reporting Program. This includes data on many thousands of homicides, such as the age, race, and sex of each victim and offender, their relationship if any, the circumstances of the crime, and the weapons. We can use these records to define the following population:

The population consisting of all one-offender and one-victim homicides committed in America during 1980, for which the offender's age is known and recorded in the FBI's *Supplementary Homicide Report*.

This population consists of 11,877 homicides.

Data describing every case in this population are available.[5] It would be a straightforward job to perform analyses with a computer, and report results based on all 11,877 cases. Imagine, though, that we are typical researchers, who do *not* have data on the entire population of interest. We must therefore draw a sample, learning what we can about the full population by studying the few cases sampled.

We will focus on a simple question: What is the mean age of the offenders in this population of homicides? Actually, age is an important issue, for age groups differ in propensity to commit crimes. Criminologists have found that for many types of crime, the distribution of offenders' ages follows a consistent pattern. Most homicides are committed by men between the ages of 18 and 40, for instance. Knowledge about age distributions and trends can therefore help predict future crime rates, as people in crime-prone age groups make up a larger or smaller fraction of the general population.[6]

The intricacies of crime rate forecasting begin with our own basic task, estimating the mean age of homicide offenders. If data on the entire population of offenders are not available, then the next-best choice—and the only option in much actual research—is to work with a random sample.

Here is one way to select a simple random sample from the population list of 11,877 homicides:

1. Assign every case on the list a five-digit identification number, from 00001 to 11877.
2. Use a table of random numbers to select cases at random, until we have a sample of the size we want.

In a random number table each five-digit combination, from 00001 to 11877, should occur with the same probability. A simple random sample can therefore be selected by reading five-digit numbers from a table like Table 8.1.

The first five-digit random number in Table 8.1 is 24679. There is no homicide case with identification number 24679; the highest is 11877, so we ignore this first random number and proceed to the next, 24655. We continue doing so until the twelfth random number, 05212: the first one that corresponds to one of the identification numbers in our population. The 5,212th homicide is therefore the first one selected for the sample. Case 5212 describes an incident in Texas, where a 31-year-old man killed a 33-year-old man with a handgun.

The second usable five-digit number in Table 8.1 is the thirteenth overall, 02348, so the 2,348th homicide becomes the second one selected. Case 02348 describes an incident in Louisiana, where a 15-year-old boy killed a 13-year-old boy with a handgun. We proceed in this fashion until we have obtained a large enough sample; for our purposes here, a sample of 20 cases ($n = 20$) suffices. Figure 8.3 includes a list of offenders' ages for this sample, with population case numbers corresponding to the first 20 five-digit random numbers between 00001 and 11877, inclusive, in Table 8.1.

Sample case #	Age	Population case #
1	31	05212
2	15	02348
3	42	03328
4	35	07491
5	21	08425
6	49	06653
7	52	08168
8	36	06494
9	36	10086
10	26	04191
11	26	05516
12	26	07466
13	23	07117
14	33	06563
15	30	05360
16	28	09937
17	21	10230
18	40	09330
19	47	02007
20	23	11059

STEM-AND-LEAF DISPLAY

```
1.  | 5
2*  | 1133      stems 10's
2.  | 6668      leaves 1's
3*  | 013
3.  | 566       5* | 2
4*  | 02        means 52 years
4.  | 79        old
5*  | 2
```

SUMMARY STATISTICS

# of cases: $n = 20$
Sample mean: $\overline{X} = 32.00$
Std. dev.: $s_X = 10.11$

Median: $Md = 30.50$

FIGURE 8.3   *Distribution of ages for a random sample of 20 homicide offenders*

PROBLEMS

The accompanying table contains data on births to teenage mothers as a percentage of all births, for the 50 U.S. states. For purposes of this exercise we will view the 50 states as a population, and consider how we might learn about this population by drawing and analyzing a random sample.

Births to Mothers Aged < 20 as a Percentage of All U.S. Births, by State (1980)

	State	Teen births		State	Teen births
1	Alabama	20.6%	26	Montana	12.4%
2	Alaska	11.8	27	Nebraska	12.1
3	Arizona	16.5	28	Nevada	15.4
4	Arkansas	21.6	29	New Hampshire	10.7
5	California	13.9	30	New Jersey	12.3
6	Colorado	13.3	31	New Mexico	18.2
7	Connecticut	11.4	32	New York	11.8
8	Delaware	16.7	33	N. Carolina	19.2
9	Florida	18.2	34	N. Dakota	10.9
10	Georgia	20.7	35	Ohio	15.7
11	Hawaii	11.5	36	Oklahoma	19.6
12	Idaho	13.1	37	Oregon	13.3
13	Illinois	15.7	38	Pennsylvania	13.9
14	Indiana	17.3	39	Rhode Island	12.3
15	Iowa	12.5	40	S. Carolina	19.8
16	Kansas	15.0	41	S. Dakota	13.5
17	Kentucky	21.1	42	Tennessee	19.9
18	Louisiana	20.1	43	Texas	18.3
19	Maine	15.3	44	Utah	11.0
20	Maryland	14.8	45	Vermont	12.0
21	Massachusetts	10.7	46	Virginia	15.5
22	Michigan	14.0	47	Washington	12.5
23	Minnesota	10.4	48	W. Virginia	20.1
24	Mississippi	23.2	49	Wisconsin	12.3
25	Missouri	16.9	50	Wyoming	15.5

10. Use Table 8.1, starting in the upper left-hand corner and reading consecutive numbers, to construct a random sample consisting of six states. Find the mean and standard deviation of the percentage of births to teenage mothers in your sample. Sample *with replacement*; if a state is selected twice, count it twice in calculating the mean and standard deviation.

11. Explain the meaning of the following symbols, as they apply to Problem 10.

    a. $\overline{X}$

b. s_X

c. n

d. μ

e. σ_X

f. N

8.5 INFERENCES ABOUT THE POPULATION

Figure 8.3 includes summary statistics for our random sample of 20 homicides. Recall our original question, "What is the mean age of offenders in our population of 11,877 cases?" We now know the answer to a related but distinctly different question, namely, "What is the mean age of offenders in our random sample of 20 cases?" The answer to the latter question is 32 years, but the answer to the former question is "We still don't know." If we have no other information to go on, our *best guess* is that the population mean is the same as the sample mean, 32 years. It would be sheer coincidence were this actually true, however. Why should we expect the mean of all 11,877 cases in the population to be identical to the mean for the 20 cases randomly picked for our sample?

Had we picked a different sample, we would probably have obtained a different sample mean. This is easily demonstrated. The last random number we chose from Table 8.1 was 11059, found towards the end of the fifth "paragraph" of numbers. The next usable number is 04042, so the 4,042nd homicide can become the first case chosen for a *second* random sample of 20 cases. Case 04042 describes another handgun killing, this one in Oklahoma. The second case in our second sample is dictated by the next usable random number, 10864, and so on.

The sample resulting from this second round of selection is shown in Figure 8.4 (page 224).[7] The mean age in the second sample is indeed different from the first: 33.7 years. Figure 8.5 shows the distribution of ages from our first (Figure 8.3) and second (Figure 8.4) samples as back-to-back stem-and-leaf displays. Although both distributions are positively skewed, they differ in many details.

This example illustrates the problem of inference. We want to estimate an unknown population parameter, like the population mean μ, on the basis of a known sample statistic, like a sample mean $\overline{X}$. But even when the sample is chosen in an impeccably random fashion, we have no reason to believe that any one sample statistic will be identical to the corresponding population parameter. If we drew another random sample, we would most likely get a different sample statistic, as Figures 8.3–8.5 demonstrate. The differences between these two samples are an example of *sample-to-sample* variation. We would see more such variation if we went on to draw more samples.

Sample case #	Age	Population case #
1	26	04042
2	19	10864
3	22	05921
4	33	06369
5	31	01297
6	23	10273
7	35	09961
8	39	10364
9	30	06370
10	28	06045
11	63	02735
12	45	05101
13	32	03390
14	58	08936
15	21	08831
16	32	09442
17	18	01621
18	29	08310
19	38	04849
20	52	03178

STEM-AND-LEAF DISPLAY

```
1.  | 89
2*  | 123        stems 10's
2.  | 689        leaves 1's
3*  | 01223
3.  | 589        6* | 3
4*  |            means 63 years
4.  | 5          old
5*  | 2
5.  | 8
6*  | 3
```

SUMMARY STATISTICS

# of cases:	$n = 20$
Sample mean:	$\overline{X} = 33.70$
Std. dev.:	$s_X = 12.52$
Median:	$Md = 31.50$

FIGURE 8.4   *Distribution of offenders' ages, for a second random sample of 20 homicide offenders*

It is important to realize that *in most studies, the researcher has only a single sample from which to draw conclusions.* That is not so here; the example of offenders' ages was especially set up for purposes of illustration, so we can select as many samples as we want. We already have two; next we will see what happens if still more samples are taken.

SAMPLE #1 (see Figure 8.3)		SAMPLE #2 (see Figure 8.4)
5	1.	89
3311	2*	123
8666	2.	689
310	3*	01223
665	3.	589
20	4*	
97	4.	5
2	5*	2
	5.	8
	6*	3

Stems digits are 10's, leaves are 1's.

5* | 2 means 52 years old.

FIGURE 8.5   *Back-to-back stem-and-leaf displays for the ages of homicide offenders in two random samples from the same population*

PROBLEM

12. In Problem 10 we selected a random sample of six states from the population of 50 states in the table. In the same manner, use Table 8.1 to select two additional samples of six states each, and find the mean and standard deviation for each one. Sample with replacement.

    a. Begin with the second "paragraph" of random numbers in Table 8.1 (08425 65259 . . .). Use $\bar{X}_2$ and s_2 to designate the mean and standard deviation calculated for this second sample.

    b. Begin with the third "paragraph" of random numbers in Table 8.1 (12779 95960 . . .) and designate the resulting sample statistics $\bar{X}_3$ and s_3.

    c. Compare the statistics obtained from all three of your random samples. What, if anything, can you guess about the true values of the population parameters μ and σ_X?

8.6 A COMPUTER EXPERIMENT

Taking more than a few samples by hand is tedious, but computers are good at such laborious tasks. If we have population data, we can easily program a computer to keep taking random samples indefinitely, calculating and recording summary statistics for each sample before replacing the selected cases and choosing a new sample. This is not the usual way to do research, for if population data are available we might as well analyze them all and skip sampling. Repeated sampling, however, lets us look at the long-run, sample-to-sample variation of the sample means, thus clarifying some of the basic theoretical ideas behind statistical inference.

With enough computing power, we can draw and analyze any number of random samples from a population data set. Table 8.7 (page 226) shows the results from 800 such samples, drawn from the homicides data. Half consisted of 20 cases each, like the samples in Figures 8.3–8.5. The remaining 400 samples were larger, 100 cases each. For every sample a mean was calculated, producing a new data set of 400 sample means from the $n = 20$ samples, plus 400 means from the $n = 100$ samples. The frequency distributions for both sets of 400 means are given in Table 8.7, along with the overall *mean of the sample means*, $\bar{\bar{X}}$, obtained by adding up the 400 means for each sample size and dividing by 400. Also shown is the *standard deviation of the sample means*, $s_{\bar{X}}$. The two distributions are shown graphically as frequency polygons in Figure 8.6.

Frequency distributions in previous chapters gave the *distribution of a variable, across many cases*. In contrast, Table 8.7 and Figure 8.6 show us the *distribution of a statistic, across many samples*. The mean of an ordinary frequency distribution is the mean value of a *variable*, and the standard deviation

TABLE 8.7

Grouped Frequency Distribution of the Mean Ages of Homicide Offenders[a]

Mean age	Small samples[b] f	Larger samples[c] f
24–24.99	1	0
25–25.99	4	0
26–26.99	7	0
27–27.99	15	0
28–28.99	28	1
29–29.99	39	12
30–30.99	53	48
31–31.99	46	104
32–32.99	49	106
33–33.99	51	81
34–34.99	38	37
35–35.99	23	11
36–36.99	16	0
37–37.99	16	0
38–38.99	3	0
39–39.99	6	0
40–40.99	2	0
41–41.99	1	0
42–42.99	2	0
Totals:	400	400
Mean of the 400 sample means:	$\overline{\overline{X}} = 32.26$	$\overline{\overline{X}} = 32.38$
Standard deviation of the 400 sample means:	$s_{\overline{X}} = 3.07$	$s_{\overline{X}} = 1.33$

[a]Drawn from a population consisting of 11,877 U.S. homicides committed during 1980.
[b]Based on 400 random samples of 20 cases each.
[c]Based on 400 random samples of 100 cases each.

measures how much the variable varies, from case to case. The means of the distributions in Table 8.7 are the means of the *sample means,* and the standard deviations measure how much these sample means vary, from sample to sample.

Three main features can be seen in Figure 8.6:

1. Both distributions are somewhat bell-shaped, but the $n = 100$ distribution is more so. (In fact, the $n = 100$ distribution is close to a normal curve.) We might guess, correctly, that for still larger samples such as $n = 500$,

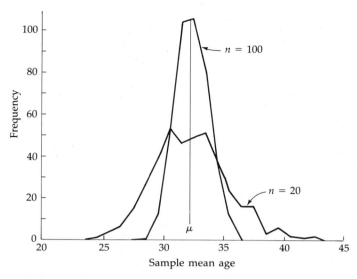

FIGURE 8.6   *Distribution of means: 400 samples of n = 20 and n = 100*

the distribution of means is even more bell-shaped, whereas for smaller samples, such as $n = 5$, it is more erratic. The relative symmetry of the polygons is also notable because the age distributions within individual samples tend to be positively skewed (see Figures 8.3–8.5). Evidently skewness in the original variable distributions does not necessarily skew these distributions of means.

2. The two distributions have almost the same centers, $\overline{\overline{X}} = 32.26$ for $n = 20$, and $\overline{\overline{X}} = 32.38$ for $n = 100$. Intuitively, it seems that the population mean must lie somewhere close to these centers. This is true: The actual population mean for all 11,877 homicide offenders is 32.28 years. A line representing this population mean, $\mu = 32.28$, is included in Figure 8.6.

3. The $n = 20$ distribution is more spread out than the $n = 100$ distribution. The means of the $n = 20$ samples range from about 24 to 43 years, with many samples well away from the overall mean of 32.26. The means of the $n = 100$ samples, on the other hand, range from only about 28 to 35, and most of them are within two years of the overall mean, 32.38. This is reflected in the smaller standard deviation, $s_{\overline{x}} = 1.33$ for the $n = 100$ samples, as compared with $s_{\overline{x}} = 3.07$ for the $n = 20$ samples. It is also reflected in the shape of the polygons in Figure 8.6, where the $n = 20$ distribution is flatter and broader, while the $n = 100$ distribution is narrower and more sharply peaked. Both distributions have about the same center, but the means from samples of $n = 100$ cluster more tightly around this center than do the means from samples of $n = 20$. *There is less sample-to-sample variation in means based on samples of 100 cases than in means based on samples of only 20 cases.*

These three features are not accidental, nor are they peculiar to one set of data. They are consistent with the *Central Limit Theorem*. The Central Limit Theorem describes what would happen if we obtained means from an infinite number of same-size samples (drawn from an infinitely large population), instead of just 400 same-size samples as shown in Table 8.7 and Figure 8.6. No one can actually calculate an infinite number of means, but mathematical theory predicts the distribution of their probabilities. The theoretical probability distribution of sample means, over infinitely many samples, is called the **sampling distribution** of the mean.

The **Central Limit Theorem** states that as sample size (n) becomes large

1. The sampling distribution of the mean becomes approximately normal, regardless of the shape of the variable's frequency distribution.
2. This sampling distribution will be centered around the population mean. That is, the mean of the sample means, $\overline{\overline{X}}$, will approach the population mean, μ.
3. The standard deviation of this sampling distribution, which is called its **standard error,** $\sigma_{\overline{X}}$, approaches the population standard deviation, σ_x, divided by the square root of the sample size:

$$\sigma_{\overline{X}} = \frac{\sigma_x}{\sqrt{n}}$$

Some of the symbols used in this section are summarized in Table 8.8.

TABLE 8.8   *Summary of Symbols Used in Chapter 8*

Symbol	Meaning
Statistics based on a single sample	
$\overline{X}$	Sample mean of variable X
s_x	Sample standard deviation of variable X
n	Number of cases in sample
$SE_{\overline{X}}$	Estimated standard error of sample mean
Population parameters	
μ	(mu) Population mean of variable X
σ_x	(sigma) Population standard deviation of variable X
N	Number of cases in population
$\sigma_{\overline{X}}$	True standard error of the sample mean
Statistics based on multiple samples	
$\overline{\overline{X}}$	Mean of sample means ($\overline{X}$) across numerous different random samples, each with n cases
$s_{\overline{X}}$	Standard deviation of sample means ($\overline{X}$) across numerous different random samples, each with n cases

The computer experiment results shown in Table 8.7 and Figure 8.6 demonstrate the Central Limit Theorem in action. Note, however, that the Central Limit Theorem refers specifically to what happens as sample size grows large. How large is large enough for this theorem to apply? We will take up this question in more detail in later chapters.

PROBLEMS

13. Use the means and standard deviations from your answers to Problems 10 and 12 to find the following statistics. Suggest an appropriate symbolic representation with each answer.
    a. The mean of the three sample means
    b. The standard deviation of the three sample means
    c. The mean of the three sample standard deviations
    d. The standard deviation of the three sample standard deviations

14. For the 50-state population of Problem 10, the mean percentage of births to teenage mothers is 15.31, with a standard deviation of 3.49. Use the appropriate symbols in answering the following questions.
    a. Is the mean of your three sample means in Problem 13, part a, the same as the population mean? Why does this difference not contradict the Central Limit Theorem?
    b. Apply the Central Limit Theorem to find the standard error of the mean for samples of size $n = 6$.
    c. According to the Central Limit Theorem, which of the statistics in Problem 13 would approach the standard error of the mean if we collected many random samples of large enough size?

15. The standard error of the mean is a measure of sample-to-sample variability in the sample mean. Other things being equal, how is sample-to-sample variability in the mean of X affected by
    a. The population standard deviation of the variable X?
    b. The number of cases in the samples?

8.7 SAMPLING DISTRIBUTIONS AND STANDARD ERRORS

From Table 8.7 and Figure 8.6, we can see that sample means have less sample-to-sample variation (corresponding to *a smaller standard error*) when the samples are larger. The Central Limit Theorem states that this *should* be true: If the standard errors are approximately equal to $\sigma_x/\sqrt{n}$, then the larger the sample size (n), the smaller the standard error must be.

The Central Limit Theorem thus provides mathematical support for our intuitive understanding that estimates based on larger samples are "better." They are better because larger samples have less sample-to-sample variation, so *the mean in any one large sample is less likely to be far from the population mean.* Unlike our computer experiment, in most real studies the researcher has only one sample, so this is an important point.

Having only one sample to work with, how can a researcher know anything about sampling distributions or standard errors? These concepts pertain to "infinitely many samples," and the standard error is defined in terms of the population standard deviation, σ_x—which like other population parameters is usually unknown. The answer is twofold:

1. The general shape of the sampling distribution may be known on the basis of statistical theory. The Central Limit Theorem tells us that the sampling distribution for the mean will be approximately normal, *given large enough samples.*
2. The standard error of this sampling distribution may be estimated using sample data.

To estimate the standard error of the mean, we substitute the sample standard deviation, s_x, for the population standard deviation, σ_x.

The **estimated standard error** of the mean, **SE$_{\bar{x}}$**, is

$$SE_{\bar{x}} = \frac{s_x}{\sqrt{n}} \qquad\qquad [8.1]$$

where s_x is the sample standard deviation and n is sample size.

For example, based on our first random sample in Figure 8.3, we estimate the standard error to be

$$SE_{\bar{x}} = \frac{s_x}{\sqrt{n}}$$

$$= \frac{10.11}{\sqrt{20}} = 2.26$$

To get the theoretical standard error, we apply the Central Limit Theorem to the population standard deviation for all 11,877 ages in this population, $\sigma_x = 13.004$:

$$\sigma_{\bar{x}} = \frac{\sigma_x}{\sqrt{n}}$$

$$= \frac{13.004}{\sqrt{20}} = 2.91$$

Our estimated standard error, 2.26, does not match the theoretical value, 2.91, bearing out our earlier observation about means: Estimates based on any one sample are unlikely to equal the population parameters. Still, one sample is often all we have to go on. The practical consequences of using the sample standard deviation to estimate the standard error show up mainly in small samples; we will examine them in Chapter 11.

In summary, a sampling distribution is the probability distribution of a sample statistic. That is, it is a distribution of the expected proportion of times possible values of the statistic would occur, if we could calculate an infinite number of statistics from an infinite number of same-size samples. Our knowledge about sampling distributions is usually based on theory, such as the Central Limit Theorem.[8]

PROBLEM

16. The standard deviation for the teenage birth percentage in the population consisting of 50 U.S. states (Problem 10) is $\sigma_X = 3.49$. If the Central Limit Theorem applies, then for samples of size $n = 6$ the true standard error of the mean should be $\sigma_{\overline{X}} = \sigma_X/\sqrt{n} = 3.49/\sqrt{6} = 1.42$. Suppose we do not know the true standard error, but must estimate it from sample data.

   a. Estimate the standard error of $\overline{X}$ based on a random sample of $n = 6$, in which the standard deviation is $s_X = 4.37$ (Problem 10).

   b. Estimate the standard error of $\overline{X}$ based on a random sample of $n = 6$, in which the standard deviation is $s_X = 2.96$ (Problem 12, part a).

   c. Estimate the standard error of $\overline{X}$ based on a random sample of $n = 6$, in which the standard deviation is $s_X = 1.85$ (Problem 12, part b).

   d. The three standard error estimates obtained from the samples of parts a–c differ widely from each other and from the true standard error, $\sigma_{\overline{X}} = 1.42$. How might these differences be explained?

8.8　THE USES OF STANDARD ERRORS

The standard error of the mean is the standard deviation of the mean's sampling distribution. Standard errors, like standard deviations, describe how spread out a distribution is. A high standard error says the mean varies a lot from sample to sample; we thus cannot be confident that the mean of any one sample is near the population mean. A second or third sample's results might differ widely. On the other hand, a low standard error indicates comparatively little variation from sample to sample. When the standard error is low, the mean from a single sample is more likely to be near the true population mean.

One use of standard errors is to compare different statistics. In the homicide example, the sample mean based on 20 cases had a higher standard error than the sample mean for 100 cases. The $n = 100$ mean is in this sense a better estimator of the population mean μ. Different kinds of statistics can also be compared in this manner. For example, it is possible to show that with normally distributed variables, the sample mean has a lower standard error than the sample median. The sample mean is therefore said to be **more efficient** than the median as an estimator of μ—at least when variable distributions are normal.

A more common use of the standard error is as a measure of **distance.** As seen in Chapter 7, we can convert statements about distances, measured in standard deviations or standard errors, into statements about normal distribution probabilities. In the chapters that follow, we will encounter several uses for such statements about probability.

PROBLEM

17. A civic group wants to estimate the mean years of schooling for all registered voters in their city. They use voting registration lists to pick a simple random sample of $n = 5$, then interview these five voters to collect data. Among the sample of five voters, the mean education is found to be $\overline{X} = 11.1$ years.

    a. Should we therefore believe that the population mean is $\mu = 11.1$? Use the concept of standard error to explain.

    b. What would provide a more convincing estimate? Refer to the Central Limit Theorem and to standard errors in your answer.

SUMMARY

This chapter began with a discussion of random numbers, which are used for selecting samples. Random selection ensures that a sample is representative of the larger population. The theories of statistical inference, which guide the process of generalizing from sample to population, all assume that sampling is random. Usually these theories rely specifically on the simple random sample.

The relationship between random sampling and sampling distributions was demonstrated with a computer experiment, repeatedly drawing random samples of cases from a population of 11,877 homicides and calculating the mean age of the offenders in each sample. The actual distribution of the 400 means for small samples ($n = 20$) and 400 for larger samples ($n = 100$) illustrates the theoretical idea of a sampling distribution. The sampling distribution of the mean is the theoretical distribution of sample means from infinitely many same-size random samples.

According to the Central Limit Theorem, for large enough samples the sampling distribution of the mean is normal and is centered on the true population mean, μ. Its variation around this center is measured by the standard error of the mean, $\sigma_{\bar{x}}$, which equals the population standard deviation divided by the square root of the sample size. When the population standard deviation is unknown, the standard error can be estimated by substituting the sample standard deviation in place of σ_x.

The computer experiment allowed us to see the Central Limit Theorem in action. The distribution of sample means from 400 smaller samples ($n = 20$) was visibly less normal and more spread out (that is, had a larger standard error) than the distribution of means from 400 larger samples ($n = 100$; Figure 8.6). Since the means varied more widely in smaller samples, we have less confidence that any one small-sample mean will be close to the population parameter.

The next two chapters examine statistical applications of normal distributions, which are models for the sampling distribution of the mean and other statistics. These distributions allow us to convert knowledge about distances, measured in standard errors, into statements of probability about how much uncertainty we face in generalizing from sample statistics to population parameters.

PROBLEMS

18. Describe and illustrate how Table 8.1 could be used to do the following.
    a. Simulate the experiment of flipping a coin ten times
    b. Simulate the experiment of drawing one of 30 names from a hat
    c. Randomly choose one page in this book
    d. Grade a stack of student papers by randomly assigning an A, B, C, D, or F

The table on pages 234–236 contains a data set in which the cases are 118 college classes. The variables are class size (number of students); percentage of these students receiving A or B as their final grade in the class; percentage of the students who rated the instructor highly on end-of-semester teaching evaluations; and the instructor's status as full-time faculty, or part-time instructor, or teaching assistant. The data were originally collected to examine whether teaching evaluations were higher in courses where the instructor gave out a higher proportion of good grades.

In Problems 19–22 treat these 118 cases as if they were a population, too large to study directly.

Data on 118 College Classes

Class	Size	A–B grades	High teacher evaluations	Teacher status[a]
1	96	75%	48%	1
2	60	45	61	1
3	86	41	43	1
4	58	53	51	0
5	29	66	61	0
6	43	74	73	1
7	75	43	56	1
8	53	68	9	0
9	63	70	23	1
10	42	60	72	0
11	51	67	68	0
12	33	52	48	0
13	57	56	31	0
14	24	51	26	0
15	16	69	93	1
16	26	65	11	0
17	21	76	95	1
18	36	46	65	1
19	44	34	51	1
20	17	65	77	1
21	56	39	58	1
22	36	91	68	0
23	86	61	70	1
24	16	69	50	0
25	38	66	28	0
26	43	63	61	1
27	52	29	38	1
28	44	57	60	1
29	72	63	39	1
30	38	53	65	1
31	89	62	52	1
32	95	38	61	1
33	53	47	51	0
34	27	37	9	0
35	97	44	73	1
36	38	63	50	0
37	89	54	67	0
38	51	67	19	0
39	39	64	32	0
40	37	51	23	1
41	43	93	81	1
42	17	52	21	1
43	62	51	31	1
44	53	75	32	0

(*continued*)

(continued)

Class	Size	A–B grades	High teacher evaluations	Teacher status[a]
45	55	67%	32%	0
46	24	66	70	0
47	33	54	37	1
48	32	56	25	1
49	28	35	78	1
50	60	36	73	1
51	91	60	0	0
52	58	60	5	0
53	123	60	16	0
54	67	79	55	1
55	68	44	35	1
56	69	68	82	1
57	24	58	100	1
58	164	68	84	1
59	87	66	30	1
60	51	41	50	1
61	82	32	33	1
62	39	33	19	1
63	35	57	56	1
64	23	83	95	1
65	36	64	45	0
66	55	42	5	1
67	47	68	90	0
68	74	50	47	1
69	77	52	60	1
70	14	57	61	0
71	96	37	77	1
72	59	51	62	1
73	25	56	33	0
74	57	86	98	0
75	53	60	36	1
76	33	70	42	1
77	59	58	33	0
78	34	62	33	0
79	44	48	33	0
80	49	39	33	0
81	19	53	33	1
82	38	74	46	1
83	31	77	17	0
84	17	65	44	1
85	31	32	13	0
86	32	53	88	1
87	80	33	45	1
88	85	61	45	1

(continued)

Data on 118 College Classes (continued)

Class	Size	A–B grades	High teacher evaluations	Teacher status[a]
89	114	67%	15%	0
90	132	66	18	0
91	90	59	55	1
92	54	35	29	1
93	15	47	73	1
94	15	67	36	0
95	15	40	25	1
96	85	47	47	1
97	52	83	74	0
98	54	35	37	1
99	66	30	20	1
100	91	59	45	1
101	22	55	78	0
102	42	50	71	1
103	75	40	56	1
104	52	73	83	0
105	48	48	29	1
106	32	41	17	1
107	50	80	81	1
108	66	79	28	0
109	58	76	37	0
110	32	69	65	0
111	32	50	53	0
112	33	67	58	1
113	35	69	58	0
114	40	42	6	1
115	24	50	87	1
116	62	18	61	1
117	30	67	89	0
118	30	73	44	1

[a]1 = full-time faculty, 0 = part-time instructor or teaching assistant.
Source: Hamilton (1980).

19. Use the random number table (Table 8.1) to select a simple random sample of $n = 10$ from the population. You will need three-digit random numbers between 001 and 118 to do this. Read the table left to right, using only the first three digits in each group of five, so that your first few numbers are 246, 246, 759, etc. The first usable random number is 118, so class 118 becomes the first one selected for your sample. Continue reading until ten cases have been selected. For the ten cases in your sample, calculate:

a. The sample mean ($\overline{X}$) and sample standard deviation (s_X) of class size.

b. The sample median (Md) of class size.

c. The sample proportion ($\hat{\pi}$) of the classes taught by full-time faculty (1).

d. Your best estimate (based only on this one sample) of the population mean (μ) and standard deviation (σ_X) of class size.

20. Use your own random numbers (for example from Problem 2), to select a second random sample of $n = 10$ from the table. Cases selected for the first sample may be chosen again for the second, but do not include the same case twice *within* one of your samples. For the ten cases in your second sample, calculate:

a. The sample mean ($\overline{X}$) and sample standard deviation (s_X) of class size.

b. The sample median (*Md*) of class size.

c. The sample proportion ($\hat{\pi}$) of the classes taught by full-time faculty (1).

d. Your best estimate (based only on this second sample) of the population mean (μ) and standard deviation (σ_X) of class size.

21. As a class exercise, you can now carry out an investigation along the same lines as the computer experiment described in this chapter. Construct a new data set consisting of all the sample means, standard deviations, medians, and percentages obtained by each student for Problem 20. Graph the distributions of these four sample statistics across all the samples, and also calculate the means of the sample statistics. The actual "population" parameters for all 118 cases are

Mean class size: $\mu = 51.74$
Standard deviation of class size: $\sigma_X = 27.06$
Median class size: 48.5
Proportion taught by full-time faculty: $\pi = .6017$

Compare these population parameters with

a. The statistics you obtained in your own single sample.

b. The distributions and means of the statistics from all samples combined.

22. Other things being equal, statistics based on larger samples allow more precise estimation of population parameters. With this in mind, how do you think the appearance of the distributions in Problem 21 would have changed if

a. Each student had constructed samples of $n = 15$ instead of $n = 10$?

b. Each student had constructed samples of $n = 5$ instead of $n = 10$?

23. (*Review question on univariate analysis*) Construct a five-stem stem-and-leaf display (stems of 0*, 0t, 0f, 0s, 0., representing 100's) for the distribution of the percentage of white household heads in the 26 housing projects of Table 8.3. Next construct a similar display for the percentage of black household heads. Describe what you see.

NOTES

1. One supposedly random generator failed dramatically in 1969, when the U.S. government held a national lottery to determine the order in which to draft young men to meet military demands during the Vietnam War. The televised selection procedure involved drawing capsules containing birthdays from a rotating drum. Draft-age men whose birthdays were drawn early would be among the first drafted. Men with birthdays drawn later would be placed farther down the list, so were increasingly unlikely to be drafted at all.

   A subsequent statistical analysis suggested that capsules from each month of the year had been added to the urn *in order,* and due to insufficient mixing the later months remained near the top—where they were more likely to be drawn first. Young men born later in the year were therefore more likely to be drafted.

2. Even simple home computers can generate pseudorandom numbers. Video games often rely on pseudorandom number generators to make their challenges less predictable. Many scientific hand calculators also feature pseudorandom number generators.

3. Could this distribution arise from a random process where each digit has a probability of .10? We will examine this question formally in Chapter 12. Many other issues might be investigated in a thorough test of randomness. We could also test whether all possible two-digit sequences are equally likely, then test three-digit sequences, etc. Beyond two digits, such analysis would require exorbitant computer time. It is also possible to look for longer cycles of repeating digits, using a method called **spectral analysis.**

   Several microcomputer random number generators are examined by Modianos, Scott, and Cornwell (1987) in a comparatively nontechnical discussion of how such generators may be studied and appraised.

4. The numbers are shown here to just seven decimal places, but the computer actually calculates them to 16 decimal places. Since there are almost 10^{16} (one with 16 zeroes after it) 16-digit random numbers between 0 and 1, each one should occur with a probability of about $1/10^{16}$.

5. These data were assembled by Kirk R. Williams and Murray A. Straus, for their project entitled "Justifiable and Criminal Homicide Among Family Members, Acquaintances, and Strangers," funded by the National Institute of Justice (85IJCX0030).

6. For research projecting future crime rates from the age distribution of criminals see Cohen and Land (1987).

7. Only 19 five-digit numbers remaining in Table 8.1 are between 00001 and 11877 inclusive. To pick the twentieth case for this second sample required generating further numbers. The first usable number so generated was 03178, so 3,178 became the last case selected for the sample in Figure 8.4.

8. Recent years have seen increased use of computer experiments as a method of learning about sampling distributions that, unlike the sampling distribution for the mean, cannot be derived from theory alone. Such experiments are called *Monte Carlo research,* after the famous gambling resort in Monaco.

Inference Using the Normal Distribution

Chapter 9

In this chapter we will start putting theoretical ideas about sampling distributions and their standard errors to work on the practical business of data analysis. *Normal distributions* are a key to this application. Normal distribution probabilities help both with *variables* that have approximately normal *frequency distributions* and, more broadly, with *statistics* that have approximately normal *sampling distributions*.

The Central Limit Theorem (Chapter 8) states that for large samples the sampling distribution of the mean is approximately normal, with a mean equal to the population mean (μ) of the variable and a standard error equal to its population standard deviation (σ_x) divided by the square root of the sample size (n): $\sigma_{\bar{x}} = \sigma_x/\sqrt{n}$. The sampling distribution is what we would theoretically obtain if we took an infinite number of same-size samples and calculated a mean for each one. Researchers cannot take an infinite number of samples; in fact, they usually have only one. The Central Limit Theorem lets them estimate the probability that their one sample mean is more than a certain distance away from the true population mean.

Starting with the definition and properties of a normal distribution, we will move on to their use. Two immediate applications are to simple summary statistics for measurement and categorical variables: the sample mean and proportion, respectively.

9.1 *DEFINITION OF THE NORMAL DISTRIBUTION*

A normal distribution is a type of symmetrical, unimodal, bell-shaped frequency distribution—but not all symmetrical, unimodal and bell-shaped distributions are normal. The term *normal distribution* is unfortunate, because the word *normal* seems to imply that variables with such distributions are usual and typical. Not so: in the many data sets examined so far, relatively few variable distributions were even symmetrical, let alone truly normal. *"Normal" does not mean "typical," nor is "normal distribution" a synonym for "bell-shaped distribution."* In some kinds of data normal distributions may well be typical, and as commonly drawn all normal distributions are bell-shaped, but neither trait defines a normal distribution.[1]

Theoretical probability distributions for continuous variables are defined by equations called *probability density functions.* When graphed, such functions show a smooth curve; areas under these curves correspond to probabilities. One probability density function defines a normal distribution, as described in the box.

If the variable X has a normal distribution with mean μ and standard deviation σ_X, the **probability density function** of X is

$$f(X) = \frac{1}{\sigma_X\sqrt{2\pi}} e^{-(X-\mu)^2/(2\sigma_X^2)} \qquad [9.1]$$

Graphing $f(X)$ against X produces the familiar shape known as the **normal curve.**

π (pi) and e are mathematical constants, approximately equal to 3.14 and 2.72, respectively.[2] Substituting these approximate values into Equation [9.1] yields

$$f(X) \approx \frac{1}{\sigma_X\sqrt{6.28}} 2.72^{-(X-\mu)^2/(2\sigma_X^2)} \qquad [9.2]$$

This substitution makes it clearer that a normal distribution has only two parameters: the population mean (μ) and standard deviation (σ_X). Consequently:

1. A different normal distribution exists for every possible combination of mean and standard deviation. Thus the normal distribution is not a single distribution, but an infinite family of distributions.
2. If we know the mean and standard deviation of a normal distribution, we can find out everything else about it. It is possible to calculate the probability of X values falling above, below, or between any given points.

Fortunately we need not make these calculations each time we encounter a normal distribution problem. The basic work has already been done, and results are available in tables like Table A.1 in the Appendix.

A **normal distribution** is a type of symmetrical, unimodal, bell-shaped frequency distribution—but not all symmetrical, unimodal, and bell-shaped distributions are normal. A normal distribution is defined by Equation [9.1].

Normal distributions are an ideal, like perfect circles, which we do not usually expect to find in nature. Smooth and symmetrical, they extend from positive infinity to negative infinity regardless of the mean and standard deviation. Like perfect circles, normal distributions are useful as simple models for many natural phenomena.

A variable whose values are the sum of many small random influences should have an approximately normal distribution. For example, the random digits in Table 8.1 are arranged in lines of 50 digits each. Adding up all 50 digits in each line would create new variables, *sums* of the original random digits. The 50-digit sum for the first line is

$$2+4+6+7+9+2+4+6+5+5+\cdots+1+5+0+5+5 = 211$$

The second line of Table 8.1 also sums to 211; the third line adds up to 235, and so forth. The 36 sums for the 36 lines of random digits from Table 8.1 are shown in Table 9.1 (page 244).

Over the long run, such sums of random digits should be approximately normally distributed. A grouped frequency distribution for the 36 sums of Table 9.1 is given in Table 9.2 and as a frequency polygon in Figure 9.1. Thirty-six sums is not a long run, but the distribution of sums already looks quite different from the nearly rectangular distribution of individual random digits (compare Figure 8.2, page 211).

If we continue the experiment of adding up 50-digit sequences of random numbers, their distribution will more and more resemble the normal curve. Figure 9.2 shows a frequency polygon for the distribution of 1,000 such 50-digit sums, obtained by generating 964 more lines of random numbers beyond the 36 lines in Table 8.1. Comparing Figure 9.2 (1,000 sums) with Figure 9.1 (36 sums), you can see that the distribution is indeed becoming more like a normal curve.

TABLE 9.1   **Sums of the 50 Random Digits per Line of Table 8.1**

Line	Sum	Line	Sum
1	211	19	211
2	211	20	231
3	235	21	212
4	224	22	234
5	241	23	250
6	235	24	224
7	276	25	232
8	223	26	209
9	233	27	242
10	241	28	238
11	239	29	265
12	243	30	201
13	232	31	239
14	200	32	228
15	233	33	227
16	234	34	192
17	220	35	261
18	276	36	236

Although individual random digits are theoretically drawn from a rectangular distribution with a mean of 4.5, sums of 50 digits each should tend toward the bell shape of a normal curve with a mean of 225 (50 × 4.5). It resembles the way the sampling distribution of the mean becomes approximately normal, regardless of how the original variables are distributed. In fact, the mean is just a special case of the general principle that sums of a large number of independent random values tend to be normally distributed.

TABLE 9.2   **Grouped Frequency Distribution for the Sums in Table 9.1**

Sum	Frequency	Proportion
190–199	1	.03
200–209	3	.08
210–219	4	.11
220–229	6	.17
230–239	13	.36
240–249	4	.11
250–259	1	.03
260–269	2	.06
270–279	2	.06
Totals:	36	1.0

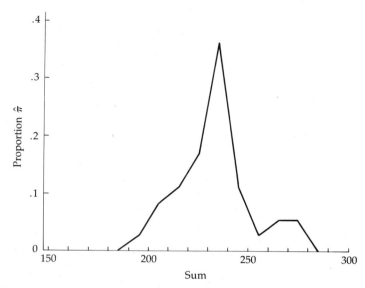

FIGURE 9.1   *Frequency polygon from grouped frequency distribution (Table 9.2)*

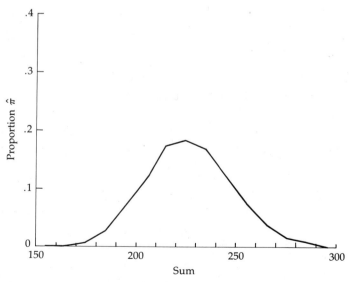

FIGURE 9.2   *Frequency polygon from grouped frequency distribution of 1,000 sums*

Recall that the mean itself is defined as a sum: the sum of all values of a variable, divided by a constant (the number of cases).

All measurement contains errors, whether they be charting errors in telescopic observations of stars or sampling errors in selecting people for a survey. If these errors result from many small random influences acting together, then the **distribution of errors** will likely be approximately normal. This is a major reason, besides its mathematical tractability, that the normal distribution is important to statistical theory.

PROBLEMS

1. Review the material on numerical summaries and distributional shape in Chapter 4.

   a. What simple tests involving mean-based and order statistics were suggested as a check on whether a variable's distribution is approximately normal?

   b. Apply these tests to the distribution of 36 sums in Figure 9.1. The summary statistics are:
   $$\overline{X} = 231.6 \qquad Md = 233$$
   $$s_X = 19.2 \qquad IQR = 18.5$$

   c. Apply these tests to the distribution of 1,000 sums in Figure 9.2. The summary statistics are:
   $$\overline{X} = 225.5 \qquad Md = 225$$
   $$s_X = 20.9 \qquad IQR = 29$$

2. Sums of 50 random digits have a theoretical "population" distribution that is approximately normal, with a mean of $\mu = 225$. How do our samples of $n = 36$ and $n = 1,000$ sums (Problem 1, parts b and c) compare with this theoretical distribution?

3. Use a calculator, computer program, or other device to produce your own table of 200 random digits. Write down these digits in groups of five, as done in Table 8.1.

   a. Construct a stem-and-leaf display to show the distribution of the 200 digits. Comment on the observed and expected shape of this distribution.

   b. Find the sums of the five digits in each of the 40 five-digit groups. Construct a new stem-and-leaf display to show the distribution of the 40 sums. Comment on the shape of this distribution, and explain why it differs from that seen in part a.

   c. What is the theoretical population mean for individual random digits? For sums of five random digits?

4. Use your own example to illustrate the distinction between the sample distribution of a variable and the sampling distribution of a statistic.

9.2 *PARAMETERS OF THE NORMAL CURVE*

To clarify the two parameters of a normal distribution, μ (*mu*, or population mean) and σ_X (*sigma*, or population standard deviation), look at their effects on the appearance of the normal curve illustrated in Figure 9.3. The mean, $\mu = 500$, is the center of symmetry and divides the curve into mirror-image halves. The standard deviation, $\sigma_X = 75$, is the distance from the mean to the **point of inflection,** the point where the curve's slope goes from convex (bulging outward) to concave (bulging inward). In Figure 9.3 the points of inflection are at $\mu + \sigma_X = 500 + 75 = 575$, and $\mu - \sigma_X = 500 - 75 = 425$.

The **normal distribution notation:**

$$X \sim N(\mu, \sigma_X)$$

is read "X is distributed normally with mean μ and standard deviation σ_X." If $\mu = 500$ and $\sigma_X = 75$, for instance, we could express this as

$$X \sim N(500, 75)$$

There is a different normal distribution for every combination of mean and standard deviation. Figure 9.4 (page 248) shows two normal curves with different means ($\mu_1 = 500$ and $\mu_2 = 650$), but identical standard deviations ($\sigma_1 = \sigma_2 = 75$). The only difference between the two curves is in the location of their centers.

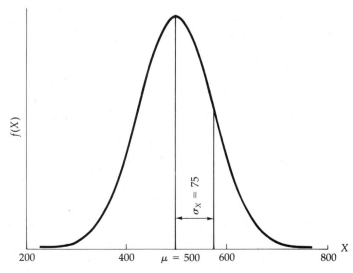

FIGURE 9.3   *Normal distribution with $\mu = 500$ and $\sigma_X = 75$, $X \sim N(500, 75)$*

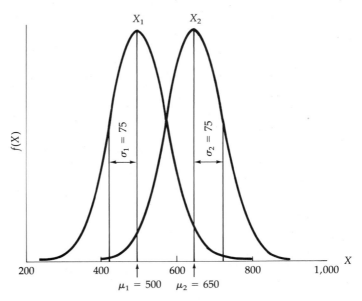

FIGURE 9.4 *Two normal distributions:* $X_1 \sim N(500, 75)$; $X_2 \sim N(650, 75)$

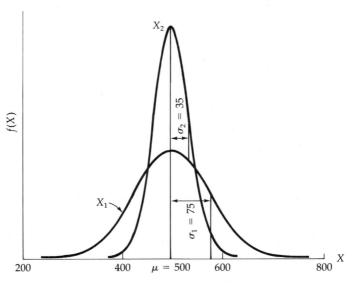

FIGURE 9.5 *Two normal distributions:* $X_1 \sim N(500, 75)$;
$X_2 \sim N(500, 35)$

Figure 9.5 shows two normal distributions with identical means, ($\mu_1 = \mu_2 = 500$), but different standard deviations ($\sigma_1 = 75$, $\sigma_2 = 35$). The distribution with the larger standard deviation is more spread out.

The third possibility is shown in Figure 9.6: a pair of normal distributions with different means and different standard deviations. The populations differ in both center and spread.

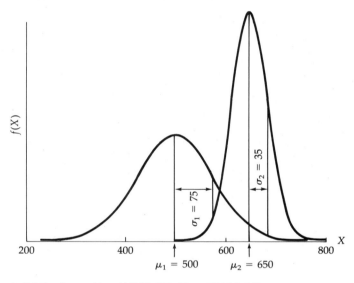

FIGURE 9.6   *Two normal distributions: $X_1 \sim N(500, 75)$; $X_2 \sim N(650, 35)$*

> A different normal distribution exists for every combination of mean (μ) and standard deviation (σ).
> The mean controls the *location* of the distribution, whereas the standard deviation controls its *spread*.

PROBLEMS

5. Sketch normal curves to represent the following trios of normal distributions. Label each mean and standard deviation on your figures.

   a. $\mu_1 > \mu_2 > \mu_3$, $\sigma_1 = \sigma_2 = \sigma_3$
   b. $\mu_1 = \mu_2 = \mu_3$, $\sigma_1 > \sigma_2 > \sigma_3$
   c. $\mu_1 < \mu_2 < \mu_3$, $\sigma_1 < \sigma_2 < \sigma_3$
   d. $\mu_1 > \mu_2 = \mu_3$, $\sigma_1 = \sigma_2 > \sigma_3$

6. Standardized academic achievement test scores often have approximately normal distributions. What can you guess about the means and standard deviations of achievement test scores among high school seniors in the following types of classes?

1. Advanced: seniors who are academically above average
2. Remedial: seniors who are academically below average
3. General: seniors chosen randomly, without regard to academic achievements

Use the symbols μ and σ, subscripted by a (advanced), g (general), or r (remedial) to summarize your answer.

9.3   *THE STANDARD NORMAL DISTRIBUTION*

Any variable can be transformed to have a mean of 0 and a standard deviation of 1. A normally distributed variable with $\mu = 0$ and $\sigma = 1$ is called a **standard normal** variable.

A normally distributed variable X is transformed into a **standard normal variable** by the **Z-score transformation:**

$$Z = \frac{X - \mu}{\sigma_X} \qquad\qquad [9.3]$$

If $X \sim N(\mu, \sigma_X)$, then Z will have a **standard normal distribution,**

$$Z \sim N(0, 1)$$

If $X \sim N(500, 75)$, for example, then the Z-score transformation is $Z = (X - 500)/75$. An X value of 575, expressed as a Z-score, is $(575 - 500)/75 = 1$, indicating that 575 is 1 standard deviation above the mean. Standard normal values are interpreted as *distance from the mean, in standard deviations.*

Figure 9.7 depicts the standard normal distribution graphically. Since the distribution is centered on $\mu = 0$, values below the mean are represented by negative Z-scores, while values above the mean are positive. A Z-score tells us exactly how many standard deviations above or below the mean that value lies. If $Z = 2.3$, we are 2.3 standard deviations above the mean; if $Z = -.5$, we are .5 standard deviation below the mean.

As its name implies, the *reverse Z-score transformation* undoes the Z-score transformation.

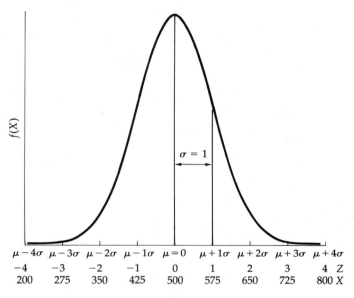

$\mu - 4\sigma$ $\mu - 3\sigma$ $\mu - 2\sigma$ $\mu - 1\sigma$ $\mu = 0$ $\mu + 1\sigma$ $\mu + 2\sigma$ $\mu + 3\sigma$ $\mu + 4\sigma$

| −4 | −3 | −2 | −1 | 0 | 1 | 2 | 3 | 4 Z |
| 200 | 275 | 350 | 425 | 500 | 575 | 650 | 725 | 800 X |

FIGURE 9.7   *The standard normal distribution:* $Z \sim N(0, 1)$

Z-scores can be expressed in terms of the original units of X by applying the **reverse Z-score transformation:**

$$X = \mu + Z\sigma_X \qquad\qquad\qquad [9.4]$$

If $X \sim N(500, 75)$, for example, then a Z-score of 2.3 corresponds to an X value of

$$X = \mu + Z\sigma_X$$
$$= 500 + 2.3(75) = 672.5$$

That is, 672.5 is the value 2.3 standard deviations above the mean in an $X \sim N(500, 75)$ distribution. Similarly a Z-score of $-.5$ corresponds to an X value of

$$X = \mu + Z\sigma_X$$
$$= 500 - .5(75) = 462.5$$

which tells us that the value 462.5 is .5 standard deviation below the mean. The correspondence between Z and X distributions is shown graphically in Figure 9.7.

PROBLEMS

7. Convert the following X values to Z-scores, and write out how the Z-score value is interpreted.
   a. $x = 800$, $X \sim N(500, 75)$
   b. $x = 275$, $X \sim N(500, 75)$
   c. $x = 22$, $X \sim N(17, 2)$
   d. $x = 1,240$, $X \sim N(2,000, 1,500)$
   e. $x = .8$, $X \sim N(.6, .2)$
   f. $x = -24$, $X \sim N(-6, 9)$

8. Return the following Z-scores to the original units of X.
   a. $z = 3$, $X \sim N(500, 75)$
   b. $z = -1$, $X \sim N(500, 75)$
   c. $z = 2.367$, $X \sim N(17, 2)$
   d. $z = .05$, $X \sim N(2,000, 1,500)$
   e. $z = -2.5$, $X \sim N(.6, .2)$
   f. $z = -1$, $X \sim N(-6, 9)$

9.4 USING TABLES OF THE STANDARD NORMAL DISTRIBUTION

A table of **standard normal probabilities** is found in the Appendix (Table A.1). Standard normal values (Z-scores) appear with their first two digits down the left-hand column, and their third digit (second decimal) across the top of the table. A Z-score of 2.34, for instance, is found by reading down the left-hand side to the row for 2.3, then horizontally to the .04 column. The value in the cell for 2.34 is .4904, which is *the probability of Z values between 0 and 2.34*. A graph at the top of Table A.1 illustrates the idea that all the probabilities in the table refer to the area between the mean (0) and some positive Z value, in the right-hand side of the standard normal distribution. Knowing that the distribution is symmetrical and its total area equals 1 lets us calculate other probabilities from the right-side values in Table A.1.

Since two-sided probabilities are just their one-sided counterparts doubled, we can readily find the area that is within ± 2.34 of the mean, above or below it. If

$$P(0 < Z < 2.34) = .4904$$

and

$$P(Z > 2.34) = .5 - .4904 = .0096$$

then

$$P(|Z| < 2.34) = 2(.4904) = .9808$$

and

$$P(|Z| > 2.34) = 1 - .9808 = .0192$$

Sketching and shading areas under the curve may help clarify the problem. A wide variety of normal curve problems can be solved in this manner.

Table A.1 readily locates probabilities for any given Z values. It also works—though awkwardly—in reverse: Look up a given probability in the body of the table, glance at the margins, and you find the corresponding Z value. Certain round-number probabilities (such as .95, .05, and .01) that are used frequently in statistics are presented more accessibly in Table A.2, an alternative view of the standard normal distribution.

Table A.2 lists **critical values** for the standard normal distribution. Its first three columns refer to three types of application, illustrated in the shaded drawings across the top: **confidence intervals, two-sided tests,** and **one-sided tests.** In this chapter we will see how to use the confidence interval column; tests are discussed in Chapter 10. Values in these three columns are probabilities, corresponding to the normal-curve areas shaded in the drawings above. Thus, the left-hand column gives the probability of values (in other words, the proportion of the total area) between $+z$ and $-z$, where z is any value of a variable with a standard normal distribution. The second column gives probabilities of values greater than $+z$ and less than $-z$. The third column gives probabilities of values greater than $+z$ only. Since the Z distribution is symmetrical, the probability of values greater than $+z$ is the same as the probability of values less than $-z$, so the third column really describes either tail of the distribution. The fourth column gives the cutoff or *critical values* of Z.

In the first row of this table, $z = .126$, or .126 standard deviation from the mean. Looking at the first column, we can see that the probability of values between .126 standard deviation below the mean ($z = -.126$) and .126 standard deviation above the mean ($z = +.126$) is .10 or 10%. That is, 10% of the cases in a normal distribution should fall within .126 standard deviation on either side of the mean. The second column shows the probability of falling *outside* of this range, or being more than .126 standard deviation from the mean. If .10 of the area is shaded in the first drawing, then the remainder, or .90, must be shaded in the second. All the way down the table, the first and second columns add up to 1.00. The third column, which refers to a single tail of the distribution, contains single-tail probabilities that are always equal to one-half of their two-tailed counterparts. Thus if .90 of the area is in both tails, as it is for $z = .126$, then each tail alone must contain .45 of the area.

Standard normal distribution tables like Table A.1 and A.2 are useful because they allow us to work back and forth between variable values and probabilities. For example, suppose an achievement test has an N(500, 75) distribution. The tables could help us answer questions such as the following.

Susan scored 612 on this test. How good is that? To find out, first convert 612 to a Z-score:

$$z = \frac{x - \mu}{\sigma_X}$$

$$= \frac{612 - 500}{75} = 1.49$$

Susan is 1.49 standard deviations above average. Table A.1 shows that the proportion between the mean and +1.49 standard deviations is .4319. The proportion below the mean in a normal distribution is .5, so the proportion below +1.49 must be .5 + .4319 = .9319. Susan's score of 612 is better than .9319 (93.19%) of the scores in an N(500, 75) population (Figure 9.8).

A school board wants to identify the most exceptional 40% of students—those scoring either in the top 20% or the bottom 20% on this exam. What scores should be used as cutoffs? Because we are starting with the probabilities, it is easier to answer from Table A.2 than A.1. A two-tailed probability of .40 corresponds to a z value of .842. The most exceptional 40% of the students will be those who are more than .842 standard deviation from the mean. To put this in terms of test scores, apply the reverse Z-score transformation. The upper cutoff point is

$$x = \mu + z\sigma_X$$

$$= 500 + .842(75) = 563.15$$

and the lower cutoff point is

$$x = \mu + z\sigma_X$$

$$= 500 + (-.842)(75) = 436.85$$

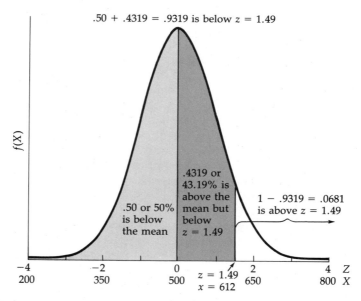

FIGURE 9.8   *Probability of X < 612 or Z < 1.49 if X ~ N(500, 75)*

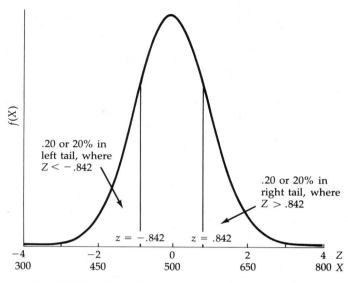

FIGURE 9.9   *Value of z such that P(|Z| > z) = .40*

Students scoring in the top 20% will be those with scores above 563.15; students scoring in the bottom 20% will be those with scores below 436.85 (Figure 9.9).

As these examples show, standard normal distribution tables work with normally distributed variables of any mean and standard deviation. We can look up a probability, given a value of Z or X; conversely, we can look up a Z or X value, given a probability. These probabilities apply to frequency distributions only if the variable has a normal population distribution, however—which many variables lack. Normal distribution probabilities apply more generally to sampling distributions, which may be normal even when variable distributions are not.

PROBLEMS

9. Table A.2 does not give a critical value of Z for an 82% confidence interval. Use Table A.1 to find this critical value and to fill out this table as if it were a row in Table A.2:

Confidence interval	Two-sided tests	One-sided tests	Critical value
.82			

10. Add further rows to the table in Problem 9 for each of the following.
    a. A critical value of .44

　　b.　A one-sided test at $P = .03$

　　c.　A two-sided test at $P = .61$

　　d.　A confidence interval at $P = .97$

11.　In Chapter 4 we encountered a measure of spread called the pseudo-standard deviation, or PSD, defined as the interquartile range divided by a seemingly arbitrary constant, 1.35. Actually this value of 1.35 was rounded off from 1.348. You are now in a position to investigate where the constant 1.35 (or 1.348) came from.

　　a.　Recall that the interquartile range is defined as the distance that spans the middle 50% of the cases. Use Table A.2 to determine the width *in standard deviations* of a normal distribution IQR.

　　b.　Using your answer to part a, write an equation for finding *the standard deviation of a normal distribution, if you know its IQR.*

　　c.　Based on your answer to part b, explain why the comparison of PSD and σ_X provides a way to check on a distribution's normality.

9.5　Z-SCORES AND NONNORMAL DISTRIBUTIONS*

The term Z-score has another context, unrelated to the standard normal distribution. It is possible to apply a version of the Z-score transformation to change *any* variable—not just those with normal distributions—so that its mean is 0 and standard deviation is 1.

For any variable X, with a sample mean $\overline{X}$ and a sample standard deviation s_X, we can calculate a **standard score** variable, X^*,

$$X^* = \frac{X - \overline{X}}{s_X} \qquad\qquad [9.5]$$

whose sample mean equals 0 ($\overline{X}^* = 0$) and sample standard deviation equals 1 ($s_{X^*} = 1$).

Equation [9.5] is the same as the Z-score transformation, Equation [9.3], except the sample mean and sample standard deviation replace their population counterparts. Standard scores are sometimes called Z-scores, but this invites confusion with the standard normal distribution.

Standard scores are convenient when variables have no natural units, or when we want to compare variables measured in different units or with different standard deviations. For example, in the 30-student sample of Table 1.1, the last student has a GPA of 3.4 and a verbal SAT score of 560. The mean GPA among these students is 2.74, so her GPA as a standard score is $X^* = (3.4 - 2.74)/.44 = 1.5$. She is 1.5 standard deviations above average (for this sample) in GPA. The mean SAT is 490, so her SAT as a standard score is $X^* =$

(560 − 490)/78 = .9, or .9 standard deviation above average. Standard scores clarify that this student is *farther above average in GPA* (1.5 standard deviations) *than in SAT score* (.9 standard deviation). They give us a way to "compare apples and oranges," by expressing both variables in standard deviations from their respective means.

One hazard of standard-score transformations is that inexperienced analysts mistake them for a way to make variable distributions normal. In fact, they cannot make a variable's distribution any more or less normal than it already was before transformation. Chapter 6 described some *nonlinear transformations* that can change distributional shape, such as the square, square root, and logarithm. Used appropriately they may make a skewed distribution more symmetrical and hence, we hope, more normal. But the standard-

TABLE 9.3   *Ordered List of 1982 per Capita GNP in 24 Countries, as Raw Data (U.S. Dollars) and Standard Scores (Standard Deviations)*

	Country	Raw GNP X	Standard score GNP[a] X^*
1	Ethiopia	150	−.746
2	Zaire	180	−.741
3	Sri Lanka	320	−.717
4	Guinea	330	−.715
5	Togo	350	−.712
6	Mauritania	520	−.683
7	Indonesia	550	−.678
8	North Korea	930	−.613
9	Mongolia	940	−.612
10	Nigeria	940	−.612
11	Nicaragua	950	−.610
12	Congo	1,420	−.530
13	Paraguay	1,670	−.488
14	Jordan	1,680	−.486
15	Taiwan	2,670	−.318
16	Greece	4,170	−.063
17	Venezuela	4,250	−.049
18	Czechoslovakia	5,540	.170
19	Austria	9,830	.899
20	Australia	11,220	1.135
21	France	11,520	1.186
22	West Germany	12,280	1.315
23	Norway	14,300	1.658
24	Brunei	22,260	3.011
Mean:		$\overline{X} = 4,540.4$	$\overline{X}^* = 0.000$
Standard deviation:		$s_X = 5,885.3$	$s_{X^*} = 1.000$

[a]$X^* = (X − 4,540.4)/5,885.3$

score transformation is an example of a **linear transformation.** *Linear transformations do not change distributional shape at all.*[3]

This property can be illustrated with the same Gross National Product data used earlier to demonstrate a nonlinear transformation (see Figure 6.6, page 160). In Chapter 6 we saw that the distribution of per capita GNP is positively skewed, but a logarithmic transformation reduces this skewness. Table 9.3 contains an ordered list of these data both raw and as standard scores. Standard scores are obtained using the sample mean ($\overline{X} = 4{,}540.4$) and standard deviation ($s_X = 5{,}885.3$):

$$X^* = \frac{X - \overline{X}}{s_X}$$

$$= \frac{X - 4{,}540.4}{5{,}885.3}$$

As expected, the standard scores have a mean of 0 and a standard deviation of 1.

Figure 9.10 shows the distribution of this variable. Before transformation, the raw data distribution is acutely skewed, with a mean ($\$4{,}540.4$) that is $\$2{,}995.4$ higher than the median ($\$1{,}545$). After transformation, the standard scores distribution remains equally skewed. The mean standard score (0) is

RAW DATA: *X*   STANDARD SCORE DATA: *X**

$$X^* = \frac{X - \overline{X}}{s_X} = \frac{X - 4{,}540.4}{5{,}885.3}$$

0*	00000000000111244	−0	77777666666544300
0.	59	0	18
1*	1124	1	1136
1.		2	
2*	2	3	0

Stems digit: $10,000	1 standard deviation
Leaves digit: $1,000	.1 standard deviation
2* | 2 means $22,000	3 | 0 means 3.0 std.dev.

SUMMARY STATISTICS

$Md = 1{,}545$	$Md^* = -.509$
$\overline{X} = 4{,}540.4$	$\overline{X}^* = 0.000$
$s_X = 5{,}885.3$	$s_{X^*} = 1.000$

FIGURE 9.10   *Per capita 1982 GNP of 24 countries, in raw data (dollars) and standard scores (standard deviations), from Table 9.3*

.509 standard deviation higher than the median standard score ($-.509$). A difference of .509 standard deviation is (within rounding error) the same as the $2,995.4 mean–median difference in the raw data:[4]

$$.509s_X = .509(5,885.3) = 2,995.6$$

The two stem-and-leaf displays show identical shapes before and after transformation.

PROBLEMS

*12. Table 9.3 shows that Nicaragua's per capita GNP, expressed as a standard score, is $-.610$. What does this value tell us?

Data are shown for the average annual water flow of the White River near Nora, Wisconsin, 1960–1971 (Wright et al., 1986). Studying the historical distribution of water flow helps predict the frequency of floods or droughts, which accompany unusually high or low flows.

Year	Average flow (cu ft/sec)
1960	845
1961	1,109
1962	1,027
1963	751
1964	1,066
1965	833
1966	441
1967	1,059
1968	1,167
1969	1,139
1970	1,117
1971	770

*13. a. Find the mean and standard deviation of the average annual White River flow. Use these statistics to express each year's flow as a standard score.

b. Interpret the standard scores of river flow for the years 1966 and 1969.

c. Find the mean and standard deviation of the standard scores. Are they as expected?

d. Obtain the mean of the standard scores by applying the standard-score transformation directly to the mean of the raw data distribution.

*14. a. Use the raw data to find the median White River flow. Compare this with the mean found in Problem 13, part a, and comment on the shape of this distribution.

b. Find the median White River flow from the standard score units. Compare this with the mean found in Problem 13, part c, and comment on distributional shape.

c. Show that the mean–median difference in the raw data (part a) is the same as the mean–median difference in the standard score data (part b). What does this tell us?

9.6 *CONFIDENCE INTERVALS FOR MEANS*

In 1984, an opinion survey was sent to a random sample of registered voters in a Vermont town where water supplies had been contaminated by chemical wastes. One goal of the survey was to find out how opinions about the contamination problems were affected by background variables such as age or education.[5] Respondents constitute a sample, from which we might draw inferences about the town's adult population.

The mean education among 155 survey respondents is $\overline{X} = 12.97$ years. What can we conclude about the mean education (μ) of the town's adult population as a whole? Probably the population mean is not the same as the sample mean, 12.97 years—it would be a great coincidence if it were. But it seems reasonable to guess that the population mean is not too far from 12.97. Statistical theory allows us to make our guess more specific:

Based on the information in this sample, we are 95% confident that the population mean education is between 12.59 and 13.35 years:

$$12.59 \le \mu \le 13.35$$

This is an example of a *95% confidence interval.*

Confidence intervals are ranges which we believe contain the population parameter.

Given a large enough sample, a **95% confidence interval for the population mean** may be constructed:

$$\overline{X} \pm 1.96(SE_{\overline{X}}) = \overline{X} \pm 1.96 \left(\frac{s_X}{\sqrt{n}}\right) \qquad\qquad [9.6]$$

where s_X is the sample standard deviation and n is sample size.

For the Vermont survey example, we have a sample mean of $\overline{X} = 12.97$, a sample standard deviation of $s_X = 2.42$, and a sample size of $n = 155$. The 95% confidence interval is therefore

$$\overline{X} \pm 1.96 \left(\frac{s_X}{\sqrt{n}}\right) = 12.97 \pm 1.96 \left(\frac{2.42}{\sqrt{155}}\right)$$

$$= 12.97 \pm .38$$

or from $12.97 - .38 = 12.59$ to $12.97 + .38 = 13.35$.

We do not claim a .95 probability that μ lies in this particular interval— either it does or it does not. But if a large number of random samples were taken, and confidence intervals were constructed using each $\overline{X}$ and s_X according to Equation [9.6], then theoretically about 95% of these intervals should actually include μ. Table 9.4 summarizes the confidence interval's construction and interpretation.

Whether a measurement variable is normally distributed or not, according to the Central Limit Theorem the sampling distribution of its mean should be approximately normal and centered on μ. (This approximation is better for large samples than for small ones.) If $\overline{X}$ is distributed normally around a mean of μ, then 95% of the time random sampling will result in $\overline{X}$ values within plus or minus 1.96 standard errors of μ—because 95% of the

TABLE 9.4

Finding and Interpreting an Approximate 95% Confidence Interval for the Mean Years of Education of Voters in a Small Vermont Town

Sample statistics

Mean	$\overline{X} = 12.97$ years
Standard deviation	$s_X = 2.42$ years
Number of cases	$n = 155$

Calculation of confidence interval for population mean (μ)
The confidence interval is the sample statistic $\pm\ z$(estimated standard error of sample statistic). That is,

$$\overline{X} \pm z(\mathrm{SE}_{\overline{X}}) = \overline{X} \pm z\left(\frac{s_X}{\sqrt{n}}\right)$$

$$= 12.97 \pm 1.96\left(\frac{2.42}{\sqrt{155}}\right)$$

$$= 12.97 \pm .38$$

so the interval is $12.59 \le \mu \le 13.35$.

Interpretation of this confidence interval

Informal: Based on our analysis of this particular sample, we are about 95% confident that the mean education among all voters in this town lies between 12.59 and 13.35 years.

Formal: If we took a large number of random samples, each with 155 cases, and calculated confidence intervals in this manner for each sample, about 95% of those confidence intervals should include the true population mean μ.

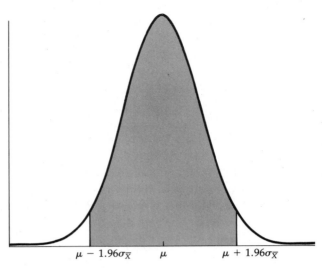

FIGURE 9.11   95% of sample means are within $\mu \pm 1.96\sigma_{\bar{x}}$

area under the normal curve is within $\mu \pm 1.96\sigma_{\bar{x}}$ (Figure 9.11). Consequently if we marked off intervals of $+1.96\sigma_{\bar{x}}$ around each sample mean, 95% of these intervals should in fact contain μ. We do not know the actual standard error, $\sigma_{\bar{x}}$, but with large enough samples we can estimate it as $SE_{\bar{x}} = s_x/\sqrt{n}$. This reasoning leads to Equation [9.6].

Appendix Table A.2 indicates that $z = 1.96$ is appropriate for a 95% confidence interval. Other intervals are constructed by using other z values. For example a 99% confidence interval requires $z = 2.576$, so the interval would be $\bar{X} \pm 2.576(SE_{\bar{x}})$. A 99.9% confidence interval is $\bar{X} \pm 3.29(SE_{\bar{x}})$.

Given a large enough sample, **any confidence interval for the population mean** may be constructed:

$$\bar{X} \pm z(SE_{\bar{x}}) = \bar{X} \pm z\left(\frac{s_x}{\sqrt{n}}\right) \qquad\qquad [9.7]$$

where z is chosen from a standard normal distribution table (Table A.2) to obtain the desired degree of confidence.

Equations [9.6] and [9.7] are specific instances of a more general formula for large-sample confidence intervals:

Sample statistic $\pm z$(Standard error of sample statistic)   [9.8]

We will later see that confidence intervals for some other statistics also follow this formula.

PROBLEMS

15. Table 9.4 showed how to find and interpret the 95% confidence interval for a mean. Using these same data,
    a. Find the 90% confidence interval for the mean years of education.
    b. Find the 99% confidence interval.
    c. How do the 90, 95, and 99% confidence intervals compare with each other? Which interval is *narrowest*? Over the long run, which type of interval *most often will contain the true parameter*?

16. Cigarettes have been found to contain measurable amounts of pesticides and heavy metals (Mussalo-Rauhamaa et al., 1986). For example, a European study of 44 different brands reported a mean cadmium content of 1.4 µg/g, with a standard deviation of .4. Assume that the 44 brands tested represent a random sample from the population of all cigarette brands.
    a. Estimate the standard error of the mean cadmium content.
    b. Construct a 95% confidence interval.
    c. Construct a 99.9% confidence interval.
    d. Describe and explain the differences between your intervals from parts b and c.

17. A study of microcomputer use in a random sample of elementary schools around the city of Chicago found wide differences between schools, but in most schools computer use was high and rapidly expanding (McGee, 1987).
    a. In a sample of 110 schools, the mean weekly microcomputer use was 39.1 hours per week, with a standard deviation of 15.7 hours per week. Find the 95% confidence interval for the mean weekly microcomputer use. Write both an informal and a formal interpretation.
    b. In the same 110 schools, the mean percentage of students who used the computers regularly was 58%, with a standard deviation of 22.6%. Find the 99% confidence interval for the mean percentage of students using computers regularly. Write both an informal and a formal interpretation of this interval.

9.7 AN ILLUSTRATION WITH REPEATED SAMPLING*

The formal interpretation of confidence intervals is awkward to put in real-world terms. It refers to a "large number of random samples," which in most research is purely hypothetical. The computer experiment in Chapter 8, however, involved hundreds of random samples, which gives us an opportunity to explore the meaning of confidence intervals.

Table 9.5 summarizes results from 400 small ($n = 20$) and 400 larger ($n = 100$) random samples. The homicide offenders in the first sample of $n = 20$ (Figure 8.3, page 221) had a mean age of 32, with a standard deviation of 10.1; in the second sample (Figure 8.4, page 224), the mean age was 33.7 and the standard deviation was 12.5. Means and standard deviations were calculated for each of the 800 samples in this experiment, enabling us to construct 800 confidence intervals. Table 9.5 shows a few examples.

Being in the unusual position of already knowing the population mean ($\mu = 32.28$ years), we can observe which of the 800 confidence intervals actually contain this value. For example, the intervals constructed from the first two samples in Table 9.5 do contain $\mu = 32.28$, but the interval constructed from the third sample does not. Figure 9.12 shows this situation graphically, for the same four small samples listed in Table 9.5.

Of the 400 confidence intervals constructed from the small samples, 363 or 90.75% actually contain the true value of μ. Among the 400 larger samples, 374 or 93.5% of the intervals contain μ. Both percentages are below the 95% accuracy we would expect, if our assumptions about the sampling distribution and standard error of X were correct. A clue to this discrepancy is the fact that our accuracy is noticeably better with samples of $n = 100$ than with smaller samples of $n = 20$.

The Central Limit Theorem states that $\overline{X}$ has an approximately normal distribution *as the sample size becomes large.* Furthermore, we are using an *estimated* standard error rather than the true standard error, $\sigma_{\overline{X}}$. The accuracy of such estimates also depends on having a large enough sample size. Just

TABLE 9.5   *Approximate 95% Confidence Intervals from 800 Random Samples[a]*

Sample number	Mean $\overline{X}$	St. dev. s_X	# of cases n	Confidence interval $\overline{X} \pm z(s_X/\sqrt{n})$	Contains $\mu = 32.28$[b]
1	32.0	10.1	20	32.0 ± 4.43	Yes
2	33.7	12.5	20	33.7 ± 5.48	Yes
3	28.5	6.8	20	28.5 ± 2.98	No
⋮	⋮	⋮	⋮	⋮	⋮
400	30.1	11.9	20	30.1 ± 5.22	Yes
1	29.8	11.2	100	29.8 ± 2.20	No
2	32.7	14.2	100	32.7 ± 2.78	Yes
3	30.8	11.8	100	30.8 ± 2.31	Yes
⋮	⋮	⋮	⋮	⋮	⋮
400	32.6	13.8	100	32.6 ± 2.70	Yes

[a]The Z value used to obtain 95% confidence intervals is $z = 1.96$. The failure to produce intervals that actually contain μ about 95% of the time, especially with $n = 20$ samples, suggests that these samples may be *too small* for the normal approximation to apply.
[b]Small-sample ($n = 20$) totals: 363 yes (90.75%),   37 no (9.25%)
Larger-sample ($n = 100$) totals: 374 yes (93.5%),   26 no (6.5%)

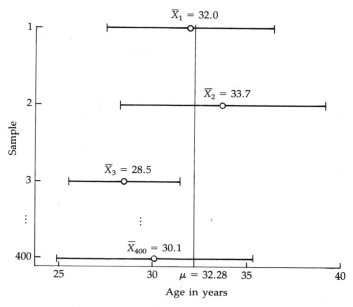

FIGURE 9.12   *Confidence intervals from four samples (size n = 20) in Table 9.5*

how large is large enough has not yet been addressed, but apparently the normal approximation with $\sigma_{\bar{x}}$ estimated by $s_x/\sqrt{n}$ does not work well with samples of only 20 cases. Chapter 10 gives more formal guidelines for deciding when a sample is "large enough" for the normal approximation to be appropriate.

9.8   CONFIDENCE INTERVALS FOR PROPORTIONS AND PERCENTAGES

If you follow news reports during political campaigns, you may notice that opinion poll results are often stated something like this:

"In our survey of 600 registered voters, we found that 54% favored candidate Smith. The margin of error in this survey is plus or minus 4%."

The *margin of error* implies that in the population of all voters (not just these 600), support for Smith could be anywhere between 50% and 58%. This margin of error is in fact a 95% confidence interval—too technical-sounding a term for network TV.

With large samples, confidence intervals for proportions (or percentages) are formed in much the same way as they are for means. Like means, sample proportions ($\hat{\pi}$) follow normal sampling distributions centered on the corresponding population parameters (π). The theoretical standard error of this sampling distribution is

$$\sigma_{\hat{\pi}} = \sqrt{\frac{\pi(1 - \pi)}{n}}$$

To form confidence intervals, we estimate this standard error by using $\hat{\pi}$ in place of π.

For large-sample confidence intervals, the **standard error of a sample proportion** is estimated by

$$SE_{\hat{\pi}} = \sqrt{\frac{\hat{\pi}(1 - \hat{\pi})}{n}} \qquad [9.9]$$

where $\hat{\pi}$ is the sample proportion and n is sample size.

The confidence interval itself is then obtained by adding and subtracting z standard errors.

Large-sample **confidence intervals** for proportions are found as

$$\hat{\pi} \pm z(SE_{\hat{\pi}}) = \hat{\pi} \pm z\sqrt{\frac{\hat{\pi}(1 - \hat{\pi})}{n}} \qquad [9.10]$$

where z is chosen from a table of the standard normal distribution (Table A.2) to give the desired degree of confidence.

For example, a 95% confidence interval requires $z = 1.96$. In the survey regarding candidate Smith, we have $\hat{\pi} = .54$ and $n = 600$, so the 95% interval is

$$\hat{\pi} \pm z\sqrt{\frac{\hat{\pi}(1 - \hat{\pi})}{n}} = .54 \pm 1.96\sqrt{\frac{.54(1 - .54)}{600}}$$

$$= .54 \pm .04$$

To express as percentages, multiply proportions by 100: 54% $\pm$ 4%. This is how the "margin of error" of plus or minus 4% was obtained for Smith's poll.

Table 9.6 shows another example of a confidence interval for a proportion, using the Vermont survey summarized in Table 9.4. The town was shocked to discover that toxic wastes from a local industry had contaminated underground water supplies as well as the air, water, and soil around the town's two public schools. Alarmed, some townspeople argued that the safety of children was paramount and that schools should be closed at once, until they

TABLE 9.6	***Finding and Interpreting an Approximate 95% Confidence Interval for the Proportion Favoring School Closings in a Small Vermont Town***

Sample statistics

Proportion favoring schools closed	$\hat{\pi} = .431$
Number of cases	$n = 153$

Calculation of confidence interval for population proportion (π)

The confidence interval is the sample statistic $\pm$ z(estimated standard error of sample statistic). That is,

$$\hat{\pi} \pm z(SE_{\hat{\pi}}) = \hat{\pi} \pm z\sqrt{\frac{\hat{\pi}(1 - \hat{\pi})}{n}}$$

$$= .431 \pm 1.96\sqrt{\frac{.431(1 - .431)}{153}}$$

$$= .431 \pm .078$$

so the interval is $.353 \le \pi \le .509$.

Interpretation of this confidence interval

Informal: Based on our analysis of this one sample we are about 95% confident that the proportion in favor of closing the schools, among all voters in this town, lies between .353 and .509.

Formal: If we took a large number of random samples, each with 153 cases, and calculated confidence intervals in this manner for each sample, about 95% of those confidence intervals should include the true population proportion π.

Source: Based on a random sample survey of registered voters living in this town (Hamilton, 1985).

could be proven safe. Other people balked at this costly step, suggesting that the dangers had been exaggerated. Among those surveyed, 66 out of 153, or 43.1% ($\hat{\pi} = .431$) favored closing the schools. The remainder thought that the schools should stay open.[6]

The calculations in Table 9.6 lead to the interval

$$.353 \le \pi \le .509$$

We can be 95% confident that the proportion of town residents in favor of school closing is between .353 and .509. To form a confidence interval for a percentage, calculate the interval for the corresponding proportion and multiply by 100. If we are about 95% confident that $.353 \le \pi \le .509$, we are also 95% confident that $35.3 \le p \le 50.9$.

PROBLEMS

18. In a New Hampshire community where the water supplies were contaminated by toxic wastes, pollsters contacted 232 adult residents. Of these 232 residents, 32 said they were still willing to drink the water. Find the 95% confidence interval for the proportion of residents in the community who might say they were still willing to drink the water. Explain what this interval means.

19. In spring of 1985, a Gallup poll of 1,021 American adults found that 63% believed America's involvement in the Vietnam War was a mistake. Construct and interpret a 95% confidence interval for the population proportion holding this belief.

20. A confidence interval's width is its upper limit minus its lower limit.
    a. What would be the width of your interval for Problem 19 (Vietnam War) if the poll had surveyed only 50 people, instead of 1,021?
    b. What would be the width if 50,000 people had been surveyed?
    c. What generalization does this suggest to you about the relationship between sample size and the width of a confidence interval?

21. Researchers studying the spread of Acquired Immune Deficiency Syndrome (AIDS) collected data from a volunteer sample of 4,955 homosexual men in Baltimore, Chicago, Los Angeles, and Pittsburgh (Chmiel et al., 1987). Of these 4,955 men, 38% were found to be infected with the HIV virus believed to cause AIDS.
    a. Construct a 99.9% confidence interval for the percentage infected with HIV. Interpret the interval *as if* these 4,955 men were a random sample of adult male homosexuals.
    b. Since the men in the sample volunteered, rather than being chosen at random, this is clearly *not* a random sample. How does this affect the interpretation of our confidence interval in part a?

9.9   *USING CONFIDENCE INTERVALS TO TEST HYPOTHESES*

Confidence intervals serve many purposes. **Hypothesis testing** is one important application. Hypotheses are *statements about the true values of population parameters.* For example, one local observer of the Vermont town's toxic waste controversy stated that the town was evenly split on school closing and that the average voter had only a high school education. These assertions can be phrased as two formal hypotheses:

1. The mean education level of the town's registered voters is 12.0 years, or $\mu = 12.0$.
2. The proportion of the town's voters favoring school closing is $\pi = .50$.

Note that each hypothesis is put in terms of a population parameter (μ and π) set equal to some specific value.

According to the confidence interval calculations in Table 9.4, we are 95% confident that the mean education of voters in the town is between 12.59 and 13.35 years. The hypothesized value, $\mu = 12.0$, does not lie in this interval. Therefore we can be *at least 95% confident that the mean education is not* 12.0. With this degree of confidence, we may *reject* (disbelieve) the hypothesis that $\mu = 12.0$.

More formally, we reason as follows. Suppose we decide to reject any hypothesis claiming that the population mean μ lies outside of the confidence interval we calculated. If 95% confidence intervals were calculated for many (large) samples, then 95% of these intervals should include the true value of the population mean μ. We would be correct about 95% of the time, over the long run, in rejecting hypothesized μ values outside of our interval. Our "95% confidence" that we are right, based on our single sample, derives from the theoretical knowledge that in the long run decisions made in this way will be correct about 95% of the time.

Our calculations in Table 9.6 indicate that we can be 95% confident that the proportion of town voters in favor of closing the schools is between .353 and .509. The hypothesis that they are equally split implies a population proportion, $\pi = .5$, that *does* lie within the confidence interval. We therefore should *not reject* the hypothesis that $\pi = .5$; we are not 95% confident that this hypothesis is false.

> We **can reject a hypothesis** (with a given level of confidence) if the hypothesized parameter value lies *outside of the confidence interval*.
>
> Conversely, we **cannot reject a hypothesis** (with a given level of confidence) if the hypothesized parameter value lies *within the confidence interval*.

Once a confidence interval has been calculated, that interval can be used with no further computation to make decisions about the plausibility of hypotheses.

The next two chapters examine hypothesis testing in greater detail. Returning to the question of "large enough" sample size, we shall see what adjustments permit inferences about mean or proportion when our samples are small.

PROBLEMS

22. Use confidence intervals to decide whether the following hypotheses can be rejected with 95% confidence.

  a. The hypothesis that the majority of Americans do not believe that the Vietnam War was a mistake (Problem 19)

  b. The hypothesis that the mean percentage of students using microcomputers in Chicago-area schools is 60% (Problem 17)

  c. The hypothesis that the mean cadmium content for all brands of cigarettes is below 1.0 μg/g (Problem 16)

23. Health inspectors in a northeastern state found that in a sample of 1,557 shellfish being sold for human consumption, 14.5% contained unsafe levels of pollutants.

  a. Assuming that these 1,557 shellfish were a random sample of those currently being sold in the state, construct a 99% confidence interval for the proportion of shellfish that are contaminated. Interpret this interval in your own words.

  b. Based on your answer to part a, consider the following hypothesis: In the population of all shellfish being sold in this state, only 10% are contaminated. Explain your decision to reject or not reject this hypothesis.

SUMMARY

The sampling distributions for many sample statistics are approximately normal. For example, the large-sample sampling distribution of the sample mean, $\overline{X}$, is approximately normal with mean μ and standard deviation (standard error) $\sigma_{\overline{X}} = \sigma_X/\sqrt{n}$. Similarly the large-sample sampling distribution of the sample proportion, $\hat{\pi}$, is approximately normal with mean π and standard deviation (standard error) $\sigma_{\hat{\pi}} = \sqrt{\pi(1 - \pi)/n}$.

A normal distribution is defined by a specific probability density function, Equation [9.1], which when graphed produces the smooth normal curve. Some properties of the normal curve are:

1. It is symmetrical around its center, μ (mean).
2. The standard deviation, σ_X, measures how spread out the curve is.
3. The curve extends from positive to negative infinity, regardless of the values of μ and σ_X.
4. Probability corresponds to area under the curve. The total area under the curve is 1.
5. About 68% of the area lies within the interval $\mu \pm 1\sigma_X$. About 95% of the area lies within the interval $\mu \pm 2\sigma_X$. About 99.75% of the area lies within the interval $\mu \pm 3\sigma_X$.

A different normal curve exists for every combination of mean and standard deviation. Any normally distributed variable can be transformed into a standard normal variable using the Z-score transformation. Standard normal variables have a mean of 0 and a standard deviation of 1.

The standard-score transformation resembles the Z-score transformation, but uses the sample mean and standard deviation in place of μ and σ_X. Since both standard and Z-score transformations are linear, they do not affect distributional shape, therefore do nothing to make a nonnormal distribution more normal.

Approximate large-sample confidence intervals are constructed for means and proportions using tables of the standard normal distribution together with the appropriate standard errors. Sample statistics such as $\overline{X}$ and $\hat{\pi}$ provide **point estimates** of the corresponding population parameters. Confidence interval methods expand point estimates into **interval estimates,** which are ranges within which we believe the parameter lies. The more confident we wish to be, the wider we must make the interval. For any given degree of confidence a larger sample allows a narrower interval than a smaller sample does.

Hypotheses are statements about the specific values of population parameters. If a hypothesized parameter value lies outside of a 95% confidence interval, then we may reject or disbelieve that hypothesis with at least 95% confidence. More elaborate hypothesis testing procedures are covered in the following chapters.

PROBLEMS

24. In the manner of Figures 9.3–9.6, sketch and label curves to show the following normal distributions.
    a. $X \sim N(0, 15)$
    b. $X_1 \sim N(25, 5)$; $X_2 \sim N(25, 10)$
    c. $X_1 \sim N(1,000, 500)$; $X_2 \sim N(2,000, 1,000)$

25. Convert the following X values to Z-scores.
    a. $x = 34$, $X \sim N(28, 16)$ .375
    b. $x = 943$, $X \sim N(1,236, 89)$ -3.29
    c. $x = -6$, $X \sim N(12, 10)$ -1.8
    d. $x = 400$, $X \sim N(400, 65)$ 0

26. Return the following Z scores to the original units of X.
    a. $z = 2.33$, $X \sim N(28, 16)$ 65.28
    b. $z = -4.91$, $X \sim N(1,236, 89)$ 799.01
    c. $z = 0$, $X \sim N(12, 10)$ 12
    d. $z = -.09$, $X \sim N(400, 65)$ 394.15

27. IQ tests are commonly designed so that scores have a normal distribution with a mean of 100 and a standard deviation of 15.
    a. Sketch a normal curve representing such a distribution.

b. Add a second horizontal scale under this curve to show the corresponding Z-scores.

c. What Z-score corresponds to an IQ of 100? 75? 145? Explain in English what these Z-scores tell us, beyond what we know from the IQ scores themselves.

28. Consider an IQ test as described in Problem 27.

a. What proportion of the population is expected to score above 124?

b. What is the probability that a single individual, selected at random from the population, will score above 124?

c. What is the probability that a single randomly selected individual will score above 115, but below 148?

d. What is the probability of scores between 64 and 94?

e. If the population consists of 200 million people, *how many* could be expected to score above 160?

f. What is the probability that a randomly selected individual will score between 85 and 130?

g. What score would place an individual in the top 5% of the population? Top 1%? Top 1/10 of 1%?

*29. Military pilots sometimes lose consciousness due to high gravity (G forces) during violent maneuvers. This danger was the subject of research resulting in the data shown in the table. The number of incidents of G-induced loss of consciousness (G-LOC), as reported by 2,459 naval aviators, is given for eleven types of aircraft.

Incidents of Gravity-Induced Loss of Consciousness (G-LOC) on 11 Types of U.S. Naval Aircraft

Aircraft type	Incidents of G-LOC	G-suits disconnected	Flight hours
AV-8	32	694	34,391
T-2	73	0	84,257
TA-4	83	823	102,908
OV-10	13	484	17,765
F-4	108	3,627	256,378
A-6	136	4,003	379,817
F-18	19	1,435	55,339
F-14	102	6,917	327,067
A-4	55	1,281	188,420
EA-6B	41	2,571	168,609
A-7	27	882	248,039

Source: Johanson and Pheeny, 1988.

a. Transform the number of incidents of G-LOC for each aircraft type into a standard score.

    b. Interpret the standard scores of G-LOC incidents for the AV-8 and A-6 aircraft.

    c. What are the mean and standard deviation of your 11 standard scores? Are they as expected?

    d. Confirm that the median of the G-LOC standard scores can be obtained by applying the standard-score transformation to the median of the original G-LOC values.

    e. Confirm that the mean–median difference for the G-LOC standard scores is the same as the mean–median difference for the original G-LOC values. What does this imply about the skewness of these two distributions?

30. A national survey of 3,813 acute-care hospitals (Alexander and Bloom, 1987) reported the following sample statistics. Find a 99% confidence interval for each and interpret it in English.

    a. The mean number of administrators per bed was .0417, with a standard deviation of .0220.

    b. The mean number of physicians per bed was .0114, with a standard deviation of .0278.

    c. The mean number of beds set up and staffed was 192.24, with a standard deviation of 186.94.

31. In a study of 180 female psychiatric patients suffering from recurrent depression, the mean age at their first depressive episode was 26 years, with a standard deviation of 11.2 years (Frank, Carpenter, and Kupfer, 1988). Assuming that these 180 cases can be viewed as a random sample, find and interpret a 95% confidence interval for the mean age of onset.

32. The most recent depressive episodes of the 180 patients (Problem 31) lasted an average of 22.1 weeks, with a standard deviation of 17.4 weeks. Find and interpret a 95% confidence interval for the mean length of the most recent episode.

33. In a random sample of 122 U.S. homicides in 1984, 44 were committed using a handgun.

    a. Find a 99% confidence interval for the proportion of all homicides during this year that were committed using a handgun.

    b. Interpret this interval in formal statistical terms, with reference to the population proportion π.

    c. Interpret the interval less formally, as you might if you were trying to explain it to an audience of nonstatisticians.

34. Find confidence intervals from the following sample results, based on the same random sample of 122 homicides in Problem 33.

    a. A 90% confidence interval for the proportion of homicide victims who are male. In the sample, 86 out of 122 are male.

b. A 95% confidence interval for the proportion of homicide offenders who are male. In the sample, 103 out of 122 are male.

35. Pollsters asked 407 New Hampshire Republican voters who they preferred in the 1988 presidential primary.[7] Their top three choices were George Bush (49%), Robert Dole (12%), and Jack Kemp (8%).

a. Find 95% confidence intervals for each percentage.

b. Which interval is widest? Which narrowest? What generalization does this suggest?

36. Refer to the presidential primary poll of Problem 35.

a. Find a 95% confidence interval for the percentage of Republicans who do *not* prefer Jack Kemp.

b. Compare the width of the interval for the percentage of Republicans not preferring Jack Kemp (part a) with the width of the interval for the percentage preferring Kemp (part a of Problem 35). What generalization does this suggest?

c. Combine your observations from part b with those from part b of Problem 35 to make a general statement about the relationship between the size of a percentage and the width of the resulting confidence interval.

37. Use the confidence intervals for the presidential primary candidates (Problem 35) to test the following hypotheses.

a. George Bush is preferred by at least half the New Hampshire Republicans.

b. Robert Dole is preferred by at least half the New Hampshire Republicans.

38. Because standard errors and confidence intervals vary with the size of the percentage, as seen in Problems 35–37, pollsters sometimes report simply a "margin of error," corresponding to the *widest possible confidence interval* for their data. For example, one poll of 300 adults sought opinions on various issues in education. Instead of reporting a confidence interval for each percentage, the pollster reported that the results have "a theoretical margin of error in the range of 5.66 plus or minus, 95% of the time."[8]

a. Given a sample of $n = 300$, what percentage will have the widest confidence interval? This is the same as asking what proportion has the largest standard error. By trial and error, can you guess what proportion results in the largest standard error?

b. Use the "largest possible" standard error, given $n = 300$, to show how the "theoretical margin of error" for this poll was obtained.

NOTES

1. The characteristic bell-shaped appearance of a normal distribution partly reflects graphing conventions. Graphs of normal distributions look most bell-like when they extend to about ± 4 standard deviations and are drawn about as high as they are wide. This is usually how they are shown, but other choices of scale can make normal distributions look less like bells. If the horizontal scale extends out much farther, such as ± 40 standard deviations, the graph will look more like a spike. A graph drawn much wider than it is high can flatten the curve into a pancake. The two mildly different normal curves shown in Figure 9.6 illustrate such possibilities.

2. π is the ratio of a circle's circumference to its diameter ($\pi = 3.14159 \ldots$), and e is the base number for natural logarithms ($e = 2.71828 \ldots$). Both constants have important mathematical properties and a broad range of scientific applications.

3. A linear transformation is one that could be written algebraically as $X^* = a + b(X)$, where X is the original variable, X^* is the transformed variable, and a and b are constants. For example, to convert distance in meters (M) to feet (F):

   $$F = 0 + 3.28(M)$$

   meaning that each meter is equal to about 3.28 feet. Likewise, to convert X to a standard score:

   $$X^* = -\frac{\overline{X}}{s_X} + \left(\frac{1}{s_X}\right)X$$

   which is a rearrangement into the form $X^* = a + b(X)$ of the usual standard-score transformation $X^* = (X - \overline{X})/s_X$, Equation [9.5].

   These and other linear transformations share the property that they have no effect on distributional shape. A distribution of heights, for example, would look the same whether measured in inches or in centimeters. The term *linear* applies because $X^* = a + b(X)$ equations can be represented graphically by a straight line. We will learn more about linear equations in Chapters 14 and 15.

4. With linear transformations, *the mean and median of the transformed data can be found by transforming the mean and median of the raw data*. If you apply the linear standard-score transformation, Equation [9.5], to the sample median (1,545) and mean (4,540.4) of the raw data in this example, you will obtain the sample median ($-.509$) and mean (0.000) of the transformed data:

   $$\text{Transformed median} = \frac{1,545 - 4,540.4}{5,885.3} = -.509$$

$$\text{Transformed mean} = \frac{4{,}540.4 - 4{,}540.4}{5{,}885.3} = 0.000$$

This does not generally apply to *nonlinear* transformations such as logarithms or powers (see Chapter 6).

5. Some of the findings are described in Hamilton (1985).

6. You may notice that the sample size is given as $n = 155$ in Table 9.4, and $n = 153$ in Table 9.6. This is not a mistake, but an example of a problem very common in survey analysis: Not all respondents answered all questions. There were actually 156 respondents in the complete data set, but one person did not answer the "years of education" question, leading to an n of only 155 in Table 9.4. Three people did not answer the opinion question in Table 9.6, reducing n to 153. Although neither reduction is large, this sort of problem can detract from our ability to generalize to the town's adult population. It seems plausible that the person who would not divulge his or her education was relatively uneducated, so the actual mean for all 156 respondents would be slightly lower than the $\overline{X} = 12.97$ found for the 155 who *did* answer. Consequently, a sample mean based on $n = 155$ would lead us to slightly overestimate the population's mean level of education. If a larger proportion had not answered, this overestimation could easily be substantial.

   Theories about sampling distributions and confidence intervals assume that the data represent simple random samples from the population. This assumption is no longer true when some cases "drop out" of the analysis for nonrandom reasons.

7. Reported in the *Boston Globe*, April 3, 1987.

8. Reported in *New Hampshire Times*, September 10, 1986.

Large-Sample Hypothesis Tests

Chapter 10

Confidence intervals and hypothesis tests help guide researchers in drawing conclusions about the values of unknown population parameters. Confidence intervals (Chapter 9) are ranges within which we believe (with a specified degree of uncertainty) the population parameter lies. Hypothesis testing has a different focus: We set up alternative hypotheses about the values of population parameters, then use sample data to evaluate their plausibility.

Like other inferential procedures, *hypothesis tests depend on the assumption that the data are a random sample.* Random sampling (Chapter 8) can be tricky in practice, but in this chapter we will generally proceed as if our examples really are simple random samples from their respective populations. While not always true, this fiction will simplify our discussion as new concepts arise. Dodging the sampling issue for now does not dismiss it: It remains a major research problem in many fields and the first area criticized when findings prove controversial.

TABLE 10.1   *Some Symbols Used in Chapter 10*

Sample statistic or estimate	Population parameter or true value	Meaning
$\hat{\pi}$	π	Proportion
$SE_{\hat{\pi}}$	$\sigma_{\hat{\pi}}$	Standard error of the sample proportion
$\hat{p}$	p	Percentage
$\overline{X}$	μ	Mean of X
s_X	σ_X	Standard deviation of X
$SE_{\overline{X}}$	$\sigma_{\overline{X}}$	Standard error of the sample mean of X

We will begin with a large sample of data on teenage drug use to illustrate basic ideas behind hypothesis testing. Later sections expand on these ideas and describe general methods for testing hypotheses with large-sample proportions and means. Small-sample methods are examined in Chapter 11.

Our discussion of hypothesis testing employs some terms and symbols from earlier chapters, which are summarized in Table 10.1. Symbols are paired: each sample statistic has a corresponding population parameter. The sample statistics provide **estimates** of the unknown population parameters.

10.1   *STATISTICAL HYPOTHESES*

A **statistical hypothesis** is a statement about the values of one or more unknown population parameters. For example, the statement "the proportion of regular cocaine users among secondary school students in River City is .06" can be viewed as a hypothesis about the proportion of cocaine users, π, in the population consisting of all River City's secondary school (junior and senior high) students. The value .06 is derived from national data; if River City students are like students nationwide, they should include about $\pi = .06$ users.

The hypothesis can be stated symbolically as

$$H_0: \quad \pi = .06$$

Hypothesis H_0 might not be true. We can also specify a second hypothesis, H_1, which must be true if H_0 is not,

$$H_1: \quad \pi \neq .06$$

Either it is true that the proportion in River City equals .06 (H_0), or else the proportion does not equal .06 (H_1). H_0 and H_1 are **complementary:** they have no overlap, and between them they exhaust all the possible values of the parameter π. Hypothesis tests begin with pairs of complementary hypotheses.

Since H_0 and H_1 are complementary, a test of whether H_0 is believable tells us indirectly whether to believe H_1 instead. We therefore need to test just one of these two hypotheses. The one tested is called the **null hypothesis**: the hypothesis that can be directly *nullified* or rejected as the result of the test. The symbol H_0 denotes the null hypothesis. Its alternative, H_1 is called the **alternative** or **research hypothesis.**

The question in this example is: Does the proportion of River City students using cocaine differ from the proportion nationwide? This possibility, for better or for worse, is represented by our research hypothesis, H_1: $\pi \neq$ .06. Data from a sample of River City students may help us decide whether to believe H_0 or H_1. No doubt the proportion of users in our sample will differ at least slightly, one way or another, from the nation's .06. But before we interpret such sample differences as support for H_1—grabbing local headlines with either "River City Students Say 'NO' to Coke!" or "River City Students Drugged Out!"—we should check carefully whether our sample results really do contradict the less dramatic possibility represented by H_0: $\pi =$ .06: "River City Students Same as Others."

Our sample proportion, $\hat{\pi}$, might differ slightly from the hypothesized value of .06 simply because of random sampling variation. That is, any one random sample could easily have a proportion greater or less than .06, even if the population it came from had a proportion of cocaine users of exactly $\pi =$ .06. **Hypothesis tests** try to ascertain whether sample results are so different from H_0 that H_0 should not be believed.

10.2   *A HYPOTHESIS TEST*

Hypothesis tests use the sample statistics that correspond to the parameters in H_0 and H_1. In the actual study on which the "River City" example is based, a sample consisting of 497 students was surveyed. Eleven percent, or $\hat{\pi} =$ .11, reported using cocaine at least once a month. Obviously this proportion is different from the hypothesized value of $\pi =$ .06. But is it so far off that we can rule out chance? Perhaps the proportion of cocaine users in the population of River City students really is $\pi =$ .06, but our sample just happened to include a higher proportion.

A sample of $n =$ 497 students is large enough that the sampling distribution of $\hat{\pi}$ should be approximately normal, centered on the population proportion π, and with standard error

$$\sigma_{\hat{\pi}} = \sqrt{\frac{\pi(1 - \pi)}{n}}$$ [10.1]

The hypothesis test begins with this theoretical knowledge and with the question: What if H_0 were true?

If H_0 is true, then the population proportion π equals .06. Our question can be restated:

What is the probability of obtaining $\hat{\pi} = .11$, or sample results even more favorable to H_1: $\pi \neq .06$, if it is true that $\pi = .06$?

We know that the sampling distribution of $\hat{\pi}$ is normal, and if $\pi = .06$ then this distribution is centered on .06. Furthermore, if $\pi = .06$, according to Equation [10.1] the standard error will be

$$\sigma_{\hat{\pi}} = \sqrt{\frac{\pi(1 - \pi)}{n}}$$

$$= \sqrt{\frac{.06(1 - .06)}{497}} = .01065$$

Thus if H_0 is true, we have $\pi = .06$, $\hat{\pi} = .11$, and $\sigma_{\hat{\pi}} = .01065$. Applying the Z-score transformation (Chapter 9) we find that the sample proportion is *4.69 standard errors away from the hypothesized population proportion:*

$$Z = \frac{\hat{\pi} - \pi}{\sigma_{\hat{\pi}}}$$

$$= \frac{.11 - .06}{.01065} = 4.69$$

This leads to another restatement of our question:

How likely is it we could randomly obtain a value as far as 4.69 standard errors from the mean, in a normal distribution?

In Table A.2 a Z value of 4.69 falls between 4.420 and 4.9, so the probability of Z values as large as 4.69 (read from the "Two-Sided Tests" column) is between .00001 and .000001 (see Figure 10.1). Formally,

$$.000001 < P(|Z| > 4.69) < .00001$$

This probability is the **obtained *P*-value.**

> The **obtained *P*-value** is the probability of obtaining the sample results, or results even more favorable to H_1, if the sample was drawn randomly from a population where H_0 is true.

The obtained *P*-value is very low, between one chance in 1,000,000 and one chance in 100,000.[1] This answers our question:

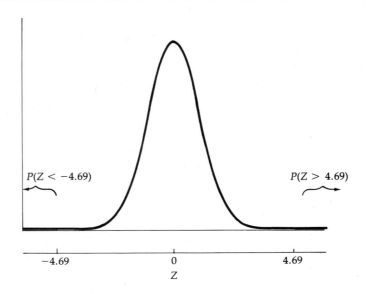

$P(Z < -4.69)$ $P(Z > 4.69)$

-4.69 0   4.69

Z

Shaded area corresponds to an obtained P-value < .0001: The probability of a Z
statistic more than 4.69 standard errors from 0

FIGURE 10.1

It is extremely unlikely ($P < .00001$) that we could obtain sample results so
favorable to H_1: $\pi \neq .06$, if H_0: $\pi = .06$ were true.

It is so unlikely that H_0: $\pi = .06$ should not be believed. We can *reject the null
hypothesis,* and *accept the alternative hypothesis H_1: $\pi \neq .06$ instead.*[2] Table 10.2
(page 282) traces the steps we have followed.

This analysis leads us to the conclusion that the proportion of cocaine
users among all River City secondary school students (not just the 497 in the
sample) is not equal to .06. That is, the proportion among River City students
is significantly different from the nationwide proportion, .06.

We will next examine some of the general principles behind this specific
example.

PROBLEM

1. The survey regarding drug use by 497 secondary school students in River
City asked about other drugs besides cocaine. For each listing, obtain a
P-value to test whether the proportion among River City students is significantly different from the proportion among secondary school students
nationwide. Follow this rule: Reject H_0 if $P < .05$.

   a. Occasional use of beer and wine: River City, .86; national, .76

TABLE 10.2	**Large-Sample Z Test for a Hypothesis About a Proportion (Two-Sided): The Proportion of Cocaine Users, Based on a Sample of 497 River City Secondary School Students**

Null hypothesis (River City is same as nation): H_0: $\pi = .06$
Research hypothesis (River City is different): H_1: $\pi \neq .06$
Sample size (large enough for large-sample methods): $n = 497$
Sample proportion of cocaine users: $\hat{\pi} = f/n = 55/497 = .11$

If this sample came from a population where $\pi = .06$ (that is, if H_0 is true), the standard error will be

$$\sigma_{\hat{\pi}} = \sqrt{\frac{\pi(1-\pi)}{n}}$$

$$= \sqrt{\frac{.06(1-.06)}{497}} = .01065$$

The test statistic is:

$$Z = \frac{\hat{\pi} - \pi}{\sigma_{\hat{\pi}}}$$

If H_0 is true, then given the sample proportion $\hat{\pi} = .11$, this test statistic becomes

$$Z = \frac{.11 - .06}{.01065} = 4.69$$

The sample value $\hat{\pi} = .11$ is about 4.69 standard errors away from the hypothesized population value $\pi = .06$ (see Figure 10.1). The probability of so large a difference (see Table A.2) is

$$.000001 < P(|Z| > 4.69) < .00001$$

This obtained *P*-value is quite low, so we should *reject* H_0. The evidence suggests that this sample of 497 cases did *not* come from a population where $\pi = .06$. The research hypothesis, that the sample came from a population where $\pi \neq .06$, is supported.

    b. Occasional use of marijuana: River City, .45; national, .25
    c. Occasional use of liquor: River City, .79; national, .61
    d. Use of illegal stimulants ("speed"): River City, .23; national, .27

Briefly summarize the conclusions that emerge from these four tests, together with the fifth test shown in Table 10.2.

10.3   *THE LOGIC OF HYPOTHESIS TESTING*

Hypothesis tests let us determine the plausibility of a particular null hypothesis (H_0). If sample evidence weighs against the null hypothesis, this strengthens the case for believing the complementary research hypothesis instead.

We say *research hypothesis* because H_1 is often the hypothesis that is of greatest interest to the researcher. It may stand for ideas (like "patients with this new treatment recovered faster" or "the failure rate of our computers is lower") for which the researcher hopes to find support. The null hypothesis then represents a negation of the researcher's hypothesis (as in "patients with this new treatment did not recover faster" or "the failure rate of our computers is as high or higher"). Although the researcher may hope that the evidence favors H_1, hypothesis tests are set up to test H_0. The research hypothesis, H_1, will be accepted only if strong evidence is found against H_0.

The nature of the research hypothesis H_1 determines whether a *two-sided test* or a *one-sided test* is appropriate. A one-sided test is called for if the researcher has specific ideas about the *direction* in which a population parameter lies. For example, a one-sided test can focus narrowly on whether the proportion of users in River City is "higher than" the national proportion. Our hypotheses would be

H_0: $\pi \le .06$

H_1: $\pi > .06$

On the other hand, the researcher may have no specific theories about the direction in which a parameter lies. A less focused research question asks simply if the population parameter is *different from* a given value. With River City, we asked first whether the proportion of cocaine users among local students is "different from" .06, the national student proportion. This called for a two-sided test:

H_0: $\pi = .06$

H_1: $\pi \ne .06$

Two-sided tests:
   Research hypothesis H_1 *does not* specify direction.
   Example: H_0: $\pi = .06$
             H_1: $\pi \ne .06$

One-sided tests:
   Research hypothesis H_1 *does* specify direction.
   Example: H_0: $\pi \le .06$
             H_1: $\pi > .06$

Decide whether to apply a one- or a two-sided test *before looking at your data*. It defeats the purpose of hypothesis testing to let the data suggest a hypothesis, then to "test" whether those same data support that hypothesis. Of course they will—but this is circular reasoning.

Whether one- or two-sided, **statistical hypothesis tests** involve the following steps:

1. We specify a pair of complementary hypotheses, H_0 (null hypothesis) and H_1 (research hypothesis).
2. Using sample data we ask: How likely are we to get sample results so favorable to H_1, if the sample comes from a population in which H_0 is true? The answer is a probability called the *obtained P-value*.
3. A high P-value says we have no reason to disbelieve H_0; we *fail to reject* the null hypothesis.
4. But a low P-value says it seems unlikely that the sample comes from a population where H_0 is true. We therefore decide to *reject* the null hypothesis.
5. We accept the research hypothesis, H_1, only if a low P-value leads us to reject H_0.

A *low P-value* is defined by some predetermined cutoff point, symbolized by the Greek letter α (**alpha**) and called the **significance level** of the test. Typically, α is set at a low value like .05 (5% chance) or .01 (1% chance).

The rule for whether to reject H_0 is:

Reject H_0 if the obtained P-value is less than α.

Example: Reject H_0 if $P < .05$.

Do not reject H_0 if the obtained P-value is not less than α.

Example: Do not reject H_0 if $P \geq .05$.

When H_0 is rejected, it may be said that the findings are **statistically significant** at level α.

These general steps are charted in Table 10.3. The chart branches into two main paths, depending on whether the obtained P-value is low enough to cast doubt on H_0. If the P-value is low, we may decide to reject H_0 and believe H_1 is true. In so doing, we risk making a *Type I error*. If the P-value is not low, then the evidence favoring H_1 is weak, so we should not reject H_0. In not rejecting H_0, we risk making a *Type II error*. Type I and Type II errors are explained in the next section.

PROBLEMS

2. For each of the following research hypotheses, state the research and complementary null hypotheses symbolically, explain what the symbols refer to, and state whether the research hypothesis calls for a one- or a two-sided test.

   a. Less than 10% of the parachutes we manufacture are defective.

   b. More than half of all U.S. college students are women.

TABLE 10.3   ***The Steps of Hypothesis Testing***

Specify two hypotheses: a **null hypothesis,** H_0, and the complementary **research hypothesis,** H_1.

$\downarrow$

Using sample data, calculate a test statistic to answer the question, "How likely are we to obtain results so favorable to H_1, if the sample was drawn randomly from a population where H_0 is true?" The answer is expressed as a probability called the **obtained P-value.**

If P-value is *very low* (e.g., below a selected cutoff value, α, of .05):	If P-value is *not very low* (e.g., not below α = .05):
$\downarrow$	$\downarrow$
H_0 may confidently be *rejected,* since it appears unlikely to be true. The evidence therefore favors the research hypothesis, H_1, instead.	There is insufficient evidence to reject H_0; the sample could have been drawn from a population where H_0 is true. The alternative, H_1, is not supported.
$\downarrow$	$\downarrow$
However, it is always possible that we are wrong in rejecting H_0. This type of mistake (rejecting a true H_0) is called a **Type I error.**	However, it is always possible that we are wrong in not rejecting H_0. This type of mistake (failing to reject a false H_0) is called a **Type II error.**

   c. Students who receive coaching do better than the national average of 18.1 on this test.

   d. The claim that 60% of small businesses pay no taxes is inaccurate.

   e. The usual average July temperature in Seattle is 64.8°F, but this year it may be warmer because of the greenhouse effect.

   f. The average recovery time for patients with this new type of surgery is less than nine weeks.

3. Imagine yourself in the role of a manufacturer who makes thousands of parachutes each month and is concerned about product quality. It is impractical to test every single parachute for defects, so you use random sampling to estimate the rate of defects. Describe in general terms the steps required to test the hypothesis that fewer than 10% of your parachutes are defective. Discuss the null and alternative hypotheses, sampling, obtained P-value, decisions regarding H_0 and H_1, and the possible types of error resulting from your decision.

10.4   *TYPE I AND TYPE II ERRORS*

Table 10.3 shows the paths in hypothesis testing that can lead to Type I or Type II errors.

TABLE 10.4 **Type I and Type II Errors: Outcomes and Probabilities for the Decision to Reject or Not Reject H_0, Given That H_0 or H_1 Is True**

	Decision[a]	
Fact	Not reject H_0	Reject H_0
H_0 is true	Correct decision: P(not reject H_0 when H_0 true) $\geq 1 - \alpha$	Type I error: P(reject H_0 when H_0 true) $< \alpha$ **significance level**
H_1 is true	Type II error: P(not reject H_0 when H_1 true) $= \beta$	Correct decision: P(reject H_0 when H_1 true) $= 1 - \beta$ **power**

[a]Based on our hypothesis test at significance level α.

Two types of error may occur in hypothesis tests:

Type I error: Rejection of H_0 when H_0 is in fact true
Type II error: Failure to reject H_0 when H_1 is in fact true

---

Table 10.4 expresses these definitions another way: The errors are possible outcomes of the decision whether to reject H_0, when either H_0 or H_1 is actually true.

The probability of making a Type I error when rejecting H_0 should be about equal to the significance level α. The probability of making a correct decision, by rejecting a false H_0, is therefore $1 - \alpha$. That is, if we reject H_0 using $\alpha = .05$, then we have at least a .95 probability of being right and less than a .05 probability of being wrong.

If we fail to reject H_0, then a Type I error is impossible. Instead, we risk making a Type II error. The probability of making a Type II error by failing to reject H_0, when H_1 is actually true, is represented by the Greek letter **β** (**beta**). The probability of making a correct decision by not rejecting H_0 is $1 - \beta$, which is called the **power** of a statistical test.

Since the probability of making a Type I error is controlled by the significance level, it may seem best to choose a very low level—$\alpha = .0001$ or $.00001$, for instance. Unfortunately there is a trade-off. Simply lowering the cutoff point at which H_0 is to be rejected does reduce the likelihood of mistaken H_0 rejection (Type I error), but it simultaneously *increases* the likelihood of mistakenly *not* rejecting H_0 (Type II error). We trade the risk of one kind of error for another.[3]

Though any kind of error is bad, Type I and Type II errors are not equally bad. Depending on the particular research question, one error usually has more serious consequences than the other. Obviously, we want to minimize the probability of whichever error is worse.

A Type I error, in the context of the River City example, would mean:

Wrongly rejecting H_0; that is, wrongly concluding that the proportion of cocaine users among River City students *is* different from the nationwide proportion.

This conclusion could alarm River City parents and educators about the unusually high level of cocaine use. Perhaps special anti-drug programs would be instituted. If the basic conclusion was erroneous, and the River City students really were no different from others, then the alarm and special programs would be overreactions.

A Type II error in the River City example would mean something quite different:

Wrongly failing to reject H_0; that is, wrongly concluding that the proportion of cocaine users among River City students *is not* different from the proportion nationwide.

This conclusion could mislead River City parents and educators about the severity of the cocaine problem in their schools. Such complacence might prevent their taking steps to improve a bad situation.

Thus Type I and Type II errors are very different mistakes. For a more dramatic example, consider the possible outcomes of a blood test for AIDS, where the null and alternative hypotheses are

H_0: The patient has AIDS

H_1: The patient does not have AIDS

The test results would be either positive (it appears that the patient has AIDS) or negative (it appears that the patient does not have AIDS). Of course, either result might be mistaken.

In this analogy a Type I error corresponds to what is called a *false negative:* a mistaken report that the patient *does not* have AIDS. This error risks fatal consequences in terms of how the patient is treated and who else is exposed. A Type II error corresponds to a *false positive:* a mistaken report that a healthy patient *does* have AIDS. This error could have serious psychological or social consequences, much different from the health consequences of a false negative report. If a trade-off is necessary, should we try to minimize the probability of a false negative or of a false positive? This is not merely a statistical question, but one that must be answered in terms of the goals of the testing itself and the consequences of different kinds of error.

Similar questions arise regarding the choice of α in any statistical research. Although $\alpha = .05$ is a widely used level for statistical significance, we can choose any level that the specific research problem warrants. A lower α, such as .01 or .001, is preferable if we consider Type I errors as costlier or more undesirable than Type II errors. If Type II errors are costlier, we may prefer to run the higher risk of Type I errors by increasing α to .10 or .20.

PROBLEMS

4. Problem 3 offered the research hypothesis that less than 10% of our parachutes are defective. In this context, discuss the nature and consequences of Type I and Type II errors.

5. As a conscientious manufacturer of parachutes, customer safety is of course your highest priority. With this in mind, is it better to choose $\alpha = .01$ or $\alpha = .10$ as the cutoff point for testing the null hypothesis H_0: $\pi \geq .10$ (Problems 3 and 4)? Why?

6. For each of the hypothesis pairs in Problem 2, parts b–f, describe the meaning of Type I and Type II errors.

10.5   LARGE-SAMPLE TESTS FOR PROPORTIONS

Given large enough samples, the sampling distribution of a proportion, $\hat{\pi}$, is approximately normal, with mean π (the population proportion) and standard error

$$\sigma_{\hat{\pi}} = \sqrt{\frac{\pi(1 - \pi)}{n}}$$

The population proportion π and hence also the true standard error $\sigma_{\hat{\pi}}$ are usually unknown. There are two ways to obtain values for this standard error, Equations [10.2] and [10.3].

For *confidence intervals* we obtain an **estimated standard error**, $SE_{\hat{\pi}}$, using the sample proportion $\hat{\pi}$:

$$SE_{\hat{\pi}} = \sqrt{\frac{\hat{\pi}(1 - \hat{\pi})}{n}} \qquad [10.2]$$

For *hypothesis tests* we find what the standard error would be if H_0 were true, using the hypothesized proportion π_0:

$$\sigma_{\hat{\pi}} = \sqrt{\frac{\pi_0(1 - \pi_0)}{n}} \qquad [10.3]$$

The symbol π_0 denotes the value for the population proportion π that is specified by the null hypothesis H_0. For example, given the null hypothesis H_0: $\pi = .06$, π_0 equals .06. If H_0 specifies more than one value for π, as in the

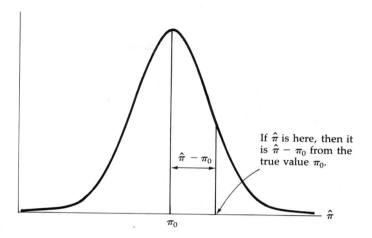

FIGURE 10.2   *Sampling distribution of the sample proportion $\hat{\pi}$, if $\pi = \pi_0$*

one-sided null hypothesis $H_0: \pi \le .06$, then π_0 is that value that is *closest to* H_1—again, this means π_0 equals .06.

Large-sample hypothesis tests for proportions are performed with a **Z statistic:**

$$Z = \frac{\hat{\pi} - \pi_0}{\sigma_{\hat{\pi}}} = \frac{\hat{\pi} - \pi_0}{\sqrt{\pi_0(1 - \pi_0)/n}} \qquad [10.4]$$

The Z statistic follows approximately a standard normal distribution when $\pi_0 = \pi$ (H_0 is true).

The hypothesis test regarding cocaine in River City schools (Table 10.2) used the Z statistic defined by Equation [10.4].

A Z statistic, Equation [10.4], tells us the distance between sample proportion $\hat{\pi}$ and null hypothesis population proportion π_0, measured in standard errors, $\sigma_{\hat{\pi}}$. Figure 10.2 shows a graphical interpretation of the distance $\hat{\pi} - \pi_0$. If the true parameter value is π_0, then the sample statistic $\hat{\pi}$ theoretically has a normal sampling distribution centered on π_0. Any one sample value of $\hat{\pi}$ will differ from the true value π_0 by the distance $\hat{\pi} - \pi_0$. Equation [10.4] converts this distance to standard errors. If the sample proportion is $\hat{\pi} - \pi_0$ from π_0, then it is z standard errors away from π_0.

Since the distribution of Z is standard normal, we can check in Table A.1 or A.2 *the probability of being so far from the center.* This probability corresponds to the area in the tails of the distribution more than $\hat{\pi} - \pi_0$, or z standard errors, from the parameter π_0. Note that Figure 10.2 shows the sampling distribution of $\hat{\pi}$ *if π_0 is the true population proportion*. We do not assume that it actually is; we merely act as if it were, to see where this leads. In Table 10.2 we saw that if H_0 is true, our sample proportion is 4.69 standard errors from the population proportion. Sample values this far out are unlikely, so H_0 is probably not true.

The normal approximation makes calculating either confidence intervals or hypothesis tests easy, *if* the sample is sufficiently large. How large the sample must be depends on the value of the population proportion π. The closer π is to 0 or 1, the more cases it takes for the normal approximation to hold. That is, we need many more cases if π is .01 or .99, than we do if π is .50. A rough guideline is that a sample is "large enough" if

$$\min\left\{\frac{n\pi}{1-\pi}, \frac{n(1-\pi)}{\pi}\right\} \geq 9 \qquad [10.5]$$

where the function "min{·, ·}" means "take the smaller of the quantities enclosed in these brackets." For example, if $\pi = .06$ is the population proportion of cocaine users, in a sample of size n we could expect about $n\pi$, or $n(.06)$ people to be users. If n equals 50 cases,

$$\frac{n\pi}{1-\pi} = \frac{50(.06)}{1-.06} = 3.19$$

and

$$\frac{n(1-\pi)}{\pi} = \frac{50(1-.06)}{.06} = 783.33$$

Then $\min\{n\pi/(1-\pi), n(1-\pi)/\pi\} = \min\{3.19, 783.33\} = 3.19$. Since 3.19 is less than 9, we conclude that a sample of size $n = 50$ is not large enough to justify the normal approximation, when $\pi = .06$.

With a sample of $n = 497$ cases (our cocaine example), Equation [10.5] gives us $\min\{n\pi/(1-\pi), n(1-\pi)/\pi\} = \min\{31.72, 7,786.33\} = 31.72$. Since 31.72 is larger than 9, this sample is large enough to justify the normal approximation.

Table 10.5 shows the minimum sample size required for proportions ranging from .00001 (one-thousandth of 1%) to .99999 (99.999%). The number of cases needed rises steeply as π approaches 0 or 1. Table 10.5 also lists the expected number of cases in the smaller category. These numbers are found by taking either π proportion of n $(n\pi)$ or $1-\pi$ proportion of n $[n(1-\pi)]$, whichever is smaller. For example, for $\pi = .10$ the minimum sample size is shown as 81 cases. Then $n\pi = 81(.10) = 8.1$; we would expect about 8.1 cases in the smaller category. The larger category would be expected to contain the other $n(1-\pi) = 81(1-.10) = 72.9$ cases.

TABLE 10.5   ***Minimum Sample Sizes for Using the "Large-Sample" Normal Approximation for a Proportion[a]***

Population proportion π	Minimum n for large sample[b]	Expected number of cases in smaller category
.00001	900,000	9.0
.0001	90,000	9.0
.001	9,000	9.0
.01	900	8.9
.05	171	8.6
.1	81	8.1
.2	36	7.2
.3	21	6.3
.4	14	5.6
.5	9	4.5
.6	14	5.6
.7	21	6.3
.8	36	7.2
.9	81	8.1
.95	171	8.6
.99	900	8.9
.999	9,000	9.0
.9999	90,000	9.0
.99999	900,000	9.0

[a]Based on the rule of thumb that a sample is "large enough" if

$$\min\left\{\frac{n\pi}{1-\pi}, \frac{n(1-\pi)}{\pi}\right\} \geq 9 \qquad [10.5]$$

[b]Larger numbers are rounded off; for example, 899,991 becomes 900,000.

The minimum sample size gets larger and larger as π approaches 0 or 1, yet the expected number of cases in the smaller category never exceeds nine. This suggests a rule of thumb, simpler than Equation [10.5], as shown in the box.

> A sample is "large enough" to use the normal approximation for a proportion if the expected number of cases in the smaller category is 9 or more; that is, if both $n\pi$ and $n(1-\pi)$ are at least 9.

Provided that a sample is sufficiently large, the methods of this chapter and Chapter 9 may be used for confidence intervals and hypothesis tests regarding the value of the population parameter π. A small sample requires other methods, described in Chapter 11.

PROBLEMS

7. Use Table 10.5 to judge, if possible, whether the normal approximation for the sampling distribution of π seems reasonable in the following situations.

   a. $\pi = .45, \quad n = 18$

   b. $\pi = .08, \quad n = 73$

   c. $\pi = .97, \quad n = 200$

   d. $\pi = .31, \quad n = 900$

   e. $\pi = .0025, \quad n = 1,500$

8. Find $\min\{n\pi/(1 - \pi), n(1 - \pi)/\pi\}$ from Equation [10.5] to confirm each of your answers to Problem 7.

9. Apply Equation [10.1] to find the standard errors of $\hat{\pi}$, based on

   a. $\pi = .2, \quad n = 50$

   b. $\pi = .2, \quad n = 150$

   c. $\pi = .2, \quad n = 250$

   d. $\pi = .2, \quad n = 350$

10. a. Use your answers to Problem 9 to illustrate the general relationship between sample size and standard errors.

    b. Why are small standard errors desirable in research?

    c. What happens to the *decrease* in standard errors, per 100 additional cases, as the sample size grows? What practical implications might this have for research?

11. Use the student data from Table 1.1 to test the following null hypothesis at the $\alpha = .01$ level: In the population of students from which this sample was drawn, 50% are male. (Remember: First convert this percentage to a proportion.)

12. In the U.S. an increasing tendency for unwed mothers to keep rather than give up their children and the lower fertility of many couples who delay childbearing until their mid-30's or later have both contributed to a seeming shortage of adoptable children. In consequence more foreign children are being admitted into America for adoption. Colombia and Korea are two major sources for these children. (Convert percentages to proportions.)

    a. In 1980 about 53% of the U.S. overseas adoptions were from Korea. In a random sample of 200 such adoptions in 1985, 120 were Korean. Test at the $\alpha = .05$ level whether the percentage of overseas adoptions from Korea changed.

    b. In 1980 about 13% of the U.S. overseas adoptions were from Colombia. The same 1985 random sample had 14 Colombians. Had the percentage from Colombia changed? Test at the $\alpha = .05$ level.

13. Apply Equation [10.5] to determine whether the sample is large enough for the normal approximation to be valid in Problem 12, parts a and b.

10.6   LARGE-SAMPLE TESTS FOR MEANS

Large-sample tests for means are similar to those for proportions. Two obvious differences are that a different standard error is used and that it is simpler to define what is meant by *large sample*; see the box.

> For tests involving a single mean, samples with at least 30 cases are generally considered "large enough" for the normal approximation to hold.

As always, "large enough" is a matter of degree, and samples much larger than 30 are even better. How close the normal approximation is for any given sample size depends partly on the underlying distribution of the variable X. The more skewed X is, the more cases it takes for the normal approximation to be reasonably good. Severe outliers can also cause problems, as they do with almost any mean-based analysis.

The Central Limit Theorem (Chapter 9) tells us that with large enough samples the sampling distribution of the mean ($\overline{X}$) is approximately normal, is centered on the population mean μ, and has a standard error $\sigma_{\overline{X}}$:

$$\sigma_{\overline{X}} = \frac{\sigma_X}{\sqrt{n}} \qquad [10.6]$$

where σ_X is the population standard deviation, and n is the sample size. In practice, if we don't know the population mean μ, we probably won't know the population standard deviation σ_X either. The population standard error must therefore be estimated using the sample standard deviation, s_X, in place of σ_X:

$$\text{SE}_{\overline{X}} = \frac{s_X}{\sqrt{n}} \qquad [10.7]$$

The estimated standard error from Equation [10.7] is used in both hypothesis tests and confidence intervals for the mean.

> Large-sample hypothesis tests for the mean employ the Z statistic:
>
> $$Z = \frac{\overline{X} - \mu_0}{\text{SE}_{\overline{X}}} = \frac{\overline{X} - \mu_0}{s_X/\sqrt{n}} \qquad [10.8]$$
>
> where μ_0 is the value of the population mean μ that is specified by the null hypothesis.

In large samples, the Z statistic has approximately a standard normal distribution.[4]

There is a family resemblance between Equations [10.8] and [10.4]. Both are for test statistics (Z) that have a standard normal distribution, and they follow the common form

$$\text{Test statistic} = \frac{\text{Sample statistic} - \text{Hypothesized parameter}}{\text{Standard error of sample statistic}}$$

Any test constructed along these lines will *measure the distance in standard errors between the statistic and the hypothesized parameter.*

The general logic of hypothesis testing is the same for means, proportions, and many other statistics. We begin by formulating two complementary hypotheses, H_0 and H_1. The test then uses sample data and asks: How likely are we to obtain a sample so favorable to H_1, if H_0 is true? The next section illustrates a large-sample hypothesis test for the mean.

PROBLEM

14. Suppose you encounter an unfamiliar sample statistic called $\overline{D}$ (*D*-bar), which is used to estimate the population parameter δ (*delta*). For large samples the sampling distribution of $\overline{D}$ is said to be normal, with a mean of δ and a standard error estimated by $\text{SE}_{\overline{D}}$.

   a. Write an equation for finding a 95% confidence interval, given values for $\overline{D}$ and $\text{SE}_{\overline{D}}$.

   b. Write an equation for testing the null hypothesis that δ equals 562, given values for $\overline{D}$ and $\text{SE}_{\overline{D}}$. Include a statement of H_0 and H_1. What value of Z is required to reject H_0 at $\alpha = .01$?

10.7 Z TEST FOR MEANS: AN EXAMPLE

The past few decades have seen great changes in the roles open to women in American society. Have these changes included shifts in women's childbearing expectations? U.S. Census research in 1975 determined that among married women aged 18 to 24, the mean of the total number of children they expected to bear was 2.17. Ten years later, another survey reported that among married women in this age group, the mean lifetime births expected was 2.18. The more recent figure is slightly higher, suggesting that the number of children expected by young married women had increased over the decade 1975–1985.

We can test whether the increase is statistically significant as follows. The null hypothesis is that the 1985 sample came from a population in which the mean is 2.17 births. Since we started by asking whether the mean had

TABLE 10.6   ***Large-Sample Test for a Hypothesis About a Mean (Two-Sided): Lifetime Births Expected by Married Women 18–24 Years Old***

Null hypothesis: H_0: $\mu = 2.17$
Research hypothesis: H_1: $\mu \neq 2.17$ (two-sided)
Sample size (large enough for normal approximation): $n = 1,000$
Sample mean births: $\overline{X} = 2.18$
Sample standard deviation of births: $s_X = 1.10$

The estimated standard error is

$$SE_{\overline{X}} = \frac{s_X}{\sqrt{n}}$$

$$= \frac{1.10}{\sqrt{1,000}} = .035$$

The test statistic is

$$Z = \frac{\overline{X} - \mu_0}{SE_{\overline{X}}}$$

Substituting the sample mean $\overline{X} = 2.18$, the hypothesized population mean $\mu_0 = 2.17$, and the estimated standard error $SE_{\overline{X}} = .035$ into this equation yields

$$Z = \frac{2.18 - 2.17}{.035} = \frac{.01}{.035} = .286$$

The sample value $\overline{X} = 2.18$ is .286 estimated standard error away from the hypothesized population value $\mu_0 = 2.17$ (see Figure 10.3, page 296). The probability of so large a difference (see Table A.2) is

$$.70 < P(|Z| > .286) < .80$$

This probability is high, so we should *not reject* H_0. The evidence suggests that this sample of 1,000 cases *could well* have been drawn randomly from a population where $\mu = 2.17$. There is insufficient support for the research hypothesis, $\mu \neq 2.17$.

"changed" (not whether it had "increased" or "decreased"), a two-sided test is appropriate. Symbolically, our null and research hypotheses are

H_0: $\mu = 2.17$

H_1: $\mu \neq 2.17$

If the 1985 sample consisted of $n = 1,000$ women, with a mean of $\overline{X} = 2.18$ and a standard deviation of $s_X = 1.10$, how strong is the evidence against H_0?

Table 10.6 shows the steps for this analysis. The sample mean $\overline{X}$ turns out to be .286 estimated standard error from the null hypothesis population mean, $\mu_0 = 2.17$. Consulting Table A.2, we find that in a normal distribution values as far as .286 standard deviation from the center have a high probabil-

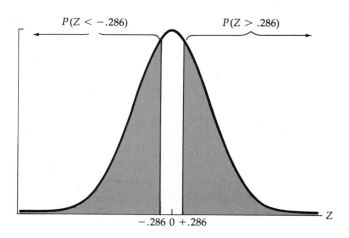

FIGURE 10.3   *Shaded area corresponds to an obtained P-value > .70: The probability of a Z statistic more than .286 standard error from 0, if $\mu = \mu_0$ (see Table 10.6)*

ity: between .70 and .80. Since this obtained *P*-value is well above the usual $\alpha = .05$ cutoff point, we have no reason to reject H_0.

The difference between the 1985 survey results and the 1975 mean of 2.17 is *not statistically significant:* It could easily be due to chance. Figure 10.3 shows the situation graphically: Sample means .286 standard error or more from the population often occur in random sampling.

PROBLEMS

15. Many elementary school educators believe that teaching young children to write their own programs in the Logo computer language fosters thinking and problem-solving skills. A study of one group of children found that their scores on a ten-point cognitive test improved by an average of $\overline{X} = 1.5$ points after instruction and experience with Logo programming. Test at $\alpha = .01$ whether this increase is significantly different from 0 (no improvement), based on $n = 60$ and $s_X = 3.0$.

16. Interpret the *Z* value obtained in Problem 15 as a *distance*.

17. The study described in Problem 15 also investigated the effects of more conventional computer instruction that did not include Logo programming. An improvement of $\overline{X} = .88$ point was observed. Test at $\alpha = .01$ whether this is significantly different from 0, based on $n = 40$ and $s_X = 2.9$.

10.8 ONE-SIDED TESTS

The examples given earlier for a proportion (Table 10.2 and Figure 10.1) and for a mean (Table 10.6 and Figure 10.3) are both two-sided tests. Whether one- or two-sided, research hypotheses should be formulated before you look at the data. With one-sided or directional hypotheses, there is always a possibility that the sample statistic will turn out to be in the opposite direction from that predicted by H_1. If so, there is no point in running the hypothesis test, since it cannot possibly reject H_0. If the sample statistic does lie in the direction predicted by H_1, however, then you can proceed with a one-sided test. Apart from this, and the use of single-tail rather than two-tail probabilities under the normal curve, one-sided tests follow the same procedures as two-sided tests. Two examples follow.

Research has shown that Americans store surprising quantities of hazardous chemicals in their homes, partly from not knowing what else to do with them. To help out, the city of Portsmouth organized a "household hazardous waste pick-up day," inviting people to bring in pesticides, paint thinners, and other unwanted chemicals they had been storing at home.

The organizers were interested in learning about the characteristics of people who participated in the pick-up. For instance, were the participants better educated than other citizens? It seems reasonable to guess that people with more education would be more conscious of the dangers posed by hazardous chemicals. According to census estimates, 30% of Portsmouth adults are college graduates. If the pick-up participants were better educated than other residents, this would imply that they came from a population in which the proportion with college educations exceeds .30. The null hypothesis is that they did not:

$$H_0: \quad \pi \le .30$$
$$H_1: \quad \pi > .30$$

A survey carried out at the pick-up site provides sample data with which to test these hypotheses: Out of $n = 99$ participants surveyed, 45 were college graduates, so the sample proportion of graduates is $\hat{\pi} = 45/99 = .455$. Table 10.7 (page 298) outlines the steps for conducting a one-sided hypothesis test. The sample is large enough to justify using the normal approximation, and the sample proportion $\hat{\pi}$ is in the direction specified by $H_1: \pi > .30$.

This Z statistic tells us that the sample proportion $\hat{\pi} = .455$ is more than 3 standard errors from the nearest possible population value if H_0 were true, $\pi_0 = .30$. The probability of a $\hat{\pi}$ value this far above the true population π is low—less than .0005 (shown in Figure 10.4). We can therefore reject H_0, and conclude in favor of H_1: The pick-up participants were significantly better educated than the Portsmouth population as a whole.

We now illustrate a one-sided test for a mean. The research hypothesis is an obvious one: that college math majors have higher math Scholastic Apti-

TABLE 10.7

Large-Sample Test for a Hypothesis About a Proportion (One-Sided): The Proportion of College Graduates, from a Sample of 99 Toxic Waste Pick-up Participants

Null hypothesis: H_0: $\pi \leq .30$
Research hypothesis: H_1: $\pi > .30$ (one-sided)
Sample size: $n = 99$
Sample proportion of college graduates: $\hat{\pi} = f/n = 45/99 = .455$

This sample proportion is in the direction specified by H_1, so we should proceed with the test.

Expected cases in smaller category: $n\pi_0 = 99(.30) = 29.7 \geq 9$, so the sample is large enough to use the normal approximation.

If $\pi = \pi_0 = .30$, the standard error is

$$SE_{\hat{\pi}} = \sqrt{\frac{\pi_0(1 - \pi_0)}{n}}$$

$$= \sqrt{\frac{.30(1 - .30)}{99}} = \sqrt{.00212} = .046$$

The test statistic is

$$Z = \frac{\hat{\pi} - \pi_0}{SE_{\hat{\pi}}}$$

Substituting the sample proportion $\hat{\pi} = .455$, and the hypothesized population proportion $\pi_0 = .30$, into this equation yields

$$Z = \frac{.455 - .30}{.046} = \frac{.155}{.046} = 3.37$$

The sample value $\hat{\pi} = .455$ is about 3.37 standard errors above the hypothesized value $\pi_0 = .30$ (see Figure 10.4). The probability of so large a difference (see Table A.2) is

$$P(Z > 3.37) < .0005$$

This probability ($P < .0005$) is low, below $\alpha = .05$, so we should *reject* H_0. The evidence suggests that this sample of 99 cases was *not* drawn randomly from a population where $\pi \leq .30$. The research hypothesis, that it came from a population where $\pi > .30$, is supported.

---

tude Test (SAT) scores than do college students in general. At one college, the mean math SAT for the entire student population was 536. In a random sample of 42 math majors at this college, the mean score was $\overline{X} = 620$, with a standard deviation of $s_X = 67.7$. These sample results can be used to test

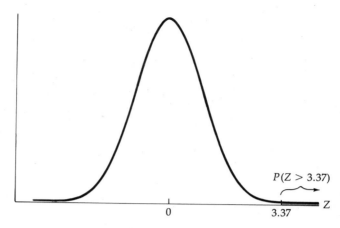

FIGURE 10.4

Shaded area corresponds to obtained P-value < .0005: The probability of a Z statistic greater than 3.37, if $\pi = \pi_0$ (see Table 10.7)

the research hypothesis that in the school's population of math majors, the mean math SAT score (μ) exceeds the mean for all students (536):

H_0: $\mu \leq 536$

H_1: $\mu > 536$

The null hypothesis H_0 states that the math majors' mean is not above 536. We will test this hypothesis using $\alpha = .001$ as the cutoff point.

Even with so low an α, Table 10.8 (page 300) shows that the null hypothesis is easily rejected. The sample mean, $\overline{X} = 640$, is more than 8 standard errors above the hypothetical population value, $\mu_0 = 536$.

As shown in Figure 10.5, there is very little area left in the tail of a normal distribution 8 standard errors above the mean. The probability of making a Type I error, if we reject H_0 here, appears from Table A.2 to be less than .0000005.[5] Since this obtained P-value is well below the $\alpha = .001$ cutoff, we can comfortably reject H_0. The difference between math majors and the student body as a whole is statistically significant at $\alpha = .001$.

In summary, for a one-sided test:

1. The null and research hypotheses are stated in directional form. Skip the testing if the sample statistic is in the direction specified by H_0.
2. Calculate the test statistic using the null hypothesis parameter value (π_0 or μ_0) that is closest to H_1. For example, if the null hypothesis is H_0: $\mu \leq 536$, calculate the Z statistic using $\mu_0 = 536$, which is the value under H_0 that is closest to H_1: $\mu > 536$.
3. Evaluate the Z test statistic for area in a single tail of the normal distribution (the "one-sided tests" column in Table A.2).

TABLE 10.8 ***Large-Sample Test for a Hypothesis About a Mean (One-Sided): Math SAT Scores from a Sample of 42 College Mathematics Majors***

Null hypothesis: H_0: $\mu \le 536$
Research hypothesis: H_1: $\mu > 536$
Sample size (marginally large enough for normal approximation): $n = 42$
Sample mean math SAT: $\overline{X} = 620$
Sample standard deviation of math SAT: $s_X = 67.7$

The sample mean is in the direction specified by H_1, so we should proceed with the test.
The estimated standard error is

$$SE_{\overline{x}} = \frac{s_X}{\sqrt{n}}$$

$$= \frac{67.7}{\sqrt{42}} = 10.45$$

The test statistic is

$$Z = \frac{\overline{X} - \mu_0}{SE_{\overline{x}}}$$

Substituting the sample mean $\overline{X} = 620$, the hypothesized population mean $\mu_0 = 536$, and the estimated standard error $SE_{\overline{x}} = 10.45$ into this equation yields

$$Z = \frac{620 - 536}{10.45} = \frac{84}{10.45} = 8.04$$

The sample mean $\overline{X} = 620$ is 8.04 estimated standard errors above the hypothesized population mean $\mu_0 = 536$ (see Figure 10.5). The probability of obtaining a sample mean so far above the population mean (see Table A.2) is

$$P(Z > 8.04) < .0000005$$

This probability is quite low, so even at $\alpha = .001$ we should *reject* H_0. The evidence suggests that this sample of 42 cases was *not* drawn randomly from a population where $\mu = 536$; we accept the alternative that it came from a population where $\mu > 536$.

PROBLEMS

18. Pick-ups of household hazardous wastes have been conducted in many different communities. Young adults are less likely than older age groups to participate. At a hazardous waste pick-up in the city of Dover, for example, 178 participants were surveyed, and 37% fit the 20–39 age group. According to census data, 49% of Dover's adult population is aged 20–39. Test at $\alpha = .05$ whether the proportion of younger adults among

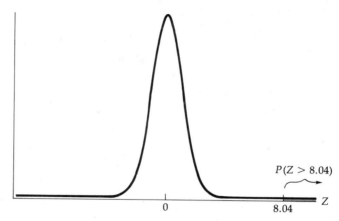

$P(Z > 8.04)$

0   8.04   Z

FIGURE 10.5   *Shaded area corresponds to an obtained P-value < .0000005: The probability of a Z statistic greater than 8.04, if $\pi = \pi_0$ (see Table 10.8)*

the pick-up participants is significantly lower than their proportion in the city's adult population as a whole. Summarize and explain your findings.

19. Among the 178 pick-up participants described in Problem 18, only 6% lived in apartments. In the city's population as a whole, 33% of the adult residents live in apartments, but it seems likely that apartment-dwellers would accumulate fewer hazardous chemicals than do people who live in houses. Test at $\alpha = .05$ whether the proportion of apartment dwellers among the pick-up participants is significantly lower than their proportion citywide.

20. During the 1987 season, 55% of all players in the National Football League were black. In a random sample of 30 players at what are considered to be "skill" positions (quarterback, wide receiver, running back, cornerback, and safety), 24 were black. Test at the $\alpha = .05$ level whether the proportion of blacks at skill positions is higher than the proportion of blacks among NFL players in general.

10.9 *SAMPLING FROM FINITE POPULATIONS**

Most of the standard procedures for statistical inference are based on the **infinite population model:** the assumption that the sample at hand was drawn from an infinitely large population. This assumption is usually untrue. The population of all voters in America, or of all mosquitoes in Maine, may be large but it is not infinite. So long as the sample size is only a tiny fraction of the population size, however, it does not matter that the population is not

TABLE 10.9   *Finite Population Correction Multiplier*

Sample as percentage of population 100(n/N)	Correction: Multiply SE by $\sqrt{(N-n)/(N-1)}$
1%	.995
5	.975
10	.949
20	.895
30	.837
40	.775
50	.707
60	.633
70	.548
80	.447
90	.316
95	.224
99	.100
100	.000[a]

[a]The sample *is* the population.

really infinite. A typical survey sample of 500 voters is only about 1/200,000th of the population of American voters, for instance. We can reasonably proceed with statistical inference as though the population were infinite.

When the sample represents a sizable fraction of the total population, however, the infinite population model is less appropriate. One remedy is the *finite population correction* (see the box).

The **finite population correction** consists of multiplying the estimated standard errors by

$$\sqrt{\frac{N-n}{N-1}}$$ [10.9]

where *n* is the sample size and *N* is the population size.

Once the standard errors have been corrected by this multiplication, we apply them to confidence intervals or hypothesis tests as usual.

Equation [10.9] will produce a number less than 1, unless the sample consists of only one case or the population is actually infinite. Otherwise, multiplying by Equation [10.9] shrinks our estimate of the standard error. If *n* (sample size) is small relative to *N* (population size), the shrinkage is unimportant. But as the fraction *n/N* enlarges, the degree of shrinkage also grows; see Table 10.9. When sample size is a large fraction of the population, less sample-to-sample variation occurs.

In Chapter 8 we dealt with samples of 20 and 100 cases, drawn from a population of 11,877 homicides. For samples of $n = 100$, the finite population correction is

$$\sqrt{\frac{11,877 - 100}{11,877 - 1}} = \sqrt{.99166} = .9958$$

Multiplying standard errors by the correction factor .9958 does not reduce them much. A sample of 100 cases is such a small fraction of 11,877 that it does little harm to regard the population as "infinite." On the other hand, the sample of $n = 42$ math majors in Table 10.8 was drawn from a population of only $N = 93$, the total number of math majors at this college. The correction factor is

$$\sqrt{\frac{93 - 42}{93 - 1}} = \sqrt{.55435} = .7445$$

The corrected standard error for Table 10.8 is then found by taking the estimated standard error, $SE_{\bar{x}} = 10.45$, and multiplying it by .7445 to obtain the corrected standard error

$$SE_{\bar{x}}^* = SE_{\bar{x}}\sqrt{\frac{N - n}{N - 1}} = 10.45(.7445) = 7.78$$

This corrected standard error, notably smaller than the uncorrected value of 10.45, would lead to an even higher Z statistic in Table 10.8.

It is possible to have a sample that includes the entire population of interest, so that $n = N$. This happens when the sample consists of data on the 50 American states, for instance. These 50 states are not a sample from some larger population of states; they *are* the population. When $n = N$, the finite population correction becomes

$$\sqrt{\frac{N - n}{N - 1}} = \sqrt{\frac{N - N}{N - 1}} = \sqrt{\frac{0}{N - 1}} = 0$$

This implies that we should multiply standard errors by 0, making the standard errors 0 themselves, if $n = N$. In other words, there is no uncertainty when we draw inferences about a population that is the same as our sample; confidence intervals or hypothesis tests are unneeded.

Although this sounds reasonable, most practicing researchers do not go so far. Instead, they continue to use standard (uncorrected) inferential procedures even when the sample constitutes most or all of the population. In this context, the inferential procedures do not really function *as inference*, to help us make guesses about an unknown population. Rather, the inferential procedures function *as description*, to help us describe what we see in the sample. Significance tests are interpreted in terms of whether a given result could be due to chance, in a vaguer sense than the strict interpretation based on random sampling from an infinite population. This vaguer usage can be justified mathematically in terms of a hypothetical infinite population, even where a real one does not exist.

The use of inferential techniques as description and the mathematical appeal to hypothetical populations also apply to nonrandom samples drawn from populations of any size. Nonrandom samples are so common as to be the rule, not the exception, in many fields. Researchers dealing with nonrandom samples, like researchers dealing with finite populations, often go ahead and use standard inferential procedures to draw descriptive, rather than strictly inferential, conclusions.

PROBLEMS

*21. A survey of 600 voters finds that 54% prefer candidate Smith. Use finite population corrections in your answers.

   a. Assume that the voting population of interest is the 1.5 million voters in the state of Colorado, and our $n = 600$ sample was drawn randomly from this population. Find a 99% confidence interval for the proportion supporting Smith. Can you be confident that Smith is supported by a majority?

   b. Assume that the voting population of interest is the 1,000 students at Lakewood High School, and our $n = 600$ sample was drawn randomly from this population. Again, find a 99% confidence interval. Can you be confident that Smith is supported by a majority?

   c. What effect does the finite population correction have on the standard errors? On the width of the resulting confidence intervals? Explain why it makes sense that your conclusions about candidate Smith's support in parts a and b differ.

*22. A sample of 36 cities is selected at random from the population consisting of 168 large U.S. cities. Apply the finite population correction to the following problems.

   a. Test at $\alpha = .05$ the null hypothesis that among large U.S. cities, the mean population growth rate over 1970–1980 was 0 or less. In the 36-city sample, the mean population growth rate was 8.6%, with a standard deviation of 19.1. (Note that this requires a test for a mean, not for a percentage or a proportion.)

   b. Construct a 95% confidence interval for mean number of homicide victims per 100,000 population per year (1980–1984). In the 36-city sample, the mean is 10.4, with a standard deviation of 9.2.

SUMMARY

Hypothesis tests ask, is it plausible that this sample could have been randomly drawn from a population characterized by certain parameter values? Unless our sample is random, this is not quite the same thing as asking whether the population *is* characterized by those parameter values. We could properly reject a null hypothesis either because

1. the null hypothesis parameter values are wrong, or
2. the parameter values are correct for some population, but not for the population our sample came from.

The first conclusion can be drawn with confidence only if the sampling is random. If sampling is nonrandom, the resulting sample might best be viewed as coming from a different population, perhaps a subset of the original population of interest. It will not necessarily tell us much then, about that original population.

Before running a hypothesis test, the researcher must decide several things. They include:

1. Identifying what kind of test is appropriate. In this chapter we have examined only tests for proportions and means, but many other kinds of tests are available.
2. Verify whether the sample is large enough that the Z statistic will have an approximately normal distribution, or if a small-sample test (Chapter 11) is required.
3. Specify the null (H_0) and research (H_1) hypotheses. At this point either a one- or a two-sided test is chosen.
4. *Optional:* Choose a cutoff point (α) for the P-value. The null hypothesis will be rejected only if the obtained P-value turns out to be below α.

The last step is optional because some researchers prefer to report only the obtained P-value and leave it to the reader to decide whether H_0 should be rejected. Their strategy avoids the objection that any cutoff point is arbitrary, and that there is little practical difference between findings where $P = .051$ and $P = .049$. With a rigid cutoff point of $\alpha = .05$, the $P = .049$ result would lead to rejecting H_0, whereas the $P = .051$ would not. Thus tiny actual differences could lead to opposite conclusions.

Once the preliminary decisions in Steps 1–4 are made, the researcher calculates the Z statistic and compares its value with the standard normal distribution. This permits estimating *the probability of obtaining sample results so favorable to H_1, if the sample came randomly from a population in which H_0 is true.* This probability is called the obtained P-value. The lower the obtained P-value, the stronger the evidence against H_0 and in favor of H_1.

A final step is to state the conclusions with reference to the original problem. If possible, conclusions should be worded so as to be comprehensible by nonstatisticians. For example, we might summarize our findings in the chemical pick-up analysis (Table 10.7 and Figure 10.4) like this:

> "Pick-up participants included a significantly higher proportion of college graduates, compared with the adult population of Portsmouth as a whole."

Clearly explaining analytical results provides a real test of our understanding.

Hypothesis tests are most straightforward in large samples, where we can rely on the Central Limit Theorem for assurance about the form of the appropriate sampling distributions. In smaller samples the Central Limit Theorem does not hold, and specific small-sample tests are needed. The small-sample analyst may also have to depend more heavily on assumptions about how variables are distributed. In this chapter, we looked at large-sample tests for percentages and for means. Their small-sample counterparts are introduced in Chapter 11.

PROBLEMS

23. In July 1986, a public opinion survey was commissioned by station WBZ, regarding attitudes toward the operation of the unfinished Seabrook nuclear power plant. Of 400 people polled, 35% said they were *in favor* of allowing the plant to operate, and 53% said that they were *opposed* to letting it operate. The remainder didn't know or had no opinion. How strong is the evidence for the research hypothesis that "a majority of the public *opposes* letting the Seabrook plant operate"? Carefully state the null and research hypotheses, carry out a one- or two-sided hypothesis test ($\alpha = .05$), and state your conclusions.

24. Suppose our research hypothesis in Problem 23 had been that "a majority of the public *favors* allowing the Seabrook plant to operate." What would the complementary null hypothesis be? Could these data possibly lead to the rejection of this null hypothesis? Explain your reasoning.

25. In the population of all 1980 homicide offenders described in Chapter 8, the mean age was $\mu = 32.28$ years. For a random sample of $n = 122$ 1984 homicide offenders, the mean was slightly higher, 32.61 years, with a standard deviation of 12.32. Test whether this evidence is enough to reject the hypothesis that the mean age of all homicide offenders was the same in 1984 as in 1980. Use the $\alpha = .01$ significance level.

26. In a sample of 7,500 male smokers, the proportion surviving to at least age 55 is found to be .88. In the population of nonsmoking males, the proportion surviving is .95. Is the proportion of smokers surviving significantly below that for nonsmokers? Conduct a test for this research hypothesis at the $\alpha = .001$ level of significance.

27. Refer to the cigarettes example in Problem 26.
    a. What would a Type I error mean, in the context of this problem? Discuss possible practical consequences of such an error.
    b. What would a Type II error mean, in the context of this problem? Discuss possible practical consequences of such an error.
    c. Which type of error is *possible,* in view of your decision regarding the null hypothesis?

28. A 1987 study (Weisman and Teitelbaum) found that in a sample of 476 married female physicians, 217 had husbands who were also physicians. Test (at the $\alpha = .001$ significance level) whether the proportion of married female doctors with physician spouses is different from the proportion of married male doctors who have physician spouses, which is $\pi = .067$.

29. The 476 female physicians in the study in Problem 28 expected to have an average of 2.6 children, with a standard deviation of 1.0. Test whether the number of children expected is significantly different, at the $\alpha = .001$ level, from the mean number of children expected by married male physicians ($\mu = 2.9$).

30. Health researchers have tracked trends in national fitness by collecting data on such variables as the amount of body fat in schoolchildren. One simple test of body fat is based on skinfold thickness: the thicker the skinfold (measured at back of arm and at shoulder blade), the more body fat. In 1960 the average skinfold thickness of 9-year-old children was 14 millimeters for boys and 17 millimeters for girls.
    a. Test (at $\alpha = .01$) whether the mean body fat of 9-year-old boys was significantly higher in 1980 than it was in 1960, based on a 1980 sample of $n = 475$ with $\overline{X} = 17$ and $s_X = 4$.
    b. Conduct a similar test for 9-year-old girls, based on a sample of $n = 502$ with $\overline{X} = 21$ and $s_X = 5$.

31. Discuss the meaning of Type I and Type II errors for the body fat analysis of Problem 30.

32. The following tests refer to the data on college course teaching evaluations and grades in Chapter 8, Problem 19 (page 236). Assume that these 118 courses constitute a random sample of all classes at a university. Use $\alpha = .05$. For each test state the null and research hypotheses and whether it calls for a one- or a two-sided test.
    a. Test whether the mean percentage of A and B grades in classes at this university is over 50%. Sample statistics: $\overline{X} = 56.9$, $s_X = 14.7$, $n = 118$.
    b. Test the hypothesis that half the classes at this school are taught by part-time instructors and teaching assistants. Sample statistics: $\hat{p} = 39.8\%$, $n = 118$.
    c. Test the hypothesis that the mean class size is 40 students. Sample statistics: $\overline{X} = 51.7$, $s_X = 27.1$, $n = 118$.

33. Use the water conservation data set of Table 2.8 (page 41) to evaluate the following null hypotheses, using the $\alpha = .05$ significance level. For each hypothesis, give the null and research hypotheses, and state whether a one- or a two-sided test, or neither, is appropriate. If a test is needed, calculate the test statistic and interpret your conclusions in English.

    a. By their own estimate, half or more of Concord's households saved no water.

    b. The mean 1981 household income in this city was the same as the national mean: $25,448.

    c. Household heads in this city have, on the average, one year of college (13 years education).

*34. The sample of 30 National Football League players at skill positions (Problem 20) was drawn from a total population of only 252 players. Since the sample is more than 10% of the population, a finite population correction is appropriate.

    a. Recalculate the test statistic using the correction.

    b. In Problem 20, without making any correction, we decided to reject the null hypothesis. Compare your results from Problem 20 with those from part a. What generalization can you make about the effect of a finite population correction, if H_0 would be rejected *without* a correction?

    c. Under what circumstances could hypothesis tests with and without corrections lead to opposite conclusions?

NOTES

1. A very low probability such as .000001 is awkward to think about; reading this number, you probably just scanned it as having many zeroes. To put small numbers in more understandable terms, try moving the decimal place. For example, a probability of .000001 corresponds to:

.000001 chance in 1
.00001 chance in 10
.0001 chance in 100
.001 chance in 1,000
.01 chance in 10,000
.1 chance in 100,000
1 chance in 1,000,000

As we said, "one chance in a million."

2. A very low P-value such as this ($.000001 < P < .00001$) should be taken with a grain of salt. The sampling distribution of a proportion is only *approximately* normal. Such approximations tend to be least reliable in the

extreme tails of the distribution, where normal probabilities are very low. We can be satisfied that the *P*-value is clearly low enough to reject H_0, without having to believe that it can be specified with such precision as between .000001 and .00001.

3. A researcher can do some things to reduce the probability of Type I or Type II error without such a trade-off. For example, certain statistical techniques are more *powerful* than others, meaning they give a lower risk of Type II error (β) for the same levels of α. Collecting more data to increase the sample size also will usually reduce the probability of both types of error.

4. When the standard error is estimated from sample data we should theoretically use the *t* distribution, introduced in the next chapter, rather than the standard normal or *Z* distribution. With large samples the *t* distribution is almost the same as the *Z* distribution, however, so it does not much matter which one we use.

5. The cautionary remarks in Note 2 apply here as well. The sampling distribution of the mean is only approximately normal, so probabilities cannot really be specified with as much precision as $P < .0000005$.

Small-Sample Inference for Means and Proportions

Chapter 11

The Central Limit Theorem says that for large enough samples, the sampling distribution of the mean is approximately normal. This holds true even if, as is often the case, the variable itself has a nonnormal distribution. Proportions too have approximately normal sampling distributions for sufficiently large samples. These properties make finding confidence intervals and conducting hypothesis tests straightforward.

The normal approximation does not work well in smaller samples, so other sampling distributions are needed. Several are introduced in this chapter. One, the *t distribution*, which resembles the normal distribution, helps with small-sample inferences involving measurement-variable summaries such as the mean. The *binomial distribution*, on the other hand, applies to categorical variables and to summaries such as proportions or percentages.

The question "When is a sample small?" was discussed in Chapter 10. It is a matter of degree: For any given variable, the smaller the sample, the worse the normal approximation is likely to be. A rule of thumb is to use the

TABLE 11.1   *Sampling Distributions for Inference Involving Means (Measurement Variable) and Percentages or Proportions*

Type of variable	Statistical summary	Large sample	Small sample
Measurement	Mean	Normal *or t* distribution if $n > 30$	*t* distribution if $n \leq 30$
Categorical	Percentage or proportion[a]	Normal distribution if $n\pi \geq 9$ *and* $n(1 - \pi) \geq 9$	Binomial distribution if $n\pi < 9$ *or* $n(1 - \pi) < 9$

[a]π refers to a population proportion of cases in a certain category. Consequently $n\pi$ is the *number* of cases we would expect in that category, in a sample of size *n*. If the population proportion π is unknown, or not supplied by a null hypothesis, then the sample proportion may be used in these calculations.

t distribution for inferences involving single means whenever the sample contains 30 cases or less; it is optional with larger samples. Strictly speaking, the *t* distribution is appropriate whenever the population standard deviation, σ_X, is unknown and estimated from sample data—as is nearly always the case. For samples of $n > 30$ the *t* and standard normal (Z) distributions are similar, however.

With proportions, a rule of thumb is that the normal approximation works as long as at least nine cases are expected in the variable's less numerous category. More detailed guidelines for proportions were given in Table 10.5 (page 291). Table 11.1 summarizes these points about sampling distributions and sample size.

We begin this chapter by introducing a small-sample problem to solve later on, after sections on the theoretical *t* distribution and how to use it. Then we will turn our attention to the binomial distribution and small-sample inference for proportions.

11.1   A SMALL-SAMPLE PROBLEM

A maker of computer chips commissioned a study to investigate reports of abnormally high rates of miscarriage among female workers exposed to chemicals in the chip manufacturing process.[1] The researchers determined that among hundreds of female workers who were not exposed to chemicals, the proportion of pregnancies ending in miscarriage was .178. Among a sample consisting of $n = 18$ exposed workers, this proportion was much higher: .389. At first glance, the proportion for exposed workers seems so high that it must reflect the chemicals' harmful properties.[2] Possibly, though, the higher rate among the exposed workers in this particular sample is due solely to chance. How likely is this?

This problem can be formulated as a test of hypotheses about the proportion of miscarriages in the population of female workers who are exposed to chemicals. If the chemicals are harmful, then this proportion will be higher than the proportion established for nonexposed workers, .178. Thus, our research hypothesis is

$$H_1: \quad \pi > .178$$

The null hypothesis is that the proportion of miscarriages among the exposed population is not higher than .178,

$$H_0: \quad \pi \leq .178$$

If we reject this null hypothesis, we will decide that the evidence favors H_1: The miscarriage rate *is* higher among exposed workers than among nonexposed workers.

A sample of only $n = 18$ pregnancies among exposed workers is too small for the normal approximation methods described in Chapter 10. (You can confirm this by applying Equation [10.5], using $\pi = .178$ and $n = 18$.) Yet the problem is urgent, and the people involved do not want to wait another year or two while more data are collected. We need statistical methods that are suitable for use with small samples.

Small-sample methods for means (measurement variables) and proportions (categorical variables) are quite distinct. We will look first at small-sample methods for means, which resemble their large-sample counterparts.

11.2 THE t DISTRIBUTION

Standard errors measure the expected sample-to-sample variation in a statistic. According to the Central Limit Theorem, for large samples the standard error of the mean is

$$\sigma_{\bar{X}} = \frac{\sigma_X}{\sqrt{n}} \qquad [11.1]$$

We usually do not know the population standard deviation (σ_X), so we use the sample standard deviation (s_X) to estimate the standard error:

$$SE_{\bar{X}} = \frac{s_X}{\sqrt{n}} \qquad [11.2]$$

In large samples, the sample standard deviation s_X provides a good enough estimate for this purpose. In small samples, however, s_X, like any other statistic, grows more variable. This increased small-sample variability increases the likelihood that we will substantially over- or underestimate the standard error. That is, our estimated $SE_{\bar{x}}$ could be considerably larger or smaller than the true standard error $\sigma_{\bar{x}}$—which partly explains why the normal approximation worsens in small samples.

A better approximation is provided by a distribution called **Student's *t*,** known more simply as the ***t* distribution.**[3] The *t* distribution resembles the standard normal distribution but has heavier tails. There are actually many *t* distributions, each characterized by a parameter called the **degrees of freedom.**

Degrees of freedom relate to sample size, so that larger samples allow more degrees of freedom.[4] As the degrees of freedom increase, the *t* distribution comes closer to resembling a normal distribution. A *t* distribution having infinite degrees of freedom is identical to the standard normal (Z) distribution. With fewer degrees of freedom, the *t* distribution grows progressively heavier tailed than the normal distribution. Figure 11.1 illustrates this comparison, showing *t* distributions with 1, 10, and infinite degrees of freedom.

Since the *t* and Z distributions are virtually indistinguishable in large samples, which offer many degrees of freedom, it matters little which one we use. It does matter with small samples, where degrees of freedom are few. There the normal approximation could let us underestimate the probability of high or low values, out in the tails of the distribution.

Critical values for the *t* distribution are given in Appendix Table A.3. Each row in Table A.3 corresponds to a certain degrees of freedom (**df**) value, from 1 to infinity (∞). The body of the table contains specific *t* values. Since the *t* distribution, like the Z distribution, is symmetrical, only positive *t*

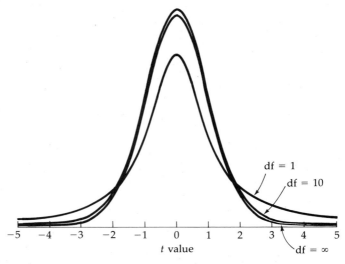

FIGURE 11.1   *t distributions with 1, 10, and infinite degrees of freedom*

values need be shown. Probabilities for confidence intervals, two-sided tests, and one-sided tests are given across the top of Table A.3. Notice that as with the Z distribution (Table A.2) these probabilities have a pattern:

1. For any given t value, the confidence interval probability (area in the center of the distribution) plus the two-sided test probability (area in the two tails of the distribution) equals 1.
2. For any given t value, the one-sided test probability (area in a single tail of the distribution) equals one-half of the two-sided test probability (area in both tails of the distribution).

Another resemblance to the normal distribution appears if you examine the critical values for a t distribution with infinite degrees of freedom: They match those for the standard normal distribution.

With large samples, one- and two-sided hypothesis tests are performed using the Z statistic,

$$Z = \frac{\overline{X} - \mu_0}{SE_{\overline{X}}} \qquad\qquad [11.3]$$

which, if H_0 is true, follows approximately a standard normal sampling distribution regardless of how X itself is distributed. For small samples, however, things are less simple.

With small samples, **one-** or **two-sided tests** of hypotheses about a mean use the **t statistic:**

$$t = \frac{\overline{X} - \mu_0}{SE_{\overline{X}}} \qquad\qquad [11.4]$$

where $SE_{\overline{X}} = s_X/\sqrt{n}$. If the population distribution of X is normal, with mean μ_0, then t will follow a **t distribution** with $n - 1$ **degrees of freedom.**

In the next section, we will look more closely at the *normality assumption* that this t test requires.

PROBLEMS

1. Use Table A.3 to find the critical value of t
   a. For a 95% confidence interval, based on a sample of $n = 25$.
   b. For a 99.9% confidence interval, based on a sample of $n = 6$.
   c. Required to reject H_0 in favor of a two-sided H_1 at $\alpha = .05$, based on a sample of $n = 14$.
   d. Required to reject H_0 in favor of a one-sided H_1 at $\alpha = .01$, based on a sample of $n = 12$.

2. Give *P*-values and state whether the null hypothesis should be rejected on the basis of the following *t* statistics.

a. $t = 2.1, n = 23, \alpha = .05$ (two-sided)
b. $t = 2.1, n = 13, \alpha = .05$ (two-sided)
c. $t = 17, n = 19, \alpha = .001$ (two-sided)
d. $t = -1.7, n = 6, \alpha = .10$ (two-sided)
e. $t = -5, n = 9, \alpha = .01$ (one-sided)
f. $t = 0.6, n = 30, \alpha = .10$ (one-sided)

11.3   *THE NORMALITY ASSUMPTION*

The *t* test is justified by the assumption that our variable's population distribution is normal. In practice, we need only assume that the population distribution is *approximately* normal. We cannot know for certain whether this **normality assumption** is true, but two sources of information help us decide whether it is believable. First, sometimes theories or findings from other studies suggest the shape of the population distribution. Second, sample distributions hold clues about the shape of population distributions.

The graphical and exploratory methods discussed in Chapters 2–5 provide important tools for checking the plausibility of the normality assumption. We can easily determine whether the *sample* distribution is symmetrical, with few outliers and approximately normal tails. The shape of the sample distribution may not match the shape of the population distribution, however. Small samples, in particular, may tell little about the shape of the population distribution. Unfortunately, the normality assumption is most crucial when samples are small.

Figures 11.2–11.4 illustrate what samples from a normal population can look like. Each figure contains box plots of 14 random samples from a standard normal population. Figure 11.2 shows samples of $n = 10$ cases each; in Figure 11.3 the samples are size $n = 50$; and in Figure 11.4 they are $n = 500$.

Among the small ($n = 10$) samples of Figure 11.2, we see much sample-to-sample variation in center, spread, and shape. Some samples are distinctly skewed, even though they came from a perfectly symmetrical population. Many of the samples have centers some distance from the true population mean (0). Such variation is to be expected in small samples, which is why it is hard to know when their source is really a nonnormal population.

Less sample-to-sample variation occurs among the medium-sized ($n = 50$) samples in Figure 11.3. The 14 shapes are more consistently symmetrical, and it is more evident that all are from the same normally distributed population. Normality becomes obvious with large samples like those in Figure 11.4 ($n = 500$). Sample-to-sample variations are slight, and each sample depicts well the true shape of the population distribution. If we saw a distinctly

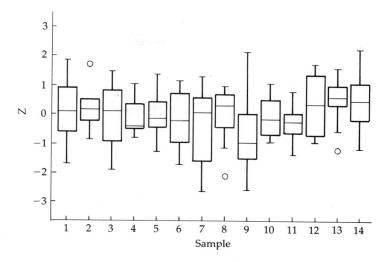

FIGURE 11.2   *Box plots of 14 random samples, size n = 10, from a standard normal population distribution*

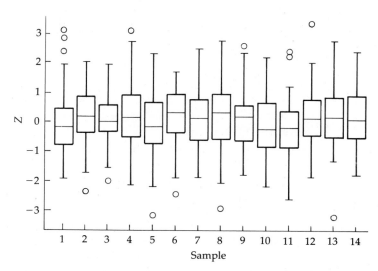

FIGURE 11.3   *Box plots of 14 random samples, size n = 50, from a standard normal population distribution*

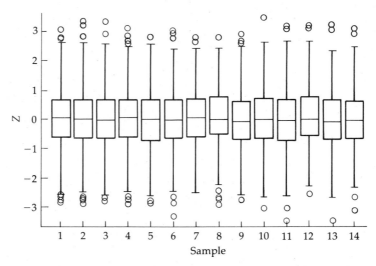

Box plots of 14 random samples, size n = 500, from a standard normal
FIGURE 11.4   population distribution

nonnormal distribution in a sample of $n = 500$, we could feel confident that
it came from a nonnormal population.

You may have noticed that as sample size increased in Figures 11.2–11.4,
so did the number of outliers. All of the $n = 500$ distributions (Figure 11.4)
contain multiple outliers, whereas only three outliers appear among the 14
$n = 10$ distributions (Figure 11.2). In normal distributions outliers occur
with a low probability, so the more cases we have the more outliers we expect. Conversely, outliers should be few in smaller samples, and the distributions are more likely to have short tails. Samples of any size should rarely
have severe outliers—that is, cases more than 3(IQR) beyond the first or
third quartile (see Chapter 5). Severe outliers occur in a normal population
fewer than three times for every one million cases.[5]

The normality assumption is less believable if we find severe (or too
many) outliers in our sample. Outliers are doubly important because

1. They may signal that the distribution is nonnormal.
2. They are a particularly troublesome *kind* of nonnormality.

Nonlinear transformation techniques (Chapter 6) may help reduce problems
with outliers and skewness. Such transformations make many distributions
more nearly normal, hence more amenable to analysis using the *t*
distribution.

PROBLEMS

3. What is a normal distribution? What sorts of nonnormality can you detect by looking at each of the following?
   a. Stem-and-leaf displays (Chapter 2)
   b. Mean–median and standard deviation–pseudo-standard deviation comparisons (Chapter 4)
   c. Box plots (Chapter 5)
4. Severe outliers occur "fewer than three times per million cases" in a normal population distribution.
   a. What is the probability that one case, drawn at random, will be a severe outlier?
   b. Show how Tables A.1, A.2, and the definition of severe outliers (Chapter 5) can be applied to derive this probability.

11.4   *CONFIDENCE INTERVALS BASED ON SMALL-SAMPLE MEANS*

Large-sample confidence intervals for means are constructed as

$$\overline{X} \pm z(SE_{\overline{X}}) \qquad\qquad [11.5]$$

where the z value is obtained from a Z distribution table to provide the desired degree of confidence. The small-sample version is quite similar. Apart from the normality assumption, the only adjustment needed is to use the t distribution in place of the Z distribution.

If the variable X has a normal distribution, then **small-sample confidence intervals** for the mean of X are obtained by

$$\overline{X} \pm t(SE_{\overline{X}}) = \overline{X} \pm t\left(\frac{s_X}{\sqrt{n}}\right) \qquad\qquad [11.6]$$

The value of t is chosen from a t distribution table with $n - 1$ degrees of freedom.

During the 1940's, 1950's, and 1960's the pesticide DDT was used extensively in U.S. agriculture, resulting in widespread contamination of wildlife and food. Laws prohibiting its use reduced new contamination after 1972, but the DDT already applied persisted in the environment. Table 11.2 (page 320) shows DDT concentrations found in lake bottom sediments along Bear Creek, Mississippi, in 1979—seven years after legal DDT applications had

TABLE 11.2   ***Pesticide Concentrations[a] in Sediments Along Bear Creek, Mississippi, in 1979***

Site	DDT at depth 0–200 mm	DDT at depth 200–400 mm
1	934.6	899.6
2	1,275.5	321.5
3	460.1	0[b]
4	673.8	200.5
5	1,275.7	1,098.4
6	362.8	41.2
7	758.3	696.7
8	557.9	922.9
9	388.7	548.1
10	226.3	253.7

[a]DDT residues in μg/kg.
[b]None detected.
Source: Cooper et al. (1987).

stopped. The scientists who collected these data were surprised that concentrations remained so high. Concentrations are given separately in Table 11.2 for sediments in the upper 200 millimeters of the lake bottom and for the next 200 millimeters down, which are older.

What is the mean DDT concentration in lake bottom sediments of this region? If the ten sites in Table 11.2 are viewed as a random sample, we can form a confidence interval for the unknown parameter μ. Small-sample methods are needed with a sample of this size ($n = 10$). Since these methods require that population distributions be approximately normal, we begin by checking the shape of the sample distributions.

Figure 11.5 shows box plots of the upper and lower depth DDT concentrations from Table 11.2. Both distributions have light tails and mild positive skewness, with no outliers. They appear to be within the range of $n = 10$ samples from a normal population (Figure 11.2) and cast no doubt on the normality assumption.

With 10 cases, we have 9 degrees of freedom (df $= n - 1$). Table A.3 indicates that with df $= 9$, the appropriate t value for a 95% confidence interval is 2.262. The confidence interval, from Equation [11.6], is therefore

$$\overline{X} \pm 2.262\left(\frac{s_X}{\sqrt{n}}\right)$$

For upper sediments (0–200 mm), the sample mean and standard deviation are $\overline{X} = 691$ and $s_X = 370$. The resulting confidence interval is

$$691 \pm 2.262\left(\frac{370}{\sqrt{10}}\right) = 691 \pm 265$$
$$\rightarrow 426 \leq \mu \leq 956$$

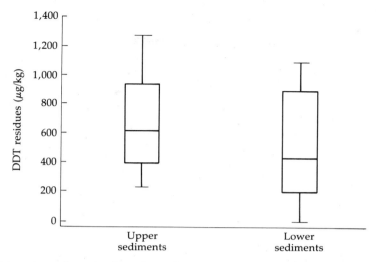

FIGURE 11.5   *DDT residues in sediments along Bear Creek, Mississippi (from Table 11.2)*

That is, we are 95% confident that in 1979 the mean DDT concentration in upper sediments of lakes along Bear Creek was between 426 and 956 micrograms per kilogram.[6]

The steps followed in constructing this interval are shown in Table 11.3. Compare Table 11.3 (page 322) with the confidence interval in Table 9.4 (page 261) to see how the large- and small-sample methods differ.

The t value used above to obtain a 95% confidence interval is 2.262. We calculated the interval by marking off a distance of 2.262 estimated standard errors on either side of the sample mean. If we had started with a larger sample and thus could use the Z distribution (or the t distribution with df = ∞), the appropriate Z or t value would be 1.96. To be 95% confident based on a large sample, we need only mark off a distance of 1.96 standard errors on either side of the mean. In a small sample we need to go farther out to obtain a given level of confidence. This difference reflects the heavier tails of the t_9 distribution (t distribution with df = 9), as compared with the normal or t_∞ distribution.

Note that even with only 9 degrees of freedom, however, the corresponding t and Z statistics are close: 2.262 vs. 1.96. A much larger difference between large- and small-sample intervals results from the fact that in smaller samples, standard errors are bigger. Standard errors measure sample-to-sample variation in the mean; a high standard error tells us to expect a great deal of variation between sample means, from one sample to the next. If the samples are small, such erratic variations are more likely than with large samples.

TABLE 11.3	***A 95% Confidence Interval for the Mean DDT Concentration in 0–200 mm Sediments (from Table 11.2)***

Sample statistics $\overline{X} = 691$

$s_X = 370$

$n = 10$

The box plot of the sample distribution in Figure 11.5 gives little reason to reject the normal-population assumption.

Calculations of confidence interval for population mean (μ)

Degrees of freedom: df $= n - 1 = 10 - 1 = 9$
t value needed for 95% confidence (Table A.3): 2.262
Sample statistic $\pm$ t(Standard error of sample statistic):

$$\overline{X} \pm t(SE_{\overline{x}}) = \overline{X} \pm t\left(\frac{s_X}{\sqrt{n}}\right)$$

$$= 691 \pm 2.262 \left(\frac{370}{\sqrt{10}}\right) = 691 \pm 265$$

The interval is therefore $426 \le \mu \le 956$.

Interpretation of this confidence interval

Informal: Based on the evidence from these ten sites we are 95% confident that in 1979 the mean DDT concentration in 0–200 mm sediments along Bear Creek was between 426 and 956 micrograms per kilogram.
Formal: If we could take a large number of random samples from this area, each with 10 sites, and calculate confidence intervals in this manner for each sample, 95% of those confidence intervals should include the true population mean μ.

This commonsense idea also follows algebraically from the definition of the standard error as a standard deviation divided by $\sqrt{n}$. With a sample of $n = 10$, the estimated standard error for upper sediment DDT levels is $SE_{\overline{x}} = s_X/\sqrt{n} = 370/\sqrt{10} = 117$ µg/kg. If we had the same mean and standard deviation, but a sample of 100 cases, then the standard error would be only $SE_{\overline{x}} = s_X/\sqrt{n} = 370/\sqrt{100} = 37$ µg/kg. The resulting confidence interval, based on 100 cases, would therefore have been only about one-third as wide as the interval needed for 95% confidence based on a ten-case sample. A narrower interval implies that with a larger sample, we could estimate more precisely the population mean μ.

PROBLEMS

5. Assuming that the variable distributions are normal, find the following confidence intervals for μ.

   a. 95% confidence interval given $\overline{X} = 34$, $s_X = 9$, $n = 26$
   b. 98% confidence interval given $\overline{X} = 0.3$, $s_X = 5$, $n = 18$
   c. 90% confidence interval given $\overline{X} = 237$, $s_X = 100$, $n = 30$

6. Construct a 95% confidence interval for the mean DDT concentration in the lower (200–400 mm) sediments, based on the data in Table 11.2. Express this interval both symbolically and in English.

11.5   A SECOND LOOK AT TABLE 9.5*

Chapters 8 and 9 described a computer experiment drawing 800 random samples, 400 each of $n = 20$ and $n = 100$. Table 9.5 (page 264) showed how to calculate confidence intervals using the means and standard deviations from each of these 800 random samples. If the Central Limit Theorem applied to this analysis, we would expect about 95% of the 800 confidence intervals to contain the population mean μ. This was not quite the case. Of the 400 intervals constructed from the larger samples, 93.5% contained μ, and for the smaller samples only 90.75% contained μ. It appears that, particularly with samples of size $n = 20$, using the normal distribution led us to construct confidence intervals that were systematically too narrow; they did not actually contain μ about 95% of the time.

There are two reasons why this analysis might have gone wrong. First, when using the sample standard deviation, s_X, to estimate the population standard deviation σ_X, it is advisable to use the t distribution instead of the standard normal. The standard normal distribution suggests that, for 95% confidence, we should construct intervals of ± 1.96 standard errors. Consulting the t distribution instead, for $n - 1$ degrees of freedom, suggests wider intervals: Intervals from $n = 20$ samples (df $= n - 1 = 19$) should be ± 2.093 standard errors wide, while intervals from $n = 100$ samples (df $= n - 1 = 99 \approx 60$) should be ± 2.0 standard errors. These corrections bring the results for the 400 samples sized $n = 100$ closer to the expected 95% accuracy: 94.25% of these 400 confidence intervals now contain μ. The slight difference between 95% (expected) and 94.25% (observed) could easily be due to random sampling variation. A problem persists with the 400 confidence intervals for samples of size $n = 20$, though: only 92.75% of them contain μ.

The second problem affects the latter samples: $n = 20$ is too small, given the nonnormal distribution of the original variable. In large samples, the t

statistic, Equation [11.4], will follow approximately a *t* distribution, regardless of the shape of the original variable's distribution. In smaller samples, this statistic follows a *t* distribution only if the distribution of the variable itself is normal. The variable in question here is the age of homicide offenders, which is positively skewed (see Figures 8.3 and 8.4, pages 221 and 224). Most offenders are relatively young, but a drawn-out tail represents the small number of older offenders in their 50's, 60's, and even 70's or 80's.

Since the $n = 20$ samples are small *and* are drawn from a nonnormal population, we cannot rely on either the Central Limit Theorem or the *t* distribution for theoretical guidance about the sampling distribution for these means. The $n = 20$ sampling distribution observed in Figure 8.6 (page 227) appears heavy tailed and positively skewed. This contrasts with the more nearly normal $n = 100$ distribution.

Table 9.5 thus shows us one of the things that can happen when the assumptions underlying an inferential procedure are untrue: We may underestimate how much sample-to-sample variation occurs in a statistic like the sample mean, $\overline{X}$. Underestimating variation makes us overoptimistic about our accuracy in estimating the population mean, μ. This mistake increases our likelihood of wrongly rejecting a true hypothesis (Type I error). Other kinds of problems can also arise.

PROBLEMS

7. Like the sample mean, sample proportions have a sampling distribution that is approximately normal when samples are sufficiently large. For large samples, 95% confidence intervals may be found as the sample proportion plus and minus 1.96 times its estimated standard error: $\hat{\pi} \pm 1.96(SE_{\hat{\pi}})$.

   A computer experiment like that in Chapter 8 was performed with the population consisting of all 1984 U.S. homicides. In this population $\pi = .171$ of the homicides were committed by strangers, and the remaining $1 - \pi = .829$ were committed by relatives or acquaintances of the victim.

   For the computer experiment 900 random samples were drawn from this population: 300 each of size $n = 5$, $n = 25$, and $n = 125$. For each of these 900 samples the proportion of homicides committed by strangers was calculated, as if to estimate the population parameter π. The formula $\hat{\pi} \pm 1.96(SE_{\hat{\pi}})$ was used to find a "95% confidence interval" for each of the 900 samples. If these intervals were valid, about 95% of them should contain the true population proportion, $\pi = .171$. The actual percentage of confidence intervals containing π follows:

   In 300 $n = 5$ samples, 189 (63%) of the confidence intervals contained the true value of π.

   In 300 $n = 25$ samples, 273 (91%) of the confidence intervals contained the true value of π.

In 300 $n = 125$ samples, 288 (96%) of the confidence intervals contained the true value of π.

What pattern do you see in these results? How do you explain it?

8. Apply Equation [10.5] to determine whether samples of size $n = 5$, $n = 25$, or $n = 125$ *should* be large enough for the normal approximation to work. Is your conclusion consistent with the results observed in Problem 7?

11.6 SMALL-SAMPLE HYPOTHESIS TESTS FOR MEANS

Like confidence intervals, small-sample hypothesis tests using the t statistic are justified by the normality assumption. Otherwise t tests, Equation [11.4], are performed much like Z tests (Chapter 10).

Table 11.4 contains data on a sample of 20 hospitals, which were among a group of hospitals claimed to be "underfunded" and in need of $120 million in state assistance. Opponents of this expenditure questioned whether these hospitals really were much worse off than others in the state, which would lose money if the aid bill passed. Table 11.4 reports 1986 occupancy rates,

TABLE 11.4   *Financial Data on 20 "Underfunded" Hospitals*

Hospital	1986 occupancy rate	1986 operating margin	1983–1986 change in operating margin
1	70.0%	6.99%	353.0%
2	73.4	1.02	24.0
3	78.6	1.78	242.9
4	52.3	−.94	−93.8
5	65.9	3.23	411.3
6	73.1	.59	−27.0
7	65.1	3.31	631.9
8	59.6	.79	132.8
9	54.0	.26	120.2
10	58.9	4.99	338.1
11	62.7	4.57	2,857.1
12	47.5	1.24	−9.2
13	69.9	2.34	278.2
14	63.9	−14.70	−1,042.5
15	62.4	1.17	−5.6
16	59.5	−.43	93.1
17	54.6	.76	131.8
18	70.1	−.03	—
19	53.5	5.51	76.2
20	50.7	2.13	22.0

Source: Data adapted from the *Boston Globe*, November 1, 1987.

operating margins, and operating margin gains or losses. We will use these data to illustrate *t* tests, beginning with a test of the hypothesis that the occupancy rate of "underfunded" hospitals is significantly below the state average. The average 1986 occupancy rate for all hospitals in this state was 65.6%, so we will test the hypotheses

H_0: $\mu \geq 65.6$

H_1: $\mu < 65.6$

With only 20 cases, small-sample procedures are required for this one-sided test.

A box plot of occupancy rates for these 20 hospitals (Figure 11.6) depicts the sample distribution as light tailed but relatively symmetrical. Its appearance does not challenge the normality assumption (compare with Figures 11.2 or 11.3). The sample mean is $\overline{X} = 62.3$ (median $Md = 62.6$), and the standard deviation is $s_X = 8.6$. The sample mean is somewhat lower than the hypothesized population mean $\mu \geq 65.6$, but might not this small difference be due to chance?

Table 11.5 shows steps for the appropriate one-sided *t* test at $\alpha = .05$. The sample mean ($\overline{X} = 62.3$) is 1.716 estimated standard errors below the hypothesized population mean ($\mu_0 = 65.6$). A negative distance this large, with $n - 1 = 19$ degrees of freedom, could arise by chance with a probability just above .05, so we cannot reject H_0 at the $\alpha = .05$ level. (A *t* value less than -1.729 would be required to reject H_0 here.) The occupancy rate for the 20 "underfunded" hospitals is not significantly lower than the state mean, 65.6%.

The claim that these hospitals are underfunded implies that their operating margins (excess of income over expenditures) are low. This idea too might

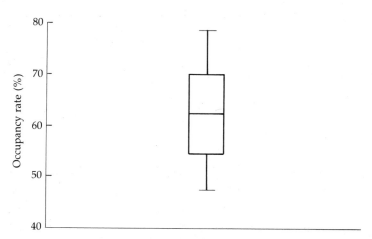

FIGURE 11.6   *Box plot of occupancy rates in 20 hospitals (from Table 11.4)*

TABLE 11.5	*Small-Sample Test at* α = .05 *for a Hypothesis About a Mean (One-Sided): Hospital Occupancy Rates (from Table 11.4)*

Null hypothesis: H_0: $\mu \geq 65.6$
Research hypothesis: H_1: $\mu < 65.6$
Sample size: $n = 20$
Sample mean: $\overline{X} = 62.3$
Sample standard deviation: $s_X = 8.6$

The sample mean is in the direction specified by H_1, so we should proceed with the test. The sample is fewer than 30 cases, calling for a t test with $n - 1 = 19$ degrees of freedom.

The sample distribution of this variable appears reasonably well behaved (Figure 11.6). If the population distribution of this variable is normal, and samples are drawn randomly from a population where $\mu = \mu_0 = 65.6$, t statistics as calculated below should follow a theoretical t distribution.

The estimated standard error is

$$\text{SE}_{\overline{X}} = \frac{s_X}{\sqrt{n}} = \frac{8.6}{\sqrt{20}} = 1.923$$

The test statistic is

$$t = \frac{\overline{X} - \mu_0}{\text{SE}_{\overline{X}}}$$

Substitution into this equation yields

$$t = \frac{62.3 - 65.6}{1.923} = -1.716$$

The sample mean $\overline{X} = 62.3$ is 1.716 estimated standard errors below the hypothesized population mean $\mu_0 = 65.6$. For df = 19 (Table A.3) this t value has a one-sided probability of $.05 < P < .10$.

Since the obtained P-value is not below $\alpha = .05$, we *cannot reject* H_0. The mean occupancy rate of these 20 "underfunded" hospitals is not significantly lower than 65.6, the statewide mean.

be tested with the data in Table 11.4. The mean operating margin for all hospitals in this state is 1.59%, so the appropriate null and research hypotheses are

H_0: $\mu \geq 1.59$

H_1: $\mu < 1.59$

For the 20 hospitals in Table 11.4, the mean operating margin is $\overline{X} = 1.23$, with $s_X = 4.3$. Consequently

$$t = \frac{1.23 - 1.59}{4.3/\sqrt{20}} = -.374$$

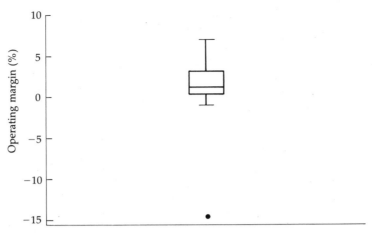

FIGURE 11.7   *Box plot of operating margins in 20 hospitals (from Table 11.4)*

From Table A.3 we obtain $P > .25$. Again we are unable to reject the null hypothesis and must conclude that the "underfunded" hospitals do not have significantly lower operating margins.

This conclusion is only tentative, however, because the sample distribution of operating margins (Figure 11.7) does not support the normality assumption. Hospital 14 operated with a large deficit of -14.70% in 1986, a value no other hospital approached, hence is a severe outlier. Severe outliers are uncommon when samples are drawn from a truly normal population.

In Figure 11.7 the normality assumption is challenged by a single case that appears inconsistent with the rest of the distribution. Outlier deletion (Chapter 6) may therefore be an appropriate strategy. Yet deleting outliers restricts the population about which we can draw inferences. We could no longer generalize about all "underfunded" hospitals in a state; rather we would be restricted to a population consisting of "underfunded" hospitals free of large operating deficits.

Often the normality assumption is challenged by the overall shape of a distribution, not just by a single case. If so, outlier deletion is inappropriate, and nonlinear transformations might be considered instead. The next section gives an example.

PROBLEMS

9. Delete Hospital 14 in Table 11.4, then recalculate $\overline{X}$, s_X, and $SE_{\overline{X}}$ for the test regarding hospital operating margin (H_0: $\mu \geq 1.59$ vs. H_1: $\mu < 1.59$) performed earlier. Is it necessary to redo the test itself? Explain the differ-

ences between the two analyses ($n = 19$ and $n = 20$) with respect to the following.

    a. Sample mean

    b. Sample standard deviation

    c. Estimated standard error

    d. Conclusion regarding H_0

10. Refer to the hospital data of Table 11.4 and test the null hypothesis that the mean operating margin change among "underfunded" hospitals is the same as the state average, 115.8%. Use a two-sided test at $\alpha = .05$.

11. Construct a stem-and-leaf display or a box plot of operating margin change for the hospitals used in Problem 10. Does this visual display appear consistent with the normality assumption? Explain your answer.

11.7 INFERENCES ABOUT MEANS OF NONNORMAL DISTRIBUTIONS*

The validity of the t distribution for small-sample inferences about means depends on the assumption that a variable's population distribution is normal. Taken literally, this assumption is very restrictive, for as we have seen many distributions are not even symmetrical, much less normal. Happily, research has shown that t tests and confidence intervals are somewhat **robust.** This means that the techniques produce reasonable results even with moderately nonnormal data. How reasonable the results are depends partly on the kind and extent of the departure from normality, however. Outliers and skewness in particular can hamper t-based inference. For example, our second look at Table 9.5 suggested that the skewed, nonnormal distribution of homicide offenders' ages led us to construct overly optimistic small-sample confidence intervals. We risk wrong conclusions when we assume nonnormal distributions are normal.

There is a second problem with drawing inferences about means in nonnormal distributions. If a distribution is at least symmetrical, with a central peak, then the mean accurately summarizes where the center of the distribution lies. Other distributional shapes may not have clear-cut centers, though. Two distinct peaks, for instance, are often more usefully described in terms of *two* centers. Such a bimodal distribution may be considered a mixture of two kinds of case, each kind more easily analyzed alone. Another type of nonnormality is skewness. In skewed distributions, different definitions of center such as median and mean disagree, so the whole concept of "center" is ambiguous. Our second problem, then, is that in nonnormal distributions the mean may be the wrong parameter to study. Unlike the first problem, this one is not improved by taking larger samples.

Neither problem will be noticed by a researcher who simply calculates a mean and a standard deviation, then proceeds to analyze without actually looking at the variable's distribution. The problems will still be there, unseen and possibly sabotaging the conclusions. If we explore the data more carefully, however, and notice evidence of nonnormality, we then face the new problem of what to do about it.

One approach, which has many advantages, is to transform the variable so that its sample distribution more closely resembles a normal curve. Chapter 6 described a family of nonlinear transformations that make skewed distributions more symmetrical. If one of these transformations works, it may overcome both of the problems with nonnormal distributions just outlined.

Table 11.6 contains data on coliform bacteria counts from samples taken in 1987 at 21 Boston-area beaches. Since actual bacteria concentrations fluctuate continuously, these counts may be viewed as a sample from the infinite population of all possible bacteria concentrations at these beaches during 1987.

The box plot in Figure 11.8 reveals that our sample bacteria count distribution does not look at all normal. Like many environmental pollution meas-

TABLE 11.6   *Coliform Bacteria Counts at Boston-area Beaches, 1987*

Beach	Average coliform count per milliliter	Highest measured coliform count per milliliter
Yirrel Beach	8	48
Short Beach	8	32
Houghton's Pond	9	40
Sandy Beach	10	40
Stacey Brook Outlet	10	> 4,000
Nantasket Beach	12	148
Winthrop Beach	13	138
Revere Beach	13	262
Lovell Island	13	268
Nahant Beach	14	375
Pearce Lake	15	95
Malibu Beach	16	275
Peckem Pond	18	150
Swampscott Beach	21	> 4,000
Kings Beach	22	> 4,000
Lynn Beach	22	2,000
Pleasure Bay	30	242
Constitution Beach	35	2,080
Carson Beach	45	4,000
Tenean Beach	52	2,260
Wollaston Beach	88	> 4,000

Source: Data from the *Boston Globe*, May 16, 1988.

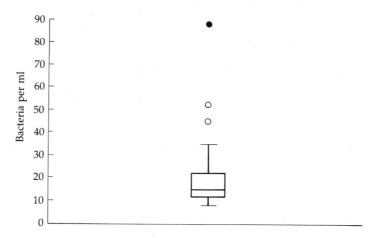

FIGURE 11.8   *Box plot of average coliform bacteria count per milliliter of water (from Table 11.6)*

ures it is positively skewed, with a few high values as outliers. This is likely to be true in the population as well, so we should not assume normality.

Although bacteria counts tend to be positively skewed, *logarithms* of bacteria counts have a more symmetrical distribution, so are routinely employed by researchers. Standard inferential procedures then apply directly to the logarithms.

Table 11.7 (page 332) shows base-10 logs of the bacteria counts, which have a less skewed distribution (Figure 11.9). Some skewness still remains; this could be eliminated by a stronger transformation like the negative reciprocal. We will stick with logs here because they are simpler, do an adequate job, and are often used with such data. Other research indicates that the *logarithm* of bacteria concentration has an approximately normal distribution.

As indicated in Table 11.7, the sample mean logarithm of the bacteria count is $\overline{X}^* = 1.252$, with a standard deviation of $s_{X^*} = .281$. These statistics may be used to form a 95% confidence interval for the population mean logarithm,

$$\overline{X} \pm t\left(\frac{s_{X^*}}{\sqrt{n}}\right) = 1.252 \pm 2.086\left(\frac{.281}{\sqrt{21}}\right)$$

so the interval is $1.124 \leq \mu^* \leq 1.38$. We are 95% confident that the mean logarithm of the beach bacteria concentrations is between 1.124 and 1.38. Since the original variable, X, was measured in bacteria per milliliter of water, the transformed variable X^* is measured in log(bacteria per milliliter). Although the transformed variable is statistically better behaved, these odd units are harder to understand.

One way to understand them is to return the statistical summaries for the transformed data to the units of the original raw data, by *inverse transfor-*

TABLE 11.7   *Raw and Logarithmic Coliform Bacteria Counts (from Table 11.6)*

Beach	Raw average coliform bacteria count X	Base-10 log of coliform count $X^* = Log_{10}(X)$
Yirrel Beach	8	.903
Short Beach	8	.903
Houghton's Pond	9	.954
Sandy Beach	10	1.000
Stacey Brook Outlet	10	1.000
Nantasket Beach	12	1.079
Winthrop Beach	13	1.114
Revere Beach	13	1.114
Lovell Island	13	1.114
Nahant Beach	14	1.146
Pearce Lake	15	1.176
Malibu Beach	16	1.204
Peckem Pond	18	1.255
Swampscott Beach	21	1.322
Kings Beach	22	1.342
Lynn Beach	22	1.342
Pleasure Bay	30	1.477
Constitution Beach	35	1.544
Carson Beach	45	1.653
Tenean Beach	52	1.716
Wollaston Beach	88	1.944
Sample statistics:	$\overline{X} = 22.57$	$\overline{X}^* = 1.252$
	$s_X = 19.20$	$s_{X^*} = .281$
	$n = 21$	

mation (Chapter 6). Since we transformed the bacteria counts by taking base-10 logarithms, we undo this transformation by taking base-10 antilogarithms of the lower and upper limits of the confidence interval. This yields values of $10^{1.124} = 13.305$ and $10^{1.38} = 23.988$ bacteria per milliliter:

$$13.305 \leq \mu_G \leq 23.988$$

This is *not* a confidence interval for the population mean concentration μ. Rather, it is a confidence interval for the antilog of the mean logarithm. We are 95% confident that the antilog of the mean logarithm is between 13.305 and 23.988 bacteria per milliliter. The antilog of the mean logarithm is actually a special type of mean, called the **geometric mean.**[7] By converting the upper and lower limits of the mean logarithm's interval back into natural

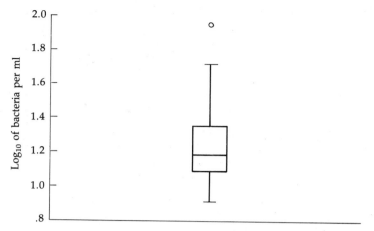

FIGURE 11.9   *Base-10 logarithms of coliform bacteria counts (from Table 11.7)*

units, we produce a 95% confidence interval for the geometric mean bacteria concentration, μ_G.

Although they make our results harder to understand, nonlinear transformations help solve many kinds of analytical problems, as later chapters will show. They let us do something about ill-behaved distributions, rather than simply not analyze them or apply inappropriate methods. They also help us realistically summarize the data, a benefit illustrated in Chapter 15.

PROBLEM

*12. Return to the data of Table 11.6. Highest measured bacteria counts are also positively skewed, so again the normality assumption is unjustified. Take base-10 logarithms of this variable and use them to construct a 95% confidence interval for the mean logarithm and for the geometric mean of highest bacteria concentrations.

11.8   THE BINOMIAL DISTRIBUTION*

In general, small-sample procedures for means are similar to large-sample procedures. The principal differences are

1. They use tables for the t distribution instead of the Z distribution.
2. They require the normality assumption.

For categorical-variable summaries such as proportions, small-sample procedures resemble their large-sample counterparts less.

The **binomial distribution** is a theoretical probability distribution for proportions. With large enough samples, the binomial distribution becomes approximately normal, justifying the large-sample methods for proportions (Chapters 9 and 10). When samples are too small to use the normal approximation, however, the binomial distribution itself must be used.

The binomial distribution is defined as follows. Assume that we have n independent cases and a categorical variable that can take on only two values. For convenience, we will refer to these two categories as 0 and 1; they could be any dichotomy such as male/female, dead/alive, or failure/success. The *binomial theorem* gives us the theoretical probabilities for all possible counts of 1's, in a sample of any given size n (see the box).

Binomial theorem: In a sample of n cases, where each case has the same probability π of falling into category 1, the probability that exactly f of these n cases fall in category 1 is

$$P(f) = \frac{n!}{f!(n-f)!}\, \pi^f (1 - \pi)^{n-f} \qquad\qquad [11.7]$$

The exclamation marks in Equation [11.7] stand for **factorials.** For example, n factorial is $n! = 1 \times 2 \times 3 \times 4 \times \cdots \times n$. If $n = 5$, this is $1 \times 2 \times 3 \times 4 \times 5 = 120$. Note that 1 factorial or 1! equals 1; 0 factorial or 0! is also defined equal to 1.

By finding the probability for every possible value of f, from 0 to n, Equation [11.7] can obtain the complete binomial distribution for a given combination of sample size (n) and probability or population proportion (π). This binomial distribution will have mean $n\pi$ and standard deviation $\sqrt{n\pi(1 - \pi)}$.

A different binomial distribution exists for each possible combination of n and π. Equation [11.7] can be used to generate each distribution, but this work gets tedious after a few cases. Alternatively, some computer programs will calculate binomial probabilities, and some textbooks provide extensive tables.[8]

To illustrate how to apply the binomial distribution for inferences about categorical variables, we will return to the study of miscarriage rates among computer chip workers. Recall that the problem was formulated in terms of whether the miscarriage rate among chemically exposed workers exceeded the rate in the population of nonexposed workers (.178):

H_0: $\pi \le .178$

H_1: $\pi > .178$

The researchers collected data on 18 pregnancies among exposed workers and found that 7 of these ($\hat{\pi} = 7/18 = .389$) ended in miscarriage. Our question is then,

How likely are we to see 7 miscarriages or more, if this sample of 18 pregnancies came randomly from a population whose overall rate is less than or equal to .178?

If the probability of this happening is low, then we can reject the null hypothesis (H_0) and conclude in favor of H_1: Miscarriage rates are systematically higher among women exposed to these chemicals. A high probability, on the other hand, would support the null hypothesis and suggest that the high miscarriage rate we see in this sample of 18 pregnancies could easily be due to chance.

Using the binomial distribution, we can calculate how likely an observed sample proportion $\hat{\pi}$ is, if the sample consists of n independent cases drawn randomly from a population whose proportion is π. To test the hypothesis about worker miscarriages, we set the population proportion π at the null hypothesis value that is closest to the research hypothesis: $\pi_0 = .178$. Our question is now more specific:

How likely are we to see a sample proportion as high as $\hat{\pi} = .389$, if the sample consists of 18 independent cases randomly drawn from a population whose proportion is $\pi = .178$?[9]

Before answering this question we will pose one that requires less computation. Suppose we saw about the same proportion of miscarriages, around $\hat{\pi} = .4$, in a sample of 5 cases instead of 18. Since 40% of 5 is 2, we will investigate the question:

What is the probability of there being two or more miscarriages, out of five pregnancies ($\hat{\pi} \geq .40$), if the population proportion is $\pi = .178$?

With $n = 5$ and $\pi_0 = .178$, the binomial probabilities can be calculated using Equation [11.7]. For example, the probability of no miscarriages in a sample of five, or $f = 0$, is

$$P(0) = \frac{5!}{0!(5 - 0)!} .178^0 (1 - .178)^{5-0} = .3753$$

(Recall that $0! = 1$, by definition; also, any number raised to the 0 power, such as $.178^0$, equals 1.) If the probability of no miscarriages is $P(0) = .3753$, then the probability of one or more must be $P(f \geq 1) = 1 - P(0) = .6247$. The probability of one miscarriage is

$$P(1) = \frac{5!}{1!(5 - 1)!} .178^1 (1 - .178)^{5-1} = .4063$$

The complete distribution of probabilities for $n = 5$ and $\pi = .178$ is given in Table 11.8 (page 336). Note that the probabilities add up to 1, as with any probability distribution.

At the bottom of Table 11.8 is a summary table, which includes the probabilities we need to test hypotheses about π. Figure 11.10 shows these same

TABLE 11.8   *Binomial Probabilities for n = 5 and π = .178*

Binomial probability that f out of n cases are in category 1:

$$P(f) = \frac{n!}{f!(n-f)!}\, \pi^f (1-\pi)^{n-f} \qquad \text{for } f = 0, 1, 2, \ldots, n$$

Given $n = 5$ and $\pi = .178$, the binomial probabilities are

$$P(0) = \frac{5!}{0!(5-0)!}\, .178^0 (1-.178)^{5-0} = .3753$$

$$P(1) = \frac{5!}{1!(5-1)!}\, .178^1 (1-.178)^{5-1} = .4063$$

$$P(2) = \frac{5!}{2!(5-2)!}\, .178^2 (1-.178)^{5-2} = .1760$$

$$P(3) = \frac{5!}{3!(5-3)!}\, .178^3 (1-.178)^{5-3} = .0381$$

$$P(4) = \frac{5!}{4!(5-4)!}\, .178^4 (1-.178)^{5-4} = .0041$$

$$P(5) = \frac{5!}{5!(5-5)!}\, .178^5 (1-.178)^{5-5} = .0002$$

Number of cases in a sample of $n = 5$ that fall in category 1	Probability of f cases falling in category 1	Probability of f or more cases in category 1
f	$P(f)$	$P(f$ or more$)$
0	.3753	1.0000
1	.4063	.6247
2	.1760	.2184
3	.0381	.0424
4	.0041	.0043
5	.0002	.0002

probabilities in graphical form.[10] We can see that the probability of two or more miscarriages out of five pregnancies, $P(f \geq 2)$, is about .2184 if it is true that $\pi = \pi_0 = .178$. Thus, the obtained P-value for this hypothesis test is $P = .2184$. This is not low enough to reject H_0: $\pi = .178$ at the usual $\alpha = .05$ level. In a sample of five pregnancies, there could well be two or more miscarriages purely by chance even if $\pi = .178$.

What if there were the same proportion of miscarriage, about $\hat{\pi} = .40$, in a sample of ten cases instead of five? A new set of binomial probabilities would have to be calculated, based on $\pi = .178$ and $n = 10$. The top part of Table 11.9 (page 338) shows these probabilities, without the details of

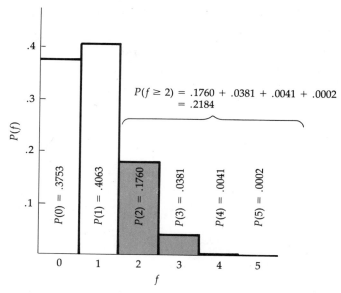

FIGURE 11.10   Binomial distribution for n = 5 and π = .178 (see Table 11.8). Right-hand tail shaded to show probability that f ≥ 2.

calculation. If it is true that the population proportion is π = .178, then in a sample of n = 10 the likelihood of four or more miscarriages equals .0854. This conclusion is shown graphically in Figure 11.11. At the usual α = .05 level of significance, an obtained P-value of .0854 is still not low enough to lead us to reject H_0: $\pi \leq$ .178.

The data actually collected for the chip manufacturing study included not five or ten, but 18 cases. Among these 18 pregnancies, seven ($\hat{\pi}$ = f/n = 7/18 = .389) ended in miscarriage. To test the hypothesis H_0: $\pi \leq$ .178 using this sample of eighteen cases, we need the binomial probabilities that can be derived from Equation [11.7] for n = 18 and π = .178. These probabilities are shown in the bottom half of Table 11.9, and graphically in Figure 11.12 (page 340). We see that the probability of seven or more miscarriages out of 18 pregnancies is only .0289. Thus, based on the study's actual results of $\hat{\pi}$ = .389 among n = 18 cases, we can reject (at α = .05) the null hypothesis that these 18 cases came from a population where $\pi \leq$ .178. Less technically, we conclude that the miscarriage rate is indeed significantly higher among exposed workers than among nonexposed workers.

This example illustrates the effect of sample size on significance tests. We tested the same null hypothesis, H_0: $\pi \leq$ .178, using three different sizes of sample. In each test, the sample proportion was similar, $\hat{\pi} \approx$ .40. Thus, in all three samples the observed proportion was more than twice as high as the

TABLE 11.9	**Binomial Probabilities for $\pi = .178$ and $n = 10$ or $n = 18$, Based on Equation [11.7]**		
f		$P(f)$	$P(f \text{ or more})$
$n = 10$:			
0		.1408	1.0000
1		.3050	.8592
2		.2972	.5542
3		.1716	.2570
4		.0650	.0854
5		.0169	.0204
6		.0030	.0035
7		.0004	.0004
8		<.0001	<.0001
9		<.0001	<.0001
10		<.0001	<.0001
$n = 18$:			
0		.0294	1.0000
1		.1144	.9706
2		.2106	.8562
3		.2432	.6456
4		.1975	.4024
5		.1198	.2049
6		.0562	.0851
7		.0209	.0289
8		.0062	.0080
9		.0015	.0018
10		.0003	.0003
11		<.0001	.0001
12		<.0001	<.0001
13		<.0001	<.0001
14		<.0001	<.0001
15		<.0001	<.0001
16		<.0001	<.0001
17		<.0001	<.0001
18		<.0001	<.0001

hypothesized population proportion. But for samples of only five or ten cases, this difference was not enough to reject the null hypothesis; it was not statistically significant. As the samples grow larger, any given difference has a lower *P*-value. This phenomenon is particularly troubling in some environmental health studies. A doubling of the rate for miscarriage or cancer is rightly quite alarming to the people involved, but if they are few in number even a tripling may not be statistically significant. That is, it does not rule out chance as the source of the difference.

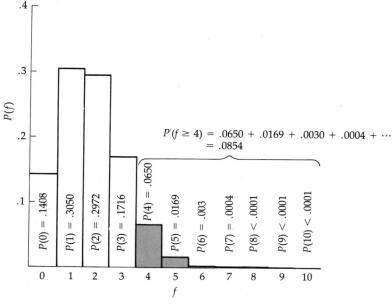

Binomial distribution for n = 10 and π = .178 (see Table 11.9). Right-hand tail shaded to show probability that f ≥ 4.

FIGURE 11.11

Calculations for the binomial probabilities in Tables 11.8 and 11.9 are laborious. A hand calculator or computer program with a factorial ($x!$) function and good numerical accuracy (since very large and small numbers are involved) is almost essential, especially for $n > 5$. There are three ways to reduce the effort further.

1. Consult texts containing tables of binomial distributions for various values of n and π.
2. Where n is large and π is small, use the Poisson distribution (see next section) as an approximation to the binomial.
3. If the large-sample normal approximation is justified (see Table 10.5), apply the standard normal distribution.

The third way is by far the easiest.

As we went from $n = 5$ to $n = 10$ and then to $n = 18$ (Figures 11.10–11.12), we saw histograms that were gradually approaching a normal curve. As n increases, more values of f (hence more histogram bars) are possible, and the distributions become more nearly continuous and normal. Figure 11.13 (page 341) shows this transition graphically, depicting the binomial distributions of Figures 11.10–11.12 and a fourth distribution for a still larger sample size, $n = 60$.

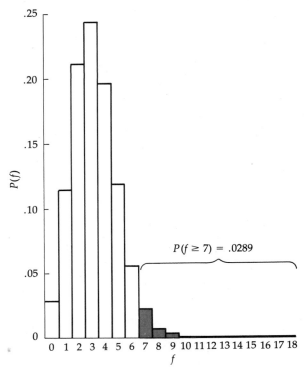

FIGURE 11.12

Binomial distribution for n = 18 and π = .178 (see Table 11.9). Right-hand tail shaded to show probability that f ≥ 7.

The rule of thumb (Chapter 10) is that the normal approximation for the binomial can be considered good enough when both $n\pi/(1 - \pi)$ and $n(1 - \pi)/\pi$ are at least nine. Applied to the four distributions in Figure 11.13, these values are

$$\text{if } n = 5, \qquad \frac{n\pi}{1 - \pi} = \frac{5(.178)}{1 - .178} = 1.08$$

$$\frac{n(1 - \pi)}{\pi} = \frac{5(1 - .178)}{.178} = 23.09$$

$$\text{if } n = 10, \qquad \frac{n\pi}{1 - \pi} = \frac{10(.178)}{1 - .178} = 2.17$$

$$\frac{n(1 - \pi)}{\pi} = \frac{10(1 - .178)}{.178} = 46.18$$

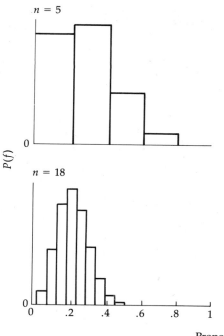
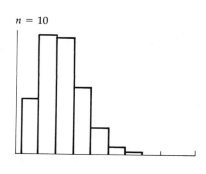
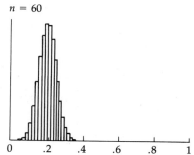

FIGURE 11.13   *Binomial distributions, $\pi = .178$*

if $n = 18$, $\dfrac{n\pi}{1 - \pi} = \dfrac{18(.178)}{1 - .178} = 3.90$

$\dfrac{n(1 - \pi)}{\pi} = \dfrac{18(1 - .178)}{.178} = 83.12$

if $n = 60$, $\dfrac{n\pi}{1 - \pi} = \dfrac{60(.178)}{1 - .178} = 12.99$

$\dfrac{n(1 - \pi)}{\pi} = \dfrac{60(1 - .178)}{.178} = 277.08$

For $n = 5$, 10, or 18, the lower of these two values is well below nine. Only the $n = 60$ distribution has both $n\pi/(1 - \pi)$ and $n(1 - \pi)/\pi$ values above nine, so only this one is large enough to justify the normal approximation. (With algebra or by trial and error, you can verify that for $\pi = .178$, at least 42 cases are needed to meet this criterion.) We can see that the $n = 60$

binomial distribution in Figure 11.13 is indeed the most normal looking. The other three are noticeably skewed, and their discreteness is apparent in the blocky outlines of their histograms.

PROBLEMS

*13. Find the binomial probabilities for $f = 0, 1, 2, 3, 4$, if $n = 4$ and $\pi = .3$. What is $P(f \geq 3)$?

*14. The Census Bureau reports that in 1980 about 6.7% of the 287,806 lawyers in America were women. If a law firm hired six lawyers that year and hired randomly with respect to gender, what is the probability that none of the six were women? If this happened, could the firm plausibly claim that it was just "by chance" that it hired no women? That is, find the probability that $f = 0$, given $n = 6$ and $\pi_0 = .067$. Reject H_0 (they hired no women "by chance") if $P < .05$.

*15. Referring to Problem 14, what is the probability of the firm hiring three or more women "by chance"? That is, find the probability that $f \geq 3$, given $n = 6$ and $\pi_0 = .067$. Again, use the $\alpha = .05$ significance level.

11.9   THE POISSON DISTRIBUTION*

The **Poisson distribution,** named after a nineteenth-century French mathematician, serves to approximate the binomial distribution when π is small and n is large. It also has other important statistical applications; for example, many variables that are *counts of random events* follow Poisson distributions.

A *Poisson variable* is a variable that follows a Poisson distribution. Such variables are discrete, with possible values $X = 0, 1, 2, 3, \ldots$.

The probability that a **Poisson variable** X takes on a specific value x is

$$P(X = x) = \frac{\mu^x e^{-\mu}}{x!} \qquad [11.8]$$

where μ is the variable's mean and e is the base of the natural logarithms ($e \approx 2.718$).

Calculating Poisson probabilities is straightforward if we know the variable's mean, μ. The value of $e^{-\mu}$ is then easily found using a calculator's e^x key, or the EXP function built into many computer programs.

The shape of a Poisson distribution depends on its mean, μ. Poisson distributions with low means have extreme positive skewness, which lessens as μ increases. Poisson distributions have the property that the mean equals the variance; hence, the square root of the mean equals the standard deviation. The higher the mean of a Poisson variable, the greater its variation.

The Poisson distribution is one of three theoretical distributions that have proved of wide use with counting variables. The other two are the binomial distribution and the **negative binomial** distribution.[11] The relationship between mean and variance provides a simple check for which of these three distributions better fits a set of counted data.

> In a *binomial distribution* the variance is less than the mean.
> In a *Poisson distribution* the variance equals the mean.
> In a *negative binomial* distribution, the variance is greater than the mean.

Binomial probabilities are difficult to calculate for large n; the numbers represented by expressions like $n!$ or $(1 - \pi)^{n-f}$ in Equation [11.7] may be too large or too small for most calculators or computer programs to handle. The normal approximation often serves, but we need huge samples to justify the normal approximation when π is small (see Table 10.5, page 291). The Poisson distribution provides an approximation to the binomial for situations where n is large but π is small.

> If $n \geq 50$, $\pi \leq .05$, and $n\pi < 5$, the binomial distribution can be approximated by a Poisson distribution with $\mu = n\pi$.

Consider an environmental health problem similar to the miscarriage study described earlier. A rare illness occurs apparently at random in the general population. There is one chance in 10,000, or a probability of $\pi = .0001$, that a person will contract this illness over a ten-year period. Water contamination has been discovered in a community of 30,000 people, where nine cases of the illness have occurred in ten years. The average number of cases for a community of this size should be $n\pi = 30,000(.0001) = 3.0$. How likely is it that the apparent outbreak in this community is simply due to chance?

This problem calls for a test of the null hypothesis that these 30,000 people were drawn randomly from a population where π (probability of contracting the illness) is not more than .0001: $H_0: \pi \leq .0001$. We will perform this test at the conservative $\alpha = .01$ level of significance. To find the probability of a sample proportion $\hat{\pi}$ as high as 9/30,000, we theoretically could use the binomial distribution, which would entail lengthy computation. We might consider the normal approximation instead, but even 30,000 cases is not a "large enough" sample; with π as low as .0001, we would need a sample of $n = 90,000$ cases (see Table 10.5).

Because π is small and n is large, the binomial distribution can be approximated by the Poisson. We first estimate the Poisson mean μ as $n\pi = 30,000(.0001) = 3.0$. To find the probability that the illness occurs nine or more times, it is easiest to find the probability it occurs 0, 1, 2, 3, . . ., 8 times, then subtract the sum of these probabilities from 1. Following Equation [11.8], the probability of exactly eight occurrences should be

$$P(X = 8) = \frac{3.0^8 e^{-3.0}}{8!} = \frac{6,561(.04979)}{40,320} = .0081$$

Similar calculations obtain the remaining probabilities of occurrence if the null hypothesis were true, as summarized in Table 11.10 along with the hypothesis test conclusions. The probability of observing nine or more occurrences of this rare illness in a sample of $n = 30,000$ is $P(X \geq 9) = .0039$, or less than one chance in 200. We can reject the null hypothesis and conclude that the occurrence of the illness in this community is significantly higher than it is in the general population.

The Poisson distribution is often applied in a different context, as a model for variables that count randomly occurring events. A classic example of this

TABLE 11.10   ***Poisson Probabilities from Equation [11.8] for Illness (Based on $\mu = n\pi = 3.0$)***

Number of occurrences	Probability of x occurrences if H_0 is true	Probability of x or more occurrences
x	$P(X = x)$	$P(X \geq x)$
0	.0498	1.0000
1	.1494	.9502
2	.2240	.8008
3	.2240	.5768
4	.1680	.3528
5	.1008	.1848
6	.0504	.0840
7	.0216	.0336
8	.0081	.0120
9	.0027	.0039

Hypothesis Test: H_0: $\pi \leq .0001$

H_1: $\pi > .0001$

The probability is $P(X \geq 9) = .0039$ that we would observe nine or more occurrences of this illness in a sample of $n = 30,000$ people, if they were chosen randomly from a population where H_0 is true. This obtained *P*-value is below the $\alpha = .01$ significance level, so we can *reject* H_0. The proportion of illnesses in this community is significantly higher than that for the general population.

application was provided by R. D. Clarke (1946), who found that the Poisson equation predicted the pattern of hits by V-1 buzz bombs launched against London during World War II. The V-1, a precursor of modern cruise missiles, was intended to terrorize and demoralize the London population with random strikes from the sky. Clarke divided South London into 576 small areas, each one-fourth of a square kilometer, and counted the number of V-1 hits in each area. A total of 535 V-1's hit this region, an average slightly under one hit per area: $\mu = 535/576 = .929$ hit per area. Of the 576 areas, 229 went untouched, but others were hit three, four, or even five times. These counts are shown in the first two columns of Table 11.11.

If the probability that any given area is hit by an incoming V-1 is independent of whether it has ever been hit before, we could expect the pattern of hits to follow a Poisson distribution. The Poisson probability that an area would not be hit at all, $P(X = 0)$, can be calculated from Equation [11.8]:

$$P(X - 0) = \frac{.929^0 e^{-.929}}{0!}$$

$$= \frac{1(.3949)}{1} \quad \text{(since by definition } .929^0 = 1 \text{ and } 0! = 1\text{)}$$

$$= .3949$$

Similar calculations obtain the other probabilities shown in the third column of Table 11.11.

TABLE 11.11   *Distribution of V-1 Buzz Bomb Hits in South London During World War II[a]*

Number of hits	Actual count of areas with x hits	Poisson probability of x hits	Poisson predicted count of areas with x hits
x		$P(X = x)$	$P(X = x)n$
0	229	.3949	227.5
1	211	.3669	211.3
2	93	.1704	98.2
3	35	.0528	30.4
4	7	.0123	7.1
5	1	.0023	1.3

Total number of hits: 535

Number of areas: 576

Mean number of hits per area: $\mu = \dfrac{535}{576} = .929$

[a]Poisson calculations are based on Equation [11.8] with $\mu = 535/576 = .929$.
Source: Based on data from Clarke (1946).

If the probability that an area has no hits is $P(X = 0) = .3949$, then out of 576 areas we should expect to see $P(X = 0)n = .3949(576) = 227.5$ areas with no hits. This prediction based on the Poisson equation is very near the mark: 229 areas were not hit. By the same reasoning we would expect to see $P(X = 1)n = .3669(576) = 211.3$ areas hit once, $P(X = 2)n = .1704(576) = 98.2$ areas hit twice, and so on. The right-hand column of Table 11.11 gives the predicted counts of areas with 0–5 hits.

The close fit in Table 11.11 between Poisson predictions and actual counts shows that the theoretical Poisson distribution provides a remarkably good model for the real-world pattern of bomb hits. This discovery was turned into a literary metaphor by Thomas Pynchon, in his strange novel *Gravity's Rainbow* (1973, Viking Press).

Other researchers have found the Poisson distribution to be a good model for phenomena as diverse as deaths by horse kick in the Prussian army and particle emissions in radioactive decay. These phenomena all involve counting an event that occurs within a given area or time interval with a small probability π. Occurrences of this event are *independent* if their probability is unchanged by any previous occurrences, as with V-1 bombs falling on London. The probability that the event occurs $x = 0, 1, 2, 3, \ldots$ times in a given area or time interval then comes from the Poisson equation, and the distribution of event counts over many areas or time intervals should be approximately Poisson.[12] Many natural and social phenomena meet this description.

PROBLEMS

*16. Demonstrate how Equation [11.8] can obtain the Poisson probabilities given in Table 11.10 for zero occurrences of the rare illness; for four occurrences of the illness.

*17. Demonstrate how Equation [11.8] can obtain the Poisson probabilities given in Table 11.11 for London areas with three buzz bomb hits; with four hits.

*18. A fire department planner calculates that the department receives an average of .51 alarm per day. Assuming that fire alarms follow a Poisson distribution, what is the probability of a day with three or more alarms?

SUMMARY

Small-sample inference regarding means requires two modifications of large-sample procedures. First, it becomes important whether the population distribution of X is normal. If so, then the t statistic should follow approximately a t distribution.

The second modification is the use of tables for the t rather than Z (standard normal) distribution. The t distributions resemble the Z distribution, but they have heavier tails. The fewer the degrees of freedom, the more heavy tailed a t distribution is relative to the normal. With many degrees of freedom t is quite similar to Z, and with infinite degrees of freedom the two distributions are identical. Degrees of freedom increase with sample size.

Apart from these differences, inference for means proceeds in the same manner whether samples are large or small. Confidence intervals are calculated by marking off a distance of t estimated standard errors on either side of the sample mean: $\overline{X} \pm t(SE_{\overline{X}})$. Hypothesis tests measure the distance between $\overline{X}$ and μ_0 in estimated standard errors: $t = (\overline{X} - \mu_0)/SE_{\overline{X}}$. A large difference (large t statistic) makes the null hypothesis look unlikely.

When the population distribution of X is not normal, standard inferential procedures can be misleading. One way to overcome this difficulty is to work with a transformed version of X, X^*, for which the normality assumption is more plausible. Confidence intervals and hypothesis tests are then performed using X^* in place of X. Interpretation is more difficult, since we do not have the natural units of X. Also the parameter being estimated or tested is no longer the mean of X; rather, it is the mean of X^*, which may be the mean logarithm of X, mean square root of X, and so on. Despite its complications this approach is popular in research, partly because positive skewness is so common.

Both the Z and t distributions are continuous, so they are appropriate with statistics like the mean that theoretically can assume infinitely many different values. Proportions, in contrast, can assume only a few different values in small samples. Their sampling distributions therefore cannot be continuous. The discrete binomial distribution is often used for small-sample inference about proportions. With larger samples binomial probabilities are hard to calculate, but the Poisson and normal distributions provide close approximations. The Poisson distribution works when n is large but the probability π is small. If n is large and π is not too near 0 or 1, the normal approximation serves.

PROBLEMS

19. Certain computer games are thought to improve spatial skills. A Mental Rotations Test, measuring spatial skills, was administered to a sample of school children after they had played one of two types of computer game.[13] Construct 95% confidence intervals based on the following mean scores, assuming that the children were selected randomly and that the Mental Rotations Test scores have a normal distribution in the population.

    a. After playing the "Factory" computer game: $\overline{X} = 22.47$, $s_X = 9.44$, $n = 19$

b. After playing the "Stellar" computer game: $\overline{X} = 22.68$, $s_X = 8.37$, $n = 19$

c. After playing no computer game (control group): $\overline{X} = 18.63$, $s_X = 11.13$, $n = 19$

(Methods for formally comparing these three groups will be introduced in Chapter 13.)

20. Researchers (Keane, Ducette, and Adler, 1985) investigated the extent of "burnout" among nurses in various hospital settings. Levels of such feelings as "I often think about finding a new job" or "I frequently get angry with my patients" were ascertained by questionnaire and measured on a Staff Burnout Scale for Health Professionals. The study focused on whether burnout levels for Intensive Care Unit (ICU) nurses differed from those for other hospital nurses. For all nurses in the large hospital studied, the burnout scale had a mean of 52.1. Among a sample of 25 ICU nurses the mean was 49.9, with a standard deviation of 14.3. Assuming that the burnout scale has an approximately normal distribution, test the hypothesis that the mean burnout level of ICU nurses is different from that for this hospital's nurse population as a whole. Include these steps.

   a. Formally state the null and alternative hypotheses, and explain whether a one- or two-sided test is appropriate.

   b. Calculate the degrees of freedom. What critical value of t is required to reject H_0 at $\alpha = .05$?

   c. Calculate the t statistic.

   d. Summarize your conclusions in English.

21. Burt and Neiman (1985) studied opinions about energy conservation policies among decision makers in city government. They constructed an index to measure approval of a variety of conservation policies. The index ranged from 0 (opposed all conservation policies) to 7 (favored all conservation policies). In a sample of 32 decision makers, the mean conservation index score was 3.03, with a standard deviation of 1.66. Among the general public, their research found a mean conservation index score of 3.91. Use this information to perform a t test ($\alpha = .01$) of the hypothesis that decision makers' opinions about conservation policy are *significantly different from* those of the general public. (Assume that the index in this example had an approximately normal distribution.)

The accompanying table contains one year's data on the number of safety failures during standby and operations at 17 U.S. nuclear power plants. If we viewed these 17 cases as a random sample from a larger population of plants and years, they could be used to form an interval estimate of the mean number of failures per year in that population. Use the table for Problems 22–27.

Safety-Related Failures at 17 Nuclear Power Plants During 1972

Reactor	Type[a]	Failures during standby	Failures during operation
Dresden 1	BWR	1	2
Yankee	PWR	8	5
Indian Point 1	PWR	7	12
Humboldt Bay 3	BWR	5	7
Big Rock Point	BWR	4	3
San Onofre 1	PWR	3	7
Haddam Neck	PWR	0	3
Nine Mile Point 1	BRW	7	13
Oyster Creek	BWR	10	19
Ginna	PWR	1	5
Dresden 2	BWR	8	20
Point Beach 1	PWR	2	4
Millstone 1	BWR	7	22
Robinson 2	PWR	3	17
Monticello	BWR	10	34
Dresden 3	BWR	4	22
Palisades	PWR	6	22

[a]BWR = boiling water reactor, PWR = pressurized water reactor.
Source: Data from the United States Atomic Energy Commission (1974), Appendix III.

22. Find the mean, standard deviation, median, and pseudo-standard deviation for the number of failures during power plant standby. Use these statistics to summarize the distribution's shape. Is the normality assumption plausible?

23. Use your statistics from Problem 22 to construct a 95% confidence interval for the mean number of nuclear power plant standby failures per year.

24. Construct a box plot or a stem-and-leaf display for the number of nuclear power plant failures during operation. Describe the distribution's shape.

*25. Take square roots of the number of failures during operation for each power plant. Construct a stem-and-leaf display of these square roots. Has symmetry been improved?

*26. Find the mean and standard deviation of the square roots of operations failures (Problem 25), and use these to construct a 95% confidence interval for the mean square root of the number of failures.

*27. Express your answer to Problem 26 in terms of the original units of the variable, number of failures during operations.

The accompanying table contains data on 14 patients who experienced accidental or intentional overdoses of the drug chlorpropamide, which is sometimes prescribed in the treatment of diabetes. The data are from a study of the mechanism causing high dosages of chlorpropamide to be harmful or fatal. When chlorpropamide dosage is too high, because of an intentional overdose (suicide attempt) or other reason, it tends to induce lowered blood glucose levels or hypoglycemia. Normally the body responds to induced hypoglycemia by reducing the production of insulin. In these 14 patients, however, insulin levels seem to be unexpectedly high. Use the table for Problems 28–31.

Chlorpropamide Toxicity and Hyperinsulinemia

Patient	Age years	Diagnosis	Serum glucose mg/dl	Serum insulin μ Units/ml	Serum chlorpropamide μg/ml
1	3	Overdose	20	90	—
2	73	Diabetic	45	38	176
3	13	Overdose	35	299	770
4	18	Overdose	29	28	690
5	16	Overdose	20	140	57
6	40	Overdose	38	118	400
7	51	Overdose	42	50	240
8	24	Overdose	33	25	523
9	14	Overdose	44	60	422
10	19	Overdose	20	240	540
11	58	Overdose	28	53	—
12	25	Overdose	40	38	—
13	71	Diabetic	16	59	171
14	78	Diabetic	28	20	162

Source: Data from Klonoff (1988).

28. a. Construct a box plot or a stem-and-leaf display for the serum glucose levels of the 14 patients.
    b. Find the mean, standard deviation, median, and pseudo-standard deviation of serum glucose levels.
    c. Use your results from parts a and b to discuss the shape of this sample distribution, compared with a normal curve.
29. The sample distribution of serum glucose levels (Problem 28) appears well enough behaved that the normality assumption is plausible. Use the t statistic to test at $\alpha = .05$ the hypothesis that the mean serum glucose level for patients with chlorpropamide toxicity is below 40 mg/dl (severely hypoglycemic).
30. Assuming normality, use the data in the table to test the hypothesis that the mean insulin level of patients with chlorpropamide toxicity exceeds the maximum normal level of 24 micro-Units per milliliter.

*31. Construct a stem-and-leaf display or box plot of the insulin levels for Problem 30. Do you see any reason to question the normality assumption? If so, what might be done about it?

*32. On July days in the high country of Rocky Mountain National Park, there is about a $\pi = .40$ probability of a thunderstorm. Assume that the occurrence of a storm one day does not change the probability of a storm the following day. A hiker is planning a seven-day trip. Use the binomial distribution to estimate the probability that she will encounter no storms. What is the probability that she will have storms on at most two days?

*33. A quiz consists of ten true–false questions, half of them false. If an unprepared student answers each question by flipping a coin, what is the probability that he will earn a passing grade (7 or more right answers)? What is the probability he will earn an A (9 or 10 right answers)?

*34. In 1986 an insurers' association released information about malpractice suits against physicians. During 1975–1983 there was a total of 2,131 malpractice suits among the 9,683 physicians insured by this association. The report stated that 3% of the physicians received 31.7% of the lawsuits.

This information seems to imply that a few physicians are especially prone to lawsuits, perhaps to the detriment of the whole profession. Before jumping to this conclusion, though, we should consider that even if lawsuits were totally random events, unrelated to physician characteristics, some physicians would have the bad luck to be sued more than once.

If lawsuits were distributed randomly among these 9,683 physicians, we could expect them to follow a Poisson distribution with a mean (μ) equal to $2{,}131/9{,}683 = 0.22$ suit per physician. Use Equation [11.8] to find the following probabilities, *if* lawsuits were indeed random.

a. $P(X = 0)$, probability a physician is not sued

b. $P(X = 1)$, probability a physician is sued only once

c. $P(X = 2)$, probability a physician is sued only twice

d. $P(X \geq 3)$, probability a physician is sued three or more times

*35. Use your answers to Problem 34 to specify the expected *number* of physicians sued never, once, twice, or three or more times, if lawsuits were a random (Poisson) process. What percentage of the physicians would be expected to receive *all* of the lawsuits?

*36. Your analysis in Problems 34 and 35 shows that even if lawsuits were random, the unluckiest 2.1% of the physicians (those sued more than one time) would be the targets of nearly 20% of them (since $2{,}131 - 1{,}710 = 421$ of the 2,131 lawsuits are against physicians who get sued more than once). Thus a completely random process in which everyone has the same chance of being sued could still produce results that superficially look

like evidence of "bad apples" among physicians. How might these percentages be different if some physicians really were more likely to be sued than others?

NOTES

1. Unpublished study by Harris Pastides and Edward Calabrese, of the University of Massachusetts, 1987; reported in the *Boston Globe*, January 27, 1987.

2. Of course, the exposed group may differ from the unexposed group in some other respect besides chemical exposure. For example, the latter might be older or include more heavy smokers; such extraneous factors could affect the miscarriage rate, making it *appear* that the exposed workers are more at risk. Researchers must carefully rule out such possibilities before drawing conclusions about the effects of chemical exposure. To keep this example simple, however, we will assume here that the two groups of workers are similar, except for their different levels of chemical exposure. Then the main statistical question is just whether the observed difference in miscarriage rates is too large to be due to chance.

3. The *t* distribution was first described in a 1908 paper published by the statistician W. S. Gosset, who was then working for Guinness Brewery in Dublin. The brewery had a policy that employees not publish their research, so Gosset wrote under the pseudonym Student, and his discovery came to be known as Student's *t* distribution.

4. This statement is a simplification. As we will see in later chapters, degrees of freedom depend not only on sample size, but also on the complexity of the analysis being done. For analyses of equal complexity, though, it is true that the larger the sample, the more degrees of freedom.

5. Outliers are somewhat more common in sample distributions because the quartiles, hence the definition of a "severe outlier" also varies. As Figures 11.2–11.4 show, such variation is greatest with small samples, so severe outliers may actually be less surprising in small samples than in medium or large samples. This is another reason it is hard to deduce nonnormality from small samples.

6. In measurement the Greek letter μ (*mu*) often stands for *micro* or *one-millionth*, so "micrograms per kilograms" is abbreviated μg/kg. This common usage of the symbol μ is unrelated to its statistical usage as the population mean.

7. You may recall that when the mean was first defined (Chapter 3) we called it the *arithmetic* mean. Among other kinds, there is also the *geometric* mean. Formally defined, the arithmetic mean is the sum of the values of *X*, times $1/n$:

$$(X_1 + X_2 + X_3 + \cdots + X_n)\left(\frac{1}{n}\right)$$

A parallel expression defines geometric mean as *the product* of the values of X, to the $1/n$ power:

$$(X_1 \times X_2 \times X_3 \times \cdots \times X_n)^{1/n}$$

To see why this definition of the geometric mean is the same as "the antilog of the mean logarithm" requires familiarity with how logarithms change multiplication into addition and powers into multiplication.

8. For example, see Bhattacharyya and Johnson (1977), Appendix, Table 2.

9. The cases in this sample are 18 pregnancies. For our purposes we assume that these are pregnancies experienced by 18 separate workers; otherwise the assumption of independent cases is not met, and the binomial distribution may not apply. Then the inferential problem becomes much more complicated.

10. Figure 11.10 shows binomial distribution probabilities as a histogram, rather than as a polygon or curve, to emphasize that the variable here is *discrete*. Miscarriages either do or do not occur, so we have no fractional counts between $f = 1$ and $f = 2$, etc. This is especially noticeable with small samples, where only a few different values of f are possible—like the six shown in Figure 11.10. A normal or t distribution, in contrast, is appropriate where the variables can be viewed as continuous, with any number of intermediate values.

11. The negative binomial is a more complicated variant of the binomial or Poisson. It describes certain types of phenomena where events are *not* independent. Contagious diseases are one possible example of such phenomena, so the negative binomial is sometimes referred to as the *contagious Poisson* distribution. This distribution is discussed in more detail in texts on probability, such as Larsen and Marx (1985).

12. If the probability *does* change systematically as a result of previous occurrences, consider the negative binomial distribution instead (see Note 11).

13. Adapted from McClurg and Chaille (1987).

BIVARIATE ANALYSIS: RELATIONSHIPS BETWEEN TWO VARIABLES

PART III

The first eleven chapters of this book dealt with univariate analysis, studying variable distributions one at a time. As the simplest kind of statistical analysis it serves best for introducing many concepts and techniques. This same simplicity also makes univariate analysis generally less interesting than **bivariate** (two variables) or **multivariate** (three or more variables) analysis. The most intriguing practical questions in statistics usually concern **causality,** the effects of one variable on another. Questions of causality are always bivariate or multivariate in nature.

In bivariate analysis, we examine variables two at a time. The two variables are often referred to as X, the **independent variable** (cause) and Y, the **dependent variable** (effect). If X is a cause of Y, then three conditions should be true.

1. *Time ordering*   The value of X at any specific time can affect Y only at some later time. Causality cannot flow backward in time, nor can it be truly simultaneous.

2. *Covariation*   The two variables vary together, in some systematic way. The variables *X* and *Y* are *not independent* as defined in Chapter 7.
3. *No third-variable explanations*   The covariation of *X* and *Y* cannot be explained by the fact that both are related to some third (or fourth, or fifth) variable.

Research seeking to demonstrate that *X* causes *Y* must address all three conditions, showing why we should believe each one is true.

Establishing Cause and Effect

The first condition, *time ordering,* is sometimes simple and sometimes impossible to meet. Many studies have a clear time sequence, based on either the nature of the variables or how the data were collected. For example, a person's gender is determined long before his or her income is; education received in fourth grade precedes achievement tests taken in fifth grade; and measurements made on November 10 obviously come before measurements made on November 11. In other studies, however, time ordering becomes a chicken-or-egg question with no definite answer. For example, do people's attitudes influence their behavior, or is it the other way around? Is a slum crime-ridden because unemployment there is high, or is unemployment high because of so much crime?

Could such questions of chronology be resolved by getting data from several different times? Not easily: The problems of analyzing data over time are surprisingly complex. Although a few simple time series are included in this book, the statistical issues of formal time series analysis are left to a more advanced text.*

Of the three conditions only the second, *covariation,* is directly addressed by bivariate analysis. This indeed is the main point of bivariate analysis: to see if and how two variables vary together. Establishing that a bivariate relationship exists is only a starting point for causal research.

The third condition, *no third-variable explanations,* is often the most troublesome. Bivariate analysis provides little help here. This condition may be the weakest link in a causal argument, since it is impossible to rule out every alternative explanation.

To narrow down the range of plausible explanations, researchers often rely on either *experimental design* or *multivariate analysis.* Experimental research designs seek to reduce the problems of "third variables" (and also of time ordering) in advance, by carefully controlling the circumstances under which the data are generated. Multivariate analysis is valuable for data that arise from uncontrollable circumstances, like the behavior of the economy or people's lives. Then it is necessary to apply a statistical control after the data are collected.

*Chapter 15, Note 1 lists some references for time series analysis.

Bivariate analysis shows if, and how, two variables vary together. Thus it can answer the question "Do they covary?" without saying anything about the two other causal conditions, "Could this covariation be explained by a third variable?" and "What is the time ordering?" As will be seen in the next four chapters, understanding a two-variable relationship is no small achievement. At the end of even the most careful and thorough bivariate analysis, however, much remains to be known. Showing that Y varies with X by no means proves that Y is *caused* by X.

The Null Hypothesis in Bivariate Analysis

In Chapters 10 and 11, we encountered the idea of a null hypothesis, a theoretical statement about the value of a population parameter. In bivariate analysis, null hypotheses often state that in the population no relationship exists between X and Y. Even when researchers believe that a relationship exists, they start out with a "no relationship" null hypothesis and hope that this pessimistic hypothesis can be rejected. This roundabout reasoning is confusing to newcomers, but it is a basic tool in most statistical research. When you hear that a research finding is *statistically significant*, it usually means that a null hypothesis of "no relationship" has been rejected because it looks improbable when compared with the sample data.

Chapter 12 introduces the hypothesis of no relationship as it applies to two categorical variables. Chapter 13 looks at relationships between one categorical variable and one measurement variable. Chapters 14 through 16 concern regression analysis, a broad family of methods for understanding relationships among variables.

Two Categorical Variables: Crosstabulation and the Chi-Square Test

Chapter **12**

A starting point for many univariate analyses is the frequency distribution table (introduced in Chapter 1), which shows how many times each value of a variable occurs. The two-variable extension of this procedure, called **crosstabulation,** is likewise the starting point for many bivariate analyses. A crosstabulation shows how many times each possible *combination* of two variables' values occurs. Like frequency distribution tables, crosstabulations work best with variables that take on only a few values. They are a particularly appropriate method for analyzing the *relationship between two categorical variables.*

Such relationships may not be obvious in the raw counts of cases in various combinations of categories. Percentages make any patterns in these counts more apparent. Taking percentages in a crosstabulation is straightforward but requires some thinking ahead: not all percentages are equally useful.

The appropriate percentages can help us see and describe how two variables are related in the sample at hand. In order to generalize to a larger population, we need tools for statistical inference. The most widely used inferential technique for crosstabulations is the *chi-square test*. This test combines basic ideas about probability and hypothesis testing with a theoretical sampling distribution called the *chi-square distribution*.

12.1   CELL FREQUENCIES AND PERCENTAGES IN CROSSTABULATION

Skiing injuries are more common among children than among adults. One study examined data on child skiers in an effort to identify contributory risk factors; Table 12.1 is a crosstabulation based on these data. Two categorical variables are shown: skiing ability and injury group. Skiing ability is the **row variable** in the table, meaning that this variable defines the rows—row 1 is Beginner, row 2 Intermediate, and row 3 Advanced–Expert. Injury group is the **column variable**. The three rows and two columns produce six **cells,** which contain **cell frequencies**—counts of cases in each combination of categories. The cell frequency in the Beginner/Injured cell is 20; that is, there are 20 injured beginners. There are 9 injured intermediates, 2 injured advanced–experts, and so forth. At the right and bottom margins of the table are the **marginal frequencies,** labeled All, which are found by adding up the cell frequencies in each column or row. The **row marginals** give the univariate distribution of the row variable, and the **column marginals** give the univariate distribution of the column variable. The number of cases ($n = 214$) is at lower right; it equals the sum of either row or column marginals.

Table 12.1 contains information about how skiing ability relates to injury, but it requires careful reading. Most of the skiers in all ability groups were uninjured. More beginners than experts were injured, but more beginners than experts were not injured, too. Converting counts to percentages brings out the patterns in this table.

TABLE 12.1   *Crosstabulation[a] of Skiing Ability by Injury Group*

	Injury group		
Skiing ability	*Injured*	*Uninjured*	*All*
Beginner	20	60	80
Intermediate	9	84	93
Advanced–Expert	2	39	41
All	31	183	214

[a]Observed cell frequencies for a sample of 214 child skiers.
Source: Data from Ungerholm and Gustavsson (1985).

TABLE 12.2   ***Skiing Ability by Injury Group: Observed Frequencies and Percentages[a]***

| | Injury group | | |
Skiing ability	Injured	Uninjured	All
Beginner	20	60	80
Row	25%	75%	100%
Column	65%	33%	
Total	9%	28%	
Intermediate	9	84	93
Row	10%	90%	100%
Column	29%	46%	
Total	4%	39%	
Advanced–Expert	2	39	41
Row	5%	95%	100%
Column	6%	21%	
Total	1%	18%	
All	31	183	214
Column	100%	100%	100%

[a]Observed cell frequencies for 214 child skiers (from Table 12.1) with row, column, and total percentages (rounded to nearest integer).

The sample includes 80 beginners, 20 of whom were injured. We could express the number of injured beginners as three different percentages. First, the percentage of beginners who are injured: $20/80 \times 100 = 25\%$. This *row percentage* is based on the number of cases (80) in the row. Second, the percentage of injured skiers who are beginners: $20/31 \times 100 \approx 65\%$. This is a *column percentage,* based on the number of cases (31) in the column. Third, the percentage of all skiers who are injured beginners: $20/214 \times 100 \approx 9\%$. This *total percentage* is based on the total number of cases (214) in the table. Table 12.2 shows the crosstabulation of ability and injury, with all three sets of percentages added to each cell. Note that the row percentages add up to 100% across each row; the column percentages add up to 100% down each column; and the total percentages add up to 100% over all cells in the table.

> **Column percentage:** cell frequency as a percentage of cases in the column
> **Row percentage:** Cell frequency as a percentage of cases in the row
> **Total percentage:** cell frequency as a percentage of all cases in the table

All three percentages are legitimate descriptions of the same cell, but they suggest different things. If we say *nearly two-thirds (65%) of the injured skiers are beginners,* that sounds like a lot of injured beginners. But it is equally true to say that *less than one-tenth (9%) of these skiers are injured beginners,* which sounds like fewer injured beginners after all. To reduce such confusion bear in mind that percentages are always calculated from some **base number.** A given number of cases, such as the 20 injured beginners, represents a large percentage relative to a small base, such as the 31 injured skiers. The same 20 cases represent only a small percentage relative to a large base, such as the 214 total cases.

We must therefore scrutinize any statement using a percentage to identify its base number. In casual reading or listening, many people fail to distinguish between such phrases as "the percentage of lawyers who are Senators" and "the percentage of Senators who are lawyers"—but each refers to a different number, the first quite small and the second large. Someone intent on persuasion can take advantage of the fact that people often neglect base numbers, noting only how large or small a percentage sounds. This is one way to "lie with statistics"—not actually lying, but exploiting the audience's statistical naiveté. Advertising and political campaigns are notorious for this practice.

Although the different percentages we can calculate for any frequency are all mathematically legitimate, not all answer specific questions equally well. The obvious question to ask of the data in Tables 12.1 and 12.2 is "How (if at all) does the risk of injury depend on skiing ability?" In this question, skiing ability is the *independent variable,* or possible cause. Injury is the *dependent variable,* or possible effect. The likelihood of injury may depend on the skier's ability, but the skier's ability before the injury cannot logically depend on whether injury occurs. In analyzing crosstabulations, *it is usually best to base percentages on the independent variable.* Since skiing ability, the row variable in Table 12.2, is independent, we should concentrate on the row percentages.

Dependent variable: possible "effect"
Independent variable: possible "cause"

Rule for choosing percentages in crosstabulations: Work with the percentage that is based on the *independent variable.* Take row percentages if the row variable is the independent variable; take column percentages if the column variable is the independent variable.

Applying this rule to Table 12.2, we get a clear description of how injury and skiing ability are related. We find 25% of the beginners in the injured group, as compared with only 10% of the intermediates and 5% of the advanced–expert skiers. Conversely, 95% of the advanced–experts were uninjured, versus 90% of the intermediates and only 75% of the beginners. Clearly, *the conditional probability of injury given beginning ability is much higher than the conditional probability of injury given advanced–expert ability.*

We draw this conclusion by first identifying the appropriate percentages (row percentages), then comparing them from one row to the next. The other percentages (column and total) in Table 12.2 provide information of possible interest in other contexts, but they are not directly useful—or may even be misleading—in describing how injury depends on skiing ability. A research report summarizing the findings of such a study would normally exclude column and total percentages. Cell frequencies and row percentages suffice to describe the relationship between skiing ability and injury group for this sample of 214 young skiers.

PROBLEMS

The accompanying table contains a crosstabulation of spouse abuse by family type. The data are from a national survey; 969 married adults who were living with at least one child are included. People reported their type of family and whether they experienced (as victim or aggressor) violent acts of spouse abuse within the past year. Problems 1–3 refer to this table.

Crosstabulation (Observed Cell Frequencies) of Spouse Abuse by Family Type

	Family Type[a]		
Spouse abuse	*Intact*	*Remarried*	*Reconstituted*
None	743	92	78
Any	36	9	11

[a]In "reconstituted" families, one or both parents brought into the present family their children from an earlier marriage ending in divorce. Any children in "remarried" families are from the current marriage only.
Source: Data from Kalmuss and Seltzer (1986).

1. Find the marginal frequencies, and use them to calculate row, column, and total percentages for each cell in the table.

2. The row percentage for the Intact/No Abuse cell in the table is 81%. We could state this in English: "Eighty-one percent of the people reporting no spouse abuse lived in intact families." Write out a similar English description for each of the following percentages.

   a. The row percentage for the Remarried/Any Abuse cell

   b. The column percentage for the Intact/No Abuse cell

   c. The total percentage for the Intact/No Abuse cell

3. Which type of percentage (row, column, or total) best clarifies the relationship between family type and recent spouse abuse? Write a short paragraph citing the appropriate percentages, to describe the relationship you see in this table.

12.2 *THE INDEPENDENCE HYPOTHESIS AND EXPECTED FREQUENCIES*

Once we have a good description of what is going on between two variables *in the sample*, the question of inference arises: What do these sample results tell us about some larger population? For the ski injury example, what does studying these 214 children tell us about the thousands of other young skiers who hit the slopes every winter? Unless we can make some generalizations, the 214 skiers alone are of limited interest.[1]

Most inference from crosstabulation starts by setting up a **null hypothesis of independence:** that the two variables involved are *not related to each other* ("independent") in the population from which the sample was drawn. The word *independent* in this context means "unrelated," which is different from the idea of an independent variable (meaning "causal"). We can state the null and alternative hypotheses as

H_0: Row and column variable are *independent* (they are unrelated)

H_1: Row and column variable are *not independent* (they are related)

If the null hypothesis (H_0) were true, we would expect to see a certain set of frequencies in the crosstabulation. We can compare the observed cell frequencies with the frequencies that are expected under H_0. If the differences are great enough, we should reject H_0.

If skiing injuries and ability are independent of each other, then the probability that a skier selected at random is both injured and a beginner should equal the probability of being injured times the probability of being a beginner. Let P(Injured) represent the probability that a skier selected at random is injured. Then

$$P(\text{Injured and beginner}) = P(\text{Injured}) \times P(\text{Beginner})$$

$$= \left(\frac{\text{Number injured}}{\text{Total}}\right) \times \left(\frac{\text{Number of beginners}}{\text{Total}}\right)$$

$$= \left(\frac{31}{214}\right) \times \left(\frac{80}{214}\right) = .1449 \times .3738 = .0542$$

This calculation follows the multiplication rule for finding $P(A \text{ and } B)$, when A and B are two independent events:

$$P(A \text{ and } B) = P(A)P(B)$$

(See Table 7.3 (page 183) and the discussion of Equation [7.6].)

If the probability of selecting an injured beginner were .0542, then out of 214 cases we would expect about 11.6 cases (.0542 × 214 = 11.6) to be injured beginners. Thus, if skiing ability and injury were independent, we would expect 11.6 injured beginners. In our sample data, we actually observe 20 injured beginners. *We observe more injured beginners (20) than we would expect (11.6) if injury and ability were independent.*

The same reasoning applies to each cell in a crosstabulation. For example, the probability of selecting an advanced or expert skier is 41/214 = .1916, and the probability of selecting an uninjured skier is 183/214 = .8551. If ability and injury were independent, then according to the multiplication rule for independent events the probability of selecting an uninjured advanced–expert would be .1916 × .8551 = .1638. If this probability is .1638 in the population, then out of the 214 cases in this sample we would expect about 35.1 cases (.1638 × 214) to be uninjured advanced–experts. The sample contains 39 uninjured advanced–experts, which is more than we would expect (35.1) if injury and ability really were independent.

The **expected frequencies** E under the independence model can be calculated as

$$E = \left(\frac{r}{n}\right) \times \left(\frac{c}{n}\right) \times n \tag{12.1}$$

where r represents the row marginal for a given cell, c represents the column marginal for that cell, and n is the total number of cases. The quantity r/n is the row probability, and c/n is the column probability. Equation [12.1] can be simplified to

$$E = \frac{r \times c}{n} \tag{12.2}$$

Table 12.3 shows the ski injury and ability crosstabulation, giving expected frequency in parentheses after the observed frequency in each cell. A comparison of the observed and expected frequencies supports our earlier observation about the percentages: There are more injured beginners, and fewer injured intermediate or advanced–expert skiers, than we would expect if injury and ability were independent of each other.

Notice that the expected frequencies in Table 12.3 add up to the same marginals as do the observed frequencies. This is always true for expected

TABLE 12.3   *Skiing Ability by Injury Group: Observed and Expected Frequencies[a]*

	Injury group		
Skiing ability	*Injured*	*Uninjured*	*All*
Beginner	20 (11.6)	60 (68.4)	80 (80)
Intermediate	9 (13.5)	84 (79.5)	93 (93)
Advanced–Expert	2 (5.9)	39 (35.1)	41 (41)
All	31 (31)	183 (183)	214 (214)

[a]Expected frequencies are in parentheses following observed frequencies and are calculated by Equation [12.2]: $E = (r \times c)/n$.

frequencies calculated under the independence hypothesis. Such expected frequencies are said to *fit the one-way (univariate) marginals.* The concept of fitting marginals becomes important in the analysis of more complex, multi-variable crosstabulations.

PROBLEMS

The accompanying table contains a crosstabulation based on a random sample survey of adolescents and their families in Buffalo, New York. The variables are the drinking habits of mothers and of their adolescent (age 12–17) children.

Crosstabulation of Drinking Habits[a] of 120 Adolescents and Their Mothers

Drinking of adolescent	Drinking of mother		
	Abstainer	Moderate	Heavy
Abstainer	7	20	7
Moderate	7	45	13
Heavy	5	5	11

[a]The "moderate" category includes infrequent, light, and moderate drinkers; "heavy" includes moderate–heavy and heavy drinkers.
Source: Barnes, Farrell, and Cairns (1986).

4. Calculate expected frequencies for each cell in this table. Write out what you learn by comparing observed and expected frequencies for the following cells.

    a. Adolescents who abstain and whose mothers also abstain

    b. Adolescents who are heavy drinkers and whose mothers are moderate drinkers

    c. Adolescents who are heavy drinkers and whose mothers are heavy drinkers

5. In which cells of the table are the observed frequencies *most different* from what we would expect if maternal and adolescent drinking habits were independent of each other?

12.3 *THE CHI-SQUARE TEST*

Independence is the usual null hypothesis for crosstabulations. Formal testing of this hypothesis is accomplished by a **chi-square test.** For the chi-square test we calculate a test statistic, X^2, to compare with the theoretical *chi-square* (χ^2) *distribution* found in Appendix Table A.4.

The X^2 **statistic** measures how closely a set of expected frequencies fits those actually observed:

$$X^2 = \sum \frac{(O - E)^2}{E} \qquad\qquad [12.3]$$

where O and E are the observed and expected frequencies, respectively. The summation (Σ) is over all cells in the table.

The greater the disparity between observed and expected frequencies, the larger the value of the X^2 statistic. If the null hypothesis is true, the X^2 statistic will follow the theoretical chi-square distribution (Table A.4). A large enough X^2 value, corresponding to a low chi-square probability, tells us to reject the null hypothesis. The theoretical rationale behind this test appears in a later section. First we will try out the test on our ski injury example.

For the first cell in Table 12.4, $O = 20$ and $E = 11.6$, so

$$\frac{(O - E)^2}{E} = \frac{(20 - 11.6)^2}{11.6} = 6.08$$

Similarly, in the Beginner/Uninjured cell, the $(O - E)^2/E$ value is

$$\frac{(O - E)^2}{E} = \frac{(60 - 68.4)^2}{68.4} = 1.03$$

These and other $(O - E)^2/E$ values are shown below the frequencies for each cell in Table 12.4.

TABLE 12.4   *Skiing Ability by Injury Group, with X² Statistic*

| | Injury group | | |
Skiing ability	Injured	Uninjured	All
Beginner	20 (11.6)	60 (68.4)	80 (80)
$(O - E)^2/E$	6.08	1.03	
Intermediate	9 (13.5)	84 (79.5)	93 (93)
$(O - E)^2/E$	1.50	.25	
Advanced–Expert	2 (5.9)	39 (35.1)	41 (41)
$(O - E)^2/E$	2.58	.43	
All	31 (31)	183 (183)	214 (214)

X^2 **statistic:**

$$X^2 = \sum \frac{(O - E)^2}{E} = 6.08 + 1.03 + 1.50 + .25 + 2.58 + .43$$
$$= 11.87$$

Equation [12.3]

The $(O - E)^2/E$ values are calculated as intermediate steps in computing the X^2 statistic, but they also have direct value for interpretation. The larger the $(O - E)^2/E$ value, the worse the fit between observed and expected frequencies in a given cell. Thus $(O - E)^2/E$ values tip us off to cells displaying an especially poor fit between the independence hypothesis and the actual data. Often one or two cells stand out markedly, like the Injured/Beginner cell in Table 12.4, whose $(O - E)^2/E$ value, 6.08, far exceeds the rest. The hypothesis that injury and skiing ability are unrelated is most strongly contradicted by the high injury rate for beginning skiers. In the other cells of this table, the discrepancy between data and hypothesis is less pronounced.

To find the X^2 statistic, add the $(O - E)^2/E$ values for all cells in a table. In Table 12.4, this gives us

$$X^2 = \sum \frac{(O - E)^2}{E}$$

$$= 6.08 + 1.03 + 1.50 + .25 + 2.58 + .43 = 11.87$$

The final step in this test is to compare the X^2 statistic with a table of probabilities for the theoretical chi-square distribution. For this we need to specify the *degrees of freedom.*

If the independence hypothesis is true, the X^2 statistic follows the theoretical **chi-square (χ^2) distribution,** with **degrees of freedom**

   df $= (K_r - 1) (K_c - 1)$ [12.4]

where K_r is the number of categories of the row variable, and K_c is the number of categories of the column variable.

Theoretical chi-square probabilities are given in Table A.4.

Since the skiing crosstabulation has three rows ($K_r = 3$) and two columns ($K_c = 2$), the degrees of freedom are

$$df = (K_r - 1) (K_c - 1)$$

$$= (3 - 1) (2 - 1) = 2$$

The X^2 statistic for this crosstabulation, 11.87, is therefore compared with the theoretical chi-square distribution for 2 degrees of freedom.

To make this comparison, turn to the chi-square table (Appendix Table A.4) and look down the left-hand margin to find the row for 2 degrees of freedom. Then scan across this row to see where our sample X^2 value of 11.87 fits into the body of the table: between the values 10.60 and 13.82. A χ^2 value of 10.60 corresponds to a P-value or probability of .005, and 13.82 corresponds to a P-value of .001 (as seen at the top of the chi-square distribution table). Sample X^2 values of 11.87 or larger would theoretically occur with a probability between .005 and .001 if the null hypothesis were true. This means that if skiing ability and injury really were unrelated in the population, we would expect to see X^2 values this large or larger in less than 5 out of every 1,000 random samples. Thus, if the null hypothesis were true, we would be quite unlikely to obtain such a sample. Since we *did* obtain such a sample, it is hard to believe that the null hypothesis is true: We should reject it.

The sample relationship between ability and injury is sufficiently strong to convince us, after the chi-square test, that a relationship also exists in the population from which this sample was drawn. That is, the relationship between injury and skiing ability is statistically significant at the $\alpha = .05$ level. This conclusion is summarized in Table 12.5.

Such chi-square tests follow the same general strategy as other hypothesis tests (Chapters 10 and 11). We begin with a null hypothesis; in this case the null hypothesis (H_0) is that the two variables are independent or unrelated. (The complementary research hypothesis, H_1, is that the two variables are not independent; they *are* related.) Next we find the expected frequencies, which are the cell frequencies we would expect to see *if* the null hypothesis

TABLE 12.5 *Skiing Ability by Injury Group: Chi-Square Test of Independence*

Skiing ability	Injury group		All
	Injured	Uninjured	
Beginner	20 (11.6)	60 (68.4)	80 (80)
$(O - E)^2/E$	6.08	1.03	
Intermediate	9 (13.5)	84 (79.5)	93 (93)
$(O - E)^2/E$	1.50	.25	
Advanced–Expert	2 (5.9)	39 (35.1)	41 (41)
$(O - E)^2/E$	2.58	.43	
All	31 (31)	183 (183)	214 (214)

$$X^2 = \sum \frac{(O - E)^2}{E} = 6.08 + 1.03 + 1.50 + .25 + 2.58 + .43$$

$$= 11.87 \qquad \text{Equation [12.3]}$$

$$\mathbf{df} = (K_r - 1)(K_c - 1) = (3 - 1)(2 - 1) = 2 \qquad \text{Equation [12.4]}$$

P-value: Compare the sample X^2 statistic, 11.87, with the theoretical chi-square distribution (Appendix Table A.4) for 2 degrees of freedom to find the obtained P-value for this test: $P < .005$. That is, there is less than a .005 chance of observing cell frequencies this far (or farther) from those expected, if the sample in fact came randomly from a population where skiing ability and injury were independent.

Conclusion: The obtained P-value is below $\alpha = .05$, so we *reject the null hypothesis* that injury and ability are independent (unrelated). We may confidently conclude that there *is a relationship* between injuries and skiing ability in the population of child skiers.

were true. The discrepancies between expected frequencies and those actually observed are measured by the X^2 statistic, which under a true null hypothesis follows a theoretical chi-square distribution.

Comparing the sample X^2 statistic with a table of the chi-square distribution gives us the **obtained *P*-value** for the test: The probability of seeing results so favorable to H_1, if H_0 were true. The lower this *P*-value, the less plausible H_0 looks. If the *P*-value is below a predetermined cutoff point, such as $\alpha = .05$, we can reject H_0 in favor of H_1. We may then say that the relationship between the two variables is *statistically significant at the .05 level*.

PROBLEMS

6. Use the chi-square distribution table (Table A.4) to decide whether you should reject the null hypothesis of independence when
   a. Testing at $\alpha = .05$, $X^2 = 2.5$, in a 2 by 2 table.
   b. Testing at $\alpha = .05$, $X^2 = 106.5$, in a 3 by 4 table.
   c. Testing at $\alpha = .01$, $X^2 = 25$, in a 9 by 3 table.
   d. Testing at $\alpha = .25$, $X^2 = 0.47$ in a 2 by 6 table.
   e. Testing at $\alpha = .001$, $X^2 = 16.9$ in a 5 by 5 table.
   f. Testing at $\alpha = .025$, $X^2 = 21.5$ in a 6 by 3 table.

7. Carry out a chi-square test to determine whether the relationship between spouse abuse and family type (Problem 1) is statistically significant at $\alpha = .05$. Which cell provides the strongest evidence against the null hypothesis? Summarize your conclusions in English.

12.4   DEGREES OF FREEDOM AND THE CHI-SQUARE DISTRIBUTION

In earlier chapters we encountered the *standard normal distribution:* a normal distribution with a mean of 0 and a standard deviation of 1. The theoretical chi-square distribution is derived from the standard normal distribution. If variable Z_1 has a standard normal distribution, then Z_1^2 is distributed as chi-square with 1 degree of freedom. If Z_1 and Z_2 are independent standard normal variables, then $Z_1^2 + Z_2^2$ is distributed as chi-square with 2 degrees of freedom. Similarly $Z_1^2 + Z_2^2 + Z_3^2$ is distributed as chi-square with 3 degrees of freedom, and so on for any number of independent standard normal variables.

> If $Z_1, Z_2, \ldots, Z_d$ are d **independently distributed standard normal variables,** then the quantity
>
> $$\chi_d^2 = Z_1^2 + Z_2^2 + Z_3^2 + \cdots + Z_d^2 \qquad [12.5]$$
>
> will follow a chi-square distribution with d degrees of freedom.

One consequence of this property is that the sum of two chi-square variables will itself follow a chi-square distribution.

If χ_d^2 is distributed as chi-square with d degrees of freedom, and χ_e^2 is distributed as chi-square with e degrees of freedom, then their sum will be distributed as chi-square with $d + e$ degrees of freedom:

$$\chi_d^2 + \chi_e^2 = \chi_{d+e}^2$$

[12.6]

This fact is particularly useful in more advanced applications of chi-square.[2]

Like the t distribution, the chi-square distribution is actually a whole family of distributions, one for every possible degrees of freedom. Unlike the symmetrical t and Z distributions, the chi-square distribution is positively skewed, most extremely for small degrees of freedom. Since chi-square is defined as a sum of squares, negative values are impossible. The mean of a chi-square distribution equals its degrees of freedom. These properties are illustrated in Figure 12.1, which shows chi-square distributions with 5 and 10 degrees of freedom. The chi-square distribution with df = 5 is more skewed and has a mean of 5; the chi-square distribution with df = 10 is less skewed and has a mean of 10.

For any given degrees of freedom, the larger a chi-square value, the lower the probability that values so large could occur by chance. For example, Table A.4 shows that with df = 3, the probability of chi-square values of 7.815 or greater is $P(\chi^2 \geq 7.815) = .05$. This probability is shown graphically as the

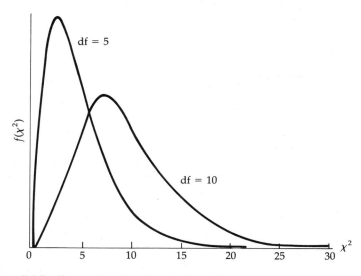

FIGURE 12.1   *Chi-square distributions with df = 5 and df = 10*

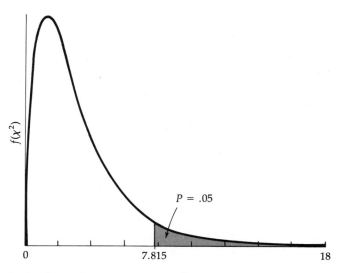

FIGURE 12.2
Chi-square distribution (df = 3) showing $P(X^2 > 7.815) = .05$

shaded area in the right-hand tail of Figure 12.2. Since $P = .05$ is a common cutoff point for statistical significance, we would ordinarily *reject a null hypothesis* if it resulted in a X^2 statistic (with df = 3) larger than 7.815.

Null hypotheses in crosstabulations imply a set of expected frequencies, like the E values in Table 12.5. Statistical theory tells us that if E is not too small, and we draw many random samples from a population where the null hypothesis is true, the quantity

$$\frac{O - E}{\sqrt{E}} \qquad\qquad\qquad [12.7]$$

(where O represents observed frequencies) will follow approximately a standard normal distribution. If $(O - E)/\sqrt{E}$ follows a standard normal distribution, then its square, $(O - E)^2/E$, should follow a chi-square distribution with 1 degree of freedom.

If the $(O - E)^2/E$ values for each cell in a crosstabulation, like those in Table 12.5, were all independent chi-square variables, then their sum would itself be a chi-square variable with degrees of freedom equal to the number of cells. This is not quite how it works, however. We do indeed calculate the sum of the individual $(O - E)^2/E$ values across all cells in the table:

$$X^2 = \sum\frac{(O - E)^2}{E}$$

This provides the X^2 test statistic of Equation [12.3]. Its degrees of freedom are not the number of cells, which would be K_r (number of rows) times K_c

(number of columns). Instead they are df $= (K_r - 1)(K_c - 1)$, as given in Equation [12.4].

When testing the independence hypothesis in a crosstabulation, we do not actually have $(K_r)(K_c)$ *independent* $(O - E)^2/E$ values to add up; rather only $(K_r - 1)(K_c - 1)$ of them are independent, and this determines the degrees of freedom. The reasoning behind this calculation is illustrated with a simple 2 by 2 crosstabulation in Table 12.6. This crosstabulation is based on an analysis of 206 articles sampled from issues of *Reader's Digest* and three "punk" magazines—*NOMAG* (Los Angeles), *The East Village Eye* (New York), and *Take It!* (Boston). At the top we see a table in which only the marginals are given: 64 of the 206 articles had violence as a major theme, while 142 did not; there were 117 articles from the punk press and 89 from *Reader's Digest*. We can calculate expected frequencies for each cell knowing only the marginals shown in this top table. The observed frequencies, and therefore the quantities $(O - E)^2/E$, could have many different values while still fitting these marginals.

TABLE 12.6   *Crosstabulation of Violence in Articles by Source: 2 by 2 with 1 df*

Marginals only:

Article content	Article source		All
	Punk press	*Reader's Digest*	
Violent			64
Not violent			142
All	117	89	206

Marginals plus one cell frequency:

Article content	Article source		All
	Punk press	*Reader's Digest*	
Violent	**41**		64
Not violent			142
All	117	89	206

Remaining cell frequencies are determined:

Article content	Article source		All
	Punk press	*Reader's Digest*	
Violent	**41**	$64 - 41 = 23$	64
Not violent	$117 - 41 = 76$	$89 - 23 = 66$	142
All	117	89	206

Source: Data from Lamy and Levin (1985).

If we fill in a single observed frequency, as done in the middle table of Table 12.6, the remaining cell frequencies are no longer free to vary. If there are 41 articles in the Violent/Punk press cell, and there are a total of 64 violent articles, then there must be $64 - 41 = 23$ cases in the Violent/*Reader's Digest* cell. Similarly, we can fill in the other observed frequencies, as shown in the bottom of Table 12.6. This is what it means to say that a 2 by 2 crosstabulation has 1 degree of freedom—because df $= (K_r - 1)(K_c - 1) = (2 - 1)(2 - 1) = 1$. If we want to fit the one-way marginals, only one observed frequency in a 2 by 2 table can vary independently; once its value is fixed, the frequencies in all other cells are fixed as well.

In a 3 by 2 table like the skiing example (Table 12.5), all frequencies are determined once the marginals and two cell frequencies are fixed, so we have 2 degrees of freedom: df $= (K_r - 1)(K_c - 1) = (3 - 1)(2 - 1) = 2$. This formula for degrees of freedom tells us the minimum number of cells, for any size crosstabulation, that can vary freely before the remaining cell frequencies are fixed by the necessity of adding up to the correct marginals.

We will later see that the X^2 statistic can be used to test other hypotheses besides independence in crosstabulations. The degrees of freedom are calculated in different ways, depending on the specific application. Regardless of application, however, the degrees of freedom reflect how many independent terms sum to obtain X^2.

PROBLEM

8. For each of the following tables, find the degrees of freedom and show how once that many cell frequencies are known, the remaining cell frequencies can be determined from the marginals.
   a. Skiers' ability and injury (Table 12.1)
   b. Spouse abuse and family type (Problem 1)
   c. Drinking habits of mothers and adolescents (Problem 4)

12.5 PARENTHOOD AND OPINIONS ABOUT WATER QUALITY

Table 12.7 contains a crosstabulation of data from residents surveyed in Acton, Massachusetts, where town water supplies were found to be contaminated by toxic wastes. Acton legislated perhaps the nation's strictest standards for water quality, mandating purification of existing contaminated supplies and shutdown of municipal wells should further contamination be found. The strict standards were costly, and the town debated whether lower, hence less expensive, standards could still adequately protect people's health. Some contended that even stricter laws were needed, prohibiting all chemical contamination. Table 12.7 shows a crosstabulation of opinions about the water

TABLE 12.7   **Opinions About Water Quality Standards, by Parenthood, for 330 Adults: Observed Frequency and Column Percentage**

| Water quality standard | Respondent has children younger than 18 in town | | |
	No	Yes	All
Make stricter	36	51	87
Column	24.7%	27.7%	
Keep as is	81	118	199
Column	55.5%	64.1%	
Make less strict	29	15	44
Column	19.9%	8.2%	
All	146	184	330
	100%	100%	

quality standards, by whether the respondent had children younger than 18 living in town.

It is possible that opinions about water quality depend on parenthood, but unlikely that parenthood depends on opinions about water quality: Parenthood is the independent variable.[3] As parenthood is the column variable in Table 12.7, column percentages are an appropriate way to look for a parenthood–opinion relationship.

Nearly one-fifth of the nonparents (19.9%) wanted lower water quality standards; only 8.2% of the parents held this view. A higher proportion of the parents of children under 18 preferred the standards to be kept as they are (64.1% vs. 55.5%) or made even stricter (27.7% vs. 24.7%). Thus people with children were more likely to approve the expensive water quality standards; fewer nonparents saw a need for such protection. These statements are only relative, for large majorities of both groups opposed lowering the standards.

To see whether this sample relationship also implies a relationship in the population, we perform a chi-square test. The null hypothesis is that the sample in Table 12.7 was drawn from a population where parenthood and water quality opinion were unrelated. Under this hypothesis, we would expect 38.5 nonparents to want stricter standards:

$$E = \frac{r \times c}{n}$$
$$= \frac{87 \times 146}{330} = 38.5$$

In the data, we observe 36 nonparents favoring stricter standards, which is slightly fewer cases in this cell than expected, were parenthood and opinions unrelated. The $(O - E)^2/E$ value for this cell is

$$\frac{(O - E)^2}{E} = \frac{(36 - 38.5)^2}{38.5} = .16$$

which is quite small, so the first cell of Table 12.7 does not challenge the null hypothesis. As shown in Table 12.8, several other $(O - E)^2/E$ values are also small. The null hypothesis runs into trouble, however, in the last row of this table: The largest $(O - E)^2/E$ value, 4.63, occurs in the cell for nonparents who want lower standards. There are significantly more such nonparents (29 observed), than we would expect (19.5) were parenthood unrelated to opinions.

To test the evidence from the table as a whole, we add up the six $(O - E)^2/E$ values:

$$X^2 = \sum \frac{(O - E)^2}{E}$$

$$= 9.60$$

TABLE 12.8   *Opinions About Water Quality Standards, by Parenthood: Chi-Square Test*

| | Respondent has children younger than 18 in town | | |
Water quality standard	No	Yes	All
Make stricter	36 (38.5)	51 (48.5)	87
$(O - E)^2/E$	.16	.13	
Keep as is	81 (88.0)	118 (111.0)	199
$(O - E)^2/E$	.56	.44	
Make less strict	29 (19.5)	15 (24.5)	44
$(O - E)^2/E$	4.63	3.68	
All	146	184	330

$$X^2 = \sum \frac{(O - E)^2}{E}$$

$$= .16 + .13 + .56 + .44 + 4.63 + 3.68 = 9.60$$
$$df = (K_r - 1)(K_c - 1) = (3 - 1)(2 - 1) = 2$$

P-value: Comparing the sample X^2 statistic, 9.60, with the theoretical chi-square distribution (Table A.4) for 2 degrees of freedom, we obtain a P-value low enough to *reject the null hypothesis* of independence: $P < .01$.

Conclusion: We conclude that parenthood and opinions about water quality standards *are related* in the population. The strongest evidence against the null hypothesis is the high number of nonparents who believe that the standards should be lowered. The $(O - E)^2/E$ value in that cell is 4.63.

and compare this test statistic with the chi-square distribution for df = $(K_r - 1)(K_c - 1) = 2$. A X^2 of 9.60 falls between 9.210 and 10.60 in the row for df = 2 (Table A.4), so we obtain $.005 < P < .01$. Since the probability of so large a X^2 is low, we reject the null hypothesis.

PROBLEMS

The accompanying table crosstabulates results from a test of two vaccines intended to prevent typhoid fever. A total of 6,907 people in Nepal were given either Vi capsular polysaccharide (Vi) or pneumococcus polysaccharide (Pneumo) vaccine. Of these 6,907 people, 71 subsequently developed symptoms of typhoid fever.

Crosstabulation of Typhoid Fever Incidence, by Type of Vaccine

| Outcome | Type of vaccine | | All |
	Vi	Pneumo	
Typhoid fever	14	57	71
No typhoid	3,443	3,393	6,836
All	3,457	3,450	6,907

Source: Data are from Acharya et al. (1987).

9. Conduct a chi-square test to determine if either vaccine is significantly more effective than the other.

10. Find the appropriate percentages for the table and use them to summarize your conclusions about the effectiveness of both vaccines.

12.6   CHI-SQUARE AS A "BADNESS-OF-FIT" TEST*

The chi-square test is called a "goodness-of-fit" test, but it might more aptly be termed a "badness-of-fit" test: The *worse* the fit between a set of observed and expected frequencies, the *higher* the X^2 statistic is. Use of X^2 is not confined to tests of the independence hypothesis in crosstabulation. The X^2 statistic can test the fit between observed and expected frequencies under a wide variety of hypotheses. Its calculation follows Equation [12.3] (or see Note 2), once we specify the expected frequencies that correspond to a particular null hypothesis.

The degrees of freedom for a test depend on the nature of the null hypothesis. Table 12.9 (page 378) lists several common applications for chi-square tests, with their degrees of freedom. In each case the degrees of freedom equal the number of *independent* $(O - E)^2/E$ values being added up to form X^2. We have already seen that there are $(K_r - 1)(K_c - 1)$ such values

TABLE 12.9	*Degrees of Freedom for Some Chi-Square Goodness-of-Fit Tests, Applied to a Variable Y^a*	

Hypothesis tested (H_0)	Degrees of freedom (df)
Univariate tests:	
Y follows a probability distribution specified from some source *outside the sample data*, such as from the U.S. Census	df = $K - 1$, where K is the number of categories of Y in the test
Y follows a *uniform distribution,* or all categories of Y are *equally probable*	df = $K - 1$, where K is the number of categories of Y in the test
Y follows a *normal distribution*	If μ and σ_Y are estimated from the sample, using either $\overline{Y}$ and s_Y or else (robust) Md and PSD: df = $K - 3$, where K is the number of categories into which Y has been grouped for this test
Y follows a *Poisson distribution*	If μ is estimated from the sample: df = $K - 2$, where K is the number of categories of Y in the test
Bivariate tests:	
X and Y are *independent*	df = $(K_X - 1)(K_Y - 1)$, where K_X and K_Y are the number of categories of X and Y respectively. An alternative notation is df = $(K_r - 1)(K_c - 1)$, referring to the row and column variables in a crosstabulation.

[a]For these tests, use either the **Pearson** X^2 statistic of Equation [12.3] or the **likelihood ratio** X_L^2 statistic (see Note 2) to test for a significant difference between a set of observed frequencies and the set of frequencies expected if the null hypothesis, H_0, were true.

for independence tests in a crosstabulation. Other hypothesis tests require other formulas to find df.[4]

One of these applications is shown in Table 12.10, which gives the distribution of a single categorical variable, gender, for 212 children pictured in a sample of advertisements for microcomputer products. In the ads examined, the researchers counted 130 boys (61%) and 82 girls (39%), so boys seem overrepresented in these ads. Many educators believe that girls may be discouraged from pursuing computer use if it is depicted as primarily a male interest.

Boys obviously predominate in this sample of ads. But can we, on the basis of this sample, conclude that computer advertisements in general are biased in their depiction of boys and girls? To test this, we need to compare

TABLE 12.10   ***A Univariate Chi-Square Test: Boys and Girls in Microcomputer Advertisements***

	Children in advertisements	
Values	*Boys*	*Girls*
Observed (Expected)	130 (108)	82 (104)
$(O - E)^2/E$[a]	4.48	4.65

$$X^2 = \sum \frac{(O - E)^2}{E} = 4.48 + 4.65 = 9.13$$

$$df = (K - 1) = (2 - 1) = 1$$

P-value: Comparing the X^2 statistic, 9.13, with the theoretical chi-square distribution (Table A.4) with 1 degree of freedom, $P < .005$.

Conclusion: This *P*-value is low enough that we may *reject the null hypothesis* on which the expected values are based: that boys and girls are depicted in the ads in the same proportions as in the U.S. population. Instead, we conclude that their proportions in the ads are *different from* their proportions in the U.S. population.

[a]The high number of boys and the low number of girls in the sampled ads provide about equally strong evidence against the null hypothesis: The $(O - E)^2/E$ values in the two cells are similar.

Source: Data from Demetrulias and Rosenthal (1985).

the observed frequencies with the frequencies to be expected if the proportions of boys and girls in the advertisements match those in the U.S. population. Census figures provide the population proportions, which in this age group are about .51 boys and .49 girls. Converting these proportions into expected frequencies gives us .51 × 212 = 108 boys, and .49 × 212 = 104 girls.

The observed frequencies, 130 boys and 82 girls, are obviously not the same as the expected frequencies, 108 boys and 104 girls. The X^2 statistic lets us assess *how likely are observed frequencies to be this different from those expected* if the sample were drawn randomly from a population where the expected proportions held true. As Table 12.10 shows, these observed and expected frequencies lead to a X^2 value of 9.13. In a univariate analysis such as Table 12.10, degrees of freedom equal the number of categories minus 1: $K - 1 = 2 - 1 = 1$. (See the first type of test listed in Table 12.9.) A chi-square value of 9.13, with 1 degree of freedom, has a probability of less than .005. Thus the data in Table 12.10 lead us to reject the null hypothesis. It is very unlikely (less than 5 chances in 1,000) that a sample with such unbalanced proportions of boys and girls could have been randomly selected from a population of advertisements where the proportions were the same as they are for U.S. children as a whole.

> A **chi-square goodness-of-fit test** can be conducted for many null hypotheses that provide expected frequencies. The test statistic
>
> $$X^2 = \sum \frac{(O - E)^2}{E}$$
>
> where O and E are the observed and expected (if H_0 is true) frequencies, is compared with a chi-square distribution.

Table 12.9 gave examples of the rules for degrees of freedom in selected tests.

PROBLEMS

*11. The accompanying table shows further results from the same study of microcomputer advertisements, enumerating adults by race. Expected frequencies are again based on census data for the nation. For example, if the ads were representative of the population, we would expect about 38 out of the 317 adults to be black; only 19 were in fact black. Are these differences significant? Apply a univariate chi-square test.

Race of 317 Adults Pictured in Microcomputer Advertisements

	Adults in advertisements		
Frequency	*White*	*Black*	*Other*
Observed	293	19	5
Expected	263	38	16

Source: Data from Demetrulias and Rosenthal (1985).

*12. As noted in Table 12.9, chi-square tests are often applied to determine how well an observed frequency distribution corresponds to a theoretical distribution such as the normal or the Poisson. Table 11.11 showed the theoretical Poisson distribution to be a remarkably good model for the spatial distribution of V-1 buzz bomb hits in London during World War II. The expected (Poisson) and observed frequencies from this example are repeated here; you may want to review the discussion in Chapter 11 of how these expected frequencies were obtained.

Distribution of V-1 Hits

	Number of hits per area					
	0	*1*	*2*	*3*	*4*	*5*
Observed	229	211	93	35	7	1
Expected (Poisson)	227.5	211.3	98.2	30.4	7.1	1.3

Although the observed and expected frequencies are close, they are not identical. Is the fit between the observed distribution of V-1 hits and the distribution predicted by the Poisson close enough that any discrepancies could easily be due to chance? Test whether the observed distribution is *significantly different* ($\alpha = .05$) from a Poisson distribution. Refer to Table 12.9 for the appropriate degrees of freedom. ·

12.7   THE PROBLEM OF THIN CELLS

The actual sampling distribution (distribution over many samples) of the X^2 statistic is approximately chi-square only if the samples are reasonably large. With very small samples, this approximation becomes inaccurate, and the X^2 statistic can no longer be interpreted in terms of the chi-square probability distribution. There is no clear point at which a sample becomes too small, however. The rule stated in the box is conservative.

> *Rule of thumb:* The chi-square test is appropriate only when *no expected frequency is less than five.*

Cells that contain few observed or expected cases are called **thin cells.** Thin cells can arise, even in large samples, if the categories are many or have very unequal marginal distributions. For example, even a sample of 100,000 cases would have many thin cells if we tried to crosstabulate a rare category, such as childhood leukemia, with environmental and family background characteristics. Thin cells pose a statistical problem in that a chi-square test may be invalid for assessing the strength of evidence about the null hypothesis. Thin cells also present the general problem of *not enough information.* If only a few cases of childhood leukemia occur in a sample of 100,000, then we have little information from which to draw such conclusions as whether it is more common in households near a toxic waste dump.

There are several ways to deal with thin cells in crosstabulations. From a theoretical standpoint, the best solution is to gather more data, increasing overall sample size to the point where even the thinnest cells have more than enough cases. Unfortunately this solution is often impractical, and a more expedient method must be found.

A second theoretically attractive approach is to use statistical tests specifically designed for small sample sizes. The best-known is **Fisher's exact test,** for 2 by 2 tables. Extensions of this test have been developed for larger tables, but they are cumbersome and not popular.

Other ways of dealing with thin cells fall under the heading of "making do." If the thin cells are caused by one uncommon category, they may disappear when that category is combined with another category. This was done with the skiing data in Table 12.1, for example. The original research report used the crosstabulation shown in Table 12.11, with four levels of skiing ability: beginner, intermediate, advanced, and expert. Only a few children were experts, however, and none of them had been injured, fitting the pattern of injury rates declining as skiing ability increased.

.   It therefore seems reasonable to combine the few experts with the larger group of advanced skiers, creating a new category of "advanced–expert" skiers. This removes the problem of thin cells found in the original 4 by 2 table. Combining categories here is defensible:

1. It makes sense theoretically that close categories ("expert" and "advanced") could be combined.
2. The basic relationship of injury to ability visible in the table is the same either way.

TABLE 12.11   ***Skiing Ability by Injury Group for Original Data on 214 Child Skiers[a]***

Skiing ability	Injury group		All
	Injured	Uninjured	
Beginner	20	60	80
Row	25%	75%	100%
Intermediate	9	84	93
Row	10%	90%	100%
Advanced	2	35	37
Row	5%	95%	100%
Expert	0	4	4
Row	0%	100%	100%
All	31	183	214

Chi-square test:

$$X^2 = \sum \frac{(O - E)^2}{E} = 12.02$$

$$\mathrm{df} = (K_r - 1)(K_c - 1) = 3$$

P-value: With 3 degrees of freedom, $X^2 = 12.02$ leads us to an obtained P-value of $P < .01$.

Conclusion: We can therefore *reject the null hypothesis* that injury and ability are unrelated. The same conclusion was reached when the expert and advanced categories were combined in Table 12.5 (page 369).

[a]Observed frequencies are given with row percentages (rounded to nearest integer).
Source: Data from Ungerholm and Gustavsson (1985).

It would have made no sense to combine the "expert" with the "beginner" category. Combining "expert" with "advanced" would also have been misleading had the original table shown any evidence that injury rates *increase* as ability rises from advanced to expert. The most obvious trend in the combined table—that injury rates decline as skiing ability increases—would then have been partly illusory: a reflection of how categories were manipulated. The principal risk in combining categories is of arbitrarily altering the conclusions drawn from the rearranged data.

Yet another way to deal with thin cells produced by rare categories is to drop those categories from the analysis. In Table 12.11, for example, we could simply drop the four expert skiers from the table, and proceed by analyzing the remaining 210 beginning, intermediate, and advanced skiers. Had we done this, our conclusions could then be generalized only to the population of "nonexpert child skiers." Systematically dropping any cases from an analysis narrows the conceptual population about which inferences can be made—which usually makes the results less interesting.

A third way to "make do" with thin cells is to go ahead and perform a standard chi-square analysis, bearing in mind that any conclusions must be tentative. Since the rule to avoid using X^2 with expected frequencies less than five is conservative, we can sometimes go below this point and still produce reasonable conclusions. This is especially true where thin cells make up only a fraction—less than one-fifth—of the total number of cells.

> If *fewer than* 20% of the cells in a table have expected frequencies below five, chi-square analysis may still be valid.

A chi-square analysis of Table 12.11, for example, leads to rejection of the null hypothesis ($P < .01$), hence to the same substantive conclusion as the analysis for Table 12.5. This does not prove that this conclusion is right, since it is possible that both analyses—combined categories (Table 12.5) and thin cells (Table 12.11)—are misleading. It is reassuring, though, to know that we reach the same conclusion either way.

PROBLEM

13. In the table for Problem 4 the crosstabulation has one thin cell: the expected number of heavy-drinking adolescents with abstinent mothers is only 3.33. Would it be reasonable to combine categories here before conducting a chi-square test? (Study the table carefully.) Explain why or why not, and describe what you think is the best way to analyze this table.

12.8 *CONTINUITY CORRECTION FOR THIN CELLS**

In tables where each variable has only two categories, combining categories is impractical—it would reduce one of the variables to a constant. For such 2 by 2 tables, a better solution to the problem of thin cells is a *continuity correction*.

The theoretical chi-square distribution is **continuous** (see Figures 12.1 and 12.2). With small cell frequencies, however, the sample X^2 statistic can take on only a few values—it is noticeably discrete. Therefore, for small samples the sampling distribution of X^2 (discrete) cannot resemble the chi-square (continuous). The continuity correction attempts to compensate for this problem.

To perform the **continuity correction** for thin cells in a 2 by 2 crosstabulation, find expected frequencies as usual, then add or subtract .5 from each observed frequency, to move them closer to the expected frequencies:

$$O^* = \begin{cases} O - .5 & \text{if } O > E \\ O + .5 & \text{if } O < E \end{cases} \qquad [12.7]$$

Calculate the X^2 statistic as usual, but replace the observed frequency, O, with O^*.

Because the continuity correction brings observed frequencies closer to the expected frequencies, it is a "conservative" procedure that makes it more difficult to reject H_0. Adding .5 to one cell in each row and subtracting .5 from the next has no effect on the marginal frequencies in a 2 by 2 table.

Table 12.12 shows a 2 by 2 crosstabulation from a study of patients who received medical radiation treatments. The 414 patients studied included 138 who later developed leukemia. Is there any connection between radiation dosage and the incidence of leukemia?

Expected frequencies are calculated as usual, following Equation [12.2]: $E = (r \times c)/n$. One cell's expected frequency is well below five, so the usual chi-square test may be invalid. Instead we apply the continuity correction by adding or subtracting .5 from each observed frequency, to move toward the expected frequencies; see Equation [12.7]. The corrected value in the Leukemia/High dosage cell is:

$$O^* = O - .5 \qquad \text{(because } O > E\text{)}$$
$$= 7 - .5 = 6.5$$

The .5 is subtracted in this cell because the observed frequency ($O = 7$) is higher than the expected frequency ($E = 3$). In the Leukemia/Low dosage cell, $O < E$, so we add .5 to obtain the corrected value of O: $O^* = O + .5 = 131 + .5 = 131.5$. These O^* values are then used in place of the observed frequencies, for calculating the $(O - E)^2/E$ values and the X^2 statistic itself.

TABLE 12.12 *Leukemia by Radiation Dose, for 414 Patients: Crosstabulation[a] and Chi-Square Test with Continuity Correction for Thin Cells*

| | Group | | |
Radiation dosage	Leukemias	Controls	All
Above 300 rads	7 (3)	2 (6)	9
O^*	$7 - .5 = 6.5$	$2 + .5 = 2.5$	
$(O^* - E)^2/E$	4.08	2.04	
Below 300 rads	131 (135)	274 (270)	405
O^*	$131 + .5 = 131.5$	$274 - .5 = 273.5$	
$(O^* - E)^2/E$	0.09	.05	
All	138	276	414

$$X^2 = \sum \frac{(O^* - E)^2}{E} = 4.08 + 2.04 + .09 + .05 = 6.26$$

$$df = (K_r - 1)(K_c - 1) = (2 - 1)(2 - 1) = 1$$

P-value: With 1 degree of freedom, $X^2 = 6.26$ yields an obtained *P*-value of $P < .025$.

Conclusion: We therefore *reject the null hypothesis* that leukemia and high radiation doses are unrelated; we conclude that there *is* a relationship.

[a]Observed frequencies are followed by expected frequencies in parentheses.
Source: Data from Gofman (1981).

Despite the conservative continuity correction, the test in Table 12.12 leads to the rejection of the null hypothesis that leukemia is unrelated to high radiation doses ($P < .025$). The strongest evidence against the independence hypothesis is provided by the seven cases observed in the Leukemia/High dose cell, where only three were expected.

PROBLEM

*14. The table at the top of page 386 presents a crosstabulation of survey data. The cases are 33 households surveyed in a city where an emergency water shortage led to a conservation campaign (the raw data are in Table 2.8, page 41). Variables are whether respondents *reported* that their household used less water and whether water meter readings showed that usage *actually* declined during the crisis. Is the relationship between self-reported and actual conservation statistically significant? Run a chi-square test with continuity correction, then use appropriate percentages (based on *uncorrected* frequencies) to help summarize your findings.

Crosstabulation of Savings During a Water Conservation Campaign

Self-reported water savings	Actual water savings		
	None	*Some*	*All*
None	10	12	22
Some	3	8	11
All	13	20	33

12.9 *SAMPLE SIZE AND SIGNIFICANCE IN CHI-SQUARE ANALYSIS*

All inferential calculations are affected by sample size. In a small sample, it may be difficult to reject the null hypothesis even when two variables are strongly related. In a large sample, on the other hand, even a weak relationship can justify rejecting the null hypothesis. This is one of the reasons why *statistical significance* should not be confused with practical importance. We can have one without the other.

Table 12.13 is a crosstabulation of survey responses by 98 college students, who were asked what was the highest degree that they hoped to complete. Twenty-eight out of 98 (28.6%) said they aspired only to complete a Bachelor's

TABLE 12.13   **Degree Aspirations by Gender, for 98 College Undergraduates: Crosstabulation and Chi-Square Test of Independence**

Gender	Highest degree aspiration		All
	Bachelor's	*Graduate*	*All*
Male	11 (11.143)	28 (27.857)	39
$(O - E)^2/E$	.00184	.00073	
Row	28.2%	71.8%	
Female	17 (16.857)	42 (42.143)	59
$(O - E)^2/E$	.00121	.00049	
Row	28.8%	71.2%	
All	28	70	98

$$X^2 = \sum \frac{(O - E)^2}{E} = .00184 + .00073 + .00121 + .00049 = .00427$$

$$df = (K_r - 1)(K_c - 1) = (2 - 1)(2 - 1) = 1$$

P-value: Given $X^2 = .00427$ and $df = 1$, $P > .50$.

Conclusion: We *do not reject* the null hypothesis that this sample came from a population of students in which gender and degree aspirations are independent. In other words, there is no significant relationship between gender and degree aspirations.

degree; the rest (71.4%) hoped for graduate degrees. Crosstabulating by gender showed little difference between men and women: 71.8% of the men aspired to graduate degrees, as did 71.2% of the women. A chi-square test confirms that this tiny difference is statistically insignificant: $X_1^2 = .00427$, with a P-value well over .50. Thus the descriptive analysis (row percentages) shows little difference between men and women in the sample, and the inferential analysis (chi-square test) indicates that this sample gives no reason to conclude that aspirations differ by gender in the larger population of students from which this sample was drawn.

Table 12.14 contains fake data that were manufactured by multiplying every cell frequency in Table 12.13 by 1,000. This simulates having a much larger sample, 98,000 instead of 98, while retaining the exact relationship between gender and degree aspirations. The row percentages in Table 12.14 are identical to those in Table 12.13: 71.8% of the men aspire to graduate degrees, as do 71.2% of the women. Whether based on 98 or 98,000 cases, a difference of only six-tenths of 1% seems too tiny to hold much practical importance.

Unlike descriptive statistics such as sample percentages, inferential statistics such as the chi-square test are drastically changed by the increase in sample size. If the observed frequencies from Table 12.13 are each multiplied by 1,000, then the expected frequencies also increase by this much. As a result, $(O - E)^2/E$ values and the X^2 statistic itself are both 1,000 times greater in Table 12.14 than they are in Table 12.13. The X^2 value is now much larger

TABLE 12.14 *Degree Aspirations by Gender, for 98,000 "College Undergraduates":[a]*
Crosstabulation and Chi-Square Test of Independence

| | Highest degree aspiration | | |
Gender	Bachelor's	Graduate	All
Male	11,000 (11,143)	28,000 (27,857)	39,000
$(O - E)^2/E$	1.84	.73	
Row	28.2%	71.8%	
Female	17,000 (16,857)	42,000 (42,143)	59,000
$(O - E)^2/E$	1.21	.49	
Row	28.8%	71.2%	
All	28,000	70,000	98,000

$$X^2 = \sum \frac{(O - E)^2}{E} = 1.84 + .73 + 1.21 + .49 = 4.27$$

$$df = (K_r - 1)(K_c - 1) = (2 - 1)(2 - 1) = 1$$

P-value: Given $X^2 = 4.27$ and df = 1, $P < .05$.

Conclusion: We *reject the null hypothesis* that this sample came from a population of students in which sex and degree aspirations are independent. In other words, we now conclude that there *is* a significant relationship between sex and degree aspirations.

[a]These data were manufactured by multiplying all cell frequencies in Table 12.13 by 1,000.

(4.27 vs. .00427) and the *P*-value correspondingly lower, $P < .05$, so at $\alpha = .05$ we should now reject the null hypothesis that gender and degree aspirations are independent.

Thus Table 12.14 leads to the conclusion that gender and degree aspirations *are* related in the population of students from which this "sample" was drawn. That is opposite to the conclusion based on Table 12.13, even though the difference between men and women has stayed the same—six-tenths of 1%. In a very large sample, this tiny difference *is statistically significant.*

Inflating cell frequencies to increase sample size is of course not a legitimate research technique. It was done here to demonstrate the role that sample size plays in determining whether a null hypothesis is rejected. In large samples, relationships can have statistical significance even when they are so weak that they have no practical importance. In small samples, even a strong relationship with great practical importance may prove statistically insignificant. This is easy to demonstrate in crosstabulations, but it is an equally great problem for other kinds of analysis as well.

PROBLEM

15. Return to the conservation crosstabulation of Problem 14. Would your inferential conclusions have been different if the same percentage distribution had been found in a table with 3,300 instead of 33 cases (that is, if each cell frequency were multiplied by 100)? Carry out this analysis and explain how your new conclusion compares to the old. (A continuity correction is not needed in your imaginary sample of 3,300 cases.)

SUMMARY

Crosstabulation is an extension of one-variable frequency tables. The examples in this chapter are bivariate, but crosstabulation can be extended to display joint distributions of three or more categorical variables; analysis then becomes more technical. Modern techniques for analyzing multivariate crosstabulations require specialized computer programs.[5]

Bivariate crosstabulations can be analyzed with comparatively simple tools: percentages and the chi-square test. Either row or column percentages, based on the independent variable, provide a readily understood summary for descriptive purposes. Choose carefully which kind of percentage to use, and state accurately what that percentage represents.

The chi-square test allows us to evaluate the independence hypothesis: that the two categorical variables are unrelated in the population. If we reject this null hypothesis, we accept that the variables *are* related. We can then use percentages or observed–expected comparisons to describe what that relationship is.

One pitfall frequently encountered with chi-square tests is thin cells. Possible solutions to this problem, ordered from theoretically "best" to "worst," include:

1. Collect more data, so no cells are thin.
2. Apply an exact test.
3. Use a continuity correction.
4. Combine categories.
5. Drop thin categories from the analysis.

Unfortunately the solutions that are best, from a theoretical point of view, may not be practical. Expedient solutions, especially combining categories, are often used in practice.

The last section of this chapter manufactured data to dramatize the impact of sample size on statistical significance. Statistical significance and practical importance are *not* the same thing, and we often find one without the other. Misunderstanding this basic point can create great confusion when statistical findings are debated.

Crosstabulation is primarily a technique to use with categorical variables. In the next four chapters we will encounter a different class of analytical techniques for measurement variables. Measurement variables provide more information than categorical variables do, and the techniques for analyzing them are enriched by this fact.

PROBLEMS

16. Use the appropriate percentages to describe the relationship between

    a. Violent content and article source (punk press vs. *Reader's Digest*— Table 12.6).

    b. Leukemia and radiation dosage (Table 12.12).

The accompanying table contains a crosstabulation of opinions about whether a controversial nuclear power plant should be permitted to begin operations, by gender. The data are from a survey of 400 nearby residents. Problems 17 and 18 refer to this crosstabulation.

Crosstabulation of Opinions About Plant by Gender

	Gender		
Should plant operate?	*Men*	*Women*	*All*
Yes	84	56	140
No	74	138	212
Don't know/No opinion	21	27	48
All	179	221	400

17. Following the rule given in this chapter, what type of percentages (row, column, or total) would best describe the relationship between opinion and gender? Explain your choice, then calculate these percentages and fill out a percentage version of the table. Use the percentages to write a paragraph describing any relationship you see between gender and opposition to plant operation.

18. Conduct a chi-square test of the crosstabulation to determine if the relationship between opinion and gender is statistically significant at $\alpha = .05$. Include the following steps.

   a. State the null and alternative hypotheses.

   b. Calculate the expected frequencies, and enter them into each cell of the table.

   c. Explain the meaning of the expected frequency you calculated for the Men/Yes cell in this table. How does this expected frequency compare with that actually observed?

   d. Find the $(O - E)^2/E$ values, and enter them into each cell as well, in the manner of Table 12.4.

   e. Calculate the X^2 statistic, and use this to perform the test. Describe your decision with respect to the null hypothesis.

   f. Summarize your final conclusions regarding gender and opposition to the nuclear power plant, drawing on your analysis both here and in Problem 17.

19. Further tabulations from the nuclear power plant survey are shown in the accompanying table, where opinions are crosstabulated by age group. Carry out a chi-square analysis of this table, and use appropriate percentages to describe your findings about the relationship between age and opposition to plant operation.

Crosstabulation of Opinions About Plant by Age

Should plant operate?	Age group				All
	18–29	*30–44*	*45–59*	*60+*	*All*
Yes	26	38	30	46	140
No	57	64	46	45	212
Don't know/No opinion	7	7	15	19	48
All	90	109	91	110	400

20. Opinion surveys commonly encounter respondents who do not know, or have no opinion, about some question. Looking at the percentages you calculated for Problems 18 and 19, do you see any relationship between "don't know" responses and either gender or age? Use these percentages to describe the pattern of "don't know" responses in this survey.

21. Using the student data in Table 1.1 (page 5), construct a crosstabulation of major by gender, omitting the three students whose majors are unknown. Calculate the appropriate percentages and enter them into the cells of your table. Discuss any patterns you see. Would a chi-square test be appropriate here? Explain.

22. Physicians often see patients complaining of chest pains. Examination later confirms that some suffer from coronary artery disease, but for others no obvious physical cause can be found for the pains. Psychiatric causes may then be suspected. The table contains data from a study of patients with chest pain (Katon et al., 1988). The presence or absence of coronary heart disease is crosstabulated by whether the patient had a history of major depression. Perform a chi-square analysis to test whether there is a significant relationship between depression and absence of coronary artery disease among chest pain patients. Calculate the appropriate percentages and use them to discuss your conclusions.

Depression and Coronary Artery Disease in 74 Patients with Chest Pains

| | Coronary artery disease | | |
Medical history	Absent	Present	All
No depression	10	35	45
Depression	18	11	29
All	28	46	74

23. The accompanying table contains a crosstabulation from a study based on interviews with preschool children who lived in two-parent families (Hamilton, 1968). Forty attended private preschools and were primarily middle-class. Others, from poorer families, were enrolled in Operation Headstart preschools. The question asked was who would listen to the child, if he or she arrived home from preschool excited about artwork done in class. Some children first named their mother; others said their father or some other person. Carry out a chi-square test (at $\alpha = .01$), and use the appropriate percentages to summarize your conclusions about the relationship between background and who will listen to the child.

Who Listens to the Child: Preschoolers' Responses by Background

| | Who will listen? | | | |
Background	Mother	Father	Other	All
Private preschool	20	5	15	40
Operation Headstart	6	12	15	33
All	26	17	30	73

*24. Table 8.1 (page 210) contains 1,800 random digits. They were actually produced by a type of equation called a pseudorandom generator, so it is open to question whether they truly have the statistical properties of random numbers. The simplest such property is that each digit, from 0 to 9, should appear with the same probability, 0.1. If this is so, we expect about 180 of the 1,800 random digits to be 0's, 180 to be 1's, etc. As shown here, none of the digits had an observed frequency exactly equal to 180. For instance, there are 185 0's, and 163 1's (see Table 8.2 and Figure 8.2). Are the observed and expected frequencies close enough that the differences could easily be due to chance?

Apply a univariate chi-square test (at $\alpha = .05$), to evaluate the null hypothesis that these numbers were generated by a process in which each digit has the same probability (df = number of categories − 1). Does our pseudorandom generator pass this elementary test of apparent randomness?

Frequencies of 1,800 Pseudorandom Digits

Digit	0	1	2	3	4	5	6	7	8	9
f	185	163	170	167	178	183	172	182	198	202

*25. During 1957 the U.S. Government conducted a series of nuclear weapons tests in Nevada, codenamed PLUMBBOB. As part of the tests servicemen were deployed near the explosions, in an effort to learn about how military units could survive and function on a nuclear battlefield. In the years since 1957, there have been persistent reports of unusually high cancer rates among these "Atomic Veterans."

Among the 13,685 PLUMBBOB participants, there were 349 known cancer deaths as of 1985. (The remaining 13,336 veterans either are still living or died of other causes.) Do 349 deaths constitute an "unusually high" incidence of cancer? In order to answer that, we have to know how many cancer deaths would normally be expected to occur in such a group. This expected frequency can only be estimated, and it has been the object of ongoing scientific controversy. Different choices of the expected number of cancer deaths can lead to opposite conclusions about the effects of the PLUMBBOB tests.

Taking into account the fact that soldiers tend to be healthy, epidemiologists have estimated that without PLUMBBOB, 284.91 cancer deaths would be expected among the 13,685 soldiers (Bross and Bross, 1987). Using this figure, the observed and expected frequencies compare as shown.

	Died of cancer?		
	Yes	*No*	*All*
Observed	349	13,336	13,685
Expected	284.91	13,400.09	13,685

Apply a univariate chi-square test to determine whether the observed number of cancer deaths among PLUMBBOB participants is significantly different from the number that would be expected if this group had not taken part in the nuclear tests.

The crosstabulation shown is from a survey of 15 college students, who were asked: Is there any area around campus where you are afraid to walk alone at night?

Afraid	Men	Women
No	6	2
Yes	1	6

26. Use percentages to describe the relationship in this sample between gender and being afraid to walk alone on campus.

*27. Run a chi-square test on the crosstabulation of Problem 26. (Because of thin cells, use the continuity correction.) At $\alpha = .05$, can we conclude that there is a relationship between gender and being afraid to walk alone at night, in the larger population of students?

*28. Does your conclusion in Problem 27 seem to contradict common sense? How does this example illustrate the difference between statistical significance and practical importance? Explain the role of sample size in this example.

NOTES

1. The inferential leap from a sample to a population depends on the assumption that the sample was selected randomly from that population. Skiers for this study were indeed chosen randomly, but only from the population of skiers at a single ski area during part of a single winter. These are important limitations on how general our conclusions can be.
2. In advanced applications a **likelihood ratio** X^2 statistic, X_L^2, is often preferred to the X^2 statistic of Equation [12.3]. Its definition is

$$X_L^2 = 2\sum O \log_e \left(\frac{O}{E}\right)$$

where $\log_e$ denotes natural or base-e logarithm. For many analyses X^2 and X_L^2 are similar and lead to the same conclusions. Either may be used to test hypotheses about the fit between a set of observed and expected frequencies. They both theoretically follow chi-square distributions with the same degrees of freedom.

For example, we might have used X_L^2 instead of X^2 to test the independence hypothesis for the skiing data of Table 12.5. Using the same observed and expected frequencies seen in Table 12.5, the likelihood ratio statistic is:

$$X_L^2 = 2\sum O\log_e\left(\frac{O}{E}\right)$$

$$= 2\left[20\log_e\left(\frac{20}{11.6}\right) + 60\log_e\left(\frac{60}{68.4}\right) + 9\log_e\left(\frac{9}{13.5}\right)\right.$$

$$\left. + 84\log_e\left(\frac{84}{79.5}\right) + 2\log_e\left(\frac{2}{5.9}\right) + 39\log_e\left(\frac{39}{35.1}\right)\right]$$

$$= 2[10.89 - 7.86 - 3.65 + 4.63 - 2.16 + 4.11] = 11.92$$

In spite of the different looking form of the likelihood ratio statistic, we end up with a value almost identical to that of the X^2 statistic in Table 12.5: $X^2 = 11.87$, $X_L^2 = 11.92$. With df = 2, either statistic leads to the conclusion that we can easily reject the independence hypothesis ($P < .005$) for these data.

The X^2 statistic defined in Equation [12.3] is sometimes called either the **Pearson** X^2 statistic or the **goodness-of-fit** X^2 statistic, to distinguish it from the likelihood ratio X_L^2 statistic.

3. Parenthood is an example of a *demographic* variable. With survey data in general, demographic variables such as gender, race, or age are almost always independent variables, since no other survey variables could "cause" these basic traits. Other variables such as parenthood, marital status, education, parental occupation, and number of siblings are also determined well before the time of the survey, hence are usually viewed as independent variables. When the chronology is not clear, as with many attitudes and behaviors, the choice of "independent" and "dependent" variables becomes more problematic.

4. As noted in Table 12.9, a chi-square test can help evaluate whether a given distribution is significantly different from a normal distribution. Since normal distributions describe continuous measurement variables, the first step in such tests is to group the measurement variable into about 6–12 categories. (The groupings are unavoidably arbitrary, which is one weakness of these tests.) Next we calculate the proportion of cases expected in each category, were the distribution normal. These proportions are then used to obtain expected frequencies, and a chi-square test is applied to compare the observed and expected frequencies in each category.

To obtain expected normal probabilities, we must specify values for the population mean and standard deviation, μ and σ_Y. These are often estimated using the sample mean ($\overline{Y}$) and standard deviation (s_Y), creating a Catch-22 sort of problem: If the distribution really is nonnormal, then the sample mean and standard deviation may be untrustworthy due to their lack of resistance to outliers. A more robust alternative is to estimate μ using the sample median (Md), and to estimate σ_Y using the pseudo-standard deviation (PSD). See Velleman and Hoaglin (1981), Chapter 9 and especially pp. 281–282.

    Either way, using sample data to estimate μ and σ_Y takes 2 degrees of freedom. If these parameters were obtained *without using* the sample (from theory or other research, for instance), degrees of freedom would be $K - 1$ instead of the usual $K - 3$.

5. The best-known method for analyzing multivariate crosstabulations is **log-linear analysis.** Log-linear analysis was first developed by Leo Goodman (Goodman, Davis, and Magidson, 1978), who also wrote and distributed computer programs for performing log-linear analysis. Log-linear programs have since been incorporated into general-purpose statistical packages like SYSTAT and SPSS. A basic introduction to log-linear methods is provided by Upton (1978).

One Categorical and One Measurement Variable: Comparisons

Chapter 13

Chapter 12 described methods for analyzing relationships between two categorical variables. Methods for two measurement variables will be described in Chapter 14. This chapter looks at ways to analyze relationships between one categorical and one measurement variable.

In general, we shall compare the measurement variable's distributions for different categories of the categorical variable. The simplest way is to compare the measurement variable's means for each category. A family of procedures has been developed to test whether differences between means are statistically significant.

The mean summarizes only the center of a distribution, however, ignoring other aspects such as variation, gaps, skewness, or outliers. Furthermore the "center" itself may have no clearcut location. These limitations naturally affect bivariate analyses where we compare several means. Before drawing conclusions from a comparison of means, we must investigate carefully the distributions involved. As with univariate analysis, graphs are essential aids.

Unlike the last chapter's single type of analysis, this chapter covers a variety of techniques. The conceptual unity underlying their seeming diversity will emerge as we proceed. We shall start with an overview of the issues.

13.1 OVERVIEW OF COMPARISON ISSUES

Most methods described in this chapter apply to data organized in either of two ways.

1. The data could be sets of values on some measurement variable, for two or more samples.
2. The data could be organized as a single sample, where one of the variables defines what category a case falls in.

Often the same information can be organized either way, as Table 13.1 illustrates.

The data in Table 13.1 are from 18 environmental monitoring sites in North America and Europe. Values are given for the average annual concentration of sulfate in precipitation measured over the years 1979–1982. Pollutants such as sulfates and nitrates, which result from burning fossil fuels, are of concern because they contribute to acid rain.

In the top half of Table 13.1 these data are organized as two independent samples: one sample consisting of 7 North American sites, and a second consisting of 11 European sites. Below each sample are summary statistics such as $\overline{Y}_1$, the mean of sample 1 (North America) and s_1, the standard deviation of sample 1. We could use these statistics to compare European and North American sulfate concentrations. Is the problem significantly worse in Europe? Such a question calls for a **two-sample test.** Further comparisons, based on samples of sites from South America, Africa, and so on, could be done with a **K-sample test,** which can encompass any number of samples, from two on up.

In the bottom half of Table 13.1, the sample data are rearranged into a single sample, with continent indicated by a categorical variable. This arrangement invites the question: Is there any relationship between sulfate concentration and continent? That is, how are one measurement and one categorical variable related, within this single sample comprising all 18 sites?

Both data formats are basically alike, and both can be analyzed by the same methods, as long as all cases are *independent* of each other. With random selection of each case from a large population, this will usually be true. If so, then it does not matter whether we view our data as consisting of one sample, or several. Because it is simpler and more general, we will use the single-sample organization with most examples in this chapter. With either format, the analytical techniques themselves are still called two-sample or *K*-sample tests.

TABLE 13.1   *Two Displays of Average Sulfate Concentrations in Precipitation in North America and Europe[a]*

Two independent samples

1. *North American sites*		2. *European sites*	
Edson, Canada	.64	Ejde, Denmark	1.40
Kelowna, Canada	.25	Sodankyla, Finland	.59
Mould Bay, Canada	.65	Monte Cimone, Italy	.69
Sable Island, Canada	.36	Trapani, Italy	.76
Wynyard, Canada	.69	Kise, Norway	1.30
Caribou, Maine, U.S.	.65	Suwalki, Poland	2.10
Huron, S. Dakota, U.S.	.92	Velen, Sweden	1.10
		Lazaropole, Yugoslavia	2.20
		Kurgan, U.S.S.R.	3.20
		Turukhansk, U.S.S.R.	1.50
		Svratouch, Czech.	2.00

$$n_1 = 7, \overline{Y}_1 = .59, s_1 = .22 \qquad n_2 = 11, \overline{Y}_2 = 1.53, s_2 = .79$$

One sample, with continent as a categorical variable

Site	Sulfate	Continent[b]
Edson, Canada	.64	1
Kelowna, Canada	.25	1
Mould Bay, Canada	.65	1
Sable Island, Canada	.36	1
Wynyard, Canada	.69	1
Caribou, Maine, U.S.	.65	1
Huron, S. Dakota, U.S.	.92	1
Ejde, Denmark	1.40	2
Sodankyla, Finland	.59	2
Monte Cimone, Italy	.69	2
Trapani, Italy	.76	2
Kise, Norway	1.30	2
Suwalki, Poland	2.10	2
Velen, Sweden	1.10	2
Lazaropole, Yugoslavia	2.20	2
Kurgan, U.S.S.R.	3.20	2
Turukhansk, U.S.S.R.	1.50	2
Svratouch, Czech.	2.00	2

[a]Sulfate concentrations are for 1979–1982, measured in milligrams per liter.
[b]1 = North America, 2 = Europe.
Source: Monitoring sites were randomly selected from Council on Environmental Quality (1986).

The most common techniques for studying relationships between one categorical and one measurement variable are based on a simple idea: Examine the means of the measurement variable within different categories of the categorical variable. For example, we could compare the mean sulfate concentrations for the North American and European sites in Table 13.1. The sample means obviously differ: $\overline{Y}_1 = .59$ mg/l (North America) vs. $\overline{Y}_2 = 1.53$ mg/l (Europe). Do the population means also differ?

Two applications of graphing are emphasized in this chapter:

1. **Exploratory:** Parallel box plots (Chapter 5) let us visually check whether mean-based analysis is reasonable.
2. **Summary:** If we decide that mean-based analysis *is* reasonable, error-bar plots can illustrate how the means differ.

Exploratory graphs such as box plots should be a first step in any analysis comparing means. Error-bar plots come later and sometimes help to understand and present conclusions.

Parallel box plots for North American and European sulfate concentrations (Table 13.1) are shown in Figure 13.1. This graph confirms that such pollution is generally worse at the European sites, and it also shows that the European sites are much more variable than the North American sites.

A family of methods called *paired-difference tests* are also included in this chapter, although they do not necessarily involve relationships between one categorical and one measurement variable. These methods are needed when the cases are *not* independent, but are naturally paired. Problems that require paired-difference techniques may superficially resemble two-sample problems, but it is important to distinguish between them.

Table 13.2 contains an example of a paired-difference analysis, using the same 18 sites from Table 13.1. Both two-sample and paired-difference tests

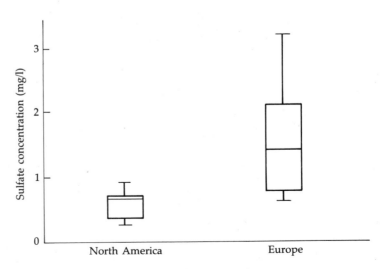

FIGURE 13.1   *Sulfate at 7 North American and 11 European sites*

TABLE 13.2

Average Sulfate Concentrations, (mg/l) in Precipitation in North America and Europe

Site	Sulfate 1975–1978	Sulfate 1979–1982	Reduction in sulfate
$n = 18$	Y_{78}	Y_{82}	$D = Y_{78} - Y_{82}$
Edson, Canada	.73	.64	.09
Kelowna, Canada	.58	.25	.33
Mould Bay, Canada	1.20	.65	.55
Sable Island, Canada	1.50	.36	1.14
Wynyard, Canada	.83	.69	.14
Caribou, Maine, U.S.	1.20	.65	.55
Huron, S. Dakota, U.S.	.97	.92	.05
Ejde, Denmark	1.60	1.40	.20
Sodankyla, Finland	.65	.59	.06
Monte Cimone, Italy	1.20	.69	.51
Trapani, Italy	.91	.76	.15
Kise, Norway	1.50	1.30	.20
Suwalki, Poland	2.70	2.10	.60
Velen, Sweden	1.40	1.10	.30
Lazaropole, Yugoslavia	2.50	2.20	.30
Kurgan, U.S.S.R.	1.50	3.20	−1.70
Turukhansk, U.S.S.R.	.63	1.50	−.87
Svratouch, Czech.	2.40	2.00	.40
Sample statistics:[a]	$\overline{Y}_{78} = 1.33$	$\overline{Y}_{82} = 1.17$	$\overline{D} = .17$
	$s_{78} = .64$	$s_{82} = .78$	$s_D = .61$

[a]In Table 13.1, sample statistics were calculated separately for two *different subsets of cases:* the $n = 7$ North American sites and the $n = 11$ European sites. Thus Table 13.1 is set up for a two-sample comparison of means. The statistics here are calculated from *the entire sample* of $n = 18$; this table is set up for a paired-difference comparison.
Source: Council on Environmental Quality (1986).

involve comparing two means. The two means compared in a two-sample test (Table 13.1 and Figure 13.1) are based on *one variable in two different subsets of the cases.* In a paired-difference test (Table 13.2 and Figure 13.2, page 402), by contrast, the two means are based on *two variables in a single sample of cases.* Paired-difference tests are often used to study change. They focus on a *differ-ence score, D,* defined as the difference between the two variables. In Table 13.2, D measures the reduction in sulfate concentration, or the 1975–1978 value minus the 1979–1982 value, for each site. These data could address the question: Was there a significant reduction in sulfate concentration in Europe and North America? One might be hoped for, given pollution controls and shifts to low-sulfur fuels.

The final section of this chapter describes methods where *ranks* replace the measured values themselves. Rank-based procedures are examples of

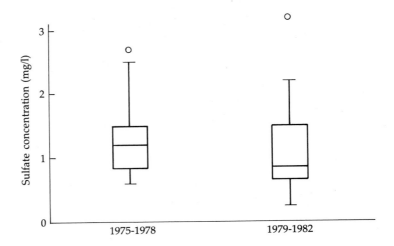

FIGURE 13.2   *Sulfate at 18 sites*

nonparametric methods, which have the advantage of being comparatively robust (insensitive to distributional problems). Rank-based tests serve well when the assumptions for a mean-based test are implausible.

PROBLEMS

1. Suggest your own examples of research questions that call for the following types of analysis. For each example, what differences between means would you expect to see?

   a. Two-sample test

   b. *K*-sample test

   c. Paired-difference test

2. In a two-sample analysis (Table 13.1 and Figure 13.1), the mean sulfate concentration was much higher for the 11 European sites than for the 7 North American sites. In a paired-difference analysis (Table 13.2 and Figure 13.2) over all 18 European and American sites, the sample mean sulfate concentration was lower in 1979–1982 than in 1975–1979. It is possible to combine both kinds of analysis to examine a third question: Was the *reduction* in sulfate concentrations larger in Europe or in North America? Find the sample statistics (mean, standard deviation, and *n*) needed to address this question, and summarize your preliminary findings.

13.2 TESTING HYPOTHESES ABOUT MEANS

The usual bivariate null hypothesis is that in the population no relationship exists between two variables. If both variables are categorical, *no relationship* corresponds to the *independence hypothesis* used in the standard chi-square test of crosstabulations (Chapter 12). With one categorical and one measurement variable, however, *no relationship* is usually translated as *no difference between means*.

The null hypothesis in tests comparing means states that the sample data come from populations with equal means. Suppose $\overline{Y}_1, \overline{Y}_2, \overline{Y}_3, \ldots, \overline{Y}_K$ are the sample means of measurement variable Y, across the K different categories of categorical variable X. The corresponding parameters are the population means of Y, namely $\mu_1, \mu_2, \mu_3, \ldots, \mu_K$.

The usual null hypothesis ("no relationship") in a comparison of K means is

$$H_0: \quad \mu_1 = \mu_2 = \mu_3 = \cdots = \mu_K$$

That is, in the population the means of Y (μ_1, μ_2, etc.) are equal across all K categories of X.

Table 13.3 (page 404) reviews the meaning of basic symbols that are used in this chapter. Similar ones were used previously, but we now need subscripts to denote the categories of X.

Measurement-variable distributions can differ in many ways, even if their means are the same. Knowing that the means are (or are not) the same is therefore not the whole story in comparing measurement variable distributions. Examine, for instance, the box plots of fictitious data in Figures 13.3 and 13.4. Both figures show the distribution of a measurement variable, Y, across three categories of X. In both plots the means have the same pattern: $\overline{Y}_1 = 75$, $\overline{Y}_2 = 125$, and $\overline{Y}_3 = 125$.

The three distributions in Figure 13.3 are all approximately normal, with standard deviations of 32: $s_1 = s_2 = s_3 = 32$. Because Y is normally distributed and has the same standard deviation within each category of X, these distributions differ only in their means. We can summarize the sample relationship between Y and X just by comparing the three means.

In Figure 13.4 it is also true that $\overline{Y}_1 = 75$, $\overline{Y}_2 = 125$, and $\overline{Y}_3 = 125$, but the three distributions differ in other obvious ways as well. In category 1 of X, Y is positively skewed with a number of severe outliers. In categories 2 and 3, the Y distributions are more symmetrical, but they differ greatly in spread. The standard deviation in category 3 is $s_3 = 52$, more than ten times larger

TABLE 13.3   *Summary of Some Symbols Used in Chapter 13*

Symbol	Meaning and examples
Y	A measurement variable *Example:* income, in dollars
X	A categorical variable, with categories 1, 2, 3, … , K *Example:* political party, 1 = Democrat, 2 = Republican, 3 = Independent
$\overline{Y}_1$	Sample mean of variable Y for all cases in the sample that fall in category 1 of variable X *Example:* mean income of Democrats in sample
$\overline{Y}_2, \overline{Y}_3, …, \overline{Y}_K$	Sample means of Y in categories 2, 3, …, K of X *Example:* mean incomes of Republicans ($\overline{Y}_2$) and Independents ($\overline{Y}_3$) in sample
n_1	Number of cases in sample that fall in category 1 of variable X *Example:* number of Democrats in sample
$n_2, n_3, …, n_K$	Numbers of cases in sample in categories 2, 3, …, K of X *Example:* numbers of Republicans (n_2) and Independents (n_3) in sample
s_1	Sample standard deviation of Y for those cases that fall in category 1 of variable X *Example:* standard deviation of incomes of sample Democrats
$s_2, s_3, …, s_K$	Sample standard deviations of Y for categories 2, 3, …, K of X *Example:* standard deviations of incomes of sample Republicans (s_2) and Independents (s_3)
μ_1	Population mean of variable Y for all cases in the population that fall in category 1 of variable X *Example:* mean income of Democrats in the population
$\mu_2, \mu_3, …, \mu_K$	Population means of Y in categories 2, 3, …, K of X *Example:* mean incomes of Republicans (μ_2) and Independents (μ_3) in the population

than that in category 2 ($s_2 = 4.8$), and twice that in category 1 ($s_1 = 26$). If we looked only at means, we might mistakenly conclude that no difference exists between the Y distributions in X categories 2 and 3.

Comparisons of means will tell the whole story about a relationship *only if the distributions differ in no other important respects, except their means.* Basing the comparison on means also makes most sense when the distributions are reasonably symmetrical. Otherwise neither the mean nor any other statistic can identify unambiguous "centers" for the distributions.

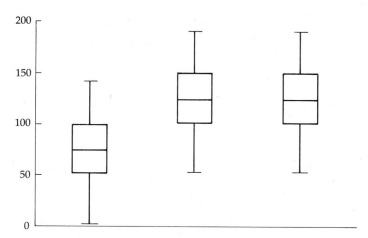

FIGURE 13.3   *Three approximately normal distributions with equal standard deviations*

The mean-based tests in this chapter handle these problems by requiring two distributional assumptions.

Two assumptions are commonly required for mean-based tests:

1. *Normality:* Within each category of X, the population distributions of Y are normal.
2. *Constant variance:* Within each category of X, Y has the same population variance or standard deviation.

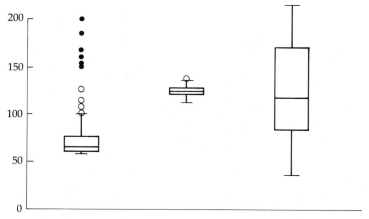

FIGURE 13.4   *Three nonnormal distributions*

If these assumptions are true, then the Y distributions are identical except for any differences in their means. (Such was the situation in Figure 13.3.) Even where the assumptions are not true, the tests sometimes produce reasonable results. Mean-based tests are not robust, however, against large departures from normality (severe outliers or skewness) or against grossly unequal variances—the kinds of problems in Figure 13.4. Comparing means is also inappropriate where, due to skewness or bimodality, the mean is a poor measure of center (see Chapter 3). Box plots or stem-and-leaf displays can help check whether such problems exist, hence whether a test of the null hypothesis

$$H_0: \quad \mu_1 = \mu_2 = \mu_3 = \cdots = \mu_K$$

is an adequate test for whether X and Y are related.

PROBLEMS

3. Assume that the 18 environmental monitoring sites in Table 13.1 constitute a sample drawn randomly from a population consisting of all nonurban points in Europe and North America. Symbolically state the null hypothesis of no relationship between continent and mean nonurban sulfate concentration, and explain the meaning of your symbols. Does the sample evidence appear to support H_0?

4. In checking sample distributions we should not expect to see perfect normality or exactly equal standard deviations, which rarely occur in practice. Rather, we hope to see distributions with no more than moderate asymmetry, few severe outliers, and reasonably similar spread.

   Chapter 5 contained numerous examples in which the distribution of a measurement variable was compared across several categories of a categorical variable. Examine the box plots listed and decide whether you see any major problems with the assumptions that (1) Y distributions are normal, and (2) Y has the same variation at all levels of X. If you see problems with outliers or skewness, describe how they might affect a comparison of means.

   a. Episodes of rough and tumble play among preschool boys and girls (Figure 5.8)

   b. Calculation test speeds of 40 personal computers (Figure 5.10)

   c. SAT scores (Figure 5.11)

   d. Nicotine concentration by airline seating (Figure 5.12)

   e. Life expectancies by continent (Chapter 5, Problem 11)

13.3   *TWO-SAMPLE PROBLEMS: DIFFERENCE-OF-MEANS TESTS*

With just two samples, or two categories of X, the null hypothesis of no difference between means reduces to

$$H_0: \quad \mu_1 = \mu_2$$

An equivalent way of stating this null hypothesis is that the **difference of means** is 0:

$$H_0: \quad \mu_2 - \mu_1 = 0 \qquad\qquad [13.1]$$

The logical alternative to Equation [13.1] is that in the population the difference of means is *not* 0:

$$H_1: \quad \mu_2 - \mu_1 \neq 0 \qquad\qquad [13.2]$$

The pair of hypotheses formulated in [13.1] and [13.2] call for a two-sided hypothesis test, since H_1 does not specify *in which direction* the means differ. With two-sample problems, as with one-sample problems (Chapters 10 and 11), we may run either one- or two-sided tests. A one-sided test is needed if H_1 specifies direction. For example,

$$H_0: \quad \mu_2 - \mu_1 \leq 0 \qquad\qquad [13.3]$$
$$H_1: \quad \mu_2 - \mu_1 > 0 \qquad\qquad [13.4]$$

The directional research hypothesis, $H_1: \mu_2 - \mu_1 > 0$, asserts that the mean in category 2 is *higher than* the mean in category 1. We could also have formulated a pair of directional hypotheses to represent the expectation that the mean in category 2 will be lower.

Difference-of-means tests proceed by comparing the sample difference of means, $\overline{Y}_2 - \overline{Y}_1$, with the population difference of means in the null hypothesis, $\mu_2 - \mu_1 = 0$; see [13.1] or [13.3]. If the sample difference of means is far enough from the hypothesized difference, 0, then we can reject the null hypothesis and accept the research hypothesis.

Table 13.4 (page 408) contains data on 34 college students, including two categorical variables (gender and degree goals) as well as four measurement variables. Difference-of-means tests might reveal whether the means of any of the four measurement variables differ significantly for men and women. Such analyses could be viewed either as two-sample problems (comparing independent samples of men and women) or as relationship problems (is gender related to GPA, and so on).

For example, is there any relationship between the variables gender (categorical) and math SAT score (measurement)? Clearly there is a relationship *in this sample.* Figure 13.5 shows box plots of the MSAT scores for 9 men and 22 women. (For one man and two women, the scores are unknown.) The box

TABLE 13.4   *Data from a Sample of 34 College Students*

Case	Gender	Degree goals	High school class rank[a]	True GPA	Reported GPA	Math SAT
1	Male	Bachelor	66	2.00	2.25	490
2	Male	Bachelor	135	2.20	2.80	550
3	Male	Bachelor	101	2.41	1.90	620
4	Male	Bachelor	111	2.44	2.30	610
5	Male	Bachelor	163	2.44	3.75	370
6	Male	Bachelor	51	2.84	2.70	670
7	Male	Bachelor	45	3.35	3.35	630
8	Male	Master	64	2.68	2.60	610
9	Male	Master	198	2.75	3.00	590
10	Male	Master	78	3.46	3.80	—
11	Female	Bachelor	175	1.57	2.30	440
12	Female	Bachelor	134	2.12	1.97	440
13	Female	Bachelor	144	2.24	2.33	650
14	Female	Bachelor	13	2.32	2.45	450
15	Female	Bachelor	67	2.44	2.45	380
16	Female	Bachelor	72	2.45	2.30	330
17	Female	Bachelor	59	2.86	2.89	590
18	Female	Bachelor	184	2.90	2.95	540
19	Female	Bachelor	5	3.33	3.25	600
20	Female	Master	52	1.75	3.13	390
21	Female	Master	51	1.91	3.55	460
22	Female	Master	148	2.39	2.80	—
23	Female	Master	33	2.41	2.44	550
24	Female	Master	24	2.52	2.50	470
25	Female	Master	225	2.58	3.00	—
26	Female	Master	79	2.66	2.70	520
27	Female	Master	2	2.67	3.20	530
28	Female	Master	27	2.82	3.40	410
29	Female	Master	15	2.97	3.07	460
30	Female	Master	84	3.57	3.50	550
31	Female	Master	7	3.65	3.83	600
32	Female	Doctorate	136	2.72	3.40	420
33	Female	Doctorate	2	2.93	2.91	550
34	Female	Doctorate	15	3.42	3.00	530

[a]Where 1 = best student in class, 2 = second best, etc.

plots indicate that the median MSAT score of the men substantially exceeds that for the women. Both distributions have similar spread, as shown by the height of the central boxes (interquartile ranges). The distribution of women's scores is symmetrical; the men's scores exhibit mild negative skewness and one low outlier.

Table 13.5 gives sample means and standard deviations. The standard deviations are similar: Among men $s_1 = 91.2$, and among women $s_2 = 83.0$.

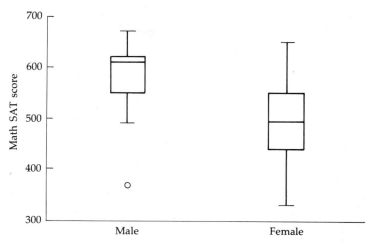

FIGURE 13.5 *Box plots of MSAT scores by gender (Table 13.4)*

Like the medians graphed in Figure 13.5, the means in Table 13.5 indicate that the men's scores tended to be higher: $\overline{Y}_1 = 571.1$ vs. $\overline{Y}_2 = 493.6$. The slight negative skewness and one low outlier in the men's distribution pull down their mean (571.1) relative to the median. For the women's more symmetrical distribution, the median (495) is nearly the same as the mean (493.6). Although the men's sample distribution does not look very normal, in fact their population distribution *is* fairly close to normal. Unless large departures from normality tip us off, it is hard to judge the normality of a population distribution from a small sample of cases.

Of course, these results are based entirely on these 31 students. If we selected another sample of 31 students, we might get much different means. The question is, what do the MSAT scores of our 31 students suggest about MSAT scores in the much larger population of students from which these 31 were drawn?

This question calls for a difference-of-means test. Its basic format resembles the one-sample tests described in Chapters 10 and 11. We calculate the

TABLE 13.5 *Sample Statistics Needed for a Two-Sample Test: Means and Standard Deviations of MSAT Scores by Gender (from Table 13.4)*

Statistic	Men	Women
Number of cases	$n_1 = 9$	$n_2 = 22$
Mean MSAT	$\overline{Y}_1 = 571.1$	$\overline{Y}_2 = 493.6$
Standard deviation	$s_1 = 91.2$	$s_2 = 83.0$

distance between a sample statistic and a population parameter in standard errors:

$$t = \frac{\text{Sample statistic} - \text{Hypothesized population parameter}}{\text{Estimated standard error of sample statistic}}$$ [13.5]

A two-sample test for a **difference of means** is performed with a *t* **statistic:**

$$t = \frac{(\overline{Y}_2 - \overline{Y}_1) - (\mu_2 - \mu_1)_0}{SE_{\overline{Y}_2 - \overline{Y}_1}}$$ [13.6]

$SE_{\overline{Y}_2 - \overline{Y}_1}$ is the estimated standard error of the difference between sample means.

If H_0 is true and certain distributional assumptions are met, this *t* statistic will follow approximately a *t* distribution (Table A.3) with degrees of freedom

$$df = n_1 + n_2 - 2$$ [13.7]

Under the null hypothesis that there is no difference between population means (that is, $\mu_2 - \mu_1 = 0$), Equation [13.6] simplifies to

$$t = \frac{\overline{Y}_2 - \overline{Y}_1}{SE_{\overline{Y}_2 - \overline{Y}_1}}$$ [13.8]

The standard error for a difference between two means, $SE_{\overline{Y}_2 - \overline{Y}_1}$, is most easily estimated in two stages:

1. Find the *pooled standard deviation.* This is an estimate of the population standard deviation of *Y*, assuming that it is the same across all categories of *X*.
2. Use this pooled standard deviation to estimate the *standard error of a difference of sample means,* $SE_{\overline{Y}_2 - \overline{Y}_1}$.

The **pooled standard deviation,** s_p, is

$$s_p = \sqrt{\frac{(n_1 - 1)s_1^2 + (n_2 - 1)s_2^2}{n_1 + n_2 - 2}}$$ [13.9]

where n_1 is the number of cases in category 1 and s_1 is their standard deviation. The corresponding statistics for category 2 are n_2 and s_2.

The **standard error of a difference of means** is estimated from

$$SE_{\overline{Y}_2 - \overline{Y}_1} = s_p \sqrt{\frac{1}{n_1} + \frac{1}{n_2}}$$ [13.10]

where s_p is the pooled standard deviation, Equation [13.9].

Table 13.6 shows the complete calculations for a difference-of-means test of the students' MSAT data. The difference is statistically significant ($P <$ .05), so we conclude that at least some difference between men's and women's mean MSAT scores exists in the population. In symbolic terms, we reject H_0: $\mu_2 - \mu_1 = 0$ and accept the alternative H_1: $\mu_2 - \mu_1 \neq 0$.

TABLE 13.6

Testing the Hypothesis of No Difference Between Mean MSATs by Gender (see Tables 13.4–13.5)

Null hypothesis:	H_0: $\mu_2 - \mu_1 = 0$	(no difference in population)
Research hypothesis:	H_1: $\mu_2 - \mu_1 \neq 0$	(some difference in population)

Sample statistics:	*Men*	*Women*
Number of cases	$n_1 = 9$	$n_2 = 22$
Mean math SAT	$\overline{Y}_1 = 571.1$	$\overline{Y}_2 = 493.6$
Standard deviation	$s_1 = 91.2$	$s_2 = 83.0$

Estimating the standard error of $\overline{Y}_2 - \overline{Y}_1$:

$$s_P = \sqrt{\frac{(n_1 - 1)s_1^2 + (n_2 - 1)s_2^2}{n_1 + n_2 - 2}} \qquad \text{Equation [13.9]}$$

$$= \sqrt{\frac{(9 - 1)91.2^2 + (22 - 1)83.0^2}{9 + 22 - 2}}$$

$$= \sqrt{\frac{211,208.52}{29}} = 85.34$$

$$SE_{\overline{Y}_2 - \overline{Y}_1} = s_P \sqrt{\frac{1}{n_1} + \frac{1}{n_2}} \qquad \text{Equation [13.10]}$$

$$= 85.34 \sqrt{\frac{1}{9} + \frac{1}{22}} = 33.77$$

Hypothesis test:

$$t = \frac{(\overline{Y}_2 - \overline{Y}_1) - (\mu_2 - \mu_1)_0}{SE_{\overline{Y}_2 - \overline{Y}_1}} \qquad \text{Equation [13.6]}$$

$$= \frac{(493.6 - 571.1) - (0)}{33.77}$$

$$= \frac{-77.5}{33.77} = -2.29$$

Conclusion: The difference between sample means, $\overline{Y}_2 - \overline{Y}_1$, is 2.29 standard errors away from the hypothesized difference between population means, $\mu_2 - \mu_1 = 0$ ($t = -2.29$). With $n_1 + n_2 - 2 = 29$ degrees of freedom, a t statistic this far from 0 has a two-sided probability (from Table A.3) of $P <$ .05. This probability is *low enough* to justify rejecting the null hypothesis at $\alpha = .05$.

Two distributional assumptions that underlie such difference-of-means tests have been mentioned. One is that Y follows a normal distribution, within the populations of cases in either of the two categories. This resembles the normality assumption required for small-sample inferences about a single mean (Chapter 11). Sample data may be used to check this assumption, which matters most when samples are small (n_1 or n_2 is less than 20).

A second assumption is that $\sigma_1 = \sigma_2$, where σ_1 is the population standard deviation for cases in category 1, and σ_2 is that for cases in category 2. The plausibility of this assumption can also be checked using sample data. A mild difference between the sample standard deviations s_1 and s_2, as in Table 13.6, should be no cause for alarm.

PROBLEMS

5. Like standardized test scores, student grade point averages tend to have well-behaved distributions, so the assumptions for a difference-of-means test are plausible. Test at $\alpha = .05$ whether there are significant differences between the mean self-reported grade point averages of the men and women students, based on these statistics for Table 13.4:

1. Men	2. Women
$n_1 = 10$	$n_2 = 24$
$\overline{Y}_1 = 2.84$	$\overline{Y}_2 = 2.89$
$s_1 = .64$	$s_2 = .48$

6. A medical journal article reported on 91 cocaine users who were treated in a hospital emergency room (Brower, Blow, and Beresford, 1988). Sixty-six of the 91 patients showed symptoms of psychosis. These 66 psychotic patients had consumed an average of 16.9 grams of cocaine during the past month, with a standard deviation of 31.9 grams. The remaining 25 nonpsychotic patients had consumed an average of 4.6 grams, with a standard deviation of 7.5 grams. Test at the $\alpha = .05$ level whether the psychotic patients used significantly more cocaine than the nonpsychotic patients (one-sided test).

7. Of the 91 patients (Problem 6), 29 were violent. The violent patients had consumed an average of 22.4 grams of cocaine in the past month, with a standard deviation of 46.6 grams. The 62 nonviolent patients had consumed an average of 9.8 grams, with a standard deviation of 11.3 grams. Test at $\alpha = .05$ whether violent patients used significantly more cocaine than nonviolent patients (one-sided test).

8. Any variable that cannot take on negative values and has a standard deviation greater than its mean must have a positively skewed distribution. From the statistics given in Problems 6 and 7, we can infer that cocaine

consumption is positively skewed in these samples. If the population distributions are also skewed, then our significance test conclusions could be called into question.

a. What does a positively skewed distribution of cocaine consumption imply, in real-world terms? Do you think that the population distribution of cocaine consumption is more likely to be skewed, or to be roughly symmetrical? Why?

b. Is there any other evidence in the description of Problems 6 or 7 that casts doubt on the *t* tests' underlying assumptions?

13.4   *CONFIDENCE INTERVALS FOR DIFFERENCES OF MEANS*

The hypothesis test in Table 13.6 is designed to answer the question, *"Is there a male–female difference in the mean population MSAT scores?"* It does not directly answer another, related question: *"What is the male–female difference in mean population MSAT scores?"* To address this second question, we will construct a *confidence interval* for the difference between means.

In Chapter 11, we constructed single-sample intervals for the mean as

(Sample statistic) $\pm$ *t*(Standard error of sample statistic)

Intervals for a difference between means share this format. Such confidence intervals require the same distributional assumptions as hypothesis tests.

A **confidence interval** for the **difference between two means** is constructed by finding

$$(\bar{Y}_2 - \bar{Y}_1) \pm t(SE_{\bar{Y}_2 - \bar{Y}_1}) \qquad\qquad [13.11]$$

where $SE_{\bar{Y}_2 - \bar{Y}_1}$ is the estimated standard error of this difference, from Equation [13.10], and the value of *t* is chosen from a table of the *t* distribution (Table A.3), with df $= n_1 + n_2 - 2$.

The standard error estimate needed to construct a confidence interval is the same as that for a difference-of-means hypothesis test. In the MSAT example (Table 13.6), we found that $SE_{\bar{Y}_2 - \bar{Y}_1} = 33.77$. For a 95% confidence interval, with $n_1 + n_2 - 2 = 29$ degrees of freedom, Table A.3 indicates that we need a *t* value of 2.045. The confidence interval is therefore

$$(\bar{Y}_2 - \bar{Y}_1) \pm t(SE_{\bar{Y}_2 - \bar{Y}_1}) = (493.64 - 571.11) \pm 2.045(33.77)$$

$$= -77.47 \pm 69.06$$

$$\rightarrow -146.53 \le \mu_2 - \mu_1 \le -8.41$$

Based on this sample, we are 95% confident that in the population of students, the male–female difference in MSAT scores is somewhere between 146.53 and 8.41 points. More formally, 95% of the intervals constructed in this way will contain the true population difference, $\mu_2 - \mu_1$.

Although closely related, the hypothesis test and confidence interval analyses yield differently flavored conclusions. The hypothesis test indicates that there *is* a difference, period. The confidence interval confirms this: Since the 95% confidence interval does not include 0, we are at least 95% confident that the difference between population means is not 0. Furthermore, the interval puts the actual difference between about 8 and 147 points, in favor of males. The great width of this interval conveys our uncertainty: The population difference could be anywhere from small (8 points) to quite large (147 points). Thus, the confidence interval is more informative than the hypothesis test.

A glance back at the box plots of Figure 13.5 helps to put these results in perspective. Although a significant difference exists between the mean scores of men and women, both distributions almost entirely overlap. Some men score much lower than the average woman; some women far outscore the average man. The best and the worst in each distribution are almost the same. In other words, *the differences within each distribution* are much larger than *the differences between the two distributions.* Knowing a person's gender is a weak basis for predicting his or her MSAT scores.

PROBLEMS

9. Problem 5 tested the hypothesis that men and women students have different mean self-reported grade point averages, based on the statistics from Table 13.4 shown here.

1. Men	2. Women
$n_1 = 10$	$n_2 = 24$
$\overline{Y}_1 = 2.84$	$\overline{Y}_2 = 2.89$
$s_1 = .64$	$s_2 = .48$

Construct and interpret a 95% confidence interval for the difference in mean GPA. In what way is this interval consistent with your conclusion in Problem 5, where the null hypothesis could not be rejected at $\alpha = .05$?

10. The table on pages 234–236 contains information on 118 college classes taught either by full-time faculty members (category 1) or by part-time instructors and teaching assistants (category 0). For each class we know the size, the percentage of A or B course grades given, and the percentage of teaching evaluations in the two highest categories. The percentages of high teaching evaluations in these 118 classes break down as shown.

0. Part-Time and TA	1. Faculty
$n_0 = 47$	$n_1 = 71$
$\overline{Y}_0 = 43.2$	$\overline{Y}_1 = 52.7$
$s_0 = 25.1$	$s_1 = 22.5$

Use this information to construct a 95% confidence interval for $\mu_1 - \mu_0$. Explain what this interval represents. Based on this interval, can we reject the hypothesis that there is no difference between the mean percentage of high evaluations for these two categories of instructor?

11. Statistics on the percentages of A and B grades in the 118 classes of the table are as shown.

0. Part-Time and TA	1. Faculty
$n_0 = 47$	$n_1 = 71$
$\overline{Y}_0 = 62.8$	$\overline{Y}_1 = 53.0$
$s_0 = 11.8$	$s_1 = 15.2$

Construct and interpret a 98% confidence interval for the difference between mean percentages of A and B grades. On the basis of this interval, can we reject at $\alpha = .02$ the null hypothesis of no difference between population means?

13.5 *PAIRED-DIFFERENCE TESTS*

The concept of a "difference-of-means" has another context: naturally paired cases. Air pollution data have already illustrated the distinction between paired-difference (Table 13.2) and two-sample (Table 13.1) problems. In brief, **paired-difference** methods *compare two variables across the same set of cases,* whereas two-sample methods compare one variable across two different sets of cases. For example, a paired-difference test could compare the means of true and self-reported grade point average (GPA) for all 34 students in Table 13.4. On the other hand, a two-sample test is appropriate to compare the mean self-reported GPA of the men with the mean of the women, as done in Problem 5.

Table 13.7 (page 416) contains data on 24 college students in a writing course that employed microcomputers for word processing—a controversial innovation at the time. One way to judge the course's success is whether the students could compose sentences faster afterward. Table 13.7 reports the number of sentences each student completed on a timed test before (X) and after (Y) taking the course.

Analysis of paired-difference problems begins by calculating a **difference score, D,** which for each case is the difference between the two variables being compared. A column of D values is given at the right of Table 13.7.

TABLE 13.7   *Test Results for a Writing Course*

Student	Sentences before course X	Sentences after course Y	Difference $D = Y - X$
1	11	23	12
2	8	9	1
3	11	24	13
4	8	20	12
5	6	16	10
6	8	18	10
7	13	21	8
8	4	24	20
9	3	20	17
10	8	23	15
11	12	27	15
12	9	33	24
13	21	42	21
14	13	33	20
15	19	25	6
16	17	30	13
17	14	47	33
18	12	24	12
19	18	34	16
20	10	36	26
21	7	22	15
22	10	29	19
23	6	25	19
24	11	28	17
Summary statistics:	$\overline{X} = 10.8$ $s_X = 4.6$	$\overline{Y} = 26.4$ $s_Y = 8.3$	$\overline{D} = 15.6$ $s_D = 6.8$

Source: Data from Nash and Schwartz (1987).

Positive difference scores indicate that the student improved, writing more sentences after taking the course. Negative difference scores are also possible, although none occur in these data—no one got worse. Sample statistics are at the bottom of the table; note that the mean difference, $\overline{D}$, is the same as the difference of means, $\overline{Y} - \overline{X}$: $15.6 = 26.4 - 10.8$.

Confidence intervals and hypothesis tests for paired differences are basically the same as one-sample methods, applied to D. We calculate the sample mean, $\overline{D}$, and standard deviation, s_D. The mean sample difference, $\overline{D}$, forms the basis for inference about the **mean population difference, δ** (*delta*). The standard error of $\overline{D}$ is estimated as

$$SE_{\overline{D}} = \frac{s_D}{\sqrt{n}}$$ [13.12]

Compare with Equation [11.2], page 313.

Confidence intervals for a **mean paired difference** (δ) are obtained from

$$\overline{D} \pm t(\text{SE}_{\overline{D}}) \qquad\qquad\qquad [13.13]$$

where $\text{SE}_{\overline{D}}$ is the estimated standard error of $\overline{D}$, from Equation [13.12], and t is found from the t table (Table A.3) with $n - 1$ degrees of freedom.

Compare Equation [13.13] with [11.6], page 319.

Hypothesis tests for **mean paired differences** are based on a t statistic:

$$t = \frac{\overline{D} - \delta_0}{\text{SE}_{\overline{D}}} \qquad\qquad\qquad [13.14]$$

where $\overline{D}$ is the mean of the sample difference, δ_0 is the null hypothesis mean of the population differences, and $\text{SE}_{\overline{D}}$ is the standard error estimated from Equation [13.12]. This t statistic is compared with a t distribution with $n - 1$ degrees of freedom.

Compare Equations [13.14] and [11.4], page 315. For small samples both confidence intervals and hypothesis tests regarding $\overline{D}$ require the assumption that D is normally distributed in the population.

Table 13.8 (page 418) shows a test of the null hypothesis that students do not improve in the number of sentences written (H_0: $\delta \le 0$). The research hypothesis is that they improve (H_1: $\delta > 0$).[1] There was some improvement in this 24-student sample; writing samples were longer by an average of $\overline{D} = 15.6$ sentences. But is this improvement large enough to rule out chance? The P-value obtained in Table 13.8, $P < .0005$, indicates that a "chance" explanation is hard to believe. The students made a statistically significant improvement.

Before accepting this conclusion we should check the variable's distribution. A box plot (Figure 13.6) indicates that D has a well-behaved, roughly symmetrical sample distribution, which does not challenge the normality assumption.[2] (Compare with the samples from normal distributions shown in Figures 11.2 and 11.3, page 317.)

We can also form a confidence interval for the mean improvement. With $n - 1 = 23$ degrees of freedom, $t = 2.069$ is required for a 95% confidence interval. Using the same standard error estimated for our hypothesis test, the interval is

$$\overline{D} \pm t(\text{SE}_{\overline{D}}) = 15.6 \pm 2.069(1.39)$$
$$\rightarrow 12.7 \le \delta \le 18.5$$

This interval applies to a hypothetical large population of students, selected like the students in Table 13.7 and taught identically. According to these

TABLE 13.8

Paired-Difference Test (One-sided) for an Increase in the Mean Number of Sentences Completed (from Table 13.7)

Null hypothesis: H_0: $\mu_Y - \mu_X \le 0$ *or* H_0: $\delta \le 0$ (no increase)

Research hypothesis: H_1: $\mu_Y - \mu_X > 0$ *or* H_1: $\delta > 0$ (some increase)

Sample statistics:

$n = 24$ $\overline{D} = 15.6$ $s_D = 6.8$

Estimating the standard error of $\overline{D}$:

$$\text{SE}_{\overline{D}} = \frac{s_D}{\sqrt{n}} \qquad \text{Equation [13.12]}$$

$$= \frac{6.8}{\sqrt{24}} = 1.39$$

Hypothesis test:

$$t = \frac{\overline{D} - \delta_0}{\text{SE}_{\overline{D}}} \qquad \text{Equation [13.14]}$$

$$= \frac{15.6 - 0}{1.39} = 11.22$$

Conclusion: The mean sample difference, $\overline{D} = 15.6$, is 11.22 standard errors away from the null hypothesis mean population difference, $\delta_0 = 0$. With $n - 1 = 24 - 1 = 23$ degrees of freedom, a t statistic of 11.22 has a one-sided probability of $P < .0005$. This is so low that we should reject the null hypothesis; such results are very unlikely if the mean population difference really is less than or equal to 0.

Students' writing test results showed a significant improvement in the number of sentences completed after taking this writing course.

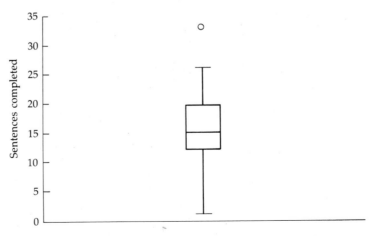

FIGURE 13.6   *Improvement in number of sentences completed (Table 13.7)*

calculations we can be 95% confident that the mean number of sentences completed by this hypothetical population would increase by somewhere between 12.7 and 18.5.

Notice again the different flavor of hypothesis test and confidence interval conclusions. The hypothesis test answers the flat question *"Is there a difference?"* The confidence interval explores further: *"How much difference is there?"*

PROBLEMS

12. Additional data on the writing course evaluation are given in the accompanying table. Conduct a paired-difference test (at $\alpha = .05$) to determine whether there was a significant improvement in the use of evidence in student writing.

Test Results for a Writing Course

	Number of paragraphs		Use of evidence[a]	
Student	Before	After	Before	After
1	2	5	2	6
2	1	5	0	1
3	1	4	0	1
4	1	6	0	5
5	1	2	1	2
6	1	5	2	1
7	4	6	3	4
8	0	5	0	1
9	2	5	0	2
10	2	4	3	4
11	1	4	2	5
12	6	6	0	4
13	2	5	3	3
14	1	6	0	4
15	4	6	3	4
16	5	9	3	3
17	2	6	3	5
18	3	6	2	4
19	2	5	1	4
20	3	8	1	2
21	3	5	1	4
22	4	8	0	4
23	3	3	1	3
24	1	6	1	4

[a]Judged on a 0–6 scale, where 6 means best or most extensive use of evidence in writing.
Source: Data from Nash and Schwartz (1987).

13. Construct a box plot for D in Problem 12 and comment on its shape. Does this box plot give you any reason to question the normality assumption?

14. Use the data in Table 13.4 to carry out a paired-difference t test ($\alpha = .05$) of the null hypothesis that true GPAs are significantly lower than self-reported GPAs.

13.6 K-SAMPLE PROBLEMS: ANALYSIS OF VARIANCE

Difference-of-means tests are hard to perform on more than two means. Three categories would require testing null hypotheses about three different pairs of means; four categories would need six different hypotheses. The two-sample methods do not readily adapt to such multiple hypotheses.

A method called **analysis of variance** is used instead for **K-sample** problems. Analysis of variance, or **ANOVA,** allows us to test the single null hypothesis that *the population mean of Y is the same, across all categories of X.* The categorical variable X may have any number of categories. With only two categories, K-sample and two-sample methods are equivalent.

Data from a survey of collegiate alcohol consumption (Table 13.9) will illustrate an ANOVA. Consumption is measured on a scale ranging from 4 to 33, where 4 indicates abstinence and 33 means heavy drinking. Year in college is treated as a categorical variable, dividing the 19 students into four groups. Means and standard deviations for the drinking scale (Y) are given for each group at the right of the table. The drinking scale mean is highest among the six sophomore students ($\overline{Y}_2 = 23.5$) and lowest among the four seniors ($\overline{Y}_4 = 13.25$). Note the use of the symbols $\overline{Y}_.$, $s_.$, and $n_.$ to represent the mean, standard deviation, and number of cases for the *whole sample*. In this context $\overline{Y}_.$ is sometimes called the **grand mean.**

Our null hypothesis is that the students were drawn randomly from a population where the mean drinking scales for freshmen, sophomores, juniors, and seniors are the same:

$$H_0: \quad \mu_1 = \mu_2 = \mu_3 = \mu_4$$

The complementary research hypothesis is most easily stated as

$$H_1: \quad \text{Population means are not all the same}$$

Of course the four *sample* means (Table 13.9) differ, but this is conceivably due to chance. Even if H_0 were true, we would expect some variation among the means of such small samples.

Analysis of variance is based on several **sums of squares.** These are the **between-groups** sum of squares (BSS), the **within-groups** sum of squares (WSS), and the **total** sum of squares (TSS$_Y$). They measure three kinds of variation.

TABLE 13.9   *Survey Data on Alcohol Consumption by Year in College*

Student	Year[a] k	Drinking scale[b] Y	Summary statistics
1	1	16	
2	1	19	
3	1	25	$n_1 = 4, \overline{Y}_1 = 19.25, s_1 = 4.03$
4	1	17	
5	2	33	
6	2	22	
7	2	32	
8	2	18	$n_2 = 6, \overline{Y}_2 = 23.5, s_2 = 7.58$
9	2	22	
10	2	14	
11	3	17	
12	3	22	
13	3	14	$n_3 = 5, \overline{Y}_3 = 15.4, s_3 = 4.45$
14	3	10	
15	3	14	
16	4	8	
17	4	18	
18	4	18	$n_4 = 4, \overline{Y}_4 = 13.25, s_4 = 5.5$
19	4	9	

All cases: $n_. = 19, \overline{Y}_. = 18.32, s_. = 6.73$

[a]Year: 1 = freshman, 2 = sophomore, 3 = junior, 4 = senior.
[b]Scale ranges from 4 to 33, where 4 indicates abstinence and 33 indicates heavy and frequent drinking.
Source: Data from a survey conducted by Sally Ward and Susan Ault at the University of New Hampshire.

1. *Between-groups sum of squares* (BSS):   How much variation is there *between the different categories or groups*? This is measured by finding the deviation of each group's mean, $\overline{Y}_k$, from the grand mean $\overline{Y}$. For each case this deviation is squared, then added to that for other cases:

$$\text{BSS} = \Sigma(\overline{Y}_k - \overline{Y}_.)^2 \qquad\qquad [13.15a]$$

2. *Within-groups sums of squares* (WSS):   How much variation is there *within each category or group*? We calculate the deviation of each Y value from the corresponding group mean, $Y - \overline{Y}_k$. Again these individual deviations are squared and summed:

$$\text{WSS} = \Sigma(Y - \overline{Y}_k)^2 \qquad\qquad [13.16a]$$

3. *Total sum of squares* (TSS$_Y$):   How much variation is there *in the sample as a whole*? As with other sums of squares, we first calculate deviations for each case, $Y - \overline{Y}_.$, then square and sum them:

$$TSS_Y = \Sigma(Y - \overline{Y}_.)^2 \qquad\qquad [13.17a]$$

The total sum of squares, TSS_Y, is also used to define the variance and the standard deviation (see Equation [4.4], page 98)—basic measures of how much a variable varies.

It can be shown algebraically that the total sum of squares equals the sum of the other two:

$$TSS_Y = WSS + BSS$$

In effect we divide the total variation (TSS_Y) into two parts: between-groups variation, which is explained by differences between the groups, and within-groups variation, which is not explained by group differences.

Table 13.10 calculates sums of squares for the student drinking data. The first student, a freshman, has a drinking scale value of 16, so she is a rela-

TABLE 13.10   **Calculation of Between-Groups, Within-Groups, and Total Sums of Squares from Table 13.9[a]**

Year	Drink scale	Between groups	Within groups	Total
k	Y	$(\overline{Y}_k - \overline{Y}_.)^2$	$(Y - \overline{Y}_k)^2$	$(Y - \overline{Y}_.)^2$
1	16	$(19.25 - 18.32)^2 = .87$	$(16 - 19.25)^2 = 10.56$	$(16 - 18.32)^2 = 5.36$
1	19	$(19.25 - 18.32)^2 = .87$	$(19 - 19.25)^2 = .06$	$(19 - 18.32)^2 = .47$
1	25	$(19.25 - 18.32)^2 = .87$	$(25 - 19.25)^2 = 33.06$	$(25 - 18.32)^2 = 44.68$
1	17	$(19.25 - 18.32)^2 = .87$	$(17 - 19.25)^2 = 5.06$	$(17 - 18.32)^2 = 1.73$
2	33	$(23.5 - 18.32)^2 = 26.88$	$(33 - 23.5)^2 = 90.25$	$(33 - 18.32)^2 = 215.63$
2	22	$(23.5 - 18.32)^2 = 26.88$	$(22 - 23.5)^2 = 2.25$	$(22 - 18.32)^2 = 13.57$
2	32	$(23.5 - 18.32)^2 = 26.88$	$(32 - 23.5)^2 = 72.25$	$(32 - 18.32)^2 = 187.26$
2	18	$(23.5 - 18.32)^2 = 26.88$	$(18 - 23.5)^2 = 30.25$	$(18 - 18.32)^2 = .10$
2	22	$(23.5 - 18.32)^2 = 26.88$	$(22 - 23.5)^2 = 2.25$	$(22 - 18.32)^2 = 13.57$
2	14	$(23.5 - 18.32)^2 = 26.88$	$(14 - 23.5)^2 = 90.25$	$(14 - 18.32)^2 = 18.63$
3	17	$(15.4 - 18.32)^2 = 8.50$	$(17 - 15.4)^2 = 2.56$	$(17 - 18.32)^2 = 1.73$
3	22	$(15.4 - 18.32)^2 = 8.50$	$(22 - 15.4)^2 = 43.56$	$(22 - 18.32)^2 = 13.57$
3	14	$(15.4 - 18.32)^2 = 8.50$	$(14 - 15.4)^2 = 1.96$	$(14 - 18.32)^2 = 18.63$
3	10	$(15.4 - 18.32)^2 = 8.50$	$(10 - 15.4)^2 = 29.16$	$(10 - 18.32)^2 = 69.15$
3	14	$(15.4 - 18.32)^2 = 8.50$	$(14 - 15.4)^2 = 1.96$	$(14 - 18.32)^2 = 18.63$
4	8	$(13.25 - 18.32)^2 = 25.66$	$(8 - 13.25)^2 = 27.56$	$(8 - 18.32)^2 = 106.42$
4	18	$(13.25 - 18.32)^2 = 25.66$	$(18 - 13.25)^2 = 22.56$	$(18 - 18.32)^2 = .10$
4	18	$(13.25 - 18.32)^2 = 25.66$	$(18 - 13.25)^2 = 22.56$	$(18 - 18.32)^2 = .10$
4	9	$(13.25 - 18.32)^2 = 25.66$	$(9 - 13.25)^2 = 18.06$	$(9 - 18.32)^2 = 86.78$

BSS $= \Sigma(\overline{Y}_k - \overline{Y}_.)^2 = 309.9$ **WSS** $= \Sigma(Y - \overline{Y}_k)^2 = 506.2$ **TSS$_Y$** $= \Sigma(Y - \overline{Y}_.)^2 = 816.1$

[a]For greater accuracy actual calculations employed $\overline{Y}_. = 18.31579$ instead of $\overline{Y} \approx 18.32$.

tively moderate drinker. The mean drinking value for all four freshmen is $\overline{Y}_1 = 19.25$, and that for all 19 students is $\overline{Y}_. = 18.32$ (see Table 13.9). Thus, the freshman group mean is .93 point above the grand mean: $\overline{Y}_k - \overline{Y}_. = 19.25 - 18.32 = .93$. The student herself is 3.25 points below her group mean: $Y - \overline{Y}_k = 16 - 19.25 = -3.25$. She is 2.32 points below the grand mean: $Y - \overline{Y}_. = 16 - 18.32 = -2.32$.

Each of these deviations is squared, then added to those for the other 18 students to obtain the three sums of squares shown at the bottom of Table 13.10. Actual calculations for the table used a more precise grand mean, $\overline{Y}_. = 18.31579$, instead of $\overline{Y}_. \approx 18.32$, to preserve accuracy: If we round off too much, the BSS and WSS will not add up to TSS_Y as they should.

The calculations in Table 13.10 illustrate how the three sums of squares are defined. Computational shortcuts help to obtain these numbers more simply if ANOVA is being done by hand. These shortcuts also let us perform ANOVA knowing only the summary statistics, in case we lack access to the raw data. A computational formula for the between-groups sum of squares is

$$\text{BSS} = \Sigma_k n_k \overline{Y}_k^2 - \frac{(\Sigma_k n_k \overline{Y}_k)^2}{n_.} \qquad [13.15b]$$

The total number of cases in the sample is $n_1 + n_2 + n_3 + n_4 = n_.$. The summations in Equation [13.15b] are over the K groups, not over all cases in the data set—hence the notation Σ_k instead of just Σ.

Applied to the data of Table 13.9, Equation [13.15b] gives us

$$\text{BSS} = n_1\overline{Y}_1^2 + n_2\overline{Y}_2^2 + n_3\overline{Y}_3^2 + n_4\overline{Y}_4^2 - \frac{(n_1\overline{Y}_1 + n_2\overline{Y}_2 + n_3\overline{Y}_3 + n_4\overline{Y}_4)^2}{n_.}$$

$$= 6,683.8 - \frac{348^2}{19} = 309.9$$

Thus, the computational formula, [13.15b], yields the same results as the definitional formula, [13.15a]; compare with Table 13.10.

The computational formula for the within-groups sum of squares is

$$\text{WSS} = \Sigma_k(n_k - 1)s_k^2 \qquad [13.16b]$$

As in Equation [13.15b], the summation is over K groups, not $n_.$ cases. Applying Equation [13.16b] to the student drinking data,

$$\text{WSS} = (n_1 - 1)s_1^2 + (n_2 - 1)s_2^2 + (n_3 - 1)s_3^2 + (n_4 - 1)s_4^2$$

$$= (4 - 1)4.03^2 + (6 - 1)7.58^2 + (5 - 1)4.45^2 + (4 - 1)5.5^2$$

$$= 506.0$$

Apart from rounding error, this WSS value of 506.0 is the same as that from the definitional formula in Table 13.10 (WSS = 506.2).

TABLE 13.11   **The Sums of Squares Used in ANOVA**

Sum of squares	Definitional formula	Computational formula
Between groups	$BSS = \Sigma(\overline{Y}_k - \overline{Y}_.)^2$ Equation [13.15a]	$BSS = \Sigma_k n_k \overline{Y}_k^2 - \dfrac{(\Sigma_k n_k \overline{Y}_k)^2}{n_.}$ Equation [13.15b]
Within groups	$WSS = \Sigma(Y - \overline{Y}_k)^2$ Equation [13.16a]	$WSS = \Sigma_k(n_k - 1)s_k^2$ Equation [13.16b]
Total	$TSS_Y = \Sigma(Y - \overline{Y}_.)^2$ Equation [13.17a]	$TSS = (n_. - 1)s_.^2$ Equation [13.17b]

Symbols used in these definitions:

$\overline{Y}_k$, s_k, n_k refer to the mean, standard deviation, and number of cases within the kth group or category of X.

$\overline{Y}_.$, $s_.$, $n_.$ refer to the mean, standard deviation, and number of cases for the sample as a whole.

Σ_k calls for summation of one value for each of the K groups or categories of X.

Σ calls for summation of one value for each of the $n_.$ cases in the sample.

Finally, we can easily obtain the total sum of squares from the overall standard deviation $s_.$:

$$TSS_Y = (n_. - 1)s_.^2 \qquad\qquad [13.17b]$$

which gives $TSS_Y = (19 - 1)6.73^2 = 815.3$ for the student drinking example. Apart from rounding error, Equation [13.17b] should give the same results as the definitional formula, Equation [13.17a]. Table 13.11 summarizes formulas for the sums of squares used in ANOVA.

In addition to the sums of squares, two further concepts are needed to carry out an analysis of variance. First, each of the three sums of squares has its specific *degrees of freedom*:

1. For the between-groups sum of squares, $\mathbf{df_B} = K - 1$, where K is the number of categories. Since there are four categories of the variable Year in Tables 13.9 and 13.10, the degrees of freedom for BSS are $df_B = 4 - 1 = 3$.

2. For the within-groups sum of squares, $\mathbf{df_W} = n_. - K$, where $n_.$ is the total sample size and K is the number of categories. Tables 13.9 and 13.10 have 19 cases and four categories, so degrees of freedom for WSS are $df_W = n_. - K = 19 - 4 = 15$.

3. The total sum of squares has degrees of freedom $\mathbf{df_T} = n_. - 1$, or $19 - 1 = 18$ in Tables 13.9 and 13.10. This is the same quantity used in one-sample tests and in the denominator of the sample variance or standard deviation.

Next, the BSS, WSS, and their respective degrees of freedom are used to form a test statistic called an *F ratio* or *F statistic.*

ANOVA hypothesis tests are based on an *F* **statistic:**

$$F_{df_w}^{df_B} = \frac{BSS/df_B}{WSS/df_w} = \frac{\Sigma(\overline{Y}_k - \overline{Y}_.)^2/(K-1)}{\Sigma(Y - \overline{Y}_k)^2/(n_. - K)}$$

[13.18]

This statistic is compared with a theoretical *F* **distribution** (Table A.5), having degrees of freedom df_B (numerator) and df_w (denominator).

Like the two-sample *t* test, ANOVA requires the assumption that population distributions of *Y* are normal, with identical standard deviations, across all categories of *X*.

Using the sums of squares from Table 13.10, we obtain the *F* statistic:

$$F_{15}^3 = \frac{309.9/3}{506.2/15} = 3.06$$

Table A.5 in the Appendix gives critical values for the theoretical *F* distribution. To use this table, we must specify the degrees of freedom, df_B and df_w, known as the **numerator degrees of freedom** and the **denominator degrees of freedom** because of their positions in Equation [13.18].

Numerator degrees of freedom are shown along the top of Table A.5 and denominator degrees of freedom down the left-hand margin. To evaluate an *F* statistic with 3 and 15 degrees of freedom, written as F_{15}^3, we find the column for $df_B = 3$ (numerator), then look for the row corresponding to $df_w = 15$ (denominator). There is no such df_w row, but the critical values for $df_w = 14$ and $df_w = 16$ are not much different. To be conservative, we round down and use $df_w = 14$. With 3 and 14 degrees of freedom, we need an *F* statistic of 3.34 or larger to reject H_0 at the usual $\alpha = .05$ level. The *F* statistic we actually obtained, 3.06, is not larger than the critical value of 3.34, so at $\alpha = .05$ we cannot reject the null hypothesis of equal population means.

Our results, however, are almost statistically significant ($P < .10$). An *F* statistic this large would occur by chance less than one time in ten, *if* the null hypothesis of equal means were true. The small sample used here may be partly to blame for our inability to reject H_0, despite substantially different sample means.

ANOVA results are typically presented in an ANOVA table like Table 13.12 (page 426). Sums of squares are from Table 13.10, with their corresponding degrees of freedom. As an intermediate step, ANOVA tables often contain **mean squares,** or **MS,** defined as sums of squares divided by their degrees of freedom. The *F* statistic is then just the ratio of two mean squares. *P*-values corresponding to this *F* statistic may be obtained from a table like Table A.5, or calculated more precisely by a computer program.

TABLE 13.12

One-way Analysis of Variance (ANOVA) for Drinking Data from Tables 13.9 and 13.10

Source	SS	df	MS = SS/df	F	Prob > F
Between groups	309.9[a]	3[b]	103.3	3.06[c]	P < .10[d]
Within groups	506.2[e]	15[f]	33.75		
Total	816.1[g]	18[h]	45.34		

[a]Between-groups sum of squares: BSS $= \Sigma(\overline{Y}_k - \overline{\overline{Y}}\,)^2 = 309.9$
[b]Degrees of freedom for BSS: $df_B = K - 1 = 4 - 1 = 3$
[c]F statistic:

$$F_{dfW}^{dfB} = \frac{BSS/df_B}{WSS/df_W} = \frac{309.9/3}{506.2/15} = \frac{103.3}{33.75} = 3.06$$

[d]Probability of $F \geq 3.06$, with 3 and 15 degrees of freedom (Table A.5)
[e]Within-groups sum of squares: WSS $= \Sigma(Y - \overline{Y}_k)^2 = 506.2$
[f]Degrees of freedom for WSS: $df_W = n. - K = 19 - 4 = 15$
[g]Total sum of squares: $TSS_Y = \Sigma(Y - \overline{\overline{Y}}\,)^2 = 816.1$ (Note that $TSS_Y = BSS + WSS$.)
[h]Degrees of freedom for TSS_Y: $df_T = n. - 1 = 19 - 1 = 18$

ANOVA tables like Table 13.12 tell us whether differences among means are significant, but they say nothing about what those differences are. Which categories have the highest and lowest means? Which categories are similar? To understand our results we must look at a table of the original group means, rather than the ANOVA table itself.

Box plots and other methods should be used where possible to check the plausibility of our assumptions of normal distributions and equal standard deviations. These assumptions are even more important in K-sample ANOVA than they are in simpler two-sample problems. Tests based on F statistics with $K > 2$ tend to be somewhat less robust than tests based on t statistics. Unfortunately, small samples, where distributional assumptions are most critical, provide little information with which to check these assumptions. Only 4–6 sample cases are in each category of the student drinking data (Table 13.9), for example. With so few cases there is no point in constructing box plots, and the shape of sample distributions will be unreliable as a reflection of the corresponding population distributions.

PROBLEMS

15. Find the obtained P-value and state whether the null hypothesis should be rejected for each of the following.
   a. $F = 8.3$, $n = 34$, X has 4 categories, $\alpha = .01$
   b. $F = 3.1$, $n = 45$, X has 3 categories, $\alpha = .05$
   c. $F = 14.6$, $n = 13$, X has 2 categories, $\alpha = .001$

d. $F = 45.3$, $n = 124$, X has 11 categories, $\alpha = .05$

e. $F = .5$, $n = 10$, X has 3 categories, $\alpha = .05$

f. $F = 4.3$, $n = 25$, X has 5 categories, $\alpha = .05$

16. The accompanying table contains insurance agency data on twelve randomly selected models of two-door cars. The frequency of injury claims and the average repair costs per vehicle are each expressed as an index. A higher index value means more frequent occupant injury claims, or higher average collision repair costs are associated with that model of car.

    Follow the steps of Tables 13.9, 13.10, and 13.12 to find BSS and WSS, and perform an analysis of variance to test (at $\alpha = .05$) the null hypothesis that the mean frequency of injury claims is the same in cars of all three sizes.

Ratings for Frequency of Injury Claims and Average Repair Costs per Insured Vehicle[a]

Car model	Frequency of injury	Repair costs	Car size[b]
Chrysler Laser	115	146	1
Mercury Lynx	137	98	1
Chevrolet Chevette	149	99	1
Chevrolet Spectrum	153	103	1
Oldsmobile Cutlass	87	85	2
Pontiac Grand Prix	88	95	2
Dodge Aries	117	90	2
Pontiac Grand Am	117	104	2
Mercury Grand Marquis	55	52	3
Ford Crown Victoria	70	68	3
Oldsmobile Ninety-Eight	73	72	3
Chevrolet Caprice	75	67	3

[a]Both are rated on scales with means of 100 for all models of 1984–1986 passenger cars.
[b]1 = small, 2 = midsize, 3 = large. All 12 models are 1984–1986 two-doors.
Source: Data from a study by the Highway Loss Data Institute, reported in the *Boston Globe*, September 15, 1987.

17. A study of American education reported on a reading achievement test taken by students at public, Catholic, and private non-Catholic high schools.[3] Mean scores are shown here.

Public	Catholic	Private Non-Catholic
$n_1 = 10,000$	$n_2 = 1,000$	$n_3 = 200$
$\overline{Y}_1 = 4.48$	$\overline{Y}_2 = 5.00$	$\overline{Y}_3 = 5.34$
$s_1 = 2.10$	$s_2 = 1.96$	$s_3 = 2.04$

Use the computational formulas for BSS and WSS in Equations [13.15b] and [13.16b] to test whether the differences between school types are significant at $\alpha = .01$.

13.7 *ERROR-BAR PLOTS**

Research reports often use **error-bar plots** to display means graphically. Figure 13.7 is an error-bar plot for the student drinking example of Tables 13.9 and 13.10. The measurement variable Y (drinking scale) defines the **vertical axis** in this graph, and categories of the X variable (year in college) are marked off along the **horizontal axis.** The location of each group mean is shown by a small circle at the appropriate Y-axis height.

The **error bars** themselves mark off a distance of *plus or minus one standard error,* or $\pm SE_{\bar{Y}_k}$, around each of the group means. ANOVA requires the assumption that the population standard deviations within each group are the same: $\sigma_1 = \sigma_2 = \sigma_3 = \cdots = \sigma_K$. If this assumption is true, then our best estimate of the common standard deviation is a pooled standard deviation, s_P.

The **pooled standard deviation** for K-sample (ANOVA) problems is

$$s_P = \sqrt{\frac{WSS}{n_. - K}} \qquad [13.19]$$

where WSS is the within-groups sum of squares, $n_.$ is the sample size, and K is the number of groups or categories of X.

The pooled standard deviation, Equation [13.19], is just the square root of the within-groups mean square, which most ANOVA tables provide (see Table 13.12). If there are only two groups ($K = 2$), then Equation [13.19] yields the same results as Equation [13.9].

The pooled standard deviation, s_P, can be used to estimate **standard errors** for each group mean:

$$SE_{\bar{Y}_k} = \frac{s_P}{\sqrt{n_k}} \qquad [13.20]$$

For the student drinking example, the pooled standard deviation is

$$s_P = \sqrt{\frac{WSS}{n_. - K}}$$

$$= \sqrt{\frac{506.2}{19 - 4}} = 5.81$$

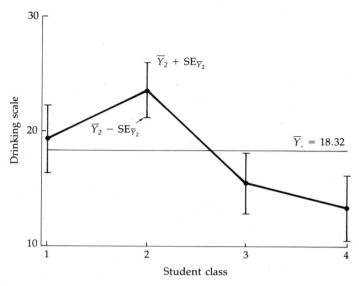

FIGURE 13.7   *Error-bar plot for student drinking data (Table 13.9)*

The estimated standard error for freshmen is obtained by dividing this pooled standard deviation by the square root of group size:

$$SE_{\bar{Y}_1} = \frac{s_p}{\sqrt{n_1}}$$

$$= \frac{5.81}{\sqrt{4}} = 2.905$$

The error bar around the freshman mean in Figure 13.7 is drawn out to a distance of $\bar{Y}_1 \pm SE_{\bar{Y}_1} = 19.25 \pm 2.905$. Similar calculations provide the error bars for the other three means.

The error-bar plot shows how the four means compare with each other and with the overall grand mean (the horizontal line). The means for sophomores and seniors depart substantially from the grand mean, in opposite directions. The individual sample means and standard errors suggest where (based on this sample) the four groups' population means probably lie. Graphs like Figure 13.7 can impart information about many different means at a glance, together with information about their uncertainty.

Our second example of an error-bar graph (Figure 13.8, page 430) is based on a study of microcomputer use in elementary schools. The researcher looked at such variables as student–computer ratios, frequency of use, extent of computer instruction, and student computer skills. All aspects were combined into a single microcomputer implementation score for each of 110 elementary schools. Means and standard deviations (s.d.) for these scores are given in Table 13.13, broken down by school socioeconomic status (SES).

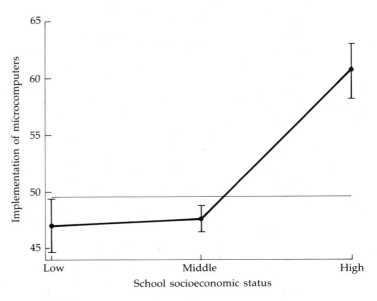

FIGURE 13.8   *Error-bar plot of microcomputer implementation by SES*

An analysis of variance reveals significant differences in microcomputer implementation among schools at three SES levels. The error-bar plot in Figure 13.8 shows them clearly: microcomputers have been implemented much more fully in the high-SES schools than in either low- or middle-SES schools. The mean implementation scores of low-SES and middle-SES schools differ little.

The sample standard deviations are similar for the three types of schools in Table 13.13, so the ANOVA assumption of identical population standard deviations seems reasonable. We can therefore use the within-groups sum of squares, WSS, to estimate standard errors (Table 13.13, bottom). These standard errors in turn provide the bars in the error-bar plot of Figure 13.8.

We connect group means in an error-bar plot by line segments only if there is a natural order to the categories of the X variable, as in Figures 13.7 and 13.8. When there is no natural order, as with "Democrat, Republican, Independent" or "treated with drug A, treated with drug B, no treatment," the group means are left separate.

Although error-bar plots make it easy to compare means at a glance, they actually require careful reading. No single rule governs what the bars represent. At least four different meanings are widely used:

1. The error bars show intervals of ± *one standard error,* as in Figures 13.7 and 13.8.
2. The bars show *95% confidence intervals.*

TABLE 13.13

Socioeconomic Status (SES) and Implementation of Microcomputers at 110 Elementary Schools

School SES[a]	Number of schools	Implementation score mean[b]	Implementation score s.d.
X	n_k	$\overline{Y}_k$	s_k
Low	19	47.0	10.8
Middle	74	47.6	9.7
High	17	60.6	10.4
All	110	49.5	11.0

Calculations for Error Bars (Figure 13.8):
Within-groups sum of squares:

$$\text{WSS} = \Sigma_k(n_k - 1)s_k^2 = (19 - 1)10.8^2 + (74 - 1)9.7^2 + (17 - 1)10.4^2 = 10{,}698.65$$

Pooled standard deviation:

$$s_p = \sqrt{\frac{\text{WSS}}{n. - K}} = \sqrt{\frac{10{,}698.65}{110 - 3}} = 10.0$$

Estimated standard errors:

$$\text{SE}_{\overline{Y}_1} = s_p/\sqrt{n_1} = 10.0/\sqrt{19} = 2.29$$

$$\text{SE}_{\overline{Y}_2} = s_p/\sqrt{n_2} = 10.0/\sqrt{74} = 1.16$$

$$\text{SE}_{\overline{Y}_3} = s_p/\sqrt{n_3} = 10.0/\sqrt{17} = 2.43$$

Error bars:
1. Low SES: $\overline{Y}_1 \pm \text{SE}_{\overline{Y}_1} = 47.0 \pm 2.29$
2. Middle SES: $\overline{Y}_2 \pm \text{SE}_{\overline{Y}_2} = 47.6 \pm 1.16$
3. High SES: $\overline{Y}_3 \pm \text{SE}_{\overline{Y}_3} = 60.6 \pm 2.43$

[a]Judged by typical parental occupation. High denotes schools where most students' parents were executives or professionals; low means most parents were unskilled workers, machine operators, clerical employees, etc.
[b]Higher scores indicate students have more access to computers, more instruction, better computer skills, etc.
Source: Data from McGee (1987).

3. The bars show intervals of $\pm$ *one standard deviation.*
4. The bars show *the entire range of Y values.*

Not all research reports specify which kind of bar they use. As some of these choices produce much narrower error bars than others—hence look more precise, even for the same data—we cannot understand an error-bar plot without knowing how the bars are defined. Examine carefully the text accompanying any error-bar plot.

PROBLEMS

*18. Problem 17 gave the means for a reading achievement test taken by samples of students at public, Catholic, and private non-Catholic high schools. They are repeated here.

1. Public	2. Catholic	3. Private non-Catholic
$n_1 = 10,000$	$n_2 = 1,000$	$n_3 = 200$
$\overline{Y}_1 = 4.48$	$\overline{Y}_2 = 5.00$	$\overline{Y}_3 = 5.34$

The within-groups sum of squares for these data is WSS = 46,609. Use the WSS to calculate a pooled standard deviation, Equation [13.19]; then estimate standard errors, Equation [13.20], for each of the three group means. Draw an error-bar plot for the means plus or minus one standard error ($\pm SE_{\overline{Y}_k}$). Explain why the group means in your plot should or should not be connected by line segments.

*19. In your error-bar plot for Problem 18, why is the error bar for private non-Catholic school students so much wider than that for public school students? What do these different widths tell us about the analysis?

*20. Draw an error-bar plot with $\pm SE_{\overline{Y}_k}$ bars for the injury claims and automobile size data in Problem 16. Include a horizontal line for the grand mean $\overline{Y}_.$. Should you connect the group means in your plot by line segments? Explain.

13.8   IQ SCORES AND READING ABILITY

ANOVA has so far been presented as a series of technical steps—assumptions, hypotheses, calculation, graphing, and so on. This section illustrates how the steps fit together as an *analytical process*. Table 13.14 contains data on 60 elementary school boys, 30 of whom were rated as poor or very poor readers—at least two years below their grade levels. The remaining 30 boys read normally, but otherwise resembled the poor readers in terms of schools, age, family background, and other variables. The 30 boys with reading problems consisted of 11 "very poor" readers and 19 who were merely "poor" readers. Reading categories are coded 1 (very poor), 2 (poor), or 3 (normal) in Table 13.14.

Reading disabilities comparable to those of the poor and very poor readers in Table 13.14 afflict an estimated 2% of elementary school children, about four-fifths of them boys. It was once believed that their difficulties reflected low intelligence, since children with reading difficulties often had below-average IQ scores. More recent research has found that the problem is not so simple, however. IQ tests typically comprise various subtests measuring different aspects of general intelligence. When children with reading disabilities

TABLE 13.14   ***IQ Subtest and Full-scale Scores for 60 Able and Reading-disabled Schoolboys***

Case	Reading ability[a]	Attention/ Concentration[b]	Spatial ability[c]	Full-scale WISC-R IQ
1	1	16	41	90
2	1	15	36	84
3	1	15	40	86
4	1	15	35	85
5	1	20	25	84
6	1	22	31	95
7	1	17	29	81
8	1	12	29	83
9	1	15	34	88
10	1	21	27	87
11	1	22	33	93
12	2	17	27	90
13	2	20	34	103
14	2	20	41	110
15	2	17	35	105
16	2	19	41	105
17	2	24	35	101
18	2	22	31	96
19	2	23	42	111
20	2	19	36	105
21	2	22	30	105
22	2	19	31	111
23	2	20	32	102
24	2	20	36	100
25	2	23	32	106
26	2	27	37	115
27	2	19	34	99
28	2	26	35	106
29	2	20	30	103
30	2	21	30	93
31	3	29	28	95
32	3	29	33	102
33	3	31	28	97
34	3	29	32	92
35	3	32	35	107
36	3	32	28	92
37	3	36	31	103
38	3	33	31	106
39	3	30	35	109
40	3	32	27	105
41	3	32	36	104
42	3	31	33	98

(continued)

TABLE 13.14   **(Continued)**

Case	Reading ability[a]	Attention/ Concentration[b]	Spatial ability[c]	Full-scale WISC-R IQ
43	3	35	32	110
44	3	31	30	94
45	3	30	29	103
46	3	34	28	102
47	3	28	31	102
48	3	29	28	98
49	3	35	29	109
50	3	34	26	96
51	3	24	28	95
52	3	35	30	101
53	3	33	29	109
54	3	30	37	112
55	3	38	25	96
56	3	33	27	93
57	3	37	34	111
58	3	30	28	96
59	3	32	35	105
60	3	30	35	105

[a]1 = very poor; 2 = poor; 3 = normal ability.
[b]Sum of WISC-R arithmetic, digit span, and coding subtests.
[c]Sum of WISC-R picture completion, block design, and object assembly subtests.
Source: Data from Treacy (1985).

are compared with normal readers on such subtests, rather than on a single overall IQ score, a complex pattern of differences emerges. Reading-disabled children are indeed well below average on some subtests, but they may equal or outscore normal readers on other subtests. Such subtest patterns shed light on the true nature of reading disabilities, which are no longer simply attributed to "low intelligence."

Table 13.14 gives both full-scale IQ and two kinds of subtest scores for each boy. The attention/concentration score is a sum of scores on three IQ subtests believed to measure attention span and concentration ability. The spatial ability score is a sum of scores on three subtests thought to measure spatial ability. Analysis of variance can test whether these two kinds of score are related to reading disability.

Before proceeding with ANOVA we must check on the usual assumptions: normal distributions and equal standard deviations. Figure 13.9 shows box plots of attention/concentration scores by reading ability. We see three reasonably well-behaved distributions with similar spreads. No outliers or severe skewness problems are apparent. The plot for very poor readers (category 1) looks slightly strange, but this is not surprising since it is based on only 11 cases. Reassuringly, full-scale IQ tests are known to have approximately normal distributions in the general population.

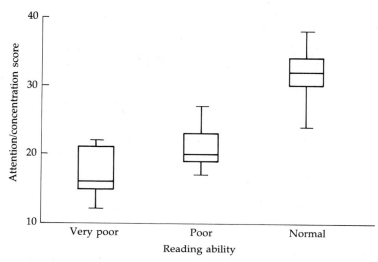

FIGURE 13.9   *Box plots of attention/concentration by reading ability*

Means and standard deviations of attention/concentration scores by reading ability group are given in Table 13.15. The means follow the same pattern as the medians (Figure 13.9): the greater the reading problems, the lower the attention/concentration score. An ANOVA based on these statistics, shown in Table 13.16 (page 436), indicates that the differences are significant ($P <$ .001). An error-bar plot (Figure 13.10) displays the nature of the group differences: As the mean attention/concentration score rises, so does reading ability. Boys of normal ability tend to score much higher than poor or very poor readers.

Our analysis has combined descriptive, inferential, and graphical methods. Table 13.15 reported the group means and standard deviations (sample description). The box plots in Figure 13.9 indicated that the three groups do not differ radically in ways other than their means (graphical exploration). The ANOVA in Table 13.16 established that statistically significant differences exist among these three means (inference). Finally, Figure 13.10 depicted how the means differ (graphical presentation).

A contrasting picture emerges from the spatial abilities scores in Table 13.14. Their summary statistics are given in Table 13.17 (page 438). Box plots

TABLE 13.15   *Attention/Concentration Ability Scores by Reading Group (from Table 13.14)*

Statistic	1. Very poor	2. Poor	3. Normal	4. All
Mean	$\overline{Y}_1 = 17.27$	$\overline{Y}_2 = 20.95$	$\overline{Y}_3 = 31.80$	$\overline{Y}_. = 25.70$
s.d.	$s_1 = 3.41$	$s_2 = 2.72$	$s_3 = 2.95$	$s_. = 6.92$
n	$n_1 = 11$	$n_2 = 19$	$n_3 = 30$	$n_. = 60$

TABLE 13.16

Analysis of Variance for Attention/Concentration Scores and Reading Ability (Tables 13.14 and 13.15)

Between-groups sum of squares:

$$BSS = \Sigma_k n_k \overline{Y}_k^2 - \frac{(\Sigma_k n_k \overline{Y}_k)^2}{n_.} \qquad \text{Equation [13.15b]}$$

$$= n_1 \overline{Y}_1^2 + n_2 \overline{Y}_2^2 + n_3 \overline{Y}_3^2 - \frac{(n_1 \overline{Y}_1 + n_2 \overline{Y}_2 + n_3 \overline{Y}_3)^2}{n_.}$$

$$= 11(17.27^2) + 19(20.95^2) + 30(31.8^2) - \frac{[11(17.27) + 19(20.95) + 30(31.8)]^2}{60}$$

$$= 2{,}326.7$$

Within-groups sum of squares:

$$WSS = \Sigma_k (n_k - 1)s_k^2 \qquad \text{Equation [13.16b]}$$

$$= (n_1 - 1)s_1^2 + (n_2 - 1)s_2^2 + (n_3 - 1)s_3^2$$

$$= (11 - 1)3.41^2 + (19 - 1)2.72^2 + (30 - 1)2.95^2 = 501.8$$

Total sum of squares:

$$TSS_Y = (n_. - 1)s_.^2 = (60 - 1)6.92^2 \qquad \text{Equation [13.17b]}$$

$$= 2{,}825.3 \qquad \text{(Allowing for rounding off, } TSS_Y = WSS + BSS.\text{)}$$

F statistic:

$$F_{df_W}^{df_B} = \frac{BSS/df_B}{WSS/df_W} = \frac{2{,}326.7/(3 - 1)}{501.8/(60 - 3)} \qquad \text{Equation [13.18]}$$

$$F_{57}^2 = 132.1$$

P-value: From Table A.5, $P < .001$

Conclusion: We can reject H_0: $\mu_1 = \mu_2 = \mu_3$; there are significant differences between the mean attention/concentration scores of elementary school boys with very poor, poor, and normal reading ability.

(Figure 13.11) show no apparent problems for the usual ANOVA assumptions. ANOVA itself (Table 13.18) shows that, like attention/concentration ability, spatial ability differs significantly by reading group. The error-bar plot in Figure 13.12 shows the surprising direction of this difference: the reading-disabled boys have *higher* mean spatial ability scores than boys who read normally. Spatial ability scores are highest among poor readers.

Spatial skills are generally controlled by the right hemisphere of the brain, whereas verbal skills are primarily a left-hemisphere function. Finding that children with reading disabilities equal or outperform normal readers at certain right-hemisphere tasks may shed light on causes and possible remedies for those disabilities.

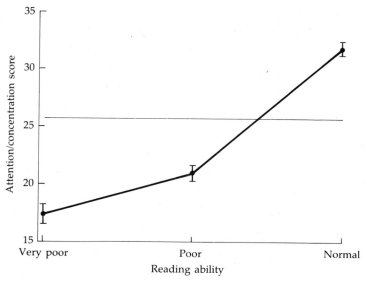

FIGURE 13.10 *Error-bar plot of attention/concentration by reading*

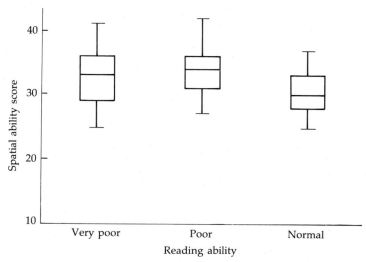

FIGURE 13.11 *Box plots of spatial ability by reading ability*

TABLE 13.17   **Spatial Ability Scores by Reading Ability Group (from Table 13.14)**

Statistic	1. Very poor	2. Poor	3. Normal	All
Mean	$\overline{Y}_1 = 32.73$	$\overline{Y}_2 = 34.16$	$\overline{Y}_3 = 30.60$	$\overline{Y}_. = 32.12$
s.d.	$s_1 = 5.12$	$s_2 = 4.11$	$s_3 = 3.27$	$s_. = 4.17$
n	$n_1 = 11$	$n_2 = 19$	$n_3 = 30$	$n_. = 60$

TABLE 13.18   **Analysis of Variance for Spatial Ability Scores and Reading Ability (Tables 13.14 and 13.17)**

ANOVA table in computer-output format:

Source	SS	df	MS	F	Prob > F
Between groups	152.48	2	76.24	4.96	0.0104
Within groups	875.35	57	15.36		
Total	1027.82	59	17.42		

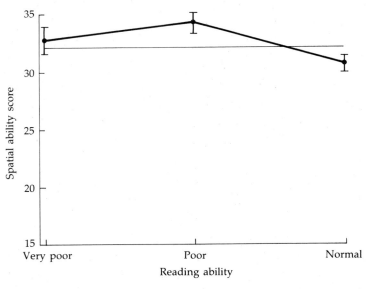

FIGURE 13.12   *Error-bar plot of spatial ability by reading ability*

PROBLEMS

21. Use computational formulas and the summary statistics in Table 13.17 to show how the following quantities from Table 13.18 can be calculated.
    a. Between-groups sum of squares (BSS)
    b. Degrees of freedom for BSS
    c. Within-groups sum of squares (WSS)
    d. Degrees of freedom for WSS
    e. Total sum of squares (TSS_Y)
    f. Degrees of freedom for TSS_Y
    g. *F* statistic

22. Use the results in Table 13.18 to find the pooled standard deviation, Equation [13.19], then estimate the standard errors, Equation [13.20], for the mean spatial ability scores in each of the three groups. In which group is the standard error highest? What does a higher standard error tell us?

23. The full-scale WISC-R IQ scores in Table 13.14 are the sums of individual subtest scores, including the attention/concentration and spatial components analyzed earlier. Are there significant differences among the three groups of boys with respect to mean WISC-R scores? Following these steps, perform an analysis:
    a. Construct box plots and comment on the distributional assumptions.
    b. Find sample means and standard deviations for the sample as a whole and for each of the three groups.
    c. Conduct an analysis of variance. Can the null hypothesis of equal population means be rejected at $\alpha = .05$?
    *d. Estimate the standard error of the mean for each group and construct an error-bar plot. Describe the relationship this plot shows between reading ability and IQ score.

13.9 *DEALING WITH DISTRIBUTIONAL PROBLEMS**

Standard inferential procedures for means are justified by making certain assumptions about population distributions. In small samples, like the student drinking data of Table 13.9, it is difficult to test these assumptions. For larger samples, informal exploratory methods like box plots or mean–median comparisons are useful. It is also possible to conduct formal tests treating distributional assumptions as null hypotheses. Chapter 12 discussed chi-square as a general "badness of fit" test (see Table 12.9, page 378). Certain versions of the chi-square test can ascertain:

1. Is a distribution significantly nonnormal?
2. Do several variances significantly differ from each other?

These tests require still further assumptions, however, and may not be very robust. For such reasons informal checks have been stressed in this book.

We have seen that if we look for distributional problems, we often find them. Where sample evidence or theoretical reasoning suggests that distributional assumptions are untrue, conclusions that follow from those assumptions are suspect. There are two general ways to deal with such problems.

One way is to apply *nonlinear transformations,* such as logarithms or square roots. Transformations can often be found that reduce several distributional problems—skewness, outliers, and unequal variances—simultaneously. The usual statistical methods are then applied to the transformed variables, as described in Chapters 6 and 11.

Nonlinear transformations work by changing the data, rendering distributional assumptions more plausible. A second approach is to use methods that bypass those assumptions. There are two general "families" of such methods. **Robust estimation** is the more modern of the two, often requiring special computer programs to do repeated calculations.[4] They typically seek statistical summaries that are highly resistant to outliers, yet have as little sample-to-sample variation (that is, small standard errors) as possible. Although promising, robust estimation is beyond the scope of a first course in statistics.

The second family of alternatives, called **nonparametric methods,** is older and computationally simpler. Like robust methods, nonparametric methods have advantages if the data involve nonnormal distributions, unequal variance, or outliers.

The next section introduces an approach combining some advantages of both nonparametric methods and nonlinear transformation.

13.10   *NONPARAMETRIC TESTS AND RANK TRANSFORMATIONS**

The term *nonparametric* applies to a variety of statistical methods. Their common feature is in not requiring strong assumptions about the shape of the population distribution. (Methods that do require such assumptions, like the mean-based procedures described earlier in this chapter, are called *parametric.*) Nonparametric methods remain valid in the presence of nonnormal distributions or nonconstant variances. A further advantage is that they are little affected by outliers.[5]

There are nonparametric alternatives to each of the comparison procedures discussed in this chapter. The two-sample *t* test has a nonparametric counterpart known by two different names (reflecting independent inventors), the **Wilcoxon rank sum test** or the **Mann–Whitney *U* test.** The nonparametric **Wilcoxon signed rank test** is analogous to the paired-difference *t* test. Finally, for *K*-sample problems, the nonparametric **Kruskal–Wallis** test provides an alternative to ANOVA. These correspondences are summarized in Table 13.19.

These three nonparametric tests share several features:

1. They are more robust than parametric methods.
2. They test hypotheses about the equality of medians, rather than means.
3. They make use of **ranks,** rather than raw data.

The process of assigning ranks is illustrated in Table 13.20 (page 442), which contains data on the average player salaries for 26 Major League baseball teams during the 1985 season. The teams are listed from lowest salary to highest. To the right of each dollar value is its rank. The team with the lowest average salary has a rank of 1, the second lowest is ranked 2, and so on. The Texas Rangers and the Minnesota Twins, tied at $258,000 each, receive the average of the ranks that they would have had otherwise. One would have been third, the other fourth, so both get rank $(3 + 4)/2 = 3.5$. The same rule resolves any other ties that occur.[6]

Table 13.20 also reports whether each team's season win–loss record placed it among the top half or the bottom half of these teams. Is there a relationship between player salaries and how well the team finished? We might test this by using a two-sample *t* test for the difference between the mean salary in the top and bottom halves.[7] Such a *t* test would require us to assume that average player salaries have a normal distribution, however. To avoid making this dubious assumption, we might prefer a nonparametric test based on ranks.

One drawback of traditional nonparametric methods is that they lack the conceptual unity of parametric methods: We seem to need a different nonparametric test, with a different sampling distribution, for every statistical problem. Furthermore, nonparametric procedures do not extend easily to more complex problems, unlike parametric methods. For these reasons, the traditional nonparametric tests are not described here. Instead, we offer a simple way to approximate such tests: We shall apply parametric techniques to ranked data.[8]

TABLE 13.19   *Classical Parametric Tests and Nonparametric Alternatives*

Type of problem	Parametric method	Nonparametric method
Two-sample	Two-sample *t* test	Wilcoxon rank sum test/ Mann–Whitney *U* test
Paired-difference	Paired-difference (one-sample) *t* test	Wilcoxon signed rank test
K-sample	Analysis of variance	Kruskal–Wallis test

TABLE 13.20   *Average Baseball Player Salary and Team Performance, 1985 Season*

	Team	Salary in dollars	Salary by rank	Team record[a]
1	Seattle	170,000	1	Top
2	Cleveland	220,000	2	Top
3	Texas	258,000	3.5	Top
4	Minnesota	258,000	3.5	Bottom
5	Montreal	315,000	5	Bottom
6	San Francisco	320,000	6	Top
7	Cincinnati	337,000	7	Bottom
8	White Sox	348,000	8	Bottom
9	Oakland	352,000	9	Top
10	Houston	366,000	10	Top
11	Kansas City	368,000	11	Bottom
12	Toronto	386,000	12	Bottom
13	St. Louis	387,000	13.5	Bottom
14	Boston	387,000	13.5	Top
15	Mets	389,000	15	Bottom
16	Pittsburgh	393,000	16	Top
17	Philadelphia	400,000	17.5	Top
18	San Diego	400,000	17.5	Top
19	Detroit	407,000	19	Bottom
20	Cubs	414,000	20	Top
21	Los Angeles	424,000	21	Bottom
22	Milwaukee	431,000	22	Top
23	California	434,000	23	Bottom
24	Baltimore	438,000	24	Bottom
25	Atlanta	541,000	25	Top
26	Yankees	546,000	26	Bottom

[a]Top or bottom half of these teams in season wins–losses.

> Nonparametric methods can be approximated by using the **rank transformation:**
>
> 1. *Rank the Y values* from all groups combined; for sets of tied cases assign the mean of the ranks they would have had if not exactly tied.
> 2. *Apply standard parametric procedures* to the ranks, rather than to the original values of Y.

This approximation approach has the advantage of requiring no new formulas or distributions beyond those already introduced in this chapter. For example, Table 13.21 shows how we can use the average salary *ranks* from Table 13.20 to test for a relationship between salary and win–loss position. Recall that the ranks were obtained after the entire sample of 26 cases was ordered. Table 13.21 splits the cases into top half and bottom half, listing

TABLE 13.21

t Test Based on Ranked Data: Average Player Salaries by Team Standing (from Table 13.20)

Top teams	Salary rank	Summary statistics
Seattle	1	
Cleveland	2	$n_1 = 13$
Texas	3.5	
San Francisco	6	
Oakland	9	$\overline{R}_1 = 12.54$
Houston	10	
Boston	13.5	
Pittsburgh	16	$s_{R_1} = 7.91$
Philadelphia	17.5	
San Diego	17.5	
Cubs	20	
Milwaukee	22	
Atlanta	25	

Bottom teams	Salary rank	Summary statistics
Minnesota	3.5	
Montreal	5	$n_2 = 13$
Cincinnati	7	
White Sox	8	
Kansas City	11	$\overline{R}_2 = 14.46$
Toronto	12	
St. Louis	13.5	
Mets	15	$s_{R_2} = 7.56$
Detroit	19	
Los Angeles	21	
California	23	
Baltimore	24	
Yankees	26	

Two-sample *t* test[a] applied to the ranks:

$$s_p = 7.738 \qquad SE_{\overline{R}_2 - \overline{R}_1} = 3.035 \qquad t = \frac{\overline{R}_2 - \overline{R}_1}{SE_{\overline{R}_2 - \overline{R}_1}} = \frac{14.46 - 12.54}{3.035} = .633$$

[a]This test applies Equations [13.9], [13.10], and [13.8] to the ranked data.

whole-sample ranks (in place of actual salary figures) from Table 13.20. Next we find the mean ($\overline{R}_k$) and standard deviation (s_{R_k}) of the ranks, within each of the two groups. Finally, we perform a difference-of-means *t* test, to see whether the mean ranks differ. This test requires calculating a pooled standard deviation, s_p, following Equation [13.9]; an estimated standard error, $SE_{\overline{R}_2 - \overline{R}_1}$, following Equation [13.10]; and finally a *t* test statistic, using Equation [13.8].

The *t* statistic we obtain is .633. With $n_1 + n_2 - 2 = 24$ degrees of freedom, Table A.3 gives a two-sided probability of $P > .50$. We should not reject the null hypothesis; the relationship between average player salary and win–loss position is not statistically significant.[9] Note that the small difference between mean ranks in Table 13.21 shows that the top half of the teams actually paid slightly *lower* salaries than the bottom half. These data provide no support for the idea that higher average salaries buy baseball success.

Here is a summary of how to use ranks for nonparametric approximations of mean-based methods.

Two-sample tests:

1. Combine both groups or categories and assign ranks, so that $n_1 + n_2$ ranks are assigned.
2. Separate the groups and calculate the means and standard deviations of the ranks within each group.
3. Perform a standard two-sample t test using the ranks rather than the raw data.

This procedure, illustrated in Tables 13.20 and 13.21, is roughly equivalent to a Wilcoxon rank sum test, also known as a Mann–Whitney U test.

Paired-difference tests:

1. Combine both groups or both variables and assign ranks, so that with n pairs we assign $2n$ individual ranks.
2. Separate the $2n$ ranks into their two original groupings, each now containing n ranks.
3. Perform a standard one-sample t test, using the set of n differences-of-ranks in place of the raw data.

This test is comparable to the Wilcoxon signed rank test.

K-sample tests:

1. Combine all K groups or categories and assign ranks.
2. Separate the groups and perform ANOVA using the ranks rather than the raw data.

This procedure is an approximation for the Kruskal–Wallis test.

Table 13.22 shows the P-values that are obtained for the main examples in this chapter, using three different kinds of test: parametric, nonparametric, and nonparametric approximations. All three tests lead to similar P-values and to essentially the same conclusions, in each of the six examples—a reassuring discovery! It shows that the conclusions themselves are robust or not dependent on particular assumptions. Had the different tests not yielded the same conclusions, we would have had to scrutinize the data and our assumptions carefully, determining which of the tests to believe.

TABLE 13.22   *Comparison of P-values Obtained with Three Test Methods*[a]

Example	Parametric (Mean-based)	Nonparametric (Median-based)	Nonparametric approximation
Student MSAT and gender (Table 13.6)	Two-sample t test: $P = .029$	Wilcoxon or Mann–Whitney: $P = .019$	Two-sample t test on ranks: $P = .016$
Sentences completed (Table 13.8)	Paired-difference t test: $P < .001$	Wilcoxon signed ranks: $P < .001$	Paired-difference t test on ranks: $P < .001$
Collegiate drinking (Table 13.12)	ANOVA F test: $P = .061$	Kruskal–Wallis X^2: $P = .114$	ANOVA F test on ranks: $P = .098$
Attention and reading ability (Table 13.16)	ANOVA F test: $P < .001$	Kruskal–Wallis X^2: $P < .001$	ANOVA F test on ranks: $P < .001$
Spatial and reading ability (Table 13.18)	ANOVA F test: $P = .010$	Kruskal–Wallis X^2 $P = .017$	ANOVA F test on ranks: $P = .014$
Baseball salaries (Table 13.21)	Two-sample t test: $P = .381$	Wilcoxon or Mann–Whitney: $P = .522$	Two-sample t test on ranks: $P = .532$

[a]The P-values were calculated with a computer, so they are more precise than those found from reference tables such as the Appendix tables.

Nonparametric approximations enable parametric methods to encompass problems otherwise inaccessible due to nonnormal distributions. Thus, they bridge parametric and nonparametric strategies. In effect, ranks are a special kind of nonlinear transformation. Like the power transformations introduced in Chapter 6, rank transformations provide a way to make ill-behaved raw data distributions more amenable to statistical analysis.[10]

PROBLEMS

This chapter began with two sets of data concerning sulfate concentrations in precipitation monitored in Europe and North America (Tables 13.1 and 13.2). Such air pollution measures often have distributions with positive skewness or outliers. Box plots indicate the presence of outliers, skewness, and nonconstant variables (Figures 13.1 and 13.2). Thus theoretical considerations and sample data both argue against the assumptions needed for a standard parametric analysis. Nonparametric or ranked-based methods provide a better alternative.

*24. Are European sulfate concentrations significantly different from North American (Table 13.1)? Since this is a two-sample problem, use the rank-based approximation to the Wilcoxon/Mann–Whitney test to test this hypothesis at the $\alpha = .05$ level.

   a. Assign ranks to all 18 sites, from 1 (lowest sulfate) to 18 (highest sulfate). For tied values, assign the mean of the ranks they would receive if not tied.

   b. Find the means and standard deviations of the ranks for the 7 North American and 11 European sites.

   c. Test whether these mean ranks are significantly different, using a standard two-sample t test.

*25. Were precipitation sulfate concentrations significantly lower in 1979–1982 than in 1975–1978 (Table 13.2)? Since this is a paired-difference problem, use the rank-based approximation to the Wilcoxon signed rank test to test this hypothesis at the $\alpha = .05$ level.

   a. You have a total of 36 sulfate measurements. Assign ranks from 1 (lowest) to 36 (highest) to each of these 36 measurements. Follow the usual rule for ties.

   b. Find the differences between ranks (1979–1982 rank minus 1975–1978 rank) for each of the 18 sites. Find the mean and standard deviation of this difference.

   c. Perform a one-sample t test to determine whether there was a significant decrease in sulfate concentration.

SUMMARY

Measurement-variable methods based on means or on squared deviations lack resistance to outliers. For inferential work, they also require normal distributions and constant variation across levels of X. One way to get around these sometimes unrealistic requirements is to use nonparametric methods, which need no such assumptions. Nonparametric methods, however, lack the conceptual unity of classical mean-based methods, and they are difficult to extend to more complex analyses. Another alternative is to change distributional characteristics by applying a nonlinear transformation. Chapter 6 described a family of power transformations useful for this purpose, which will reappear in a new context in Chapter 15.

In this chapter, we used a different kind of nonlinear transformation: the rank transformation. Rank transformations also help cope with ill-behaved distributions. Applying standard parametric procedures to rank-transformed data provides a reasonable approximation of certain nonparametric tests, without the burden of learning a new set of techniques.

Even without going into true nonparametric methods, this chapter has introduced what may seem like a bewildering variety of techniques. Most addressed the same general question: How can we test hypotheses about differences among means? At a deeper mathematical level, there is less variety here than meets the eye. Two-sample tests are just a simpler kind of K-sample test; running an ANOVA or a difference-of-means t test on a two-sample problem yields exactly the same conclusion. Paired-difference methods, which at first seem to be separate techniques, are actually just versions of one-sample methods. In a sense, one-sample problems are K-sample problems where $K = 1$, just as two-sample problems are K-sample problems where $K = 2$.

From this perspective, ANOVA is the most general technique described in this chapter. The other techniques are just special cases of ANOVA. The remainder of this book explores a related technique called *regression analysis*—an even more general method than ANOVA. Just as one- and two-sample methods are special kinds of K-sample ANOVA, ANOVA itself is a special kind of regression. That is, any ANOVA problem can be reformulated as a regression problem, but regression also encompasses many other kinds of problem that ANOVA cannot address. The more you learn about these seemingly separate statistical methods, the more their basic similarities emerge.

PROBLEMS

26. Which general type of statistical method (two-sample, K-sample, or paired-difference) would be appropriate for each of the following research questions? Identify specific statistical techniques that might be used.

    a. Among automobile accident victims, are medical costs significantly higher for those who were not wearing seat belts?

    b. A high school begins an experimental SAT preparation course. Do graduates of this course earn significantly higher SAT scores than they did before taking the course?

    c. Do graduates of the SAT preparation course in part b earn significantly higher SAT scores than their fellow students who did not take the course?

    d. Are there significant differences in the average length of prison sentences set by six criminal court judges?

27. In a study 66 hospital nurses were told a story about the difficult behavior of a five-year-old girl (Bordieri, Solodky, and Mikos, 1985). They were asked to assess whether the child likely suffered from emotional problems, using a 9-point scale, with 9 meaning highest likelihood. Half of the 66 nurses were shown a photograph of a physically attractive five-

year-old girl, and the other half were shown a photograph of an unattractive child. Among those shown the attractive picture, the mean "likelihood of emotional problems" assessment was 2.94, with a standard deviation of 1.56. Among those shown the unattractive picture, the mean and standard deviation were 4.03 and 2.23, respectively. Conduct a two-sample t test to establish whether there is a significant difference (at $\alpha = .05$) between these means. What can you conclude? Upon what distributional assumptions does this conclusion depend?

28. Refer back to the writing course data of Problem 12. Test whether there was a significant improvement (at $\alpha = .05$) in the number of paragraphs students wrote in a fixed amount of time.

29. Construct a box plot for D in Problem 28, and comment on its shape. Does the normality assumption seem reasonable?

30. Problem 16 contained an index representing the average collision repair cost for 12 types of small, midsize, and large cars. Use ANOVA to test whether the relationship between car size and collision repair cost index is statistically significant at $\alpha = .05$.

31. What is the meaning of the three sums of squares in ANOVA? Using the example in Problem 30, explain what these quantities measure in your own words, without simply repeating the formulas.

*32. Construct an error-bar plot for car repair cost means from Problem 30. Briefly summarize the relationship this shows between car size and average repair costs.

The data in the accompanying table are from an experiment testing a drug's efficacy in preventing motion sickness. Astemizole was administered to half of the 20 subjects; the rest got a placebo. The experimenters recorded how many mechanically controlled head movements subjects could endure before becoming nauseated. The numbers given in the table are the change in number of head movements tolerated after taking astemizole or the placebo. A positive value means that the subject tolerated more head movements before nausea, so susceptibility to motion sickness apparently decreased after treatment. A zero value means unchanged susceptibility, and a negative value means that susceptibility actually increased. Use the information in the table to solve Problems 33 and 34.

33. Conduct a two-sample t test to determine whether astemizole is significantly different from a placebo in its effects on susceptibility to motion sickness ($\alpha = .05$). What do you conclude about the effectiveness of this drug?

Susceptibility to Motion Sickness

Subject	Group	Change in number of movements tolerated
1	Placebo	−130
2	Placebo	−80
3	Placebo	−45
4	Placebo	0
5	Placebo	5
6	Placebo	5
7	Placebo	30
8	Placebo	40
9	Placebo	65
10	Placebo	70
11	Astemizole	−85
12	Astemizole	−55
13	Astemizole	−50
14	Astemizole	0
15	Astemizole	0
16	Astemizole	15
17	Astemizole	25
18	Astemizole	70
19	Astemizole	75
20	Astemizole	80

Source: Data from Kohl et al. (1987).

34. Construct box plots for the astemizole and placebo groups in Problem 33. Do the box plots show anything that would make you doubt the conclusion reached in Problem 33?

35. A study examined whether rings affect the amount of bacteria on health care workers' hands (Jacobsen et al., 1985). The subjects were college microbiology students. Bacteria counts were made for each student's hands after a careful washing, once when the student was wearing rings and a second time when the student wore no rings. Difference scores were defined as each student's bacteria count with rings minus the same student's bacteria count without rings. The mean difference was $\overline{D} = 842$ with a standard deviation of $s_D = 4{,}215$ ($n = 32$). That is, when the students wore rings their bacteria counts were higher by an average of 842 bacteria (per milliliter of rinse solution). Is this difference statistically significant? Apply a one-sided t test at $\alpha = .05$.

36. An article reported on how much different advertisement wordings appeal to women with low, moderate, and high desires to work at a career (Barry, Gilly, and Doran, 1985). An imaginary magazine advertised as a "homemaker's helper" appealed most strongly to women with low de-

sires to work. When the same magazine was advertised as a "woman's best friend," it appealed more strongly to women with moderate desires to work. Mean ratings are shown by desire to work (DW) category for how strongly this imaginary magazine appealed to women when advertised as a "career helper."

1. *Low DW*	2. *Moderate DW*	3. *High DW*	*All*
$n_1 = 62$	$n_2 = 126$	$n_3 = 61$	$n_. = 249$
$\overline{Y}_1 = 1.75$	$\overline{Y}_2 = 2.09$	$\overline{Y}_3 = 2.52$	$\overline{Y}_. = 2.11$
$s_1 = .876$	$s_2 = .948$	$s_3 = .813$	$s_. = .936$

Are the differences statistically significant at $\alpha = .01$? What does "statistical significance" in this instance mean?

*37. Estimate standard errors and construct an error-bar plot for the advertisement analysis in Problem 36.

38. The statistics shown describe mean *self-reported* grade point averages (GPA) by degree goals, for the college student sample of Table 13.4.

1. *Bachelor*	2. *Master*	3. *Doctorate*	*All*
$n_1 = 16$	$n_2 = 15$	$n_3 = 3$	$n_. = 34$
$\overline{Y}_1 = 2.62$	$\overline{Y}_2 = 3.10$	$\overline{Y}_3 = 3.10$	$\overline{Y}_. = 2.88$
$s_1 = .51$	$s_2 = .45$	$s_3 = .26$	$s_. = .52$

Test whether the relationship between degree goals and self-reported GPA is statistically significant ($\alpha = .05$).

*39. Use the pooled standard deviation to estimate the standard errors for the three group means in Problem 38, and use these standard errors to draw an error-bar chart. Why is the standard error for students with doctorate degree goals so much larger than the other two standard errors?

40. The statistics shown are for *true* mean GPA for the same 34 students of Problems 38–39.

1. *Bachelor*	2. *Master*	3. *Doctorate*	*All*
$n_1 = 16$	$n_2 = 15$	$n_3 = 3$	$n_. = 34$
$\overline{Y}_1 = 2.49$	$\overline{Y}_2 = 2.72$	$\overline{Y}_3 = 3.02$	$\overline{Y}_. = 2.64$
$s_1 = .47$	$s_2 = .54$	$s_3 = .36$	$s_. = .51$

Is the relationship between degree goals and true GPA significant ($\alpha = .05$)? Can you think of a real-world explanation for the discrepancy between the conclusions in Problems 38 and 39?

41. Probation officers make recommendations to judges regarding the sentencing of convicted criminals. A recent study categorized probation officers according to their judicial philosophies as either "liberal" or "con-

servative," and used a scale to measure the severity of actual sentences given to offenders (Walsh, 1985). Among a random sample of $n_1 = 67$ cases processed by liberal probation officers, the mean sentence severity was 290 with a standard deviation of 496. Among $n_2 = 179$ cases processed by conservative probation officers, the mean sentence severity was 551, with a standard deviation of 767. Use a two-sample t test to determine whether sentences are significantly more severe in cases processed by conservative probation officers ($\alpha = .05$).

42. The sentence severity scale in Problem 41 takes on only positive values. From the information available here, do we have any reason to doubt either of the distributional assumptions that underlie our analysis in Problem 41? What might be done about this?

43. When there are only two categories in X ($K = 2$), K-sample ANOVA produces the same results as a two-sample t test. Their obtained P-values are identical, and the ANOVA F statistic (with $df_B = K - 1 = 1$ and $df_w = n - K = n - 2$ degrees of freedom) equals the t statistic (with $n - 2$ degrees of freedom) squared:

$$F^1_{n-2} = (t_{n-2})^2$$

Problem 19 in Chapter 8 presented data on grades, teaching evaluations and instructor status in a sample of 118 college courses. In this chapter (Problem 11) you constructed a confidence interval for the difference between mean percentages of A and B grades given by full-time faculty and by part-time instructors. The relevant sample statistics are shown.

0. *Part-Time*	1. *Full-Time*	*All*
$n_0 = 47$	$n_1 = 71$	$n_. = 118$
$\overline{Y}_0 = 62.8$	$\overline{Y}_1 = 53.0$	$\overline{Y}_. = 56.9$
$s_0 = 11.8$	$s_1 = 15.2$	$s_. = 14.7$

a. Use this information to carry out a two-sample t test of the null hypothesis $H_0: \mu_1 - \mu_0 = 0$.

b. Now perform an ANOVA test of the null hypothesis $H_0: \mu_0 = \mu_1$. (Note that this null hypothesis is equivalent to that tested in part a.)

c. Compare your results from the t test (part a) and ANOVA (part b). Are they as expected with respect to t and F statistics, P-values, and conclusions?

44. Researchers compared IQ test scores for a sample of junior high and high school students (Reilly, Wheeler, and Etlinger, 1985). All were classified as emotionally disturbed, mentally retarded, learning disabled, or juvenile delinquent—four distinct groups that share poor IQ test performance. This similarity sometimes leads to misdiagnoses. Results for the full-scale IQ tests are given at the top of page 452.

	Emotionally Disturbed	Mentally Retarded	Learning Disabled	Juvenile Delinquent	All
	$n_1 = 21$	$n_2 = 20$	$n_3 = 40$	$n_4 = 40$	$n_. = 121$
	$\overline{Y}_1 = 84.05$	$\overline{Y}_2 = 57.95$	$\overline{Y}_3 = 76.40$	$\overline{Y}_4 = 75.70$	$\overline{Y}_. = 74.45$
	$s_1 = 18.42$	$s_2 = 7.41$	$s_3 = 9.64$	$s_4 = 12.76$	$s_. = 14.55$

Use ANOVA to test whether there are significant differences (at $\alpha = .001$) among these four groups.

*45. Construct an error-bar plot for the IQ data of Problem 44, and briefly describe what you see.

The accompanying table contains data on a small Vermont town where chemical wastes seeped into local water supplies. (Other data from this study appear in Chapters 10 and 12.) Alarmed residents formed a Health and Safety Committee to work on this problem. The survey items shown are for 19 respondents and include: whether they became active members of the Committee; years of residence in town; years of education completed; and a scale of how important it was to study the causes and effects of the contamination.

Survey Data for a Town with Water Contamination

Subject	Active in committee	Years lived in town	Years of education	Importance of problem[a]
1	No	14	16	1.6
2	No	15	12	3.6
3	No	34	12	1.6
4	No	24	12	2.3
5	No	54	12	3.3
6	No	30	12	1.6
7	No	45	13	1.3
8	No	39	14	1.0
9	No	36	9	3.0
10	No	53	12	1.3
11	No	65	9	1.6
12	Yes	5	12	3.3
13	Yes	25	14	6.0
14	Yes	3	17	1.6
15	Yes	5	16	3.0
16	Yes	10	12	3.3
17	Yes	27	12	3.3
18	Yes	13	12	2.3
19	Yes	3	13	3.3

[a]Higher values mean that studying the contamination problem was more important.
Source: This study is described in Hamilton (1985).

46. Based on the data from the table, are Committee members significantly different from others in this small town, with respect to length of residency? Conduct a *t* test for the difference in mean residency between these two groups. Apart from the question of statistical significance, suggest how this difference might have had practical importance to those involved in the controversy.

*47. The scale of the contamination problem's importance likely has a nonnormal population distribution, positively skewed with severe outliers. Apply the rank-based approximation for the Wilcoxon/Mann–Whitney nonparametric test to determine whether Committee activists attached significantly higher importance than other residents to studying the contamination's causes and effects ($\alpha = .05$).

The accompanying table contains data from well water tests. The wells are classified by distance from the nearest road salted in winter: *near* means less than 100 feet from the road; *far* means more than 100 feet. This categorization of the distance measurements in the original data is appropriate because of the markedly bimodal distribution of that variable (see Chapter 6, Problem 24, page 172).

Data on Ten Private Water Wells in Lee, New Hampshire

Well	Chloride concentration in mg/l	Distance from salted road
1	10	Far
2	10	Far
3	10	Far
4	10	Near
5	17	Far
6	21	Far
7	110	Near
8	150	Near
9	620	Near
10	680	Near

Source: Data courtesy of New Hampshire Water Supply and Pollution Control Commission.

*48. Is chloride concentration significantly higher in wells that are near a salted road (at $\alpha = .05$)? Chloride concentration has an extremely nonnormal distribution, so a standard difference-of-means *t* test would not be valid here. Instead, perform the rank-based approximation of the Wilcoxon/Mann–Whitney test. What do you conclude?

*49. Apply the rank-based approximation for the Kruskal–Wallis test to the car model and injury ratings data of Problem 16.

    a. First, assign ranks to the injury ratings variable for all 12 models.

    b. Perform an analysis of variance on the *ranks*, to determine whether there are significant differences between small, midsize, and heavy cars.

    c. When a parametric ANOVA is performed with these data (Problem 16), we obtain a *P*-value of $P < .001$, so H_0 (no difference) could be rejected. Do your nonparametric approximation results also support the parametric conclusion that H_0 can be rejected?

NOTES

1. These hypotheses could be stated either in terms of the difference of population means, $\mu_Y - \mu_X$, or the mean of population differences, δ, since both will be the same. To avoid confusion with two-sample methods, we emphasize δ.

2. Strictly speaking, a discrete variable like the number of sentences completed cannot be normally distributed, since normal distributions are continuous. This variable takes on enough different values (ranging from 1 to 33 in the sample) that its distribution could be "approximately" normal, however. The same can be said of many other measurement variables.

3. The means and standard deviations are from Coleman, Hoffer, and Kilgore (1982). The sample sizes given in this example are less than those of the actual study, to simplify calculations.

4. Hoaglin, Mosteller, and Tukey (1983) give an introduction to robust methods.

5. The term *nonparametric* is actually a misnomer; a better, but less often used, term is *distribution-free.*

6. If there are no ties at all, the distribution of ranks is *uniform* (see Chapter 8) regardless of the shape of the original Y distribution. Uniform distributions have the desirable properties of symmetry with no outliers or gaps. The ranks distribution is still approximately uniform if the data tie only occasionally. But if many cases have the same value of Y, the ranks distribution may be no better-behaved than the original Y distribution. Then the rank-based methods described here offer no improvement over methods using the raw data.

7. For purposes of this example, we view the data in Table 13.20 as a "random sample" from a larger population of teams and years.

8. See Koopmans (1987, pages 397–407), for further examples and explanation of these approximations.

9. The actual null hypothesis being tested here has been left indistinct.

Since we are working with mean *sample* ranks, it might seem natural that our hypotheses would refer to mean *population* ranks—but there is no such thing as a "mean population rank." (Why?) In their traditional form, rank-based tests test the equality of population medians (of the original variable, not of ranks). Although the tests described here are only approximations, we may think of them in this way too.

10. Rank transformations differ from power transformations in the important respect that the rank transformation is not reversible. You cannot recover the original Y values from a set of ranks, as you can by squaring, for example, if Y was transformed by taking square roots. The fact that the rank transformation is irreversible means that *some information has been thrown away* in making the transformation. This is one reason why power transformations are preferable to rank transformations, if either works equally well to overcome distributional problems in a specific set of data.

Two Measurement Variables: Regression Analysis

Chapter 14

Bivariate analyses where both variables are categorical can be done by cross-tabulation (Chapter 12). With one categorical and one measurement variable, we can compare means of the measurement variable across categories (Chapter 13). This chapter explores the third possibility in bivariate analysis: two measurement variables. A graphical tool called the *scatter plot* and a family of numerical methods known as *regression analysis* will be introduced.

Regression is the core technique of statistical analysis in many fields. One reason for its importance is that regression adapts readily to a wide variety of analytical problems, from simple to highly complex. For example, every comparison of means in Chapter 13 could be reformulated as a regression problem. Regression can also handle more complicated problems involving three, four, or twenty different variables at once (Chapter 16).

Regression furthermore provides a straightforward description of *how* variables are related. If variable X is suspected of being a cause of variable Y, such questions arise as: If X increases, does Y increase or decrease? By how much? Is the change in Y the same at high levels of X as it is at low levels?

How accurately can we predict values of *Y*, if we know values of *X*? Regression analysis addresses these questions directly.

Scatter plots are a versatile way to graph two-variable distributions. Like regression analysis, scatter plots can be adapted to many analytical purposes. This chapter uses them as a first step to performing a regression and, later, to interpret the regression results. Chapter 15 offers a third use of scatter plots: to check and improve the validity of our analyses.

We begin here by showing what a scatter plot is, then introduce regression in the context of this plot. Statistical details will be filled in gradually, over the final chapters.

14.1 *SCATTER PLOTS*

Table 14.1 contains data on the first 25 flights of the U.S. space shuttle. The twenty-fifth flight was the *Challenger* disaster, in which seven crew members died. Subsequent investigations pinpointed the cause as the burn-through of an O-ring seal at a joint in one of the shuttle's solid-fuel rocket boosters. Many of the 24 previous shuttle flights had also experienced heat damage to the booster rocket joints. The number of incidents of damage to field joints for each flight is shown in Table 14.1, together with estimates of the temperatures in the joints at the time of launching. The joint temperatures depend mainly on weather.

Could the joint problems be related to cold weather? The rubber O-rings were known to be less resilient, hence to seal less effectively, at low temperatures. This problem was an immediate focus of the crash investigation because *Challenger* had endured a night of freezing weather prior to launch. Figure 14.1 (page 460) shows a *scatter plot* for the data of Table 14.1, set up to investigate this issue. A temperature scale drawn across the bottom forms the *X-axis* of this plot. The vertical scale, here the number of incidents of O-ring damage, is called the *Y-axis*. By convention the independent ("cause") variable goes on the horizontal or *X-axis*, while the dependent ("effect") variable goes on the vertical or *Y-axis*. They are known as the *X variable* and *Y variable*, respectively.

> A **scatter plot** is drawn with two axes:
>
> The *horizontal* **X-axis** is a scale for the **X variable** or "cause."
> The *vertical* **Y-axis** is a scale for the **Y variable** or "effect."

Each shuttle flight for which we know *X* (temperature) and *Y* (incidents of O-ring damage) values is located as a dot in the scatter plot of Figure 14.1. The first flight, STS-1, was launched with a joint temperature of 66° F and experienced no O-ring damage. This flight is represented by a dot in the bottom center of the plot, with the (*X, Y*) **coordinates** (66, 0). The second

TABLE 14.1 <u>*Status of Booster Rocket Field Joints on the First 25 Space Shuttle Flights*</u>

Flight	Launch date	Temperature at launch (°F)	Incidents of damage
		X	Y
STS-1	4/12/81	66	0
STS-2	11/12/81	70	1
STS-3	3/22/82	69	0
STS-4	6/27/82	80	—
STS-5	1/11/82	68	0
STS-6	4/4/83	67	0
STS-7	6/18/83	72	0
STS-8	8/30/83	73	0
STS-9	11/28/83	70	0
STS 41-B	2/3/84	57	1
STS 41-C	4/6/84	63	1
STS 41-D	8/30/84	70	1
STS 41-G	10/5/84	78	0
STS 51-A	11/8/84	67	0
STS 51-C	1/24/85	53	3
STS 51-D	4/12/85	67	0
STS 51-B	4/29/85	75	0
STS 51-G	6/17/85	70	0
STS 51-F	7/29/85	81	0
STS 51-I	8/27/85	76	0
STS 51-J	10/3/85	79	0
STS 61-A	10/30/85	75	2
STS 61-B	11/26/85	76	0
STS 61-C	1/12/86	58	1
STS 51-L	1/28/86 (*Challenger*)	31	—

Source: Data from the Presidential Commission on the Space Shuttle *Challenger* Accident (1986).

flight, STS-2, launched at 70° F and having one damage incident, is shown at coordinates (70, 1). Five pairs of flights, such as STS-2 and STS 41-D, have the same combination of X and Y values. These pairs of cases are each represented by a single dot, though for some purposes it might be better to show them as closely spaced but distinct.

The number of damage incidents is unknown for two flights: STS-4, which lost a booster rocket at sea, and *Challenger's* last flight, STS 51-L. Neither is included in Figure 14.1.

To an experienced data analyst, an alarming conclusion almost leaps off the page from this scatter plot: *All* flights below 65° F experienced booster rocket joint damage. The most known incidents occurred on the coldest flight (STS 51-C), whereas only three out of 16 warmer weather flights suffered joint damage. Temperature may not be the sole cause of the joint problems, but Figure 14.1 strongly suggests it is important.

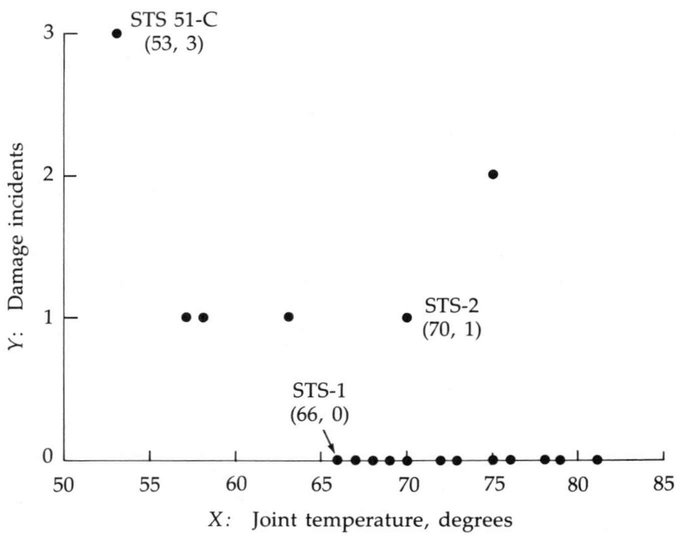

FIGURE 14.1 *Scatter plot of damage incidents vs. joint temperature for 23 space shuttle flights*

Because of storms and cold that left three-foot icicles on the launch tower, the booster joint temperature on *Challenger*'s last launch was 31° F, far below any shown in Figure 14.1. Before the launch, engineers discussed whether these cold conditions were dangerous. Among evidence they examined was a scatter plot like Figure 14.1, but with one crucial difference: It included only flights that had sustained some damage. Their plot looked like Figure 14.2, with no obvious trend to the scattered dots and no alarming conclusion that leaps off the page. The incomplete scatter plot gives the false impression that there is no link between temperature and joint damage. A more complete graph like Figure 14.1 was not examined until it was too late—after the *Challenger* explosion.

The scatter plot in Figure 14.1 is exceptional not only because of its dramatic history, but also because its conclusions can be grasped so readily. Often the visual pattern in a scatter plot is less clear-cut, and its interpretation seems more subjective—a bit like reading tea leaves. We may need to summarize the pattern numerically, for purposes of description, comparison, or inference. Regression analysis provides ways to do this systematically.

PROBLEMS

1. Although low temperatures were one cause of booster rocket joint damage, since even some warm weather shuttle flights experienced damage other causes must exist. A number of changes in design and procedures were made during the course of the shuttle program; some were intended to

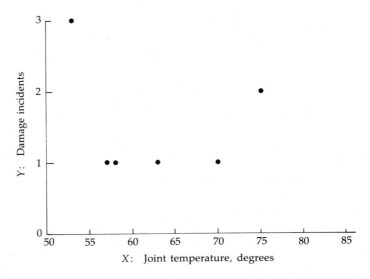

Scatter plot of damage incidents vs. joint temperature, omitting undamaged flights

FIGURE 14.2

reduce the likelihood of joint damage. Use the data of Table 14.1 to construct a scatter plot of the number of damage incidents (Y) against year of launch (X) and describe what you see. Was the problem getting better or worse?

2. The accompanying table contains data on eight professional basketball players. The variables are height, number of rebounds, and number of points scored during the 1986–1987 season. Construct a scatter plot of players' rebounds (Y) against height (X). The X-axis scale in your graph can begin at 70 inches, since all are taller. Describe any patterns you see in the plot.

Data on Eight Basketball Players, 1986–1987 Season

Player	Height in inches	Season rebounds	Season points	Minutes played
Willis	84	704	1,010	2,300
Bird	81	805	2,115	3,113
Walton	83	544	606	1,546
Leavell	73	67	583	1,190
Lee	82	351	431	1,197
Brewer	76	53	211	570
Sampson	88	781	1,597	2,467
Reid	80	301	986	2,157

Source: Data courtesy of Steve Tullar.

14.2   REGRESSION LINE AND REGRESSION EQUATION

Table 14.2 presents data from an era far removed from the *Challenger* disaster. The cases are ten river valley regions in France. Over 15,000 years ago these valleys were inhabited by cave dwelling humans who left a rich legacy of cave art, primarily animal pictures. Archaeologists have excavated areas where these people lived, and bones in the debris reveal what animals were actually caught and eaten.

The data in Table 14.2 reflect the prevalence of two types of animal, reindeer and mammoths, in the paleolithic hunters' paintings and in their bone piles. Four measurement variables are shown: the prevalence of reindeer bones, of reindeer art, of mammoth bones, and of mammoth art. The study's authors wondered what purpose the cave paintings served—for example, hunting magic, storytelling, totemism, or "art for art's sake."

A scatter plot based on these data is shown in Figure 14.3. Its X-axis represents the mammoth "bone prevalence index," a percentage scale describing how common mammoth bones are, compared to other animals' bones among the cave dwellers' debris in each region. The Y-axis is the amount of mammoth art as a percentage of the total cave art in each region. In the Northern Garonne River region, for instance, mammoth bones were moderately prevalent (compared with other regions) and pictures of mammoths made up 14% of the total cave art. There are only eight dots in Figure 14.3, because three of the ten cases share coordinates.

In the space shuttle scatter plot of Figure 14.1, the scatter had a down-to-right trend—fewer damage incidents occurred at warmer temperatures. In contrast, the scatter of points in Figure 14.3 has an up-to-right trend: Regions with a higher prevalence of mammoth bones tend also to have more mam-

TABLE 14.2   *Paleolithic Cave Art and Animal Bone Prevalence[a]*

Region	Reindeer bone	Reindeer art	Mammoth bone	Mammoth art
1. N. Garonne	21%	4%	4%	14%
2. Dronne	30	0	2	4
3. Isle	42	18	0	0
4. Vezere	34	13	2	7
5. Dordogne	24	3	0	21
6. Lot	19	4	6	21
7. Aveyron	29	0	0	0
8. Gers	28	11	1	3
9. S. Garonne	31	7	0	3
10. Ariege	31	10	0	0

[a]Bone prevalence is expressed as a percentage, based on a "bone prevalence index" averaged for the many excavation sites and levels within each river valley region.
Source: Data from Rice and Paterson (1985).

moth art. It seems plausible that the percentage of cave art devoted to mammoths reflects how important mammoths were to the hunters of each region. If so, this is indirect evidence that the cave art is related to hunting success.

We can summarize such trends in the data by **fitting a line** to the scatter, like the one shown in Figure 14.4 (page 464). It is usually impossible to draw a straight line that connects all the points in a scatter plot, but we may be able to draw one that summarizes the general pattern. Such lines are called **regression lines.** The process of fitting such a line is called *the regression of Y on X.* In Figure 14.4, we see the *regression of mammoth art prevalence on mammoth bone prevalence.*

A regression line with upward trend like the one in Figure 14.4 is said to have a *positive slope,* which means that there is a *positive relationship* between the two variables.

> A **positive relationship** or a **positive slope** means that low values of the *Y* variable tend to occur with low values of the *X* variable, while high values of *Y* tend to occur with high values of *X*. A line with a positive slope rises to the right.

A positive relationship indicates that in regions where mammoth bones are prevalent, mammoth art tends to be prevalent too. In regions where mammoth bones are scarcer, mammoth art is scarcer as well.

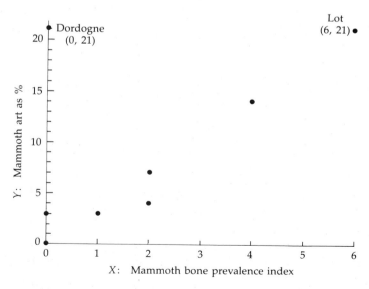

FIGURE 14.3   *Scatter plot of mammoth art vs. mammoth bones*

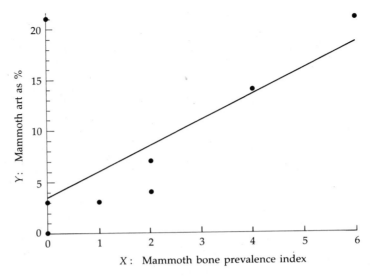

FIGURE 14.4   *Mammoth art vs. mammoth bones, with regression line*

We can be more specific in describing the regression line. Any straight line corresponds to an algebraic equation (see box).

Straight lines have equations of the form

$$\hat{Y} = a + b(X) \qquad \text{[14.1]}$$

where X and $\hat{Y}$ are variables; a and b are constants. The constant a is called the Y-intercept, and the constant b is called the slope.

The equation for the line in Figure 14.4 is

$$\hat{Y} = 3.5 + 2.53(X) \qquad \text{[14.2]}$$

In Equation [14.2] $a = 3.5$ and $b = 2.53$. X represents the mammoth bone prevalence, and $\hat{Y}$ ("Y-hat") represents the *predicted* percentage of cave art devoted to mammoths. The predicted percentage of mammoth art is not the same as the actual percentage for any given case. Rather, it represents the percentage we would obtain if we substituted a region's actual X value into Equation [14.2].

Figure 14.5 shows the graphical meaning of the Y-intercept and slope.

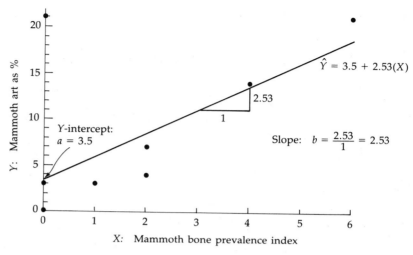

FIGURE 14.5   *Meaning of regression line Y-intercept and slope*

> The **Y-intercept** (a) is the predicted value of Y ($\hat{Y}$) when $X = 0$. Graphically, this is *the height at which the line crosses the Y-axis.*
>
> The **slope** (b) is the change in predicted Y ($\hat{Y}$) for each 1-unit increase in X. Graphically, this is the *steepness of the line.*

Since $a = 3.5$ in Equation [14.2], this line crosses the Y-axis at the point (0, 3.5). For a region with no mammoth bones ($X = 0$), we predict that about 3.5% of the cave paintings will be of mammoths. Since $b = 2.53$, for every additional percentage point on the mammoth bone prevalence index, the predicted percentage of mammoth paintings rises by 2.53.

One region stands apart from this pattern. In the Dordogne River region, no mammoth bones were found ($X = 0$), yet mammoth pictures made up 21% of the cave art. Other regions have X values as low as 0 and Y values as high as 21, so the Dordogne is not an outlier with respect to either variable alone. It *is* an outlier with respect to both variables at once, however. No other region has a similar *combination* of very low X value and very high Y value. The Dordogne is a **bivariate outlier.**

Figure 14.6 (page 466) shows another regression line, this one fit to the space shuttle data. Its equation is

$$\hat{Y} = 4.79 - .06(X) \tag{14.3}$$

where $\hat{Y}$ is the predicted number of booster joint damage incidents and X is joint temperature. The scale in Figure 14.6 is drawn all the way down to 0° F

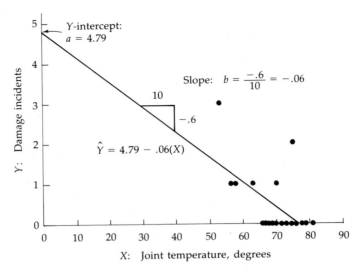

FIGURE 14.6   *Regression of joint damage incidents on joint temperature*

to show the graphical meaning of the intercept, $a = 4.79$. At 0° F, we predict 4.79 joint damage incidents—but this temperature is far outside the range of the actual data (53–81° F), so the prediction is very tentative. When X values of 0 do not occur in the data, the Y-intercept may be unrealistic: It is simply an algebraic property of any straight line. The slope in Figure 14.6 ($b = -.06$) tells us that the predicted number of damage incidents decreases by .06 with each one-degree increase in temperature.

The cave paintings example in Figure 14.5 showed a positive relationship, or a regression line with a positive slope ($b = 2.53$). Figure 14.6, in contrast, shows a *negative relationship*, a regression line with a *negative slope* ($b = -.06$).

> A **negative relationship** or a **negative slope** means that low values of Y tend to occur with high values of X, while high values of Y tend to occur with low values of X. A line with a negative slope descends to the right.

The negative relationship implies that warmer weather flights tended to have fewer damage incidents.

To draw any line, we need only two points. A regression equation can be used to find two points that lie on the regression line. For example, *Challenger* was launched at the lowest temperature of any flight in the shuttle program: 31° F. Substituting $X = 31$ into Equation [14.3] gives us

$$\hat{Y} = 4.79 - .06(X)$$
$$= 4.79 - .06(31) = 2.93$$

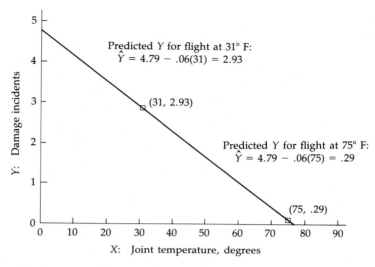

Predicted Y for flight at 31° F:
$\hat{Y} = 4.79 - .06(31) = 2.93$

(31, 2.93)

Predicted Y for flight at 75° F:
$\hat{Y} = 4.79 - .06(75) = .29$

(75, .29)

X: Joint temperature, degrees

FIGURE 14.7   *Drawing a line by connecting two (X, Y) points: (31, 2.93) and (75, .29)*

That is, based on the cold launch temperature we would have predicted 2.93 damage incidents on *Challenger*'s last flight.[1] Since it satisfies the regression equation, the point (31, 2.93) lies on the regression line, as shown in Figure 14.7. We next substitute a second X value, such as $X = 75$, to find a second point:

$$\hat{Y} = 4.79 - .06(75) = .29$$

so the point (75, .29) is also on the regression line. To draw this line, we need only connect the two points (31, 2.93) and (75, .29). Any two points that satisfy the equation will do, but for ease of drawing they should be well apart.

The regression line in Figure 14.5 provides a closer fit to the cave data than does the line for the space shuttle data in Figure 14.6. The following sections take up the problem of summarizing how well a regression line fits and the related problem of how to find or define regression lines in the first place.

PROBLEMS

3. Think up your own examples of pairs of measurement variables that have the following types of relationship. Explain your reasoning for each example.

   a. Positive relationship

   b. Negative relationship

4. In a sample of 143 college students, the regression equation for predicting weight in pounds (Y) from height in inches (X) was determined to be $\hat{Y} = -238 + 5.6(X)$.

   a. Explain what the slope (b) in this equation tells us.

   b. What is the literal meaning of the intercept (a) in this equation? Why should we not expect a meaningful intercept in an equation for predicting college students' weights from their heights?

5. Table 2.1 (page 25) contains data on nicotine concentrations in airplanes. The regression of nicotine concentration (Y, in micrograms per cubic meter of air) on number of passengers in the smoking section (X) is described by the regression equation $\hat{Y} = 9.07 + .99(X)$.

   a. Interpret the value of the slope.

   b. Interpret the value of the intercept.

   c. Construct a scatter plot of the data and draw in the regression line. Indicate the graphical meanings of a and b.

6. The regression equation for predicting reindeer art prevalence (Y) from reindeer bone prevalence (X) in Table 14.3 is $\hat{Y} = -10.8 + .62(X)$.

   a. Interpret the slope and the Y-intercept in English.

   b. Construct a scatter plot of reindeer art vs. reindeer bones. Find two widely spaced points on the regression line and use them to draw the line.

14.3   SUMMARY STATISTICS FOR TWO MEASUREMENT VARIABLES

Summary statistics for one measurement variable often make use of the *sum of squared deviations from the mean,* also known as the total sum of squares (TSS):

$$TSS_X = \Sigma(X_i - \overline{X})^2 \qquad [14.4]$$

where $\overline{X}$ is the mean of X. The subscript i denotes the ith case, and we are summing over all values of i (all cases) in the data. Similarly, for the variable Y the total sum of squares is

$$TSS_Y = \Sigma(Y_i - \overline{Y})^2 \qquad [14.5]$$

where $\overline{Y}$ is the mean of Y. The TSS, which measures *how much a variable varies,* is the basis for the variance and standard deviation (Chapter 4).

Table 14.3 shows squared deviations and their sums for the cave paintings data. For example, the mammoth bone prevalence value for the Northern Garonne River region is 4. The mean is 1.5, so Northern Garonne's squared X deviation is $(X_1 - \overline{X})^2 = (4 - 1.5)^2 = (2.5)^2 = 6.25$. Similarly, mammoth paintings make up 4% of the cave art in the Dronne River region; the mean is 7.3%, so the Dronne's squared Y deviation is $(Y_2 - \overline{Y})^2 = (4 - 7.3)^2 =$

TABLE 14.3 — *Squared Deviations and Cross Products for Mammoth Bone Prevalence (X) and Mammoths as a Percentage of All Cave Art (Y), from Table 14.2*

Case i	Bone X_i	Squared deviation $(X_i - \bar{X})^2$	Art Y_i	Squared deviation $(Y_i - \bar{Y})^2$	Cross product $(X_i - \bar{X})(Y_i - \bar{Y})$
1	4	6.25	14	44.89	16.75
2	2	.25	4	10.89	−1.65
3	0	2.25	0	53.29	10.95
4	2	.25	7	.09	−.15
5	0	2.25	21	187.69	−20.55
6	6	20.25	21	187.69	61.65
7	0	2.25	0	53.29	10.95
8	1	.25	3	18.49	2.15
9	0	2.25	3	18.49	6.45
10	0	2.25	0	53.29	10.95

Sums:

$\Sigma X_i = 15$ TSS$_X$: $\Sigma(X_i - \bar{X})^2 = 38.5$ $\Sigma Y_i = 73$ TSS$_Y$: $\Sigma(Y_i - \bar{Y})^2 = 628.1$ SCP: $\Sigma(X_i - \bar{X})(Y_i - \bar{Y}) = 97.5$

Means:

$\bar{X} = \Sigma X_i/n = 15/10 = 1.5$ $\bar{Y} = \Sigma Y_i/n = 73/10 = 7.3$

$(-3.3)^2 = 10.89$. The sums of the squared deviations for X (TSS$_X$) and for Y (TSS$_Y$) are given at the bottom of Table 14.3.

A related two-variable idea is the **cross product.** The cross product (CP) is the deviation of X_i from the mean of X, times the deviation of Y_i from the mean of Y:

$$CP_i = (X_i - \bar{X})(Y_i - \bar{Y}) \qquad [14.6]$$

For example, region 4 in Table 14.3 (Vezere River) has a mammoth bone prevalence of 2% and has 7% mammoth art. Its cross product is therefore

$$(X_4 - \bar{X})(Y_4 - \bar{Y}) = (2 - 1.5)(7 - 7.3) = (.5)(-.3) = -.15$$

Cross products have the following properties:

1. If for the ith case X_i is below the mean of X, and Y_i is below the mean of Y, then both deviations are negative and *the cross product is positive* (regions 3, 7, 8, 9, and 10 in Table 14.3). If both X_i and Y_i are above their respective means, then both deviations are positive and again *the cross product is positive* (regions 1 and 6).
2. On the other hand, if X_i and Y_i are on opposite sides of their respective means—high X_i and low Y_i (regions 2 and 4), or low X_i and high Y_i (region 5)—then *the cross product is negative.*

Figure 14.8 summarizes these properties of cross products in terms of the cave paintings scatter plot. Cases with positive cross products lie in either the lower left (low X, low Y) or the upper right (high X, high Y) quadrants of this plot. Cases with negative cross products lie in either the upper left (low X, high Y) or the lower right (high X, low Y) quadrants.

Cross products thus can be either positive or negative (or zero, if either X_i or Y_i equals its mean). To summarize the pattern of cross products for the entire sample, we add them up to form a *sum of cross products*.

The **sum of cross products** (SCP) is

$$\text{SCP} = \Sigma(X_i - \overline{X})(Y_i - \overline{Y}) \qquad [14.7]$$

The summation is over all cases in the data.

The sum of cross products will be positive if high X/high Y and low X/low Y cases predominate, as they do in the cave paintings data. The sum of cross products will be negative if high X/low Y and low X/high Y cases predominate. If neither type of case predominates, then positive cross products will be canceled out by negative cross products, and the sum of cross products will be near zero.

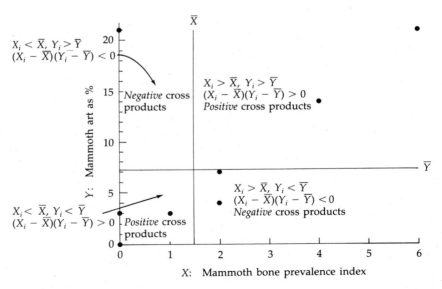

FIGURE 14.8   *Cross product signs relative to X and Y means*

> The **sign** of the sum of cross products (SCP) corresponds to the sign of the relationship—positive, negative, or zero (no relationship).

The sum of cross products is a basic building block of bivariate statistics, much like the total sum of squares in univariate statistics. These building blocks can be put together to define numerous other statistics, including the regression slope and the Y-intercept.

> The **slope** (b) for *the regression of Y on X*, also called the **regression coefficient** *on X*, is found from the sum of cross products divided by the total sum of squares of X:
>
> $$b = \frac{\text{SCP}}{\text{TSS}_x}$$
>
> $$= \frac{\Sigma(X_i - \overline{X})(Y_i - \overline{Y})}{\Sigma(X_i - \overline{X})^2}$$
>
> [14.8a]

> The **Y-intercept** (a), also called the **regression constant,** is found from the slope and the means of X and Y:
>
> $$a = \overline{Y} - b(\overline{X})$$
>
> [14.9]

The sum of cross products also serves to define another bivariate statistic, the *correlation coefficient.*

> The **correlation coefficient** (r) is found from the sum of cross products and the total sums of squares of X and Y:
>
> $$r = \frac{\text{SCP}}{\sqrt{\text{TSS}_x}\,\sqrt{\text{TSS}_Y}}$$
>
> $$= \frac{\Sigma(X_i - \overline{X})(Y_i - \overline{Y})}{\sqrt{\Sigma(X_i - \overline{X})^2}\,\sqrt{\Sigma(Y_i - \overline{Y})^2}}$$
>
> [14.10]
>
> A correlation coefficient measures *how well the regression line fits the data.*

Several kinds of statistics are called "correlations." The one defined in Equation [14.10], by far the most often used, is the **Pearson product–moment correlation.**

The correlation between X and Y is the same as the correlation between Y and X. It makes no difference in Equation [14.10] if we replace each X with a Y and each Y with an X; the results are still the same. By contrast, a similar replacement in Equations [14.8a] for slope, b, or [14.9], for intercept, a, would produce entirely different results.

The equations for slope, [14.8a], and correlation, [14.10], both have the sum of cross products, which may be positive, negative, or zero, in their numerators. The denominators of both equations involve sums of squares, which can only be positive. Thus the *signs* of the regression slope and the correlation coefficient depend only on the sign of the sum of cross products. If positive and negative cross products cancel each other out so the sum of cross products is zero, then the slope and correlation must be zero as well. A zero slope, correlation, or sum of cross products means that no linear upward or downward trend was detected in the data.

A regression line derived using Equations [14.8a] and [14.9] will always pass through the point $(\overline{X}, \overline{Y})$, which is the intersection of the $\overline{X}$ and $\overline{Y}$ lines in Figure 14.8. If it has a positive slope, the line will go through the lower left and the upper right quadrants in this figure, the same regions that contribute positive cross products. Likewise, a negative slope line will go through upper left and lower right, the same regions that contribute negative cross products.

For the cave paintings data of Figure 14.5 the regression line was earlier given, without explanation, as $\hat{Y} = 3.5 + 2.53(X)$. We now have the information that was used to obtain this equation. Substituting the SCP and TSS_X values from Table 14.3 into Equation [14.8a], we get

$$b = \frac{\text{SCP}}{\text{TSS}_X} = \frac{97.5}{38.5} = 2.53$$

This value, $b = 2.53$, is the slope used in Equation [14.2] and graphed in Figure 14.5.

The slope is then used with the values of $\overline{Y}$ and $\overline{X}$ (from Table 14.3) to obtain the Y-intercept, following Equation [14.9]:

$$a = \overline{Y} - b(\overline{X}) = 7.3 - 2.53(1.5) = 3.5$$

This value, $a = 3.5$, is the Y-intercept seen earlier in Equation [14.2] and Figure 14.5.

The third statistic defined above, the correlation coefficient, can also be found from the sums in Table 14.3:

$$r = \frac{\text{SCP}}{\sqrt{\text{TSS}_X}\ \sqrt{\text{TSS}_Y}} = \frac{97.5}{\sqrt{38.5}\ \sqrt{628.1}} = .627$$

The correlation coefficient tells us how well the regression line fits the scatter. For positive relationships, it has a minimum of 0 and a maximum of 1. A correlation of $r = .627$ indicates that the fit between line and scatter in Figure 14.5 is moderately good. We will look more closely at the properties of correlations later in this chapter.

PROBLEMS

7. The authors of the French cave art study later analyzed data from four regions in Spain (Rice and Paterson, 1986). Too few mammoth bones were found for statistical analysis, but prevalences for reindeer and ibex are shown.

Region	Reindeer bone	Reindeer art	Ibex bone	Ibex art
East	7.26%	3.61%	23.11%	8.43%
Central	3.07	.67	13.67	9.11
West	.31	2.47	24.73	11.11
Far West	1.01	0	25.25	11.54

a. Using Equation [14.8a], find the slope (b) for the regression of reindeer art prevalence (Y) on reindeer bone prevalence (X).
b. Using Equation [14.9], find the intercept (a) for this regression.
c. Using Equation [14.10], find the correlation coefficient (r).
d. Write out the regression equation, and explain what it says about the relationship between reindeer art and reindeer bone prevalence.

8. Repeat parts a–d in Problem 7, using the ibex art and ibex bone prevalence variables.

14.4 PREDICTED VALUES AND RESIDUALS

A regression equation like [14.2] or [14.3] is an equation for the **predicted values** of Y, not the actual values. Predicted values can be calculated for any value of X. Suppose we wish to estimate the percentage of mammoth art in a region where the bone prevalence value is 5%. To find predicted Y when $X = 5$, just substitute this number into the regression equation, Equation [14.2]:

$$\hat{Y} = 3.5 + 2.53(X)$$

If $X = 5$ then

$$\hat{Y} = 3.5 + 2.53(5) = 16.15$$

In a similar fashion we can find predicted Y, $\hat{Y}$, for any other value of X.

The Northern Garonne River region, the first case in Table 14.2, has a mammoth bone prevalence value of $X_1 = 4$. Based on this bone prevalence, the percentage of mammoths in cave art is predicted to be

$$\hat{Y}_1 = 3.5 + 2.53(X_1) = 3.5 + 2.53(4) = 13.62$$

In fact, Table 14.2 shows that the actual percentage of mammoth art in Northern Garonne is close to this prediction: $Y_1 = 14$. The region's *predicted* mammoth art percentage is $\hat{Y}_1 = 13.62$, but its *actual* mammoth art percentage is $Y_1 = 14$. The difference between actual and predicted values is called the *residual*.

> The **residual** for the *i*th case, e_i, is the actual minus the predicted value of Y:
>
> $$e_i = Y_i - \hat{Y}_i \qquad [14.11]$$
>
> The residual tells us *the amount of prediction error* for a given case.

The residual for Northern Garonne is $e_1 = Y_1 - \hat{Y}_1 = 14 - 13.62 = .38$. The fact that the residual is positive means that the prediction for this case was too low.

> A **positive residual** occurs when the actual value of Y is *higher than pre-dicted:* Y_i is greater than $\hat{Y}_i$.
>
> A **negative residual** occurs when the actual value of Y is *lower than pre-dicted:* Y_i is less than $\hat{Y}_i$.
>
> A **zero residual** occurs when there is no prediction error: Y_i equals $\hat{Y}_i$.

Table 14.4 shows predicted values ($\hat{Y}_i$) and residuals (e_i) for mammoth art in each of the ten French valley regions. For the Dordogne River region, based on the mammoth bone prevalence, we predict that 3.5% of the art will involve mammoths. In fact 21% does, giving Dordogne the largest residual in the table. A large residual indicates that the prediction is particularly poor for that case. For nine of the ten regions in Table 14.4, simply knowing the mammoth bone prevalence lets us predict the percentage of mammoth art within ±5 percentage points. Dordogne, however, has much more mammoth art than we would expect, given the absence of mammoth bones.

Graphically, predicted values *lie on the regression line,* and residuals are *the vertical distance between the regression line and the actual data points.* This idea is illustrated in Figure 14.9. In the Gers River region, for example, the residual is $e_8 = Y_8 - \hat{Y}_8 = 3 - 6.03 = -3.03$, which tells us that the actual Y is 3.03 units *lower* than the predicted Y.

Figure 14.10 (page 476) shows three kinds of vertical distance that are important in regression. First, there is the vertical distance of each actual Y value from the mean of Y: $Y_i - \overline{Y}$. When these distances are squared and summed, they form the total sum of squares, TSS_Y. The second graph in Figure 14.10 shows vertical distances between the predicted values of Y and the mean: $\hat{Y}_i - \overline{Y}$. When squared and summed, they form the *explained sum of squares,* or ESS.

TABLE 14.4   *Mammoth Art and Bones Data from Table 14.2*

Case	Region	Prevalence of bones	Actual art	Predicted art	Residual
i		X_i	Y_i	$\hat{Y}_i$	$e_i = Y_i - \hat{Y}_i$
1	N. Garonne	4	14%	13.62%	.38
2	Dronne	2	4	8.56	−4.56
3	Isle	0	0	3.50	−3.50
4	Vezere	2	7	8.56	−1.56
5	Dordogne	0	21	3.50	17.50
6	Lot	6	21	18.68	2.32
7	Aveyron	0	0	3.50	−3.50
8	Gers	1	3	6.03	−3.03
9	S. Garonne	0	3	3.50	−.50
10	Ariege	0	0	3.50	−3.50

Explained and residual sums of squares:

$$\text{ESS} = \Sigma(\hat{Y}_i - \overline{Y})^2 = 246.43$$

$$\text{RSS} = \Sigma e_i^2 = \Sigma(Y_i - \hat{Y}_i)^2 = 381.18$$

[a]Predicted values are from the regression equation [14.2]: $\hat{Y} = 3.5 + 2.53(X)$.

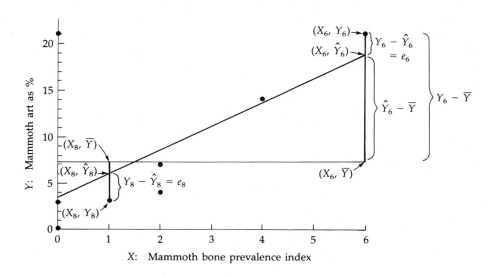

FIGURE 14.9   *Vertical distances between actual Y (Y_i), predicted Y ($\hat{Y}_i$), and mean Y ($\overline{Y}$)*

The **explained sum of squares** or **ESS** is

$$\text{ESS} = \Sigma(\hat{Y}_i - \overline{Y})^2 \qquad\qquad [14.12]$$

where $\hat{Y}_i$ is the predicted value of Y for case i, $\overline{Y}$ is the mean of Y, and the summation is over all cases in the data.

The ESS measures the total variation of the predicted values of Y. The explained sum of squares for the cave paintings example is shown at the bottom of Table 14.4.

The third graph in Figure 14.10 shows the residuals, or distances between actual and predicted values: $e_i = Y_i - \hat{Y}_i$. Squaring and summing the residuals produces the *residual sum of squares* or RSS.

The **residual sum of squares** or **RSS** is

$$\begin{aligned}\text{RSS} &= \Sigma(Y_i - \hat{Y}_i)^2 \qquad\qquad [14.13]\\ &= \Sigma e_i^2\end{aligned}$$

where Y_i is the actual value of Y for case i, $\hat{Y}_i$ is the predicted value for that case, and the summation is over all cases in the data.

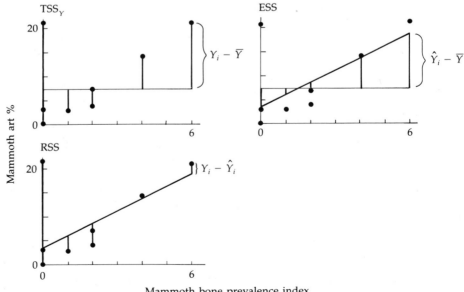

FIGURE 14.10   *Variation of Y (TSS_Y), predicted Y (ESS) and residuals (RSS)*

The RSS measures the variation of the residuals or prediction errors. This value is also given for the cave paintings example at the bottom of Table 14.4.

PROBLEMS

9. This regression equation describes the relationship between cumulative grade point average (Y) and average number of classes missed per month (X) in a sample of college students: $\hat{Y} = 3.04 - .064(X)$.

    a. What is the predicted GPA for a student who never misses class?

    b. What is the predicted GPA for a student who misses ten classes per month?

10. In the regression of reindeer art (Y) on reindeer bone prevalence (X) in Problem 7, we obtain the equation $\hat{Y} = .76 + .32(X)$.

    a. Use this equation to find the predicted values and residuals for reindeer art in each region in the Spanish data of Problem 7.

    b. Which region has the single largest residual (positive or negative)? Describe what the residual tells us about this region.

    c. Draw a scatter plot for reindeer art (Y) and bones (X) in the four Spanish regions. Draw the regression line on this scatter. Show the graphical meanings of predicted values and residuals in this plot.

11. Use the residuals and predicted values from Problem 10 to calculate the explained and residual sums of squares (ESS and RSS).

12. Find the residuals and predicted values for the regression of ibex art on ibex bone prevalence (Problem 8): $\hat{Y} = 6.66 + .16(X)$.

    Which region has the largest (positive or negative) residual? Explain what the predicted Y and the residual for this region tell us.

14.5 ASSESSING FIT IN REGRESSION

Residuals are prediction errors, so the residual sum of squares (RSS) measures how poorly the regression line fits the data. The better the fit, the lower the RSS. This leads to a criterion for selecting the "best" of the infinitely many lines we could draw on a scatter plot: The **best-fitting line** is the line that produces the lowest residual sum of squares. This is called the *least squares* criterion.

> The **least squares** line has a lower residual sum of squares, RSS = $\Sigma(Y_i - \hat{Y}_i)^2 = \Sigma e_i^2$, than any other line.

We use the sum of *squared* residuals partly because squaring gets rid of the negative values. Otherwise, positive residuals would be canceled out by negative residuals, since the residuals from any line passing through the point $(\overline{X}, \overline{Y})$ add up to zero. You can easily verify that $\Sigma e_i = 0$ (within rounding error) by checking the mammoths data in Table 14.4. The sum of unsquared residuals is therefore useless as a measure of fit.

The definitions of regression slope, Equation [14.8a], and intercept, Equation [14.9], meet the least squares criterion. That is, for any set of data a line based on these equations will have a lower residual sum of squares than any other straight line. No other values of *a* or *b* yield an RSS as low as the 381.18 seen in Table 14.4, for example. Because of this property, this kind of regression analysis is known technically as **ordinary least squares**, or OLS.

The explained sum of squares (ESS) can also be used to describe *how well* the best-fitting line fits. Recall that the ESS measures total variation of the predicted values around the mean of *Y*. The ratio of this predicted or explained variation to the total variation of *Y*, TSS$_Y$, provides a statistic called the *coefficient of determination* or R^2.

The **coefficient of determination** (R^2) is

$$R^2 = \frac{\text{Explained variation of } Y}{\text{Total variation of } Y}$$

$$= \frac{\text{ESS}}{\text{TSS}_Y} \qquad\qquad [14.14]$$

$$= \frac{\Sigma(\hat{Y}_i - \overline{Y})^2}{\Sigma(Y_i - \overline{Y})^2}$$

R^2 is interpreted as *the proportion of the total variation of Y that is explained by X.*

We can also interpret R^2 as the proportion of *variance* explained, referring to the sample statistic $s_Y^2 = \text{TSS}_Y/(n - 1)$. The coefficient of determination varies between 0 (none of *Y*'s variation or variance is explained) and 1 (all of *Y*'s variation or variance is explained).

The coefficient of determination for the mammoth paintings data can be calculated using the ESS from Table 14.4 and TSS$_Y$ from Table 14.3:

$$R^2 = \frac{\text{ESS}}{\text{TSS}_Y} = \frac{246.43}{628.1} = .39$$

This tells us that 39% of the total variation in mammoth art is explained by mammoth bone prevalence.[2]

If 39% of the variance is explained, clearly the remaining 61% is unexplained. The proportion of variance not explained can be found either as $1 - R^2$ or as the ratio of residual to total *Y* variation:

$$1 - R^2 = \frac{RSS}{TSS_Y} \qquad\qquad [14.15]$$

Calculated either way, this value is .61 for the cave paintings example. So 61% of the variation in mammoth art must be explained by things other than bone prevalence. Such things may include:

Omitted variables   Some of the remaining variation might reflect the effects of other, unanalyzed variables, such as local beliefs or artists' preferences.

Measurement error   We surely have not found every cave painting done, so we cannot know the true percentage of mammoth art in each region. Measurement error is really a special kind of omitted variable.

Sampling variability   Variation around the regression line in our sample may be greater (or less) than it is in the population.

Misspecification   The true mammoth art–bones relationship may not be a straight line, as assumed for this regression.

Regression almost always leaves some variation unexplained, for one or more of these reasons.

The coefficient of determination summarizes how well a regression line fits the data. If R^2 is a low value, such as .1 or .2, we know that the data are scattered widely around the line. On the other hand, a high R^2 like .8 or .9 confirms a close fit between line and data. The next section provides further examples of how we can interpret R^2 and its relative, the correlation coefficient (r).

PROBLEMS

13. Below are sums of squares for several regressions. Use them to find the coefficients of determination (R^2), and interpret these coefficients in English.

   a. Booster rocket joint temperature (X) and number of joint damage incidents (Y) on 23 space shuttle flights (Figure 14.6): ESS = 4.302, TSS_Y = 13.652

   b. Heights (X) and weights (Y) of 143 college students (Problem 4): ESS = 50,026, TSS_Y = 92,104

   c. Passengers in smoking section (X) and nicotine concentration (Y) in airliner cabins (Problem 5): ESS = 865.5, TSS_Y = 17,867.6

14. For the basketball player sample of Problem 2, the following summary statistics were obtained concerning height in inches (X) and number of rebounds (Y): $\overline{X}$ = 80.875, $\overline{Y}$ = 450.75, TSS_X = 152.875, TSS_Y = 645,209.5, SCP = 8,366.75, ESS = 457,907, RSS = 187,327. Find and interpret the following values.

   a. The slope, b

   b. The Y-intercept, a

   c. The coefficient of determination, R^2

   d. The proportion of variance *not* explained

15. Write out the regression equation for predicting basketball rebounds from height (Problem 14). For each player calculate the predicted number of rebounds and the residual. Which player has the largest residual? What does this tell us about him?

16. (*Class project*) Your work in Problems 14 and 15 gave you the equation for the least squares regression line. In theory, this line should have a lower residual sum of squares (RSS) than any other line. That is, we should be unable to find any values of a and b that, applied to the basketball data, yield an RSS less than 187,327. This property can be confirmed mathematically by the use of calculus, or experimentally as in the following class project.

    Each student or work group should be assigned a possible slope value b, such as integers extending above and below the least squares b value obtained in Problem 14b (e.g., $b = \ldots 51, 52, 53, 54, 55, 56, 57, \ldots$). For each assigned value of b, calculate a value of a as usual following Equation [14.9]. Using Equation [14.9] is an optional restriction, but it keeps things simpler by ensuring that your line, like the least squares line, will go through the point $(\overline{X}, \overline{Y})$.

    Use your own a and b values to find "predicted values" and "residuals" for each of the eight cases in the table for Problem 2. Then calculate the "residual sum of squares" from your equation. Pool your results with those of other students to make up a graph, putting the residual sum of squares on the vertical axis and your chosen value of b on the horizontal axis. What does this graph show?

14.6   CORRELATION COEFFICIENTS

The correlation coefficient, Equation [14.10], is another way to summarize how well the regression line fits the data. Correlations range from -1 to $+1$. The closer the correlation is to -1 or $+1$, the more nearly perfect is the fit between line and data. Positive correlations occur with positive slopes and positive relationships; negative correlations go with negative slopes and negative relationships. A correlation of 0 means that there is no linear relationship.

    Table 14.5 gives guidelines for interpreting correlation coefficients. We earlier found that the correlation between mammoth bones and mammoth art is $r = .627$, which could thus be described as a moderate positive relationship. These guidelines are just rules of thumb and do not apply to every field.

TABLE 14.5

Rules of Thumb for Interpreting the Bivariate Correlation Coefficient and the Coefficient of Determination

Correlation coefficient	Interpretation: Linear relationships	Coefficient of determination[a]
r		R^2
1.0	Perfect positive relationship	1.00
.8	Strong positive relationship	.64
.5	Moderate positive relationship	.25
.2	Weak positive relationship	.04
.0	No relationship	.00
−.2	Weak negative relationship	.04
−.5	Moderate negative relationship	.25
−.8	Strong negative relationship	.64
−1.0	Perfect negative relationship	1.00

[a]In two-variable analysis the coefficient of determination is simply the square of the correlation coefficient: $R^2 = r^2$. The coefficient of determination is interpreted as the proportion of the variation in Y that is explained by X.

Some social scientists call correlations of $\pm.3$ strong, while physical scientists may view anything less than $\pm.7$ as weak.

Figure 14.11 shows examples of scatter plots for correlations of different strengths. At upper left is a perfect positive relationship, $r = +1.0$, where all points lie on a regression line of positive slope. In a perfect negative relationship ($r = -1.0$, lower right) all points lie on a line with a negative slope. The closer the correlation is to 0, the more scattered the points are. At $r = 0$, the regression line has a horizontal slope ($b = 0$).

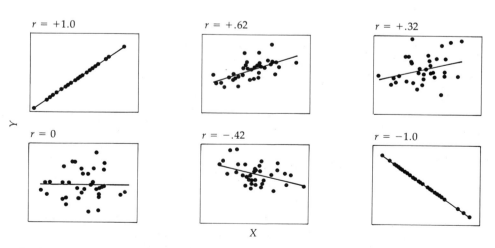

FIGURE 14.11   Scatter plot examples of correlations

In two-variable analysis, a simple relationship links the correlation coefficient and the coefficient of determination (see box).

For *bivariate regression only,* the coefficient of determination, R^2, equals the square of the correlation coefficient, r:

$$R^2 = r^2$$ [14.16]

With the mammoths example, R^2 can thus be found either from Equation [14.14],

$$R^2 = \frac{ESS}{TSS_Y} = \frac{246.43}{628.1} = .39$$

or from Equation [14.16],

$$R^2 = r^2 = .627^2 = .39$$

This property of correlation coefficients adds another dimension to their interpretation, shown at the right of Table 14.5: By squaring a correlation, we find what proportion of the variation in Y is explained by X.

PROBLEMS

17. For each of these correlation coefficients, interpret the correlation itself; then calculate and interpret the corresponding coefficient of determination.

   a. VSAT and GPA, Table 1.1 (page 5): $r = .17$

   b. Weight and GPA, Table 1.1: $r = -.10$

   c. Serum chlorpropamide level and serum insulin level, Chapter 11, Problem 28 (page 350): $r = .45$

   d. Number of safety-related failures during nuclear power plant standby and operation, Chapter 11, Problem 22 (page 349): $r = .68$

   e. Attention/concentration ability and spatial ability, Table 13.14 (page 433): $r = -.34$

18. Use the summary statistics for basketball players' height and number of rebounds in Problem 14 to find the correlation coefficient for these data. Show how the coefficient of determination can be found from the correlation coefficient and from the explained and total sums of squares.

19. Without calculation, what can you guess about the sign of the correlation between the percentage of gross sales returned to state, and the percentage paid in prizes, for the ten state lotteries in Chapter 3, Problem 27 (page 88)? Explain your reasoning.

14.7 PHYSICIAN PROBLEMS AND HOSPITAL SIZE

Table 14.6 contains a small data set in which the cases are 14 Massachusetts hospitals. There are three measurement variables: number of patient admissions, number of beds, and number of physicians who had been sued, disciplined, or charged with serious violations. A scatter plot of two of these variables, number of physicians accused (Y) and number of patient admissions (X), is shown in Figure 14.12 (page 484). Admissions values are missing for four of the 14 hospitals listed in Table 14.6, so we will focus on the remaining ten. Each of the ten hospitals is represented by a dot in Figure 14.12.

Figure 14.12 shows a strong linear pattern. Even before calculating any statistics, we can make some guesses:

1. The relationship between number of physicians accused and number of patient admissions is positive; both correlation (r) and slope (b) will be positive.
2. The regression line will intersect the Y-axis somewhere between $Y = 0$ and $Y = 10$; that is, the Y-intercept (a) will be between 0 and 10.
3. The regression line should be able to fit these data closely; we might expect the correlation, r, to be .8 or higher.

A more formal analysis is begun in Table 14.7, where the total sums of squares and sum of cross products for these ten hospitals are calculated. Based on these values, the regression summary statistics are found, as shown

TABLE 14.6　　　*Data on 14 Massachusetts Hospitals, 1984*

Hospital	Patient admissions	Beds	Doctors accused[a]
1. Ludlow	2,041	—	10
2. Brookline	2,126	98	20
3. Amesbury	2,364	63	14
4. N. Shore Children's	2,398	50	11
5. Southwood	2,425	—	12
6. J. B. Thomas	2,857	99	16
7. Fairlawn	—	104	15
8. Sancta Maria	3,916	—	18
9. Anna Jacques	—	156	22
10. Milton	—	161	24
11. Cambridge	5,527	182	27
12. Choate Symmes	8,560	—	43
13. Mercy	—	311	49
14. Newton Wellesley	12,129	351	64

[a]Number sued, disciplined, or charged with violations, 1983–1986.
Source:　Data from the *Boston Globe*, June 19, 1986.

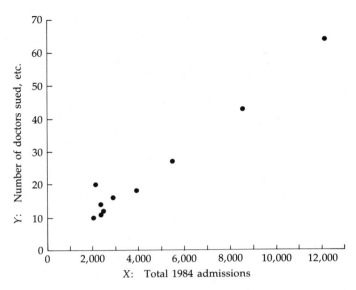

FIGURE 14.12   *Number of physicians accused vs. number of patient admissions at ten hospitals*

TABLE 14.7   ***Hospitals Data from Table 14.6 with Squared Deviations and Cross Products***

Case	Patient admissions	Squared deviation	Doctors accused	Squared deviation	Cross product
i	X_i	$(X_i - \overline{X})^2$	Y_i	$(Y_i - \overline{Y})^2$	$(X_i - \overline{X})(Y_i - \overline{Y})$
1	2,041	5,727,885	10	182.25	32,309.55
2	2,126	5,328,249	20	12.25	8,079.05
3	2,364	4,286,142	14	90.25	19,667.85
4	2,398	4,146,518	11	156.25	25,453.75
5	2,425	4,037,287	12	132.25	23,106.95
6	2,857	2,487,875	16	56.25	11,829.75
8	3,916	268,645	18	30.25	2,850.65
11	5,527	1,193,993	27	12.25	3,824.45
12	8,560	17,021,400	43	380.25	80,451.15
14	12,129	59,208,408	64	1,640.25	311,635.35

Sums:

		TSS$_X$: $\Sigma(X_i - \overline{X})^2 =$	$\Sigma Y_i =$	TSS$_Y$: $\Sigma(Y_i - \overline{Y})^2 =$	SCP: $\Sigma(X_i - \overline{X})(Y_i - \overline{Y}) =$
	$\Sigma X_i = 44,343$	103,706,392	235	2,692.5	519,208.5

Means:

$\overline{X} = \Sigma X_i/n$ $\overline{Y} = \Sigma Y_i/n$

$\quad = 44,343/10$ $\quad = 235/10$

$\quad = 4,434.3$ $\quad = 23.5$

TABLE 14.8   *Regression of Number of Doctors Accused on Number of Patient Admissions*

Regression equation:

Slope: $b = \dfrac{SCP}{TSS_x}$ Equation [14.8a]

$$= \frac{519,208.5}{103,706,392} = .005$$

Intercept: $a = \overline{Y} - b\overline{X}$ Equation [14.9]

$$= 23.5 - .005(4,434.3) = 1.3$$

Regression equation: $\hat{Y} = a + bX$ Equation [14.1]

$$= 1.3 + .005(X)$$

Interpretation: (Graphed in Figure 14.13.) For each one-patient increase in admissions, the average number of doctors accused increases by .005, or by five doctors for every 1,000 patient admissions. A hospital with no patient admissions ($X = 0$) is predicted to have 1.3 doctors accused, but this X value is far outside the range of the data, and the prediction for $X = 0$ makes no real-world sense.

Correlation:

Correlation coefficient: $r = \dfrac{SCP}{\sqrt{TSS_x}\,\sqrt{TSS_Y}}$ Equation [14.10]

$$= \frac{519,208.5}{\sqrt{103,706,392}\,\sqrt{2,692.5}} = .98$$

Coefficient of determination: $R^2 = r^2$ Equation [14.16]

$$= .98^2 = .96$$

Interpretation: There is a near perfect positive relationship between doctors accused and patient admissions. The number of patient admissions explains 96% of the total variation in the number of doctors accused.

in Table 14.8. The resulting regression line, graphed in Figure 14.13 (page 486), shows the strong positive relationship ($r = .98$) we expected.

     A strong positive relationship means that the number of physician problems is closely tied to the number of admissions. The busier hospitals are the ones where more physician problems occur. This finding is unsurprising because we would expect counts of almost any hospital activity—not only counts of doctors in trouble, but also counts of births, deaths, lives saved, sheets used, or whatever—to be higher in the larger, busier hospitals.

     For someone in hospital administration, the most interesting aspect of this analysis probably is not the regression equation itself, but the residuals, given in Table 14.9. They show which hospitals have substantially more or

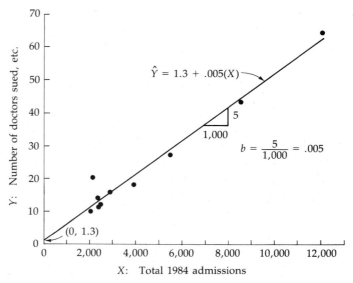

FIGURE 14.13   *Regression of physicians accused on number of patient admissions*

TABLE 14.9   *Hospitals Regression[a] from Tables 14.7–14.8*

Case	Hospital	Patient admissions	Accused: Actual	Accused: Predicted	Residual
i		X_i	Y_i	$\hat{Y}_i$	$e_i = Y_i - \hat{Y}_i$
1	Ludlow	2,041	10	11.5	−1.5
2	Brookline	2,126	20	11.9	8.1
3	Amesbury	2,364	14	13.1	.9
4	N. Shore Children's	2,398	11	13.3	−2.3
5	Southwood	2,425	12	13.4	−1.4
6	J.B. Thomas	2,857	16	15.6	.4
8	Sancta Maria	3,916	18	20.9	−2.9
11	Cambridge	5,527	27	28.9	−1.9
12	Choate Symmes	8,560	43	44.1	−1.1
14	Newton Wellesley	12,129	64	61.9	2.1

[a]Predicted values and residuals in this table are based on the regression equation $\hat{Y} = 1.3 + .005(X)$, with the values of a and b rounded off. For certain statistical calculations, more precise values of a and b are preferable.

With these data, rounding off introduces enough inaccuracy to obscure some of the theoretical properties of predicted values and residuals. For example, the sum of residuals is not 0, and r^2 does not equal ESS/TSS_Y, unless a more precise version of the regression equation is used: $\hat{Y} = 1.299677 + .0050065(X)$.

fewer physician problems than would be expected from size alone. Ludlow's negative residual, $e_1 = -1.5$, says Ludlow Hospital had *1.5 fewer* physician problems than we would expect based on admissions. Brookline's positive residual, $e_2 = 8.1$, means Brookline had *8.1 more* physician problems than predicted. This residual is much larger than any other in the table, implying that other factors besides size affect Brookline. A more detailed study of this hospital might help identify those factors.

PROBLEMS

20. Carry out the regression of number of patient admissions (as the Y variable) on number of hospital beds (as X), using the data of Table 14.6. (Because of missing values only six cases are in this analysis.) Write out and interpret the following.

    a. The regression equation
    b. The coefficient of determination

Fourteen hospitals are in Table 14.6, but missing values reduce the sample available for any two-variable analysis. Such difficulties are common in research, and they worsen as more variables are considered. Your regression analysis in Problem 20 offers a solution, however. Among the six hospitals with complete data, bed counts and the number of patient admissions are so closely related that we can predict one almost perfectly from the other. On the reasonable assumption that a similar relationship holds true for the other hospitals, we can use a technique called **missing value replacement.**

21. Use your regression equation from Problem 20 to calculate the *predicted* number of patient admissions for each of the four hospitals in Table 14.6 where the actual number admitted is unknown. Assemble a new set of data on admissions and number of physicians accused, in which the original data are unchanged except that the four missing admissions values have been replaced by the four predicted values. These new data should contain 14 cases with no missing values.

22. Based on the ten cases with known patient admissions values, we earlier (Table 14.8) found the regression equation for predicting physician problems (Y) from patient admissions (X):

    $$\hat{Y} = 1.3 + .005(X) \qquad R^2 = .96$$

    Repeat this analysis using your new $n = 14$ data set from Problem 21. Has replacing the missing values caused much change in the regression equation or R^2?

23. The largest negative residual in the $n = 14$ regression of Problem 22 belongs to one of the cases brought in by missing value replacement. Identify this hospital and explain what the residual tells us.

14.8 PREDICTING STATE SAT SCORES

Table 14.10 contains data from the 21 U.S. states in which the Scholastic Aptitude Test (SAT) is the predominant college admissions test. (In the remaining 29 states, other standardized tests are used more often than the SAT.) The variables are median education level of each state's adult population, percentage of the population living below the official poverty line, mean salary of public school teachers, percentage of high school seniors who take the SAT, and mean composite (math plus verbal) SAT score. Composite SAT scores have a theoretical range of 400–1,600 points.

When state-by-state average SAT scores are published each year, many people check how their own state compares. If the state average is high, local

TABLE 14.10   ***Data on U.S. States Favoring SAT for College Admissions[a]***

State	Median education in years	Pop. below poverty	Mean teacher salary	Pupils taking SAT	Mean SAT
1. California	12.7	11.3%	$23,614	31.6%	899
2. Connecticut	12.6	8.7	21,036	70.7	896
3. Delaware	12.5	11.9	20,625	57.2	897
4. Florida	12.5	13.0	18,275	26.8	889
5. Georgia	12.2	16.4	13,040	55.0	823
6. Hawaii	12.7	10.0	24,319	52.9	857
7. Indiana	12.4	9.8	20,347	50.4	860
8. Maine	12.5	12.9	16,248	44.2	890
9. Maryland	12.5	9.9	22,800	51.4	889
10. Massachusetts	12.6	9.8	21,841	68.2	888
11. New Hampshire	12.6	8.7	16,549	52.1	925
12. New Jersey	12.5	9.7	21,536	71.3	869
13. New York	12.5	13.7	25,000	60.2	896
14. N. Carolina	12.2	14.6	17,585	47.4	827
15. Oregon	12.7	11.3	21,746	39.8	908
16. Pennsylvania	12.4	10.5	21,178	55.2	885
17. Rhode Island	12.3	10.3	23,175	59.6	887
18. S. Carolina	12.1	15.9	16,523	50.0	790
19. Texas	12.4	14.8	19,550	32.2	868
20. Vermont	12.6	11.4	16,299	50.5	904
21. Virginia	12.4	11.5	18,535	54.3	888

[a]Education and poverty values are for 1980; teacher salaries, 1983; SAT variables, 1982.

politicians and educators often take pride and credit. If it is low, editorials ask "What's wrong with our schools?" and we hear excuses or blame. Both reactions may be premature: To compare states fairly on their SAT scores is trickier than it first appears (see Wainer, 1986).

One problem is that students who take the SAT are a self-selected sample, rather than a randomly selected sample from which to infer about a state's student population. In Table 14.10 participation rates range from 26.8% (Florida) to 71.3% (New Jersey) of the high school seniors. If the SAT is taken by only the best students in Florida, but by most students in New Jersey, it is hardly surprising that the average Floridian score is higher.

Figure 14.14 shows the regression of average SAT scores on the percentage of students taking the test. The regression line, $\hat{Y} = 879 - .024(X)$, indicates that average SAT declines only slightly as the percentage tested increases. Every increase of one percentage point in students tested leads to an average decline of only .024 SAT point; even going from 0% to 100% would drop the predicted SAT by merely 2.4 points! This is a trivial effect. An X value of 0 lies well beyond the data, so we should attach no substantive meaning to the Y-intercept, $a = 879$.

The dots are scattered randomly around the nearly horizontal line in Figure 14.14, so the correlation is near 0: $r = -.01$. Squaring this correlation we find the coefficient of determination, $R^2 = .0001$. Apparently, the percentage of seniors taking the SAT explains almost none of the variation in mean scores.

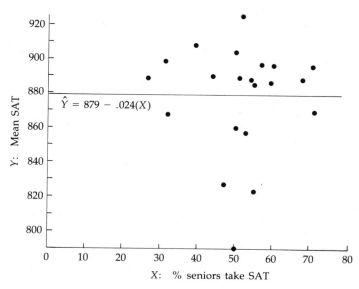

FIGURE 14.14   *Regression of mean SAT score on percentage of seniors tested for 21 states*

Some of the other variables in Table 14.10 are better predictors of mean SAT scores. Figure 14.15 shows the regression of SAT scores on the states' median levels of education.[3] Unlike Figure 14.14, Figure 14.15 shows a scatter with a relatively strong, positive relationship: $r = .77$. The regression line is $\hat{Y} = -946 + 146(X)$, where Y is mean SAT and X is median adult education. For every 1-year increase in median education, a state's predicted mean SAT score increases by 146 points. The Y-intercept, $a = -946$, makes no substantive sense: It is the predicted mean SAT score for a state where the adult population's median education is 0. Since median education is above 12 years in all 21 of these states, we should not be surprised by an unreasonable prediction for $X = 0$.

Note that the three states with the lowest median educations all lie below the regression line in Figure 14.15. It looks as if a curve, dropping towards the lower left, might fit these data better than a straight line. Curvilinear regression is discussed in Chapter 15.

Unlike most scatter plots seen so far, Figure 14.15's X-axis does not range all the way to 0. Consequently the Y-intercept cannot be shown. To extend the education scale from 0 to 13, when all the actual X values are between 12.1 and 12.7, would make the graph nearly unreadable. For the same reason, the Y-axis scale covers only 790–930 SAT points.

If mean SAT scores are related to adult education levels, then educators and politicians should be cautious about taking credit for high scores, or blame for low ones. Perhaps mean test scores largely reflect the demographic characteristics of a state's population. This possibility shifts the question from

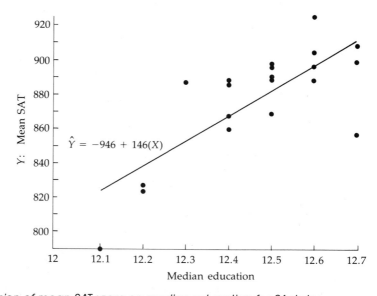

FIGURE 14.15   *Regression of mean SAT score on median education for 21 states*

"How do our students' scores compare with those in other states?" to "How do our students' scores compare with those we would expect, given the demography of our state?" The latter question is less simple and less satisfying, but potentially more useful. Individual states' predicted values and residuals can help to address it.

Median education explains about 59% of the variation in mean state SAT scores: $R^2 = .59$. Although the overall fit of the regression line is good, several states are substantially over- or under-predicted. Rhode Island and New Hampshire have the two largest *positive* residuals: Actual SAT scores exceed those predicted from their education levels. In Rhode Island, the median education is 12.3 years and the mean SAT score is 887. From the regression equation, we would predict Rhode Island's mean SAT to be

$$\hat{Y}_{17} = -946 + 146(X_{17}) = -946 + 146(12.3) = 850$$

Thus Rhode Island's actual mean SAT is 37 points higher ($e_{17} = Y_{17} - \hat{Y}_{17} = 887 - 850 = 37$) than we would predict, based on its median education.

Both South Carolina and Hawaii, on the other hand, have substantial *negative* residuals. South Carolina has the lowest median education in this sample, $X_{18} = 12.1$, so we would predict it to have the lowest mean SAT as well:

$$\hat{Y}_{18} = -946 + 146(X_{18}) = -946 + 146(12.1) = 821$$

The actual mean SAT, $Y_{18} = 790$, is even lower, giving South Carolina a residual of $e_{18} = Y_{18} - \hat{Y}_{18} = 790 - 821 = -31$.

The residuals show us the variation in mean state SAT scores, after the effect of education levels has been subtracted. Other influences besides general education levels must be examined to account for the unexpectedly high or low scores in such states as Rhode Island, New Hampshire, Hawaii, and South Carolina.

PROBLEMS

Statistics on median education (X) and average teacher salary (Y) for the 21-state sample of Table 14.10 are shown.

X	Y
$\overline{X} = 12.47$	$\overline{Y} = 19{,}991.5$
$\text{TSS}_X = .5629$	$\text{TSS}_Y = 197{,}774{,}971.6$
$\text{SCP} = 5{,}110.49$	

24. Use these statistics to find the regression equation for predicting average teacher salary from a state's median education level.

25. Find the correlation coefficient and the coefficient of determination for median education and average teacher salary. Describe the relationship between these two variables.

26. In what state is the average teacher salary the farthest below what we would predict, based on median education? In what state is the average teacher salary the farthest above what we would predict?

14.9   *OUTLIERS AND INFLUENCE IN REGRESSION ANALYSIS*

We earlier saw how outliers can influence the mean. They affect even more strongly statistics based on sums of squares. Recall that the "best" straight line is chosen by the least squares criterion, and that the definitions of slope, intercept, correlation, and coefficient of determination all involve sums of squares. Consequently, like the mean and the standard deviation, ordinary least squares regression is not resistant to outliers.

The mammoths data seen earlier contain a bivariate outlier, the Dordogne River region, where there are many mammoth paintings but no mammoth bones. Based on the absence of bones, we would predict mammoths in only 3.5% of the paintings; in fact they comprise 21%, leaving a large positive residual (Table 14.4).

Squaring a large number enlarges it vastly, which is why statistics based on squared deviations have such low resistance to outliers: Outlier influence magnifies. In Table 14.4 the sum of squared residuals for all ten regions is shown to be 381.18. The squared residual from the Dordogne region alone is $17.50^2 = 306.25$. More than 80% of the RSS for the entire sample thus comes from a single case! Since least squares regression seeks a line for which the residual sum of squares (RSS) will be a minimum, the regression line is *drawn toward* the outlying case.

To see this, we can redo the regression leaving out the Dordogne region (outlier deletion, Chapter 6). The resulting nine-case regression is compared with the original ten-case line in Figure 14.16. With Dordogne deleted, the regression equation becomes

$$\hat{Y} = .21 + 3.34(X) \tag{14.17}$$

The slope of Equation [14.17] exceeds the slope in Equation [14.2] ($3.34 > 2.53$), so deleting the outlier makes the best-fitting line steeper. The intercept in Equation [14.17] is lower than the intercept in Equation [14.2] ($.21 < 3.5$), so with the outlier deleted the best-fitting line intersects the Y-axis at a lower point. The outlying location of Dordogne pulls up the lower end of the ten-case regression line, producing a higher intercept and a shallower slope.

This illustrates how a single case can influence the regression equation. Measures of fit are affected as well. The coefficient of determination for the ten-case analysis is $R^2 = .39$, meaning that 39% of the variance of paintings is explained by bone prevalence. In the nine-case analysis, R^2 increases dramatically, from .39 to .96. Likewise the correlation coefficient goes from moderate, $r = .627$, to nearly perfect, $r = .979$. These statistics reflect the fact that the second regression line fits the nine-case scatter much better than the first regression line fits the ten-case scatter. The outlier's influence pulled the original regression line away from the trend followed by the rest of the data.

   An improved fit is not in itself sufficient reason to accept the second line and reject the first. *We can always achieve a better fit by throwing away data that disagree with our theories.* The way in which Dordogne differs from the rest suggests something may be unusual about that valley. Perhaps its people cached all their mammoth bones somewhere not yet found. If so, then it is an outlier due to measurement error, and keeping Dordogne in the analysis can only obscure the true art–bones relationship. On the other hand, the measurements may indeed be accurate. If we collected more data, we might find other legitimate instances where the prevalence of mammoths in cave art looms in disproportion to their actual remains in bone piles. Mammoths were, after all, impressive animals and a formidable challenge to stone-age hunters—doubtless more easily painted than caught.[4]

   The point is that the statistical evidence alone is not enough to justify deleting Dordogne from the analysis. Other information would be needed to support this decision. In the absence of such information, it is a good idea to report the analysis *both ways:* give the results with and without this one influential case.

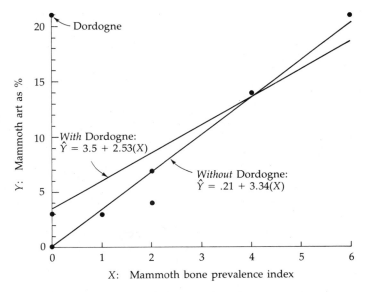

FIGURE 14.16   *Regression of mammoth art on mammoth bones with and without Dordogne*

More generally, outliers pose essentially the same problems in regression as in univariate analysis. An influential case may distort the regression results, just as a severe outlier can distort the mean or standard deviation. When this happens, the analyst faces a decision about what should be done, following the same considerations discussed in Chapter 6. Scatter plots have many uses in regression, but one of the most important is to help detect and deal with outliers and influential observations.

PROBLEMS

It seems unlikely that a strong causal relationship exists between alcohol sales and the marriage rate among U.S. states. Nonetheless, a regression of marriage rate (Y) on per capita alcohol consumption (X) for the 13 western states in the accompanying table yields the equation $\hat{Y} = -69 + 27.4(X)$. The R^2 for this equation is .81.

**Yearly Alcohol Sales and Marriage Rates
in 13 Western U.S. States**

State	Alcohol sales, gal/person	Marriage rate/ 1,000 pop.
1. Alaska	3.96	13.34
2. Arizona	3.15	11.12
3. California	3.38	8.91
4. Colorado	3.23	12.08
5. Hawaii	3.26	12.29
6. Idaho	2.55	14.23
7. Montana	3.09	10.60
8. Nevada	6.88	142.83
9. New Mexico	2.90	12.77
10. Oregon	2.79	8.74
11. Utah	1.69	11.61
12. Washington	2.97	11.55
13. Wyoming	3.37	14.63

27. Describe what these regression results seem to imply about the relationship between per capital alcohol sales and marriage rates.

28. Construct a scatter plot for marriage rate (Y) vs. alcohol sales (X) in the 13 western states to check for possible problems with the regression results in Problem 27. Draw in the regression line.

    a. What do you see?

    b. What should be done about this problem?

    c. If corrective action is taken, what do you expect will happen to the regression line?

14.10 NOTES ON CALCULATION*

The equation given for the regression slope, [14.8a], is the **definitional formula** for this statistic. If you are calculating by hand, a *computational formula* is available that achieves the same numerical result with less effort.

> The **computational formula** for the regression **slope** is
>
> $$b = \frac{n\Sigma XY - (\Sigma X)(\Sigma Y)}{n\Sigma X^2 - (\Sigma X)^2}$$
>
> [14.8b]

It calculates the slope directly from the sum of XY products (ΣXY), the sum of X values (ΣX), the sum of Y values (ΣY), and the sum of X^2 values (ΣX^2).

Table 14.11 contains a small data set we can use for illustration. The variables are the year and two measures of global atmospheric pollution: the carbon dioxide concentration measured at the South Pole and the cumulative global release of chlorofluorocarbons. Carbon dioxide, which is produced by the burning of wood, oil, gas, coal, and other fuels, is linked to a worldwide "greenhouse effect" that may change climates and sea levels over coming decades. Chlorofluorocarbons are widely used in aerosol sprays, refrigerators, and certain plastics. They are suspected of causing a decrease in the earth's protective ozone layer.

Table 14.12 (page 496) shows the computational formula calculations for regressing carbon dioxide concentration (Y) on year (X). We need columns for X^2 and for X times Y, in addition to columns for X and Y themselves. The sums of these four columns provide the quantities we need to find the slope with Equation [14.8b]. The regression equation for predicting CO_2 concentration from year is

TABLE 14.11   *Two Measures of Air Pollution, 1971–1985*

Year	Air CO_2 concentration, parts/million	Global release of chlorofluorocarbon, millions of kg
1971	325	4,615.1
1973	327	5,900.5
1975	329	7,355.7
1977	332	8,737.9
1979	335	9,963.8
1981	338	11,136.1
1983	340	12,309.1
1985	343	13,588.9

Source: Council on Environmental Quality (1987).

TABLE 14.12   *Obtaining the Regression Slope by the Computational Formula*

Year X	X^2	CO_2 Y	XY
1971	3,884,841	325	640,575
1973	3,892,729	327	645,171
1975	3,900,625	329	649,775
1977	3,908,529	332	656,364
1979	3,916,441	335	662,965
1981	3,924,361	338	669,578
1983	3,932,289	340	674,220
1985	3,940,225	343	680,855

Sums:

$\Sigma X =$ 15,824	$\Sigma X^2 =$ 31,300,040	$\Sigma Y =$ 2,669	$\Sigma XY =$ 5,279,503

Regression calculations:

Slope: $b = \dfrac{n\Sigma XY - (\Sigma X)(\Sigma Y)}{n\Sigma X^2 - (\Sigma X)^2}$ Equation [14.8b]

$$= \frac{8(5,279,503) - (15,824)(2,669)}{8(31,300,040) - (15,824)^2} = 1.315$$

Intercept: $a = \overline{Y} - b\overline{X}$ Equation [14.9]

$$= 333.6 - 1.315(1978) = -2,267.47$$

Equation: $\hat{Y} = -2,267.47 + 1.315(X)$ (Figure 14.17)

$$\hat{Y} = -2,267.47 + 1.315(X) \tag{14.18}$$

This tells us that the CO_2 concentration rose an average of 1.315 parts per million each year during this period. As the scatter plot in Figure 14.17 shows, the fit is extremely good. R^2 equals .996, indicating that nearly all of the variation in CO_2 concentration is explained by this steady upward trend.

Regression is sometimes used with time series, like the CO_2 data, to find a **trend line** which can then be projected into the future. We could use Equation [14.18], for instance, to predict the CO_2 concentration in 1995:

$$\hat{Y} = -2,267.47 + 1.315(X)$$

so if $X = 1995$,

$$\hat{Y} = -2,267.47 + 1.315(1995) = 355.96$$

If CO_2 concentration continues to increase at the 1971–1985 rate, then by 1995 it will be about 355.96 parts per million. This prediction is shown by the extension of the regression line in Figure 14.17.

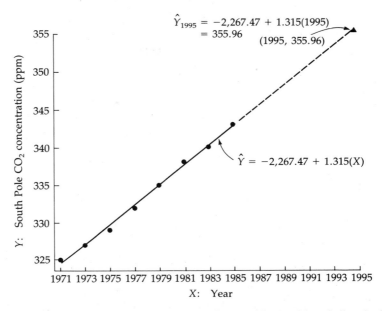

$\hat{Y}_{1995} = -2{,}267.47 + 1.315(1995)$
$= 355.96$
$(1995,\ 355.96)$

$\hat{Y} = -2{,}267.47 + 1.315(X)$

X: Year

FIGURE 14.17   *Regression of CO_2 concentration on year, with trend line extrapolation to 1995*

Although such trend line extrapolation is a popular use of regression, it demands caution. One problem is that the standard inferential procedures for hypothesis tests and confidence intervals, described in Chapter 15, are often invalid with time series data. Secondly, a straight line may not be the best way to describe the data. Time series often have patterns of up and down cycles, or change at ever-steepening or declining rates. The third limitation is familiar from some of the other regression examples we have seen: Extrapolating into the future asks the regression equation for predictions beyond the X range of the original data. The data themselves cannot tell us whether such predictions are any good.

Although Equation [14.8b] somewhat eases hand calculation in regression, such calculations are still laborious. Partly for this reason, regression used to be considered an advanced analytical technique, beyond the scope of an undergraduate statistics course. The computer revolution has changed that situation, since programs that zip through regressions are now available for every type of computer. The spread of such programs has brought a boom in the popularity of regression analysis among researchers.

PROBLEMS

29. Find the regression equation for predicting cumulative chlorofluorocarbon release from year, using the data in Table 14.11. Calculate the cumu-

lative release expected by the year 2000, if the trend seen over 1971–1985 continues. Can you think of specific reasons why this prediction might not prove accurate?

30. Construct a scatter plot for your chlorofluorocarbon analysis in Problem 29, showing how the regression line can be projected out to the year 2000. From the scatter plot, what can you guess about the values of r (correlation coefficient) and R^2 (coefficient of determination)?

SUMMARY

This chapter introduced regression analysis in the context of scatter plots. Although regression can be done without ever looking at a scatter plot, that is the statistical equivalent of flying blind. Many problems that can bedevil regression show up clearly in a scatter plot. Scatter plots also make it easier to understand what the numerical regression results mean.

Regression provides an equation for the best-fitting straight line. This equation, and especially its slope, can be interpreted directly in terms of the average or predicted change in Y, for every 1-unit increase in X. The equation also implies for each case a predicted value and a residual, which may themselves be the research focus. Finally, in correlation (r) and coefficient of determination (R^2) we have summary statistics that tell us how well the best-fitting line fits.

Our discussion of regression so far has treated it as only a *descriptive* tool, a method for analyzing sample data. In Chapter 15 we will turn attention to *inference* in regression analysis: how it works to form confidence intervals or test hypotheses about population parameters. Like other inferential procedures, those for regression require a number of assumptions about the population. The final sections of Chapter 15 describe how residuals can check on a regression's validity and suggest how that validity might be improved.

PROBLEMS

31. Construct a scatter plot for the regression of number of employees (Y) on gross sales in millions (X), for the state lotteries data of Chapter 3, Problem 27 (page 88). Draw in the regression line, $\hat{Y} = 100 + .1269(X)$. Indicate the residuals graphically in your plot.

32. Use the regression equation in Problem 31 to calculate predicted values and residuals for the number of employees of each state lottery. Does one lottery stand out as having a particularly large positive or negative residual? Discuss what this residual means. Could it be interpreted as evi-

dence that that particular state lottery is efficiently or inefficiently run? Explain.

33. Suppose you were to collect data on the age and the year of birth for everyone in your class. What can you guess about the correlation between these two variables?

34. Find the correlation between the actual and Poisson predicted counts of London areas with x buzz bomb hits in Table 11.11. Based on this correlation, how would you describe the fit between the mathematical model (Poisson distribution) and real-world data?

35. Temperature in degrees Celsius can be converted to approximate degrees Fahrenheit by a linear transformation: Multiply by 1.8, then add 32.
    a. Write this transformation in $\hat{Y} = a + b(X)$ form, using $\hat{Y}$ to represent degrees Fahrenheit and X to represent degrees Celsius.
    b. Explain the meaning of the values of a and b in your equation.

36. Carry out a regression analysis to estimate the values of a and b from these lowest recorded temperatures, in Celsius (X) and Fahrenheit (Y).

Region	X	Y
U.S. lower 48	− 56.5	− 70
Antarctica	− 89	− 129
Australia	− 22	− 8

Is the regression equation close to that theoretically expected (Problem 35)?

37. Problem 5 in Chapter 5 (pages 129–130) contains data on calculation test times (in seconds) and clock speeds (in megaHertz, MHz) for 11 personal computers. Construct a scatter plot with calculation time as Y and clock speed as X. Is the relationship positive or negative? What does this imply?

38. The statistics shown describe clock speed and calculation times for the 11 personal computers of Problem 37.

Clock Speed (MHz)	Calculation Time (sec)
$\overline{X} = 6.97$	$\overline{Y} = 107.95$
$TSS_X = 35.95$	$TSS_Y = 15{,}297.39$
$SCP = -602.28$	

Use these statistics to find and interpret:
a. The regression equation.
b. The correlation and the coefficient of determination.

39. Data on the estimated decommissioning costs for five nuclear power plants are given in the accompanying table. Regress the estimated decommissioning costs on years in use. Find and interpret:

a. The regression equation.

b. The correlation coefficient.

c. The coefficient of determination.

Estimated Decommissioning Costs for Five Inoperative Nuclear Power Plants

Plant	Capacity (megawatts)	Closing costs (millions)	When operated	Years of use
Elk River	24	$14	1962–1968	6
Sellafield/Windscale	33	64	1963–1981	18
Humboldt Bay-3	65	55	1963–1976	13
Shippingport	72	98	1957–1982	25
Dresden-1	210	95	1960–1978	18

Source: Data from Brown et al. (1986).

40. Regress decommissioning costs on capacity for the five nuclear plants in Problem 39.

a. What would you predict as decommissioning cost for a 100-megawatt plant?

b. Construct a scatter plot for these data, and draw in your regression line.

c. Describe what you see in the scatter plot.

41. The accompanying table shows data for ten technical rock climbing routes on granite cliffs in the Sierra Nevada of California. Each route was named

Data on Ten Technical Rock Climbing Routes

Route	Year first climbed	Difficulty rating[a]
Great White Book	1962	6
Chartres	1968	9
The Coming	1968	9
Far West	1969	6
Mosquito	1971	7
Botch	1972	9
Sticks and Stones	1972	10
Foolish Pleasures	1976	11
Slipstream	1976	11.50
Hyper Space	1977	10.25

[a]Higher ratings mean more difficult.
Source: Data from Reid and Falkenstein (1983).

and given a difficulty rating by its first climbers. For each route, the table gives the difficulty rating (higher ratings mean harder) and the year it was first climbed.

a. Is there an upward trend in the difficulty of new routes? Find the regression slope, and use it to address this question.

b. If this upward trend continued, what would you predict the average difficulty rating will be in 1993?

42. The accompanying table contains data on municipal waste generation, in thousands of metric tons per person per year, for 21 industrial countries. The second variable included is per capita gross national product (GNP).

Per Capita Municipal Waste Generation and Gross National Product (GNP) in 21 Industrial Nations, 1980

Country	GNP dollars/person	Municipal waste kg/person
Canada	11,334	526
United States	13,159	703
Austria	9,832	208
Belgium	10,537	313
Denmark	12,334	399
Finland	10,847	290
France	11,525	289
West Germany	12,283	338
Greece	4,167	259
Ireland	5,030	188
Italy	6,754	246
Netherlands	10,781	382
Norway	14,296	415
Portugal	2,479	152
Spain	5,372	215
Sweden	13,858	301
Switzerland	16,836	337
United Kingdom	9,584	282
Japan	10,047	344
Australia	11,218	681
New Zealand	8,040	488

Source: Data from Council on Environmental Quality (1987).

a. Find the slope and intercept, and write out the regression equation for the regression of municipal waste generation (Y) on the per capita GNP (X).

b. Construct a scatter plot, and draw in your regression line.

c. Find the correlation coefficient and R^2.

    d. Explain what your results from parts a–c tell you about the relationship between municipal waste generation and per capita GNP.

43. Using your regression equation from Problem 42, calculate the predicted values and residuals for each of the 21 industrial nations.

    a. Which country has the largest positive residual? Which country has the largest negative residual? Discuss what these residuals mean.

    b. Sweden and the United States have similar per capita gross national products. Are they also similar in terms of (1) predicted waste generation, and (2) residuals? Discuss.

NOTES

1. Because *Challenger*'s X value (31° F) lies outside of the range of data used to obtain the regression line (X values from 53° F to 81° F), the prediction of "2.93 incidents" for *Challenger* should be taken as only an approximation. It is more supportable just to say that based on this analysis, *Challenger* was more likely to encounter booster joint problems than any previous shuttle flight.

2. The term *explained* here is used in a technical sense. It does not necessarily imply causality. We are not claiming that mammoth bone prevalence *caused* mammoth art, only that there is a statistical relationship between the two, such that knowing one helps to predict what the other might be.

3. Median levels of education for adults over age 25 were obtained using a grouped-data procedure similar to Equation [3.2]. Using the simpler individual-level definition gives a median in each state of 12 years.

4. Rice and Paterson (1985, 1986) conclude that the *size* of an animal is the best predictor of how often that type of animal appears in cave paintings. Table 14.2 shows that reindeer bones are much more prevalent than mammoth bones, but reindeer art is not correspondingly more prevalent. Other smaller animals were also underrepresented in paintings. Possibly this is because larger animals are more dangerous to hunt and yield much more food. As the biggest animal around, mammoths were relatively popular subjects for the cave painters.

Inference and Criticism in Two-Variable Regression*

Chapter 15

Regression analysis provides a variety of ways to describe relationships in sample data. We can focus on the regression equation, slope, scatter plot, sums of squares, correlation coefficient, coefficient of determination, predicted values, or residuals. Inferences about the population can similarly focus on any aspect of regression.

Basic inferential procedures for regression are introduced in this chapter. Their details are new, but they should look somewhat familiar: They bear a family resemblance to other measurement-variable inferential methods, especially comparison-of-means techniques (Chapter 13). For example, we will see that, given an estimate of the standard error for a regression slope, we can construct confidence intervals and run one- or two-sided t tests on the same lines as for a difference of means. The F statistic used with ANOVA also has a counterpart in regression.

The resemblance between comparison-of-means and regression methods is not superficial. The two are mathematically related, and comparison-of-means procedures can be viewed as just a special kind of regression—as will be shown later in this chapter and in Chapter 16. Experienced analysts use regression as a single, unified approach for many different research problems.

Table 15.1 summarizes symbols from Chapter 14 used also in this chapter. You may want to refer to this table as new terms are defined.

TABLE 15.1   ***Some Symbols from Chapter 14***

Symbol	Meaning	Definition
Regression equation:		
	$\hat{Y}_i = a + b(X_i)$ or $Y_i = a + b(X_i) + e_i$	
X_i, Y_i	Actual values of X and Y for case i	
$\hat{Y}_i$	Predicted value of Y for case i	$\hat{Y}_i = a + b(X_i)$
e_i	Residual or prediction error for case i	$e_i = Y_i - \hat{Y}_i$
b	Sample slope or regression coefficient	$b = \dfrac{\text{SCP}}{\text{TSS}_X} = \dfrac{\Sigma(X_i - \overline{X})(Y_i - \overline{Y})}{\Sigma(X_i - \overline{X})^2}$
a	Sample Y-intercept or regression constant	$a = \overline{Y} - b\overline{X}$
Sums of squares:		
TSS_X	Total sum of squares, variable X	$\text{TSS}_X = \Sigma(X_i - \overline{X})^2$
TSS_Y	Total sum of squares, variable Y	$\text{TSS}_Y = \Sigma(Y_i - \overline{Y})^2$
ESS	Explained sum of squares	$\text{ESS} = \Sigma(\hat{Y}_i - \overline{Y})^2$
RSS	Residual sum of squares	$\text{RSS} = \Sigma e_i^2 = \Sigma(Y_i - \hat{Y}_i)^2$
SCP	Sum of cross products	$\text{SCP} = \Sigma(X_i - \overline{X})(Y_i - \overline{Y})$
Correlation and R^2:		
r	Correlation coefficient	$r = \dfrac{\text{SCP}}{\sqrt{\text{TSS}_X}\sqrt{\text{TSS}_Y}}$
R^2	Coefficient of determination	$R^2 = r^2 = \dfrac{\text{ESS}}{\text{TSS}_Y}$

15.1   *INFERENCE IN REGRESSION*

The regression equation $\hat{Y} = a + b(X)$ describes a relationship in sample data. To describe a *population* relationship we use the notation

$$\mu_{Y|X} = \alpha + \beta(X) \qquad\qquad\qquad [15.1]$$

Here α (*alpha*) represents the **population intercept,** and β (*beta*) represents the **population slope.** The symbol $\mu_{Y|X}$ ("mu-sub-Y given X") represents the **conditional mean of Y:** the population mean of Y given the value of X.

The population parameters α and β correspond to the sample statistics a and b. The intercept, α, is *the population mean of Y when X is 0:*

$$\alpha = \mu_{Y|X=0}$$

The slope, β, is *the change in the population mean of Y for every 1-unit increase in X:*

$$\beta = \mu_{Y|X_0+1} - \mu_{Y|X_0}$$

where X_0 is any specific value of X. Since α and β are generally unknown, we view the sample statistic a as an *estimator* of the population parameter α, and b as an estimator of β.

The sample regression equation is often written for the *predicted* values of Y, as $\hat{Y} = a + b(X)$. It can also be written for the *actual* values of Y, by including the sample residuals e: $Y = a + b(X) + e$, where e is simply the difference between actual and predicted Y. Similarly the population regression equation, Equation [15.1], for the *mean* values of Y, can also be written for the actual values of Y by including a **disturbance term, ϵ** (*epsilon*):

$$Y = \alpha + \beta(X) + \epsilon \qquad\qquad [15.2]$$

The disturbance term represents the effects of "everything else" (besides X) that influences the measured values of variable Y. This covers a lot of ground, and inference can proceed only by making strong assumptions about the unknown properties of ϵ. Sample residuals assist in checking some of these assumptions.

Sample-to-population inference may focus on either the right- or the left-hand side of a regression equation. That is, we can either

1. run hypothesis tests or construct confidence intervals regarding the unknown parameters α and β; or
2. construct confidence intervals around the predicted values of Y.

The most common inferential procedure is a test of the hypothesis that β equals 0, meaning that X is unrelated to Y in the population.

Table 15.2 (page 506) summarizes symbols used for the population parameters in bivariate regression.

15.2   STANDARD ERRORS IN REGRESSION

We estimate the parameters in a regression equation by calculating a and b from sample data. As with any sample statistics, it is unlikely that the calculated values exactly equal the corresponding parameters α and β. *Standard*

TABLE 15.2   ***Symbols for Population Parameters in Bivariate Regression***

Symbol	Meaning	Definition
Regression equation:		

$$\mu_{Y|X} = \alpha + \beta(X) \quad \text{or} \quad Y = \alpha + \beta(X) + \epsilon$$

Symbol	Meaning	Definition
X, Y	Actual values of X and Y	
$\mu_{Y|X}$	Conditional mean of Y given X	$\mu_{Y|X} = \alpha + \beta(X)$
ϵ	Disturbance term; effects of all other variables that influence Y	$\epsilon = Y - \mu_{Y|X}$
β	Population slope; change in population mean of Y per 1-unit increase in X. Estimated by b.	
α	Population intercept; population mean of Y when X is 0: $\mu_{Y|X=0}$. Estimated by a.	

errors describe the sample-to-sample variability of regression statistics. The more variability, the higher the standard errors, and the less precise our estimates of regression parameters.

In previous work with measurement variables, we saw that confidence intervals and hypothesis tests follow a common pattern, yet each different application requires its own standard error. The same is true in regression analysis.

Standard errors in regression analysis use a statistic called the *standard deviation of the residuals,* or s_e.

The **standard deviation of the residuals,** s_e, is defined as

$$s_e = \sqrt{\frac{\text{RSS}}{n - K}} \qquad [15.3]$$

where RSS is the residual sum of squares, n is the sample size, and K is the number of coefficients (intercept and slopes) in the regression equation.

It is like an ordinary standard deviation, except that its denominator is $n - K$ instead of the usual $n - 1$. The quantity $n - K$ is called the *residual degrees of freedom* or df_R.

The **residual degrees of freedom,** df_R, are

$$df_R = n - K \qquad [15.4]$$

where n is sample size and K is the number of coefficients.

In bivariate regression (Chapters 14 and 15) there are two coefficients, *a* and *b*, so $K = 2$ and $df_R = n - 2$. Multiple regression (Chapter 16) can include any number of coefficients.

The standard error of a regression slope measures the sample-to-sample variation in *b*.

The **standard error of a bivariate regression slope, SE_b,** is estimated as

$$SE_b = \frac{s_e}{\sqrt{TSS_X}} \qquad [15.5]$$

where s_e is the standard deviation of residuals, Equation [15.3], and TSS_X is the total sum of squares for *X*.

Besides the standard error of the slope, SE_b, three other standard errors will be introduced:

1. The standard error of an estimated conditional mean of *Y* given a specific value of *X*, X_0: $SE_{\hat{\mu}_{Y|X_0}}$

2. The standard error of predicted *Y* for an individual case when $X = X_0$: $SE_{\hat{Y}_0}$

3. The standard error of the intercept, SE_a

Table 15.3 gives equations for estimating these standard errors. Standard errors for the estimated conditional mean of *Y*, Equation [15.6], and for the

TABLE 15.3 *Standard Errors for Bivariate Regression*

Standard deviation of the residuals, e:

$$s_e = \sqrt{\frac{RSS}{n - K}} \qquad [15.3]$$

Standard error of the slope, b:

$$SE_b = \frac{s_e}{\sqrt{TSS_X}} \qquad [15.5]$$

Standard error of an estimated conditional mean of Y, $\hat{\mu}_{Y|X_0}$, given $X = X_0$:

$$SE_{\hat{\mu}_{Y|X_0}} = s_e \sqrt{\frac{1}{n} + \frac{(X_0 - \overline{X})^2}{TSS_X}} \qquad [15.6]$$

Standard error of an individual predicted value of Y, $\hat{Y}_0$, when $X = X_0$:

$$SE_{\hat{Y}_0} = s_e \sqrt{\frac{1}{n} + \frac{(X_0 - \overline{X})^2}{TSS_X} + 1} \qquad [15.7]$$

Standard error of the intercept, a (derived from [15.6] by substituting 0 for X_0):

$$SE_a = s_e \sqrt{\frac{1}{n} + \frac{(\overline{X})^2}{TSS_X}} \qquad [15.8]$$

predicted Y of an individual case, Equation [15.7], both depend on the *specific value of X*, denoted X_0, for which the prediction is made. The standard error of the intercept, Equation [15.8], is simply the standard error of the conditional mean when $X_0 = 0$.

We next turn to concrete examples to illustrate the calculation and use of these standard errors, focusing particularly on SE_b and inferences about the regression slope.

PROBLEM

*1. Use the sums of squares in Tables 14.3 and 14.4 (pages 469 and 475) to calculate these statistics for the mammoth bones and cave art data.

  a. Standard deviation of the residuals, s_e
  b. Residual degrees of freedom, df_R
  c. Standard error of the slope, SE_b
  d. Standard error of the intercept, SE_a
  e. Standard error of the estimated conditional mean, $SE_{\hat{\mu}_{Y|X_0}}$, for $X_0 = 5$
  f. Standard error of an individual predicted value of Y, $SE_{\hat{Y}_0}$, for $X_0 = 5$

15.3   INCOME AND HOMICIDE RATE

Table 15.4 gives yearly homicide rates and median family incomes for 12 U.S. cities (seen earlier in Table 1.4). Because these cities were chosen randomly from the population consisting of all U.S. cities, we can draw inferences from them about the relationship between income and homicide rate in that population.

It is widely thought that homicide rates are related to poverty and are higher in poor areas than in rich ones. This implies a negative relationship between homicide rate and income. We can state this idea as a research hypothesis:

$$H_1: \quad \beta < 0$$

The corresponding null hypothesis is

$$H_0: \quad \beta \geq 0$$

H_1 asserts that in the population of all U.S. cities, the regression of homicide on income has a negative slope (mean homicide rate falls as income rises). The null hypothesis (H_0) claims a positive or zero slope (mean homicide rate rises or stays the same as income rises). Because the hypotheses are directional, a one-sided test is appropriate.

TABLE 15.4   ***Family Income and Homicide Rate in 12 U.S. Cities***

Case	City	Family income[a]	Homicide victims[b]
i		X_i	Y_i
1	Warren, MI	26.5	2.61
2	Pueblo, CO	18.3	4.13
3	Raleigh, NC	21.8	6.79
4	Fort Wayne, IN	19.6	6.97
5	Tucson, AZ	19.4	7.32
6	Anchorage, AK	30.7	8.48
7	Toledo, OH	20.2	9.53
8	Portsmouth, VA	16.8	14.53
9	Memphis, TN	16.9	16.93
10	Hartford, CT	14.0	19.65
11	Savannah, GA	15.4	20.23
12	Birmingham, AL	15.2	27.07

Summary statistics:

$$\overline{X} = 19.57 \qquad\qquad \overline{Y} = 12.02$$
$$s_X = 4.87 \qquad\qquad s_Y = 7.55$$
$$n = 12$$

[a]1980 median family income in thousands of dollars per year.
[b]1980–1984 mean homicide rate in victims per 100,000 pop. per year.

When homicide rate (Y) is regressed on median family income (X) for the 12 cities of Table 15.4, we obtain

$$\hat{Y} = 32.224 - 1.0326(X) \qquad\qquad [15.9]$$

The sample regression slope, $b = -1.0326$, and Y-intercept, $a = 32.224$, are found by applying Equations [14.8a] and [14.9]. (The slope is given to four decimal places to ensure that inferential calculations work out properly.)

Equation [15.9] tells us that on the average, a \$1,000 increase in median family income is associated with a decline in the homicide rate of slightly more than one (1.0326) victim per 100,000 population *in this sample*. The negative sign of the sample regression slope ($b = -1.0326$) supports our research hypothesis that homicide rates tend to be higher in poor cities and lower in wealthy ones. If the median income in a city were 0 (a nonsensical possibility), we would predict a homicide rate of 32.224 victims per 100,000 population. Figure 15.1 (page 510) superimposes the regression line on a scatter plot.

Table 15.5 repeats the original data along with predicted homicide rates, $\hat{Y}_i = 32.224 - 1.0326(X_i)$, and residuals, defined as the differences between actual and predicted rates ($e_i = Y_i - \hat{Y}_i$). For example, Pueblo has a median family income of \$18,300 per year. Substituting this X value into Equation [15.9] gives us Pueblo's predicted homicide rate:

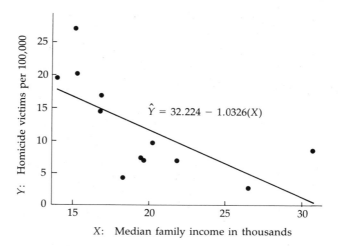

FIGURE 15.1   *Homicide rate vs. income for 12 U.S. cities*

TABLE 15.5   ***Regression of Homicide Rate on Median Family Income***

Case	City	Income	Homicide	Predicted values $\hat{Y}_i = 32.224 - 1.0326(X_i)$	Residuals
i		X_i	Y_i		$e_i = Y_i - \hat{Y}_i$
1	Warren, MI	26.5	2.61	4.86	-2.25
2	Pueblo, CO	18.3	4.13	13.33	-9.20
3	Raleigh, NC	21.8	6.79	9.71	-2.92
4	Fort Wayne, IN	19.6	6.97	11.99	-5.02
5	Tucson, AZ	19.4	7.32	12.19	-4.87
6	Anchorage, AK	30.7	8.48	.52	7.96
7	Toledo, OH	20.2	9.53	11.37	-1.84
8	Portsmouth, VA	16.8	14.53	14.88	$-.35$
9	Memphis, TN	16.9	16.93	14.77	2.16
10	Hartford, CT	14.0	19.65	17.77	1.88
11	Savannah, GA	15.4	20.23	16.32	3.91
12	Birmingham, AL	15.2	27.07	16.53	10.54

Sums of squares:

Total (X): $\text{TSS}_X = \Sigma(X_i - \overline{X})^2 = 261.23$

Total (Y): $\text{TSS}_Y = \Sigma(Y_i - \overline{Y})^2 = 627.01$

Explained: $\text{ESS} = \Sigma(\hat{Y}_i - \overline{Y})^2 = 278.53$

Residual: $\text{RSS} = \Sigma(Y_i - \hat{Y}_i)^2 = \Sigma e_i^2 = 348.48$

$$\hat{Y}_2 = 32.224 - 1.0326(X_2)$$

$$= 32.224 - 1.0326(18.3) = 13.33$$

The actual homicide rate in Pueblo for this period is $Y_2 = 4.13$, so the residual is $Y_2 - \hat{Y}_2 = 4.13 - 13.33 = -9.20$. From its relatively low median income, we might expect a much higher homicide rate. The other $\hat{Y}_i$ and e_i values in Table 15.5 were calculated similarly.

Four sums of squares appear at the bottom of Table 15.5. TSS_Y (total sum of squares for Y) is the sum of squared deviations of the individual Y values, Y_i, around the mean of Y:

$$TSS_Y = \Sigma(Y_i - \bar{Y})^2$$

$$= (2.61 - 12.02)^2 + (4.13 - 12.02)^2 + \cdots + (27.07 - 12.02)^2$$

$$= 88.55 + 62.25 + \cdots + 226.50 = 627.01$$

The standard deviation of Y (see Table 15.4) is calculated from TSS_Y:

$$s_Y = \sqrt{\frac{TSS_Y}{n - 1}}$$

$$= \sqrt{\frac{627.01}{11}} = 7.55$$

The total sum of squares for X, TSS_X, resembles TSS_Y, except that we square and sum deviations of each X_i from the mean of X.

The third sum of squares in Table 15.5 is the explained sum of squares, or ESS. This is the sum of squared deviations of the predicted values around the mean of Y:

$$ESS = \Sigma(\hat{Y}_i - \bar{Y})^2$$

$$= (4.86 - 12.02)^2 + (13.33 - 12.02)^2 + \cdots + (16.53 - 12.02)^2$$

$$= 51.27 + 1.72 + \cdots + 20.34 = 278.53$$

Finally, Table 15.5 gives the sum of squared residuals, RSS:

$$RSS = \Sigma(Y_i - \hat{Y}_i)^2 = \Sigma e_i^2$$

$$= (-2.25)^2 + (-9.20)^2 + \cdots + (10.54)^2$$

$$= 5.06 + 84.64 + \cdots + 111.09 = 348.48$$

The explained sum of squares plus the residual or "unexplained" sum of squares equals the total sum of squares for Y:

$$TSS_Y = ESS + RSS$$

$$627.01 = 278.53 + 348.48$$

The residual sum of squares is used to find the standard deviation of the residuals, Equation [15.3]:

$$s_e = \sqrt{\frac{\text{RSS}}{\text{df}_R}}$$

For the data in Table 15.5, RSS = 348.48 and its degrees of freedom are $\text{df}_R = n - K = n - 2 = 10$. The standard deviation of the residuals is therefore

$$s_e = \sqrt{\frac{348.48}{10}} = 5.903$$

Like any standard deviation, s_e measures spread or variation. Since the residuals, e, are the vertical distances between actual Y values and the regression line, s_e measures the variation of Y *around the regression line* (Figure 15.2). The higher the s_e, the farther the dots scatter away from the line. A low s_e indicates that the data fit the line closely. Because all of the standard errors in regression increase as s_e increases (see Table 15.3), a low s_e permits more precise estimates of population parameters. *High* and *low* are relative terms here. The standard deviation of the residuals can be as low as 0 (all points lie on the line) or almost as high as the standard deviation of the Y variable itself.

The standard deviation of the residuals (s_e) and the total sum of squares for X (TSS_X) are used to estimate the standard error of the regression slope, SE_b, Equation [15.5]:

$$\text{SE}_b = \frac{s_e}{\sqrt{\text{TSS}_X}} = \frac{5.903}{\sqrt{261.23}} = .365$$

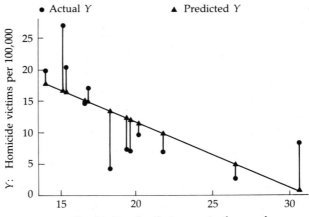

FIGURE 15.2   *Residuals are vertical distances from line*

Like any standard error, SE$_b$ measures sample-to-sample variability. A high SE$_b$ means high variability, so the slope of any one sample could easily be far from the true parameter β. A low SE$_b$ means the reverse: Most likely, the *b* for our sample is not very far from β.

Our estimate of the standard error of the regression slope relating homicide rate to household income is SE$_b$ = .365. Our estimate for the slope itself, in Equation [15.9], is *b* = −1.0326. These estimates are the basic elements required for confidence intervals and significance tests.

PROBLEMS

*2. The standard deviation of the residuals was 5.903 homicides/100,000 population in our example. How might the appearance of the scatter plot (Figure 15.1) differ if this standard deviation were 7? If it were 1?

*3. Examine Equation [15.5] for SE$_b$. Other things being equal, how would increasing sample size affect this standard error? What does this imply, for a researcher trying to estimate β from *b*?

*4. The data shown are prices and processing speeds for a sample of five personal computer systems. Higher values on the speed index mean faster computers. All five designs are fairly similar; price and performance are their main selling points (data from *Infoworld*).

System	Price	Speed Index
Mitac	$4,895	339
PC's Limited	4,799	400
Dale	5,300	373
Acer	3,995	339
PC Designs	4,015	413

If you pay more, do you get a faster computer? Carry out the regression of speed (*Y*) on price (*X*). Find the following.

a. The regression equation for predicting speed (*Y*) from price (*X*)

b. The residual sum of squares, RSS

c. The standard deviation of the residuals, s_e

d. The estimated standard error of *b*, SE$_b$

15.4   *t TESTS FOR REGRESSION COEFFICIENTS*

Many hypothesis tests for measurement variables are based on a *t* statistic, defined as the distance between sample statistic and hypothesized population parameter, in standard errors:

$$t = \frac{\text{Sample statistic} - \text{Hypothesized population parameter}}{\text{Estimated standard error of statistic}}$$

Confidence intervals are found as:

Sample statistic $\pm$ t(Estimated standard error of statistic)

These formulas apply to individual means (Chapter 11), to pairs of means (Chapter 13), and to regression analysis as well. Any of the standard errors seen in Table 15.3 may be used to construct confidence intervals or perform hypothesis tests along these lines.

Most often tests or intervals are constructed for the regression coefficient or slope, b, as shown in the box.

Hypotheses about a regression slope may be tested using the t statistic:

$$t = \frac{b - \beta_0}{\text{SE}_b} \qquad [15.10]$$

where SE_b is the standard error of b and β_0 is the value of the population slope supplied by the null hypothesis H_0.

If H_0 is true and certain assumptions are met, this t statistic will follow a theoretical t distribution (Appendix Table A.3) with $\text{df}_R = n - K$ degrees of freedom.

For homicide rate and income we earlier calculated a sample regression slope of $b = -1.0326$ and a standard error of $\text{SE}_b = .365$. The hypotheses to be tested, at the $\alpha = .05$ level, are

H_0: $\beta \geq 0$

H_1: $\beta < 0$

If H_0 were true, the population slope β could be as close to H_1 as $\beta_0 = 0$. This value of β_0 can be used in Equation [15.10] to obtain the t statistic:

$$t = \frac{b - \beta_0}{\text{SE}_b} = \frac{-1.0326 - 0}{.365} = -2.83$$

The usual **regression null hypotheses** are H_0: $\beta = 0$, H_0: $\beta \geq 0$, or H_0: $\beta \leq 0$. Under these hypotheses $\beta_0 = 0$, and the t test is just the sample regression slope divided by its standard error:

$$t = \frac{b}{\text{SE}_b} \qquad [15.11]$$

| TABLE 15.6 | *A One-Sided t Test in Regression: Testing the Hypothesis of a Negative Relationship Between City Homicide Rate and Median Family Income* |

Null hypothesis: H_0: $\beta \geq 0$ (0 or + relation in population)
Research hypothesis: H_1: $\beta < 0$ (− relation in population)

Sample statistics:

Number of cases:	$n = 12$
Regression slope:	$b = -1.0326$
Total sum of squares, X:	$TSS_X = 261.23$
Residual sum of squares:	$RSS = 348.48$

Estimating the standard error of b:

Standard deviation of residuals: $s_e = \sqrt{\dfrac{RSS}{df_R}}$ Equation [15.3]

$$= \sqrt{\dfrac{348.48}{10}} = 5.903$$

Standard error of b: $SE_b = \dfrac{s_e}{\sqrt{TSS_X}}$ Equation [15.5]

$$= \dfrac{5.903}{\sqrt{261.23}} = .365$$

Hypothesis test: $t = \dfrac{b - \beta_0}{SE_b}$ Equation [15.10]

$$= \dfrac{-1.0326 - 0}{.365} = -2.83$$

Conclusions: The sample regression coefficient, $b = -1.0326$, is 2.83 standard errors below the hypothesized population coefficient, $\beta_0 = 0$. With $df_R = n - K = 12 - 2 = 10$ degrees of freedom, a t statistic of -2.83 has a one-sided probability (Table A.3) of $P < .01$. This probability is below $\alpha = .05$, so we should reject the null hypothesis. We therefore conclude that there *is* a negative relationship (H_1: $\beta < 0$) between homicide rate and median income in U.S. cities.

We use the t table (Table A.3) in regression just as in Chapter 13. In this example, the t statistic is $t = -2.83$, with degrees of freedom $df_R = n - K = 12 - 2 = 10$. This t statistic falls between the values of 2.764 (one-sided $P = .01$) and 3.169 ($P = .005$) in the row for 10 degrees of freedom in Table A.3. Our obtained P-value is therefore between .01 and .005:

$$.005 < P < .01$$

Since this P-value is well below $\alpha = .05$, we may reject the null hypothesis that the population slope β is positive or zero. There is less than one chance in one hundred of randomly drawing a sample so favorable to H_1: $\beta < 0$ if H_0: $\beta \geq 0$ were true. By rejecting the null hypothesis, we indirectly conclude that there *is* a negative relationship between homicide rates and median income, in the population consisting of all U.S. cities. The sample relationship is statistically significant at the $\alpha = .05$ level.

Table 15.6 outlines these steps. Compare it with Tables 13.6 or 13.8 (pages 411 and 418) to see the similarities between different *t*-test applications.

PROBLEMS

*5. In a sample of 89 Mathematics majors, the regression of college grade point average (Y) on *verbal* SAT score (X) yields the regression equation $\hat{Y} = 1.459 + .0025(X)$. The estimated standard error of b is $SE_b = .0007$. Test at the $\alpha = .01$ level whether Math majors' grades rise significantly with VSAT (one-sided test). State the null and research hypotheses, and summarize your conclusions in English.

*6. In a sample of 436 English majors, the regression of college GPA (Y) on *math* SAT score (X) is $\hat{Y} = 2.156 + .001(X)$, with $SE_b = .0003$. Is the relationship between GPA and MSAT score among English majors significant at $\alpha = .01$ (one-sided test)?

*7. Is the relationship between computer speed and price statistically significant? Use your results from Problem 4 to conduct a two-sided test at $\alpha = .10$.

15.5 *t TESTS FOR HYPOTHESES OTHER THAN* $\beta = 0$

The t statistic is most often employed to test the null hypotheses $\beta = 0$, $\beta \geq 0$, or $\beta \leq 0$, all of which lead to β_0 values of zero (in which case [15.10] reduces to [15.11]). If we reject such hypotheses, we conclude that there is a relationship as specified by H_1—but this relationship could still be so weak it has no practical importance. The t test also applies to other null hypotheses, however. We might test whether a slope is large enough to be of *practical importance*, rather than just testing whether it is *not zero*.

For example, suppose we regress 1988 income in dollars on years of education beyond high school, for a sample of people born in 1958. The slope b will tell us how much predicted income rises for each additional year of education. But with a large enough sample, any nonzero sample slope is statistically significant. (Recall the discussion of sample size and statistical significance for Tables 12.13 and 12.14, pages 386 and 387). Higher education is expensive, however, and we might be disappointed to find that in the population mean income rises only $2 ($\beta = 2$) with each additional year in college. Rejecting the null hypothesis that $\beta \leq 0$ does not answer our real concern: Does college pay off financially?

Suppose that an economic analysis had convinced us that college expenses are a break-even investment if mean income at age 30 increases by \$2,000 with each year of college ($\beta = 2,000$). To test whether education pays off, we should thus test the hypotheses

H_0: $\beta \leq 2,000$

H_1: $\beta > 2,000$

Such a one-sided test would follow the steps outlined earlier for tests of $\beta = 0$. Our t statistic, following Equation [15.10], is

$$t = \frac{b - \beta_0}{SE_b} = \frac{b - 2,000}{SE_b}$$

This t statistic could be compared with the t distribution table in the usual manner, with $df_R = n - 2$ degrees of freedom. If we reject the null hypothesis $\beta \leq 2,000$, we could conclude that college does indeed pay off.

PROBLEM

*8. Although the relationship between English majors' grades and their MSAT scores (Problem 6) is statistically significant at $\alpha = .01$, it is not very strong. Can we reject at $\alpha = .01$ the null hypothesis that mean GPA rises by no more than .0005 (on the usual four-point grade scale where $4.0 = $ A, $3.0 = $ B, etc.) for every one additional point on the MSAT? This requires a one-sided test of the hypotheses H_0: $\beta \leq .0005$ vs. H_1: $\beta > .0005$.

15.6   CONFIDENCE INTERVALS FOR REGRESSION COEFFICIENTS

The same standard errors used in regression t tests are also employed for confidence intervals. These are intervals that we believe, with a specified degree of confidence, contain the population parameter β. Confidence intervals in regression follow the usual formula of taking the sample statistic plus and minus t times its estimated standard error. The value of t is chosen from the theoretical t distribution (Table A.3) with $df_R = n - K$ degrees of freedom. In bivariate regression, $df_R = n - 2$.

> **Confidence intervals for regression coefficients** are formed by
>
> $b \pm t(SE_b)$ [15.12]

Table 2.8 (page 41) gave data from a study of household responses to an emergency water conservation campaign. The variables include household income and the amount of water used during the summers of 1980 and 1981, before and after the shortage was declared. In this sample, average household water use rose with income. The regression of 1980 water use on income yields the equation

$$\hat{Y} = 1{,}665.97 + 37.992(X) \tag{15.13}$$

The regression slope, $b = 37.992$, tells us that *in this sample*, predicted water use rose by 37.992 cubic feet with each $1,000 increase in income. By how much did water use increase *in the population* of households, from which these 33 were chosen? This question calls for a confidence interval.

Table 15.7 shows some of the predicted values and residuals from this regression. Summary statistics and sums of squares, given at the bottom, are used to construct a confidence interval in Table 15.8.

The confidence interval will be $b \pm t(SE_b)$, Equation [15.12], so we need to know the estimated standard error of b, SE_b. That requires first finding the standard deviation of the residuals, s_e, done at the top of Table 15.8.

TABLE 15.7

Some Predicted Values and Residuals from the Regression of Water Use on Household Income

Household	Income ($1,000)	1980 usage (ft³)	Predicted usage[a]	Residuals
i	X_i	Y_i	$\hat{Y}_i$	$e_i = Y_i - \hat{Y}_i$
1	15	5,700	2,235.85	3,464.15
2	30	3,300	2,805.73	494.27
3	25	1,200	2,615.77	−1,415.77
.	.	.	.	.
.	.	.	.	.
.	.	.	.	.
32	15	3,700	2,235.85	1,464.15
33	30	2,100	2,805.73	−705.73

Summary statistics:

$$\overline{X} = 26.82 \qquad \overline{Y} = 2{,}684.85$$
$$s_X = 18.94 \qquad s_Y = 1{,}641.10$$
$$n = 33$$

Sums of squares:

Total (X): $TSS_X = \Sigma(X_i - \overline{X})^2 = 11{,}479$

Total (Y): $TSS_Y = \Sigma(Y_i - \overline{Y})^2 = 86{,}182{,}424$

Explained: $ESS = \Sigma(\hat{Y}_i - \overline{Y})^2 = 16{,}568{,}747$

Residual: $RSS = \Sigma(Y_i - \hat{Y}_i)^2 = \Sigma e_i^2 = 69{,}613{,}677$

[a]Predicted values are from the regression equation $\hat{Y}_i = 1{,}665.97 + 37.992(X_i)$.

TABLE 15.8	**Finding and Interpreting a 95% Confidence Interval for the Slope in the Regression of Water Usage on Income**

Sample statistics:

Regression equation: $\hat{Y} = 1{,}665.97 + 37.992(X)$

Standard deviation of residuals:

$$s_e = \sqrt{\frac{RSS}{n-2}} \qquad \text{Equations [15.3], [15.4]}$$

$$= \sqrt{\frac{69{,}613{,}677}{33-2}} = 1{,}498.533$$

Estimated standard error of b:

$$SE_b = \frac{s_e}{\sqrt{TSS_X}} \qquad \text{Equation [15.5]}$$

$$= \frac{1{,}498.533}{\sqrt{11{,}479}} = 13.987$$

Confidence interval for the population slope (β):

$b \pm t(SE_b)$ Equation [15.12]

$37.992 \pm 2.042(13.987)$

$9.431 \le \beta \le 66.553$

Interpretation: We are 95% confident that in the population of households from which this sample was drawn, the parameter β is between 9.431 and 66.553. That is, mean summer water consumption rose between 9.431 and 66.553 cubic feet, for every \$1,000 increase in household income.

The appropriate t value is selected from the t distribution table, for $df_R = n - K = 31$ degrees of freedom. We can round this down to 30 degrees of freedom. With 30 df, a 95% confidence interval requires $t = 2.042$. Since the regression slope is $b = 37.992$, and its estimated standard error is $SE_b = 13.987$, the 95% confidence interval is

$$b \pm t(SE_b)$$

$$37.992 \pm 2.042(13.987)$$

$$\rightarrow 9.431 \le \beta \le 66.553$$

Based on this sample of 33 households, we are 95% confident that in the population, average water use rose somewhere between 9.431 and 66.553 cubic feet with each additional thousand dollars of income. (Stated more formally, 95% of the intervals constructed in this manner should contain the true value of β.)

This confidence interval may seem wide, but two things should be noted. First, it is based on a sample of only 33 households. For a more precise estimate we would need a larger sample. Second, the interval *does not include zero*. We saw in Chapter 9 that confidence intervals can be used to perform two-sided hypothesis tests. If we are at least 95% confident that the parameter β is not zero, as our confidence interval suggests, then we know that a two-sided hypothesis test would also reject H_0: $\beta = 0$ in favor of H_1: $\beta \neq 0$ at the $\alpha = .05$ significance level. We can be confident that water use has at least some relation to income: Wealthier households do not use the same amount of water as poor households do.

As observed in earlier chapters, the confidence interval conveys more specific information than does the hypothesis test alone. Not only do we conclude that "some" relationship exists; we can describe how strong we think that relationship is and indicate the uncertainty of our estimate.

PROBLEMS

* 9. Using the sample slope ($b = 37.992$) and estimated standard error ($SE_b = 13.987$) for income and water use in Table 15.8, find and interpret:

   a. A 90% confidence interval for β.

   b. A 99% confidence interval for β.

   Comment on the relationship between 90%, 95% (Table 15.8), and 99% confidence intervals.

*10. For income and water use we noted that the 95% confidence interval, $9.431 \leq \beta \leq 66.553$, was wide—partly due to the relatively small sample ($n = 33$). This small sample is actually a randomly selected subset of a larger sample of $n = 496$ households. For those 496 households, the regression of 1980 water use on income results in a sample regression coefficient $b = 45.61$, with an estimated standard error of $SE_b = 5.68$. Use this information to construct and interpret a 95% confidence interval for β, based on all 496 cases. How does the width of this interval compare with that in Table 15.8? What general principle about precision and sample size does this comparison illustrate?

*11. Using the sample slope ($b = -1.0326$) and estimated standard error ($SE_b = .365$) for median incomes and homicide rates (Table 15.6), find:

   a. An 80% confidence interval for β.

   b. A 95% confidence interval for β.

   c. A 99.9% confidence interval for β.

*12. Find and interpret a 95% confidence interval for β in the regression of the number of 1986–1987 season points (Y) on minutes played (X), based on the sample of eight basketball players in Chapter 14, Problem 2 (page 461).

*13. Examine your regression of season points on minutes played in Problem 12.

a. Is the relationship between points and minutes statistically significant at $\alpha = .05$? Use the confidence interval from Problem 12 to answer.

b. How strong is the relationship between points and minutes? Find the correlation coefficient (r) and the coefficient of determination (R^2) to answer.

15.7 CONFIDENCE INTERVALS FOR REGRESSION PREDICTIONS

So far we have concentrated on inferences about the population slope β. Similar methods can be applied to draw inferences about the population intercept α, a conditional mean, $\mu_{Y|X}$, or an unknown individual Y value, using the formulas for estimating standard errors given in Table 15.3.

Both $\mu_{Y|X}$ and unknown individual Y values are estimated by the regression prediction $\hat{Y}$. For example, we can use Equation [15.13] to estimate the population mean water use for households with incomes of $60,000—that is, to estimate $\mu_{Y|X_0}$ for $X_0 = 60$. Our estimate, $\hat{\mu}_{Y|X_0}$, is

$$\hat{\mu}_{Y|X_0} = 1,665.97 + 37.992(X_0)$$

$$= 1,665.97 + 37.992(60) = 3,945.49 \text{ cubic feet}$$

To construct a 95% confidence interval for this estimate of the conditional mean, we find

$$\hat{Y}_0 \pm t(SE_{\hat{\mu}_{Y|X_0}})$$

where $\hat{Y}_0$ denotes the predicted value of Y for a specific value of X, $X = X_0$. With 31 degrees of freedom the necessary t value is 2.042. The estimated standard error of $\hat{\mu}_{Y|X_0}$ is calculated from Equation [15.6], given $X_0 = 60$:

$$\hat{Y}_0 \pm t(SE_{\hat{\mu}_{Y|X_0}}) = 3,945.49 \pm 2.042 \, s_e \sqrt{\frac{1}{n} + \frac{(X_0 - \overline{X})^2}{TSS_X}}$$

$$= 3,945.49 \pm 2.042(1,498.533) \sqrt{\frac{1}{33} + \frac{(60 - 26.82)^2}{11,479}}$$

$$= 3,945.49 \pm 2.042(532.37)$$

$$\rightarrow 2,858.4 \le \mu_{Y|X_0} \le 5,032.6$$

Based on our analysis of this sample, we can be 95% confident that in the population of all Concord households with $60,000 incomes, the mean summer 1980 water use was between 2,858.4 and 5,032.6 cubic feet.

A different use of regression prediction is to answer questions like: What is our best prediction of water use for the Smith household, which earned $60,000 in 1980? The regression equation yields the same prediction we used earlier in estimating the mean for *all* households earning $60,000:

$$\hat{Y}_0 = 1{,}665.97 + 37.992(X_0)$$
$$= 1{,}665.97 + 37.992(60) = 3{,}945.49 \text{ cubic feet}$$

Thus, the predictions for an individual household $(\hat{Y}_0)$ and a conditional mean $(\hat{\mu}_{Y|X_0})$ are the same for any given X value, X_0.

The 95% confidence interval for our estimate of the Smiths' actual water use, denoted Y_0, is constructed with the standard error of $\hat{Y}_0$, Equation [15.7]:

$$\hat{Y} \pm t(\text{SE}_{\hat{Y}_0}) = 3{,}945.49 \pm 2.042 s_e \sqrt{\frac{1}{n} + \frac{(X_0 - \overline{X})^2}{\text{TSS}_X} + 1}$$

$$= 3{,}945.49 \pm 2.042(1{,}498.533)\sqrt{\frac{1}{33} + \frac{(60 - 26.82)^2}{11{,}479} + 1}$$

$$= 3{,}945.49 \pm 2.042(1{,}590.29)$$

$$\rightarrow 698.1 \le Y_0 \le 7{,}192.9$$

Based on our analysis of this sample, we can be 95% confident that the Smiths used between 698.1 and 7,192.9 cubic feet of water.

Comparing Equations [15.6] and [15.7] in Table 15.3 you will see that the standard error for an individual predicted value always exceeds the standard error for the conditional mean at the same level of X. This is because group means can be predicted more precisely than the values of individual cases. Note also that the standard errors depend on the specific values of X, X_0. Standard errors are smallest, hence our estimates are most precise, when $X_0 = \overline{X}$. Confidence intervals for regression predictions are consequently narrowest with $\overline{X}$, and progressively widen as we move to higher or lower values of X.

Figure 15.3 illustrates these properties. The 95% confidence intervals for predicted water use are shown as bands around the regression line. The

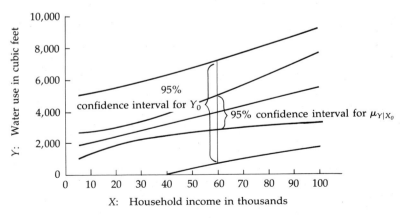

FIGURE 15.3   *95% confidence bands, predicted water use*

narrower inner band represents the 95% limits for the conditional mean of water use, while the outer band represents the 95% limits for individual case predictions. At any given value of X we can be 95% confident that the population mean usage falls within the inner band and that an individual household's usage falls within the outer band.

PROBLEMS

*14. The Osgood family lived in Concord in 1980 on a household income of $19,000.

    a. What is our best point estimate of how much water they used?

    b. Construct a 95% confidence interval for the Osgoods' water use.

    c. What is our best point estimate of the mean water use among *all* Concord households earning $19,000?

    d. Construct a 95% confidence interval for the mean water use among all Concord households earning $19,000.

*15. What X value will result in the narrowest confidence interval for the conditional mean of summer water use (Table 15.8)? Calculate the 90% confidence interval at this value.

*16. Refer to the data on hospital admissions and physicians sued, disciplined or charged, in Table 14.6 (page 483).

    a. How many physician problems would you predict for a hospital with 4,000 patient admissions?

    b. Construct and interpret a 99% confidence interval for the number of physician problems at an individual hospital with 4,000 patient admissions.

15.8   F TESTS IN TWO-VARIABLE REGRESSION

There are close parallels between regression and ANOVA. The total sum of squares for Y has the same meaning in both methods: It is the sum of squared deviations of the Y value for each case in the sample, from the overall mean of Y. That is,

$$TSS_Y = \Sigma(Y_i - \overline{Y})^2$$

The explained sum of squares (regression) is a more general version of the between-groups sum of squares (ANOVA). The explained sum of squares is the sum of squared deviations of the *predicted value* of Y for each case from the overall Y mean:

$$ESS = \Sigma(\hat{Y}_i - \overline{Y})^2$$

In ANOVA, the between-groups sum of squares is defined as the sum of squared deviations of the *group mean* for each case from the overall mean:

$$BSS = \Sigma(\overline{Y}_k - \overline{Y})^2$$

Both definitions are similar in that the group means $(\overline{Y}_k)$ used in ANOVA are a kind of predicted value: In ANOVA, we "predict" that each case will have a value equal to the mean for that case's group.

Finally, the regression residual sum of squares is a more general version of ANOVA's within-groups sum of squares. The residual sum of squares is based on deviations of the actual values for each case from the *predicted values:*

$$RSS = \Sigma(Y_i - \hat{Y}_i)^2$$

The within-groups sum of squares is based on deviations of the actual values for each case from the *group means:*

$$WSS = \Sigma(Y_i - \overline{Y}_k)^2$$

Since the group means in ANOVA have the same function as predicted values in regression, the ANOVA WSS is equivalent to the regression RSS. Both represent the variation in Y that is left over after subtracting the variation related to X.

The three sums of squares are used in ANOVA to calculate an F statistic, which tests the null hypothesis that all groups have the same population mean. This hypothesis implies that no relationship exists between the categorical variable that defines the groups and the measurement variable whose means are being studied. A similar F statistic can be calculated using the three sums of squares in regression. In regression, the F statistic again tests a null hypothesis of *no relationship*: that no relationship exists between the Y (dependent) variable and all of the X (independent) variables in the analysis. For now we will consider only bivariate regression, with one X variable.

Table 15.9 illustrates the calculation of an F statistic for household income and water use. Its key elements are the three sums of squares from Table 15.7. Each sum of squares has certain degrees of freedom:

df$_E$ = $K - 1$ for the explained sum of squares
df$_R$ = $n - K$ for the residual sum of squares
df$_T$ = $n - 1$ for the total sum of squares

where n is the number of cases and K equals the number of coefficients. Dividing each sum of squares by its degrees of freedom produces a **mean square**. The F statistic equals the explained mean square divided by the residual mean square. If H_0 is true and certain assumptions are met, this F statistic will follow the theoretical F distribution (Table A.5) with df$_E$ = $K - 1$ and df$_R$ = $n - K$ degrees of freedom, where K is the number of coefficients (2 in bivariate regression).

TABLE 15.9   **Regression F Test for Water Use and Income**

Source	SS	df	MS = SS/df	F	Prob > F
Explained	16,568,747[a]	1[b]	16,568,747	7.38[c]	< .05[d]
Residual	69,613,677[e]	31[f]	2,245,602		
Total	86,182,424[g]	32[h]	2,693,201		

[a]Explained sum of squares: ESS = $\Sigma(\hat{Y}_i - \overline{Y})^2$ = 16,568,747
[b]Degrees of freedom for ESS: $df_E = K - 1 = 2 - 1 = 1$, where K is the number of coefficients (2 in bivariate regression).
[c]F statistic: $F_{df_R}^{df_E} = \dfrac{ESS/df_E}{RSS/df_R} = \dfrac{16,568,747/1}{69,613,677/31} = 7.38$
[d]Probability of $F \geq 7.38$, with 1 and 31 degrees of freedom (Table A.5)
[e]Residual sum of squares: RSS = $\Sigma(Y_i - \hat{Y}_i)^2$ = 69,613,677
[f]Degrees of freedom for RSS: $df_R = n - K = 33 - 2 = 31$
[g]Total sum of squares (Y): $TSS_Y = \Sigma(Y_i - \overline{Y})^2$ = 86,182,424 (Note that $TSS_Y = RSS + ESS$.)
[h]Degrees of freedom for TSS_Y: $df_T = n - 1 = 33 - 1 = 32$

An F test of the null hypothesis of no relationship ($\beta = 0$) between Y and all of the X variables in a regression is

$$F_{df_R}^{df_E} = \frac{MS_E}{MS_R}$$

$$= \frac{ESS/df_E}{RSS/df_R}$$ [15.14]

$$= \frac{\Sigma(\hat{Y}_i - \overline{Y})^2 / (K-1)}{\Sigma(Y_i - \hat{Y}_i)^2 / (n-K)}$$

Compare Table 15.9 with Table 13.12 (page 426) to see the similarities between F tests in regression and ANOVA. Although the regression F test is sometimes called an *analysis of variance,* the analysis is basically a regression, not ANOVA.

In bivariate regression, the F test always leads to the same conclusion as a two-sided t test. Specifically:

1. Both F and t tests test the null hypothesis H_0: $\beta = 0$.
2. The obtained P-values from F and two-sided t tests are the same.
3. The F statistic equals the t statistic squared, $F = t^2$.

By comparing Tables A.3 and A.5 you can see that two-sided P-values for a t distribution with df_R degrees of freedom are identical to P-values for an $F = t^2$ distribution with 1 and df_R degrees of freedom.

Table 15.9 shows that for the water use and income data, we obtain an F statistic of 7.38, with 1 and 31 degrees of freedom. Its probability is $P < .05$, so we can *reject* the null hypothesis of no relationship between Y (household water use) and X (household income) in the population. This conclusion is consistent with the earlier observation that the 95% confidence interval for β (found in Table 15.8) does not contain 0.

In bivariate regression, the F and two-sided t tests are redundant. The F statistic is less versatile than t, since it is not used to construct confidence intervals or one-sided tests. In multivariate analysis (Chapter 16), however, F statistics serve to test hypotheses that cannot be addressed by individual t statistics.

PROBLEMS

*17. The water conservation data set in Table 2.8 includes a rough measure of the amount of water saved during the water shortage: 1980 use minus 1981 use. When water savings (Y) are regressed on household income (X) for these 33 households, we obtain: $\hat{Y} = 1{,}079 - 26(X)$. Since b is negative, this regression suggests that wealthier households *saved less water* than poorer households did. Use an F test to determine whether the relationship is significant ($\alpha = .05$), using these sums of squares: $\text{TSS}_Y = 46{,}169{,}091$; $\text{ESS} = 7{,}757{,}530$; $\text{RSS} = 38{,}411{,}561$.

*18. The water conservation data set also records the head of household's education. The regression of water savings (Y) on education uncovers another negative sample relationship: $\hat{Y} = 1{,}315 - 63(X)$. Is this relationship significant at $\alpha = .05$? Use $\text{TSS}_Y = 46{,}169{,}091$, $\text{ESS} = 1{,}255{,}761$, and $\text{RSS} = 44{,}913{,}330$.

*19. a. Find and interpret the coefficient of determination, R^2, for the regression of water savings on income (Problem 17). Show how the correlation (r) can be obtained using the values of R^2 and b.

   b. Find and interpret R^2 and r for the regression of water savings on education (Problem 18).

15.9   ASSUMPTIONS AND PROBLEMS IN REGRESSION ANALYSIS

Like other inferential procedures, regression t tests, confidence intervals, and F tests depend on assumptions. Several of these assumptions can be checked against sample data, to see whether they are reasonable. Evaluating other assumptions requires information from beyond the sample: how the data were collected, or what causal processes relate X to Y. Certain statistical problems can also invalidate regression even as a summary of sample data. Some key assumptions and potential problem areas are outlined in this section.

Table 15.10 lists three kinds of problem—influential cases, nonlinear relationships, and omitted variables—that can undermine both descriptive and

TABLE 15.10   *Some Problems Invalidating Regression for Description and Inference*

1. Influential cases: They disproportionately influence the values of b and other sample statistics, making conclusions reflect only a small fraction of the data. Such cases also cast doubt on the normality assumption (see 4 in Table 15.11). *Informal checks:* Examine scatter plots of Y vs. X. If suspicious data points appear, redo the regression without those cases, and see whether results differ much.

2. X–Y relationship not linear: If the X–Y relationship is strongly curved, then trying to summarize it with a straight line is misleading. *Informal checks:* Examine scatter plots of Y vs. X or of e vs. $\hat{Y}$. Curvilinearity may disappear when you subject one or both variables to a nonlinear transformation reducing skewness.

3. Omitted third variables: Leaving out some important X variable(s) may make the sample estimate of b higher or lower than the true causal effect of X on Y. This problem arises only in causal interpretation. *Informal checks:* Omitted variables are difficult to detect and deal with. Experimental design is the best way to rule them out. In nonexperimental research, multivariate analysis offers a way to statistically control for the effects of some additional variables. Theories about the variables being studied may also provide guidance.

inferential aspects of regression. The table also suggests simple informal checks to help decide whether these problems affect your data. Often these informal checks involve scatter plots, either of Y vs. X or of the sample residuals against the predicted values of Y (such e vs. $\hat{Y}$ plots are discussed later). First, we will look briefly at each of the possible problems.

1. *Influential cases.* The problem here is that a small fraction of the data controls our conclusions. (Figure 14.16, page 493, showed the effect of an influential case in the mammoth bones and cave paintings data.) Influence is not inherently bad, but it is something to be aware of. If certain cases are influential, examine them closely and decide whether it is appropriate for them to dominate your final conclusions.

2. *Nonlinear relationships.* So far we have seen only straight-line regressions, but sometimes the true relationship is better described by a curve. Curvilinearity often shows up in scatter plots, which is one reason to study such plots early in any regression analysis. If the relationship is not linear, then it is a mistake to push ahead with line-fitting. Instead, we might try fitting an appropriate curve. Simple ways to do so are shown later in this chapter.

3. *Omitted variables.* They are a problem especially in causal research (see introduction to Part III). Failing to control for the effects of some variable that causes both X and Y may make us over- or underestimate how much X itself affects Y. In practice, this problem does not require *no* omitted variables, but rather no *important* omitted variables. If the variables left out

have only weak relationships with X and Y, then omission should not distort our analysis of the X–Y relationship.

The omitted variables problem is difficult to solve; more discussion and an example are presented later. Experimental design, multivariate statistics, and theory may help. But if we are not interested in causality, wanting only to describe a statistical relationship between X and Y, omitted variables matter less.

Some additional problems that affect inferential (but not descriptive) aspects of regression are listed in Table 15.11. The inferential methods in this chapter assume that none of these problems is present. A mathematical proof, the **Gauss–Markov Theorem,** asserts that ordinary least squares (OLS) regression is indeed the best estimation method *if* these and certain other assumptions are met. These assumptions are often not justified in practice, however, and we should view them with some skepticism. We shall discuss each assumption in turn.

4. *Nonnormality.* We justify using t and F distributions in regression by making a normality assumption about the population: We assume Y has a normal distribution at every level of X. Scatter plots of Y vs. X, or of e vs. $\hat{Y}$, provide visual evidence about how reasonable this assumption is. Box

TABLE 15.11	*Some Problems Invalidating Regression for Inference*

4. Nonnormal distribution of Y: Using the theoretical t and F distributions is justified by the assumption that the population distribution of Y is normal for every possible value of X. This assumption is most important in small samples, although outliers, bimodality, or severely skewed distributions can cause difficulties in samples of any size.
Informal checks: Examine e vs. $\hat{Y}$ scatter plots and box plots of e. Mean–median and standard deviation–pseudo-standard deviation comparisons can also be applied to e. If residuals show marked skewness or severe outliers, consider nonlinear transformation or outlier deletion.

5. Nonconstant variance of Y (heteroscedasticity):   We assume that Y is equally variable at all levels of X. Such homoscedasticity is required for the usual standard errors, significance tests, and confidence intervals to be valid.
Informal checks: Scatter plots of e vs. $\hat{Y}$ should reveal whether heteroscedasticity is present in the sample. Look especially for a pattern of residuals that fans out (widens) from left to right or right to left across the plot. Nonlinear transformations may reduce this problem.

6. Cases not independent: Standard errors, significance tests, and confidence intervals calculated in the usual manner may be misleading if cases are not independent. Autocorrelation often occurs in time series data.
Informal checks: Think about how the data were collected, and whether there are any plausible relationships between the cases. Sort by case number to see if the residuals show systematic runs of positive or negative values.

plots of e or Y are also helpful. Mean–median and standard deviation–pseudo-standard deviation comparisons (Chapters 3 and 4) can also help in checking the normality assumption. In large samples, this assumption is somewhat less important. Severe outliers or skewness, however, can cause problems in samples of any size.

5. *Nonconstant variance.* Constant variance implies that the population variance (or its square root, the standard deviation) of Y is the same at all levels of X. This assumption is required for regression t and F tests just as for difference-of-means tests or analysis of variance. The technical term for constant variance is **homoscedasticity**. If Y's variation is *not* constant, it shows **heteroscedasticity**. Examples are provided in later sections.

6. *Dependence.* The cases are assumed to be independent of each other. This should be true if sample cases were individually and randomly selected from a large population. It is typically *not* true when the variables are time series, like the camping data in Table 2.6 (page 36). The regression methods described in this book may be misleading when used to analyze relationships between time series. Better methods and the issues involved are discussed in the many specialized books about time series analysis.[1]

The requirements that the cases be independent (6) and no important variables are omitted (3) raise difficult statistical and theoretical issues that trouble even sophisticated analysts. On the other hand, influence, linearity, normal distributions, and constant variation *are* all relatively easily checked, at least informally. The next section describes a simple graphical approach to identifying such problems.

15.10 *SCATTER PLOTS FOR REGRESSION CRITICISM*

The problems summarized in Tables 15.10 and 15.11 are often easier to detect after an initial regression has been done. This process is called statistical **criticism**: Results from an analysis are checked for possible problems. If problems do turn up, criticism may also suggest how the situation could be improved. In recent years sophisticated techniques have been developed for regression criticism.[2] A simple yet incisive graphical approach is described here.

Earlier we confirmed a positive, statistically significant relationship of household water use to income (Tables 15.7–15.9). Figure 15.4 (page 530) shows the scatter plot and regression line this conclusion rests on. The plot's most conspicuous feature is the single high use, high income household at the upper right. Does this one household distort the regression line, steepening its slope and making the use–income relationship seem stronger? The easiest way to test this possibility is to redo the regression using only the other 32 households. If we reach essentially the same conclusion either way, then outlier influence is not a problem. On the other hand, it is possible that

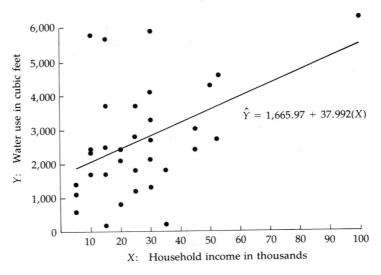

$$\hat{Y} = 1,665.97 + 37.992(X)$$

FIGURE 15.4   *Regression of water use on income*

our conclusion about water use and income rests mainly on this one household, which happens to have both high use and high income.

Potentially influential cases show up well in Y vs. X scatter plots. For
some other problems in Tables 15.10 and 15.11, an **e vs. $\hat{Y}$ plot** is more revealing. This is a plot graphing the residuals (e) on the vertical axis and
the predicted values of Y ($\hat{Y}$) on the horizontal axis.

Figure 15.5 shows an e vs. $\hat{Y}$ plot for water use and income. The "Y"
variable in Figure 15.5 is the regression residuals (e), from the last column of
Table 15.7. The "X" variable in Figure 15.5 is the regression predicted values
($\hat{Y}$), the next column in Table 15.7. Thus, household 1 is graphed at ($\hat{Y}$, e)
coordinates (2,235.85, 3,464.15), household 2 is at (2,805.73, 494.27), and so
on. Box plots for e and $\hat{Y}$ in the margins of Figure 15.5 show distributional
shapes. A horizontal line drawn at $e = 0$, which is the mean of the residuals,
also represents the regression line. Cases plotted above this line (like household 1) have positive residuals, meaning that their water use was *higher* than
income would predict. Cases plotted below this line have negative residuals,
meaning that their water consumption was *lower* than predicted from income.

Compare the e vs. $\hat{Y}$ plot in Figure 15.5 with the Y vs. X plot in Figure
15.4 to see the point-by-point correspondence between them. The high use,
high income household we suspected of being influential does not have a
large residual, so it does not stand out as clearly in this plot as in Figure 15.4.
The horizontal box plot at the top of Figure 15.5 reveals that this household
is a severe outlier in the $\hat{Y}$ distribution, however. Because $\hat{Y}$ is a linear function of X, the $\hat{Y}$ distribution has exactly the same shape as the X distribution;
hence this household must be a severe outlier on X as well. The vertical box

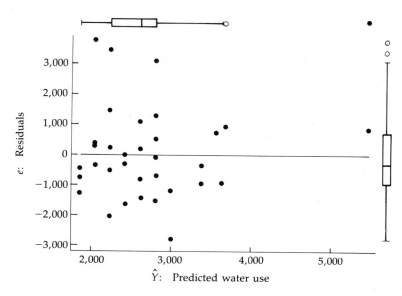

FIGURE 15.5   *e vs. Ŷ plot for regression of Figure 15.4*

plot at right (*e* distribution) shows that the residuals are reasonably symmetrical, which is consistent with the normality assumption.

In two-variable regression, the *e* vs. $\hat{Y}$ plot resembles the *Y* vs. *X* plot, but with the points scattered around a regression line brought horizontal at $\bar{e} = 0$.

Figure 15.6 (page 532) shows an *e* vs. $\hat{Y}$ plot for the city homicide rate and income data, graphed earlier in Figure 15.1. The relationship between homicide rate and median family income is negative, so the rotation to a horizontal regression line is not the only difference between Figures 15.1 and 15.6. The positively skewed *X* distribution in the *Y* vs. *X* plot, Figure 15.1, transforms to a negatively skewed $\hat{Y}$ distribution in the *e* vs. $\hat{Y}$ plot, Figure 15.6. The vertical box plot in Figure 15.6 gives us no reason to doubt the normality assumption, and there are no severe outliers. Otherwise the scatter is hard to describe. We will return to it later, after viewing more examples of what *e* vs. $\hat{Y}$ plots might reveal.

In bivariate regression the $\hat{Y}$ and *X* distributions have the same shape. If the regression slope *b* is negative, however, the $\hat{Y}$ distribution resembles a left-to-right (low-to-high) reversal of the *X* distribution.

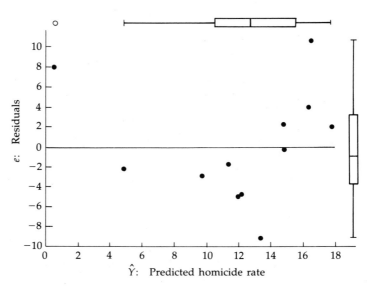

FIGURE 15.6   *e vs. Ŷ plot for regression of Figure 15.1*

The best situation in regression criticism is finding nothing to criticize. Figure 15.7 depicts an e vs. $\hat{Y}$ plot that sounds an "all clear" signal:

1. Both the e and $\hat{Y}$ distributions are approximately normal, with no outliers or skewness. This reassures us about points 1 and 4 in Tables 15.10 and 15.11.
2. The scatter of dots around the $\bar{e} = 0$ line does not fan out in either direction, suggesting that Y has the same variability at all levels of X. This implies that heteroscedasticity (point 5) is not a problem.
3. There is no indication that we are systematically over- or underpredicting Y anywhere along the regression line, as could happen if the relationship were not linear (point 2).

Based on this fictitious data plot, we have no reason to question the validity of the original regression.

Figure 15.8, in contrast, contains a rogue's gallery of e versus $\hat{Y}$ plots that mean trouble. At the upper left we see how an influential case might appear in such a plot. The rightmost case controls the regression line, so has a very small residual, but it shows up here as a severe outlier on $\hat{Y}$ (and therefore, X).

The upper right plot of Figure 15.8 illustrates nonlinearity. Instead of being scattered randomly above and below the 0 line, the residuals are consistently negative at low values of $\hat{Y}$, become consistently positive, then turn negative again at high values of $\hat{Y}$. Such U or inverted-U patterns in residuals indicate that we should try fitting a curve, rather than a straight line, to the original X–Y data.

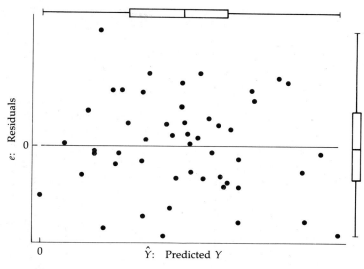

FIGURE 15.7   "All clear" e vs. Ŷ plot

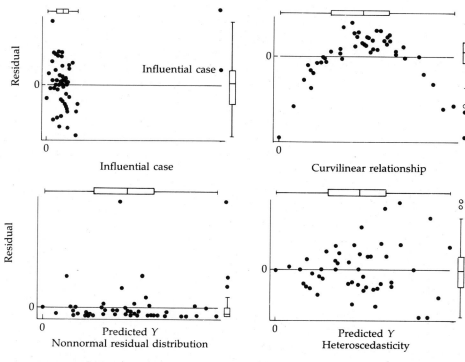

FIGURE 15.8   Problems e vs. Ŷ plots reveal

At the lower left in Figure 15.8 the residuals have a grossly nonnormal distribution. They are highly skewed: Most residuals lie below the 0 line, with a handful of moderate and severe outliers above it. The assumption that Y has a normal distribution at all levels of X (point 4 in Table 15.11) is refuted by these data.

The fourth plot in Figure 15.8 shows the most common kind of heteroscedasticity: The scatter of residuals fans out from 0 like a megaphone. This tells us that Y is less variable, hence more predictable, at low levels of X than at high ones. More generally, heteroscedasticity (point 5) shows up when the scatter of dots around $\bar{e} = 0$ systematically widens or narrows as $\hat{Y}$ increases.

Influential cases are less clear in e vs. $\hat{Y}$ plots than in ordinary Y vs. X plots. For example, the upper left graph in Figure 15.8 depicts a severe influence problem, but the only evidence is an unusually high value of $\hat{Y}$. Figure 15.9 plots Y vs. X for the same data, exposing the influential point at the upper right: It is a case with high X and high Y values. The steeper of the two regression lines is based on all 50 points in the data. It has a statistically significant positive slope—but that slope is simply tracking the influential case. The flatter regression line shows what we get if we delete this outlier and redo the regression with only the remaining 49 cases: a nonsignificant, weakly negative slope.

Among 49 of the 50 cases in Figure 15.9, Y is unrelated to X. But a very high Y and a very high X coincide once, and this single case has enough influence to create an apparent relationship. Had we not looked at these graphs, we might have reported a "significant positive relationship," unaware that this conclusion reflects only a single case.

The data in Figures 15.7–15.9 were invented to provide clear-cut examples. Real data often present more ambiguous pictures that call for judgment

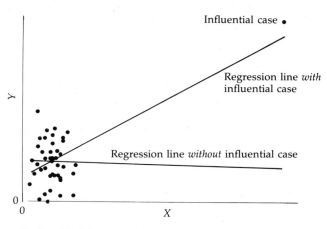

FIGURE 15.9   *Influential case in Y vs. X plot*

and careful thought. You will have several chances to test your own judgment in the next problem set. In the next section we will examine other real data examples and consider what to do when our critical graphs warn of trouble.

PROBLEMS

*20. Make up e vs. $\hat{Y}$ plots of your own, different from those shown in Figure 15.8, to illustrate each of the following problems.

   a. Nonconstant variance

   b. Nonlinearity

   c. Influence

   d. Nonnormality

*21. Examine the e vs. $\hat{Y}$ plot for city homicides and income (Figure 15.6). How does this plot compare with the examples in Figures 15.7 and 15.8? Is there any sign of trouble? Consider the original Y vs. X plot (Figure 15.1) in formulating your answer.

*22. Use the mammoth bones and cave art data from Table 14.4 (page 475) to construct your own e vs. $\hat{Y}$ plot (marginal box plots are an optional effort). In Chapter 14, we concluded that the Dordogne River region was a mildly influential case. How does it appear in the e vs. $\hat{Y}$ plot?

15.11   COPING WITH PROBLEMS IN REGRESSION

The preceding section showed how to detect some common regression pitfalls. If criticism uncovers no major problems, you can proceed with analysis more confidently. But the presence of any of the problems shown in Figure 15.8 warns that the initial analysis may be invalid. What can you do then?

Chapter 6 introduced two tactics for dealing with ill-behaved univariate distributions: outlier deletion and nonlinear transformation. Both are of great help in regression.

Outlier deletion was illustrated with the mammoth bones and cave paintings regression (Figure 14.16) and the fictitious example of influence (Figure 15.9). No special techniques are involved; just do the regression with and without the suspicious case. If the two results are substantially different, we indeed have an influence problem. We must then decide whether to remove the influential case or whether it contains important information worth retaining.

We have used nonlinear transformations to make skewed univariate distributions more symmetrical. The transformations do the same in regression, bringing either X or Y closer to the ideal normal curve. This often reduces problems of influence or nonnormality.

Two more properties of nonlinear transformations are useful in regression. First, they may reduce heteroscedasticity, making the variation of Y more nearly the same at all levels of X. Second, they provide a simple way to fit curves, instead of straight lines. When either X or Y has been transformed, the application of linear regression methods results in a **curvilinear regression** with respect to the original untransformed variables.[3]

Nonlinear transformations thus help with four of the six problems described: influence, nonlinearity, nonnormality, and heteroscedasticity. Furthermore, the same transformation that makes the univariate distributions of X or Y more symmetrical often improves the regression with respect to influence, linearity, normality, and homoscedasticity as well. These multiple advantages make nonlinear transformations a powerful supplement to basic regression techniques. Transformation is not a magic wand; many ill-behaved distributions are hard to improve and nonlinear transformations cannot help. But they often do succeed, as the following examples demonstrate.

Table 15.12 contains gross national products (X) and life expectancies (Y) of the 24 countries first seen in Table 1.6. Figure 15.10 is a scatter plot of these data, with the regression line

$$\hat{Y} = 55.9 + .0012(X) \qquad [15.15]$$

The line's positive slope informs us that in countries with higher per capita GNP, life expectancy also tends to be higher. This conclusion sounds reasonable, but note the poor fit between line and data.

Figure 15.11 (page 538), an e vs. $\hat{Y}$ plot, displays an inverted fishhook pattern—evidence of nonlinearity. (Compare Figure 15.8, upper right.) Another problem is signalled by the horizontal box plot: The predicted values, hence also X (GNP) itself, are positively skewed. We suspect an outlier of being an influence point: Brunei, a small oil-exporting sultanate on the island of Borneo, whose people have a very high per capita GNP ($22,260) but only a middling life expectancy (66 years).

A third problem to consider is bimodality (two peaks). The sample distribution of life expectancy is bimodal (Figure 3.2, page 61). If the regression residuals are also bimodal, this is further evidence against the normality assumption. (Problem 50 explores bimodality more.)

Positively skewed distributions like GNP can often be made more symmetrical by a nonlinear transformation such as the logarithm or square root (Chapter 6). For GNP, logarithms work well. Table 15.13 (page 539) gives the base-10 logarithm of per capita GNP for each country.[4] By transforming, we create a new variable, $X^* = \log(X)$. If we regress life expectancy on the *log* of per capita GNP (X^*) instead of on per capita GNP itself (X), we obtain the line

$$\hat{Y} = 14.21 + 14.47(X^*)$$

or

$$\hat{Y} = 14.21 + 14.47[\log(X)] \qquad [15.16]$$

TABLE 15.12   *Per Capita Gross National Product and Life Expectancy, 1982*

Case	Country	GNP in dollars	Life expectancy in years
i		X_i	Y_i
1	Nicaragua	950	58
2	Paraguay	1,670	65
3	Venezuela	4,250	68
4	France	11,520	74
5	West Germany	12,280	73
6	Greece	4,170	73
7	Norway	14,300	75
8	Czechoslovakia	5,540	71
9	Austria	9,830	72
10	Jordan	1,680	61
11	Sri Lanka	320	67
12	Brunei	22,260	66
13	Indonesia	550	50
14	North Korea	930	66
15	Mongolia	940	64
16	Taiwan	2,670	72
17	Australia	11,220	74
18	Congo	1,420	48
19	Ethiopia	150	41
20	Guinea	330	44
21	Mauritania	520	44
22	Nigeria	940	49
23	Togo	350	48
24	Zaire	180	48

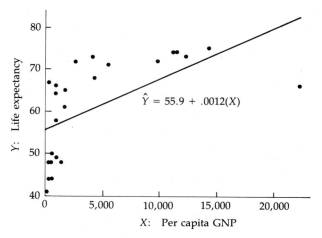

FIGURE 15.10   *Regression of life expectancy on GNP*

This line is shown with a scatter plot of Y vs. X^*—that is, Y vs. log (X)—in Figure 15.12.

Comparing Figure 15.12 with Figure 15.10, we find the log of GNP has wrought improvements:

1. Nonlinearity is now much reduced.
2. The log of GNP has a nearly symmetrical distribution, with no apparent outliers or influence points.

The nonlinear transformation that makes the univariate distribution of GNP more symmetrical also has the bivariate benefits of reducing problems with nonlinearity and outlier influence. For both descriptive and inferential purposes, the regression of Equation [15.16] and Figure 15.12 is better than that of Equation [15.15] and Figure 15.10.

Although nonlinear transformations can improve the validity of regression analysis, the resulting equations take more effort to interpret. The straight line of Equation [15.15] implies that for every additional $1,000 per capita GNP average life expectancy rises 1.2 years. Not so! As the scatter plot of Figure 15.10 testifies, life expectancy does *not* rise by a constant amount per dollar of GNP. Nonetheless, the temptation to use the linear regression is strong because its terms, unlike log terms, are familiar: dollars and years.

The curvilinear regression equation, [15.16], suggests instead that for every 1-unit increase in the *logarithm* of per capita GNP, predicted life expectancy goes up 14.47 years. Many people find the statement in log terms baffling, for they do not see what it means in dollars.

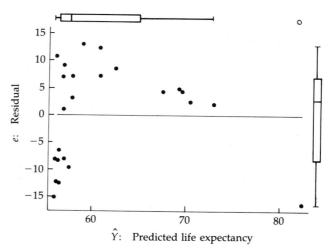

FIGURE 15.11   *e vs. $\hat{Y}$ plot for Figure 15.10 regression*

TABLE 15.13   *Per Capita GNP with Its Base-10 Log and Life Expectancy, 1982*

Case	Country	GNP in dollars	Log of GNP	Life expectancy in years
i		X_i	$X_i^* = \log(X_i)$	Y_i
1	Nicaragua	950	2.98	58
2	Paraguay	1,670	3.22	65
3	Venezuela	4,250	3.63	68
4	France	11,520	4.06	74
5	West Germany	12,280	4.09	73
6	Greece	4,170	3.62	73
7	Norway	14,300	4.16	75
8	Czechoslovakia	5,540	3.74	71
9	Austria	9,830	3.99	72
10	Jordan	1,680	3.23	61
11	Sri Lanka	320	2.51	67
12	Brunei	22,260	4.35	66
13	Indonesia	550	2.74	50
14	North Korea	930	2.97	66
15	Mongolia	940	2.97	64
16	Taiwan	2,670	3.43	72
17	Australia	11,220	4.05	74
18	Congo	1,420	3.15	48
19	Ethiopia	150	2.18	41
20	Guinea	330	2.52	44
21	Mauritania	520	2.72	44
22	Nigeria	940	2.97	49
23	Togo	350	2.54	48
24	Zaire	180	2.26	48

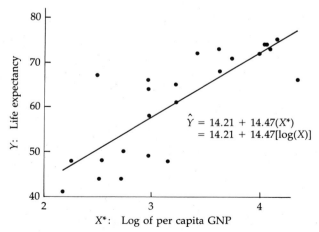

$$\hat{Y} = 14.21 + 14.47(X^*)$$
$$= 14.21 + 14.47[\log(X)]$$

FIGURE 15.12   *Regression of life expectancy on log(GNP)*

Equation [15.16] describes a straight line relationship between Y and $\log(X)$, hence implies that the relationship between Y and X itself is curved. Curvilinear relationships can be hard to describe in words; they are more easily grasped by viewing graphs.

PROBLEMS

In the 1980's many people discovered in alarm that they lived near toxic waste dumps. Such dumps have been found in every state. Citizen response is often to form activist groups; one of the first was the Love Canal Homeowners Association. Hundreds now exist, with more formed weekly. Below are data on a random sample of nine U.S. states. The variables are number of federally recognized hazardous waste dumps (X) and number of anti-dump groups known in each state (Y).

State	X	Y
Colorado	15	9
Delaware	14	0
Hawaii	6	1
New York	65	63
Oklahoma	5	3
South Dakota	1	1
Tennessee	8	20
West Virginia	6	7
Wisconsin	30	5

*23. Regress number of anti-dump groups (Y) on number of dumps (X), and construct an e vs. $\hat{Y}$ plot. What statistical problems do you see?

*24. Take base-10 logarithms of the number of dumps and number of anti-dump groups. *First add 1 to each case to eliminate 0 values.* Then plot $Y^* = \log(Y + 1)$ vs. $X^* = \log(X + 1)$. How does taking logs affect the distributional problems?

*25. Construct a stem-and-leaf display of Y to explain why it is reasonable to use the transformation $Y^* = \log(Y + 1)$ rather than $Y^* = Y^2$ in Problem 24. Why is it necessary to "eliminate 0 values" by adding 1's before we take logarithms?

*26. Regress your transformed Y, Y^*, on your transformed X, X^*, from Problem 24. Interpret the intercept and slope in terms of the transformed variables.

*27. Is there a significant positive relationship between log(number of dumps) and log(number of anti-dump groups)? Apply a one-sided t test at $\alpha = .05$.

15.12 *UNDERSTANDING CURVILINEAR REGRESSION*

A regression equation defines the predicted value of Y for any value of X. If we find the $\hat{Y}$ values that correspond to each X value in our sample, we can plot a series of points $(X_i, \hat{Y}_i)$ that lie on the regression line. Connecting these points shows us the line itself. The same principle can be used to show regression curves. Any two points plus a ruler are enough to draw a straight line, but it may take a number of $(X_i, \hat{Y}_i)$ points to approximate well the shape of a curve.

Table 15.14 (page 542) shows X_i, log(X_i), and $\hat{Y}_i$ values for life expectancy and GNP. The predicted values ($\hat{Y}_i$) are found by substituting each country's X value into Equation [15.16]. Since Y itself was not transformed, the predicted values are measured in the same units as Y—years of life expectancy. When these $\hat{Y}$ values are plotted against X, they follow a curve (Figure 15.13). Figure 15.14 superimposes this curve of predicted values on the original scatter plot of (X_i, Y_i) data points (Figure 15.10). Note that Figure 15.14 is otherwise an ordinary Y vs. X plot, life expectancy in years vs. GNP in dollars.

The curve in Figure 15.14 fits the data visibly better than does the straight line in Figure 15.10. We can compare their fit quantitatively as well. The coefficient of determination for the straight line is $R^2 = .37$. A comparable statistic for curves is *the squared correlation between observed and predicted Y values,* which is here $r^2 = .66$. Thus, the curvilinear relationship with GNP explains about 66% of the variance in life expectancy, whereas the linear relationship explains only about 37%.[5]

Figure 15.14 shows that average life expectancy rises steeply from very poor to moderately poor countries. The lowest life expectancies are in countries with a per capita GNP below $600, such as Ethiopia, Guinea, or Mauritania, where life expectancy is in the 40's. At $1,000 per person—only a few hundred dollars more—only one country, Congo, has a life expectancy below 60 years. It seems that a few dollars (and the concomitant national infrastructure for safe water, health care, etc.) make a lot of difference at the low end of the GNP scale. At the high end, on the other hand, life expectancy rises very little with GNP.

This curvilinear description of the life expectancy–GNP relationship is more true than the straight-line notion that every dollar increase in GNP adds b years to life expectancy. The nonlinear transformation thus led to a more realistic conclusion. We got there, ironically, by taking the seemingly unrealistic step of converting measurements into logarithms of dollars.

TABLE 15.14   *Log per Capita GNP and Predicted Life Expectancy*

Case	Country	GNP in dollars		Predicted life expectancy in years
i		X_i	$log(X_i)$	$\hat{Y}_i = 14.21 + 14.47[log(X_i)]$
1	Nicaragua	950	2.98	57.33
2	Paraguay	1,670	3.22	60.80
3	Venezuela	4,250	3.63	66.74
4	France	11,520	4.06	72.96
5	West Germany	12,280	4.09	73.39
6	Greece	4,170	3.62	66.59
7	Norway	14,300	4.16	74.41
8	Czechoslovakia	5,540	3.74	68.33
9	Austria	9,830	3.99	71.95
10	Jordan	1,680	3.23	60.95
11	Sri Lanka	320	2.51	50.53
12	Brunei	22,260	4.35	77.15
13	Indonesia	550	2.74	53.86
14	North Korea	930	2.97	57.19
15	Mongolia	940	2.97	57.19
16	Taiwan	2,670	3.43	63.84
17	Australia	11,220	4.05	72.81
18	Congo	1,420	3.15	59.79
19	Ethiopia	150	2.18	45.75
20	Guinea	330	2.52	50.67
21	Mauritania	520	2.72	53.57
22	Nigeria	940	2.97	57.19
23	Togo	350	2.54	50.96
24	Zaire	180	2.26	46.91

Finding $\hat{Y}$ values to connect for the curve was easy because the Y variable was not transformed. If Y is transformed, then one additional step is required. This will be illustrated using the data in Table 15.15 (page 544) on geographic variations in motor vehicle fatality rates. The cases are 17 Nevada counties. Variables are population density (X, in people per square mile), and average annual motor vehicle fatality rate, 1979–1981 (Y, in deaths per 100,000 population per year).

Figure 15.15 shows the regression of fatality rate on population density. The fatality rate is lower in more densely populated counties:

$$\hat{Y} = 156 - .83(X) \qquad\qquad [15.17]$$

A t test for the null hypothesis H_0: $\beta = 0$ leads to a P-value of $.20 < P < .50$, however: The relationship is not statistically significant. Failure to reject H_0

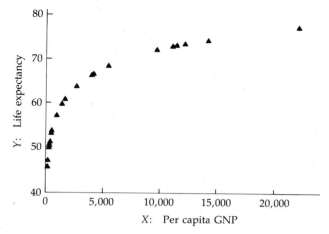

FIGURE 15.13   *Predicted Y from regression of Y on X**

suggests that the negative relationship seen in Figure 15.15 could well be due to chance.

Figure 15.16, an e vs. $\hat{Y}$ plot for the fatalities–density regression, exposes several glaring statistical problems. The X variable, population density, is severely skewed. The box plot for $\hat{Y}$, which has the same shape as the X distribution, is crowded into one corner, with a long tail of outliers. The worst outlier is Ormsby–Carson City, whose population density is nearly one hundred times higher than the median. The residuals distribution is also

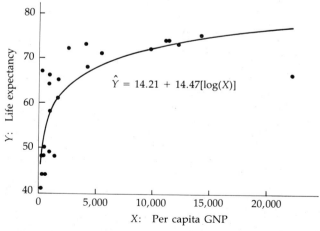

FIGURE 15.14   *Curvilinear regression of life expectancy on GNP*

TABLE 15.15   ***Motor Vehicle Fatality Rate and Population Density for Nevada by County***

Case	County	Density (pop./mi²)	Fatality rate (Deaths/100,000/yr)
i		X_i	Y_i
1	Washoe	30.7	17.56
2	Ormsby–Carson City	219.9	26.02
3	Clark	58.8	27.35
4	Douglas	27.4	53.21
5	Mineral	1.7	64.34
6	Churchill	2.8	69.46
7	White Pine	.9	93.87
8	Elko	1.0	94.58
9	Lyon	6.8	95.63
10	Storey	5.7	110.89
11	Lander	.7	130.85
12	Lincoln	.4	133.98
13	Nye	.5	136.31
14	Humboldt	1.0	173.13
15	Pershing	.6	185.84
16	Eureka	.3	389.54
17	Esmeralda	.2	557.70

Source: Data courtesy of Susan Baker; see Baker, Whitfield, and O'Neill (1987).

skewed, dominated by the exceptionally high fatality rates in Eureka and
Esmeralda Counties. The *e* vs. $\hat{Y}$ plot further suggests that heteroscedasticity
is a problem: The residuals are most variable at high values of $\hat{Y}$, which
correspond to low values of X. (Remember that with a negative relationship,

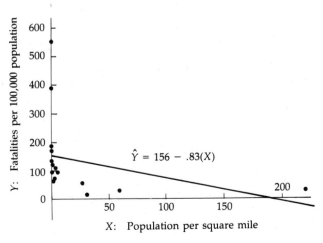

FIGURE 15.15   *Vehicle fatality rate vs. density*

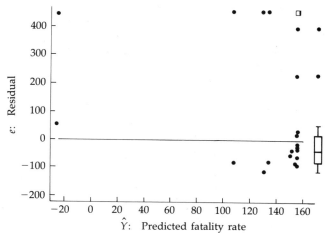

FIGURE 15.16   *e* vs. *Ŷ* plot for Figure 15.15 regression

the $\hat{Y}$ distribution resembles a left-to-right reversal of the X distribution.) We thus see potential problems with influence, nonnormality, and heteroscedasticity. A t or F test should not be trusted, since the basic assumptions are probably false. Even as a purely descriptive statement, Equation [15.17] is suspect because of the obviously influential position of Ormsby–Carson City.

Like Figures 15.10 and 15.11, Figures 15.15 and 15.16 are a fine illustration of when *not* to trust regression. Yet simply taking logarithms of both fatality rate (Y) and population density (X) makes the statistical problems mostly vanish. Logarithms for the 17 county densities and fatality rates are given in Table 15.16 (page 546), and Figure 15.17 shows the regression of $\log(Y)$ on $\log(X)$:

$$\log(\hat{Y}) = 2.14 - .383[\log(X)] \qquad [15.18]$$

The points in Figure 15.17 are spread evenly around the regression line, and the outliers and heteroscedasticity are gone. Unlike Equation [15.17], the negative relationship in Equation [15.18] *is* statistically significant, and we have less reason to disbelieve the t or F test. The appropriate use of nonlinear transformations has again overcome several problems at once.

Equation [15.18] may be valid, but what does it mean? A graph will help us interpret it, but we cannot simply plug X values into the equation—our predicted values will be in logarithms too, since Y was also transformed for this analysis. To return Y to its original units, fatalities per 100,000 people, we need to take antilogarithms of the predicted values from Equation [15.18]. Since the logs here are base-10, we just raise 10 to the power of the predicted values. Table 15.17 shows the results.

TABLE 15.16
Nevada Motor Vehicle Fatality Rate and Population Density, with Base-10 Logarithms

Case	County	Density[a]	Fatality rate[b]	Log density[c]	Log fatality rate[d]
i		X_i	Y_i	$log(X_i)$	$log(Y_i)$
1	Washoe	30.7	17.56	1.49	1.24
2	Ormsby–Carson City	219.9	26.02	2.34	1.42
3	Clark	58.8	27.35	1.77	1.44
4	Douglas	27.4	53.21	1.44	1.73
5	Mineral	1.7	64.34	.23	1.81
6	Churchill	2.8	69.46	.45	1.84
7	White Pine	.9	93.87	−.05	1.97
8	Elko	1.0	94.58	0	1.98
9	Lyon	6.8	95.63	.83	1.98
10	Storey	5.7	110.89	.76	2.04
11	Lander	.7	130.85	−.15	2.12
12	Lincoln	.4	133.98	−.40	2.13
13	Nye	.5	136.31	−.30	2.13
14	Humboldt	1.0	173.13	0	2.24
15	Pershing	.6	185.84	−.22	2.27
16	Eureka	.3	389.54	−.52	2.59
17	Esmeralda	.2	557.70	−.70	2.75

[a]Units: people per square mile (1980).
[b]Units: motor vehicle occupant fatalities per 100,000 pop. per year (1979–1981).
[c]Units: log(people per square mile).
[d]Units: log(fatalities per 100,000 people).

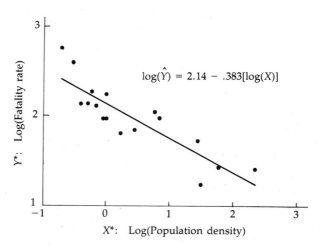

$$log(\hat{Y}) = 2.14 - .383[log(X)]$$

FIGURE 15.17   *Log(fatality rate) vs. log(density)*

TABLE 15.17   *Predicted Values and Their Antilogs from Regression of Log Vehicle Fatality Rate on Log Population Density*

Case	County	Log density[a]	Predicted log fatality[b]	Antilog predicted log fatality[c]
			$\log(\hat{Y}_i) =$	
i		$\log(X_i)$	$2.14 - .383[\log(X_i)]$	$10^{\log(\hat{Y}_i)}$
1	Washoe	1.49	1.569	37.07
2	Ormsby–Carson City	2.34	1.244	17.54
3	Clark	1.77	1.462	28.97
4	Douglas	1.44	1.588	38.73
5	Mineral	.23	2.052	112.72
6	Churchill	.45	1.968	92.90
7	White Pine	−.05	2.159	144.21
8	Elko	0	2.140	138.04
9	Lyon	.83	1.822	66.37
10	Storey	.76	1.849	70.63
11	Lander	−.15	2.197	157.40
12	Lincoln	−.40	2.293	196.34
13	Nye	−.30	2.255	179.89
14	Humboldt	0	2.140	138.04
15	Pershing	−.22	2.224	167.49
16	Eureka	−.52	2.339	218.27
17	Esmeralda	−.70	2.408	255.86

[a]Units: log(people per square mile).
[b]Units: log(fatalities per 100,000 population).
[c]Units: fatalities per 100,000 population.

To sketch the curve implied by any two-variable regression equation in which the Y variable has been transformed to Y^*:

1. Substitute X_i values from the data into the regression equation to obtain the corresponding predicted values $\hat{Y}_i^*$, which will be in *transformed units* (logs, square roots, etc.).
2. Apply the appropriate inverse transformation (see Table 6.8, page 168) to those predicted values, to return them to the *original units of Y* (years, fatalities, etc.).
3. Plot the resulting $(X_i, \hat{Y}_i)$ points on a Y vs. X scatter plot, and connect them with line segments.

In Washoe County, for example, the population density is 30.7 people per square mile, so the log density is $\log(30.7) = 1.49$. Substituting this value into the regression equation, [15.18], gives us the predicted *log* fatality rate:

$$\log(\hat{Y}_1) = 2.14 - .383[\log(X_1)]$$

$$= 2.14 - .383(1.49) = 1.569$$

To see what *fatality rate* this corresponds to, we take the antilog of 1.569:

$$\hat{Y}_1 = 10^{\log(\hat{Y}_1)} = 10^{1.569} = 37.07$$

If we predict that the *log* of the fatality rate is 1.569, this is equivalent to predicting that the fatality rate itself will be 37.07.

In taking antilogs of the predicted Y values, we undid the original transformation of Y. This is necessary whenever Y itself has been subjected to a nonlinear transformation. The Y transformation is undone using an appropriate inverse transformation, like those shown in Table 6.8 (page 168). Inverse transformations take us back to the original units of Y. If Y was originally measured in fatalities per 100,000 population, then the antilog of the predicted log of Y will also be in fatalities per 100,000 population. Thus we predict 37.07 fatalities per 100,000 in Washoe County, 17.54 fatalities per 100,000 in Ormsby–Carson City, and so forth.

Figure 15.18 shows the curve traced out by plotting $(X_i, \hat{Y}_i)$ values on a scatter plot of fatality rate versus population density. To quantify how well the curve fits we can find the squared correlation between the original Y values (Y_i column, Table 15.16) and the predicted Y values in natural units (right column, Table 15.17). This squared correlation is $r^2 = .65$, a big improvement over $R^2 = .10$ for the straight-line regression in Figure 15.15.

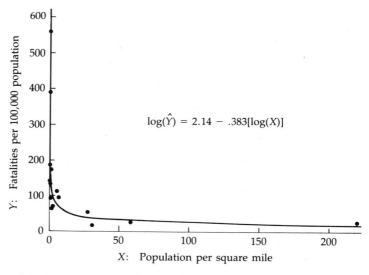

FIGURE 15.18   *Vehicle fatality rate vs. density*

> To summarize how well a curve fits the raw data, calculate the squared correlation between Y and $\hat{Y}$ values. If Y has been transformed, use an inverse transformation to return $\hat{Y}$ to the units of the original data before finding this correlation.

The squared correlation between Y and $\hat{Y}$ is analogous to the coefficient of determination, R^2. In a linear regression, the squared correlation between Y and $\hat{Y}$ *is* R^2.

The curve in Figure 15.18 fits the data better, and it also brings out a new aspect of the negative relationship between fatality rate and population density. The fatality rate plummets as we go from the sparsest counties to those slightly more dense. In sparsely populated counties such as Esmeralda or Eureka, the vehicle death rate is many times higher than it is even in uncrowded counties like Elko (one person per square mile).[6] At densities above 40 people per square mile, the fatality rate declines much less steeply. This contrasts with the straight-line implication that fatality rates fall steadily as density goes up. As Figure 15.15 shows, if that were true Ormsby–Carson City should have about -27 fatalities per 100,000 people! Its actual fatality rate, not negative 27 but positive 26.02, is closer to the curved-line prediction of 17.54 (see Table 15.17). Thus, another advantage of curvilinear regression is that it results in more reasonable predictions.

In the examples of both life expectancy–GNP and fatality rate–population density, criticism uncovered major statistical problems with the initial straight-line regression. Nonlinear transformations provided a way to solve these problems. The resulting regression curves not only are more valid statistically, but make better real-world sense as well.

PROBLEMS

*28. Return to the toxic waste dump analysis of Problems 23–27. Construct a Y vs. X scatter plot, calculate $(X_i, \hat{Y}_i)$ points, and sketch in the implied regression curve.

*29. Since there are fewer individuals to bring in and spread infection, epidemics tend to be less frequent in small, isolated communities. Bartlett (1978) demonstrated this for measles epidemics in 19 English and Welsh towns over 1940–1956. A scatter plot of his data shows that the mean number of weeks between epidemics (Y) and town population (X) have a curvilinear relationship, and both variables are positively skewed. These problems are reduced if we work with the negative reciprocal of weeks between epidemics, $Y^* = -(Y^{-1})$, and the base-10 logarithm of town population, $X^* = \log_{10}(X)$. The following regression equation is obtained:

$$\hat{Y}^* = .00245 - .00268(X^*)$$

or, equivalently,

$$-(\hat{Y}^{-1}) = .00245 - .00268[\log_{10}(X)]$$

Calculate and plot a set of $(X, \hat{Y})$ points for populations (X values) of 1,700, 22,000, 269,000, and 1,046,000 people. Since Y has been transformed, you will have to apply the appropriate inverse transformation (Table 6.8) to obtain $\hat{Y}$ values in the original units, weeks between epidemics. Interpret the curve implied by your plotted points.

15.13   *ALTERNATIVE EXPLANATIONS*

The finding that motor vehicle fatality rates are highest in the least populated counties (Figure 15.18) contradicts the popular belief that driving is safer in rural areas. Since this belief underlies proposals to raise rural speed limits, our contrary conclusions deserve further scrutiny. Why do fatality rates seem to decline with density? Could the findings be a statistical fluke, "explained away" once some third variable is taken into account? If not, what is there about rural driving that makes it more dangerous?

The authors of the highway fatalities study found a similar fatality rate–population density relationship nationwide, not just in Nevada. They considered several omitted variables that might account for this consistent pattern. For example:

1. Could the high death rates in rural areas result from *travel by people from other areas*? Two pieces of evidence argued otherwise. First, there is a moderately strong correlation ($r = .67$) between county fatality rates based on where crashes occurred and on where victims lived. Second, no correlation was found between fatality rates and major travel routes such as interstate highways.
2. Could highway death rates reflect *different age structures* of rural and urban populations? About 40% of the people killed in motor vehicle accidents are 15 to 24 years old, so if rural ares have a higher proportion of people in this age group that could explain the higher rural fatality rates. Areas of high and low fatality proved to have similar proportions in this accident-prone age group, however.
3. Are we simply seeing the effects of *greater travel distances* in rural areas? The authors adjusted their rates to control for the amount of travel. They found that death rates per million vehicle miles were still lowest in the crowded Northeast and highest in the rural West.

These analyses led the researchers to conclude that the fatality rate–population density relationship cannot easily be "explained away." The evidence against alternative explanations strengthens the conclusion that there really are more hazards in rural driving.

This still leaves open the question of what the hazards are. The authors suggest six possibilities:

1. In rural areas emergency facilities are farther apart, and there is less access to major hospital trauma centers.
2. Rural roads are often worse.
3. People drive faster on rural roads, and speed is known to affect death rates.
4. Utility vehicles and pickup trucks, which are linked to higher death rates, are more common in rural areas.
5. Seat belts are used less often in rural areas.
6. Vehicles that are older, are less well-maintained, and have fewer safety features are more common in rural areas.

All six of these factors may contribute to the relationship between motor vehicle fatality rate and population density.

The process of considering alternative explanations, and where possible bringing evidence to bear on them, is a tremendously important step in science. Both data collection (especially experimental design) and data analysis have important roles in this process. Research is called **well-designed** if it is done in a way that limits the number of competing explanations and makes it easier to choose between them. Scientific articles often conclude with a section in which the authors discuss omitted variables and possible alternative explanations. If a study's original authors don't think about alternative explanations, very likely their critics will—especially if the study's conclusions are important or controversial.

PROBLEMS

*30. Earlier in this chapter we found a statistically significant positive relationship between household income and water use.

    a. How might higher income cause an increase in water use?

    b. Suggest an alternative explanation for how the water use–income relationship could arise even if income did *not* in any way cause water use.

*31. We found that for a sample of U.S. cities, the higher the median income, the lower the homicide rate.

    a. Suggest some ways in which low income could be a cause of higher homicide rates.

    b. Suggest an alternative explanation: a way this relationship could have arisen *without* income being a cause of homicide rate.

SUMMARY

We started out this chapter by looking at the standard deviation of residuals, which reflects variation around a regression line. Besides being a measure of how well the line fits, the standard deviation of residuals is useful in estimating standard errors for regression coefficients and predicted values. If certain assumptions are met, such standard errors may be used with the t distribution to construct confidence intervals or hypothesis tests. An alternative way to test certain hypotheses about regression slopes is by using the F test, in a manner similar to its role in analysis of variance.

Confidence intervals can be constructed for regression predictions. These predictions may apply to either the population mean of Y, at any given level of X, or the value of Y for an individual case, given a known value of X. There is more uncertainty in predicting individual values than population means, and this greater uncertainty is reflected by larger standard errors and hence wider confidence intervals.

The last half of this chapter introduced the idea of regression criticism: using analytical results (namely, the residuals) to help decide whether the original analysis should be believed. One critical technique, the e vs. $\hat{Y}$ plot, provides a simple way to look for problems.

Some kinds of problem (nonnormality, heteroscedasticity) affect only the inferential aspects of regression analysis. Other kinds of problem (influence, nonlinearity) affect descriptive aspects as well, so that the regression provides a poor summary of either sample or population. Distributional problems of both kinds are fairly common. Nonlinear transformations may reduce such problems, but they also make interpretation harder. Their use in any given analysis should be weighed in terms of this trade-off. Graphs help convey what curvilinear regression equations imply.

Many of the examples in this book have illustrated a general approach called **interactive data analysis,** in which the results from one analysis guide the next. Regression criticism is the newest stop on this route. Other examples from earlier chapters are mean–median comparisons, stem-and-leaf displays, box plots, outlier deletion, and nonlinear transformations. Interactive data analysis can sometimes be done by hand, but its modern popularity is largely due to the ease of computer calculation.

At one time, the effort of completing one regression was so great that an analyst might view it as the final goal. If regression is so difficult, the idea of regression criticism may seem masochistic. With a computer, though, the effort of doing a regression is trivial. This frees us to view any result as provisional, rather than final: If this is the result we get when we do such-and-such, what happens if we do the analysis in a different way? It is easy to find out.

Through interactive analysis we can see whether certain cases are influencing the results, whether nonlinear transformations improve normality and fit, or if some third variable changes the outcome. We can also get a feel for how *sensitive* our conclusions are: Do they change when we take a slightly different analytical approach? Or are they robust, holding firm no matter how we look at the data? There may be no one "final" analysis, but by experimentation we can get a feel for what works and what does not, selecting an approach that provides the best balance of realism, statistical adequacy, and understandable simplicity.

PROBLEMS

*32. Carry out the regressions needed for the following problems, and use t or F statistics to test the null hypothesis H_0: $\beta = 0$ at $\alpha = .10$. To keep things manageably simple, assume that the usual requirements for valid inference are met.

   a. Regress expected statistics grade (Y) on grade point average (X), using the data on 30 college students in Table 1.1 (page 5).

   b. Find the equation for predicting true VSAT score from self-reported score, using the data on 16 students in Chapter 2, Problem 10 (page 42).

   c. Find the number of incidents of G-LOC and of G-suit disconnects *per 10,000 flight hours,* from the data on naval aircraft in Chapter 9, Problem 29 (page 272). Regress the number of G-LOC incidents per 10,000 hours (Y) on the number of disconnects per 10,000 hours (X). Is there a significant relationship between these two rates?

   d. Regress serum insulin level on serum chlorpropamide, using the patients in Chapter 11, Problem 28 (page 350).

   e. Is the perceived importance of a toxic waste problem related to respondent's education (Chapter 13, Problem 46, page 452)?

   f. Are years of education (part e) related to years lived in a Vermont town?

The table at the top of page 554 contains data from a study of ethnicity and real estate practices in Denver, Colorado. The percentage of residents who are Anglo (white non-Hispanic) is given for ten neighborhoods, along with the percentage of Anglo realtors with real estate listings in that neighborhood. Use these data for Problems 33 and 34.

*33. Regress percentage of Anglo realtors with listings in a given neighborhood (Y) on percentage of Anglo residents (X) for the ten neighborhoods. Write out the regression equation and describe what it implies.

*34. Construct an e vs. $\hat{Y}$ plot for the neighborhoods regression of Problem 33, and describe the problem this plot reveals.

Percentages of Anglo Residents and Anglo Realtors for 10 Denver Neighborhoods

Neighborhood	Anglo residents	Anglo realtors with area listings
Aurora	88.1%	36%
Arvada	82.0	34
City Park West	47.2	10
Curtis Park	19.5	6
Englewood	93.5	26
Hampden	91.3	12
Highlands	37.5	10
Littleton	95.8	28
Valverde	48.6	10
Wheat Ridge	95.5	34

Source: The data are a random sample from Palm (1986).

*35. Table 11.2 (page 320) contains data on pesticide residues at 10 sites along Bear Creek, Mississippi. When DDT residues at depth 200–400 mm (Y) are regressed on residues at 0–200 mm (X), we obtain

$$\hat{Y} = 122.28 + .54(X)$$

The standard deviation of the residuals is $s_e = 356.27$, the mean X is $\overline{X} = 691.4$, and $\text{TSS}_X = 1{,}233{,}756$.

   a. Is the relationship between residues at these two depths significant? Apply a one-sided t test at $\alpha = .10$.

   b. If we examined an eleventh site and found upper sediments (0–200 mm) had a DDT concentration of 1,200 μg/kg, what is our best prediction of the lower sediment (200–400 mm) concentration?

   c. Construct a 90% confidence interval for the lower sediment DDT concentration of the added site ($X_0 = 1{,}200$) in part b.

*36. In the regression of weeks between measles epidemics and town size (Problem 29), we saw the regression equation

$$-(\hat{Y}^{-1}) = .00245 - .00268[\log_{10}(X)]$$

This equation is based on data from 19 towns. The estimated standard error for b in this equation is $\text{SE}_b = .00067$. Construct and interpret a 95% confidence interval for β.

*37. In the straight-line regression of vehicle fatality rates on population densities for Nevada counties (Figure 15.15), the equation is $\hat{Y} = 156 - .83(X)$ with $\text{SE}_b = .63$. Perform a t test at the $\alpha = .05$ significance level to verify the claim made earlier, that this straight-line relationship is "not statistically significant." Why does the evidence in the e vs. $\hat{Y}$ plot (Figure 15.16) cast doubt on this conclusion?

We have looked at the regression of homicide rate on median income for a random sample of 12 U.S. cities (Tables 15.4–15.6). If you studied Figures 15.1 and 15.6 closely, you may have noticed evidence that the distributions of X or $\hat{Y}$ are somewhat skewed. There is also evidence of curvilinearity: Note that residuals in Figure 15.6 are positive at low levels of $\hat{Y}$; negative at middle levels of $\hat{Y}$; and positive again at high levels of $\hat{Y}$. This same curvilinearity is visible in the Y vs. X plot of Figure 15.1. Use this information for Problems 38–41.

*38. Take logarithms to reduce the skewness of median income in Table 15.4. Carry out the regression of homicide rate on log(income); write out the regression equation and interpret the values of a and b.

*39. Find the predicted values and residuals for each of the 12 cities, using your regression equation from Problem 38. Explain the meaning of the residual and predicted values for the city of Anchorage.

*40. Use the residuals from your curvilinear regression in Problem 39 to obtain the following.

a. The standard deviation of the residuals, s_e

b. The estimated standard error, SE_b

c. A one-sided t test of $H_0: \beta \geq 0$ vs. $H_1: \beta < 0$

*41. Evaluate your regression curve from Problems 38–40.

a. Draw a scatter plot of homicide rate versus income, and use your $\hat{Y}$ values from Problem 39 to sketch in the regression curve.

b. The coefficient of determination for the straight-line fit in Figure 15.1 is $R^2 = .44$. Find the squared correlation between Y and $\hat{Y}$ values in Problem 39. Does the curve fit measurably better than the straight line?

c. What does the curve tell us about the relationship between homicide rates and median family incomes in these cities?

The table at the top of page 556 contains measurements of the aluminum concentration in forest litter (organic debris) at eight ridgetop sites near an aluminum reduction plant in western Tennessee. Use these data for Problems 42–45.

*42. Regress aluminum concentration (Y) on distance from plant (X).

a. Write out and interpret the regression equation.

b. Calculate and interpret the coefficient of determination, R^2.

c. Construct a scatter plot and draw in your regression line.

Aluminum Concentration on Forest Floor and Distance from an Aluminum Reduction Plant

Site	Distance to plant (km)	Aluminum concentration (mg/kg)
1	1.2	6,100
2	1.4	4,400
3	3.6	2,800
4	4.4	2,700
5	4.5	2,600
6	8.8	2,800
7	32.0	2,500
8	33.0	2,700

Source: Data from Beyer, Fleming, and Swineford (1987).

*43. Find predicted values and residuals for the aluminum and distance data and use them to construct an e vs. $\hat{Y}$ plot. What problems do you see?

Up to this point we have focused mainly on *statistical* motivations for curvilinear regression, such as improving fit or reducing skewness. Curvilinear regression can also be motivated by *theoretical* considerations. Scientists may have theoretical reasons to expect the relationship between two variables to follow a particular type of curve. For example, many physical phenomena tend to decrease with the inverse square of the distance from their source. In other words there is a negative linear relationship between Y and $1/X^2$. We can estimate parameters for such a relationship by regressing Y on a transformed X variable, $X^* = X^{-2} = 1/X^2$. The aluminum concentration data provide an example of this.

*44. Create a transformed X variable, $X^* = X^{-2}$, for each of the eight Tennessee sites. Regress Y on X^*.

   a. Write out the regression equation.

   b. Draw a scatter plot of Y versus X (same as in Problem 42c), and use predicted Y values to sketch in the regression curve.

   c. Comment on what this curve implies.

*45. Find the squared correlation between Y and $\hat{Y}$ values in Problem 44. Use this statistic and the scatter plot of Problem 44b, in contrast to your answers to Problems 42 and 43, to discuss how the curvilinear regression improves our ability to describe the aluminum–distance relationship.

The table on page 162 contains data on 15 species of primate, including average adult body weight and an index of dietary quality. In Problems 5 and 6 of Chapter 6 we found that the body weight distributions are positively skewed, but their logarithms are much more symmetrical. We will use this univariate finding as a starting point to study whether an ecological relationship called the Jarman–Bell Principle (the larger the body weight, the lower the quality of diet) holds true among these primates.

*46. Regress female dietary quality (Y) on the base-10 logarithm of female body weight, $X^* = \log(X)$.

   a. Write out the regression equation, and interpret the value of b.

   b. Find and interpret the coefficient of determination, R^2.

*47. Test whether the relationship between female primate dietary quality and log body weight (Problem 46) is significant at $\alpha = .05$, using a one-sided t test.

*48. Construct an e vs. $\hat{Y}$ plot for the female primate analysis of Problems 46 and 47. Do you see any major problems?

*49. Draw a scatter plot of female primate dietary quality (Y) versus body weight (X). Use the predicted values from your regression of female primate dietary quality on log body weight (Problem 46) to draw in the regression curve. What does this curve tell us?

*50. We earlier noted that the distribution of life expectancy is bimodal (two-peaked) in our sample of nations (Figure 3.2, page 61). If the regression residuals are also bimodal, this casts doubt on the normality assumption. Stem-and-leaf displays of the residuals can help us investigate this possibility.

   a. Find residuals from the linear regression of life expectancy on GNP, using Equation [15.15] and the data of Table 15.12.

   b. Find residuals from the curvilinear regression, life expectancy with log(GNP), using Equation [15.16] (or Table 15.14) and the data of Table 15.12.

   c. Graph both sets of residuals in a back-to-back stem-and-leaf display with the stems shown.

Residuals from linear regression		Residuals from curvilinear regression
	−1.	
	−1*	
	−0.	
	−0*	
	0*	
	0.	
	1*	
	1.	

Stems digits 10's, leaves digits 1's; 1.6 means a residual of 16 years' life expectancy.

d. Is bimodality still apparent in the linear regression residuals? In the curvilinear regression residuals? Summarize your conclusions.

(*Further material: Dummy variable regression*) Comparison-of-means procedures like the two- or K-sample tests in Chapter 13 can be viewed as simple kinds of regression. For example, a two-sample t test is equivalent to a regression of measurement variable Y on a two-category variable X, where the two categories of X are assigned values of 0 and 1. This method is called **dummy variable regression.** The predicted values of Y in a dummy variable regression are actually group means. That is, the mean of Y given $X = 0$ is

$$\overline{Y}|(X = 0) = \hat{Y}_0$$

$$= a + b(X)$$

$$= a + b(0)$$

$$= a \qquad (Y\text{-intercept})$$

and the mean of Y given $X = 1$ is

$$\overline{Y}|(X = 1) = \hat{Y}_1$$

$$= a + b(X)$$

$$= a + b(1)$$

$$= a + b \qquad (\text{intercept plus slope})$$

Since the slope b is the difference between the two means $\overline{Y}_0$ and $\overline{Y}_1$, a significance test for b tests whether there is *a significant difference between the two means.* The null hypotheses $H_0: \mu_1 - \mu_0 = 0$ (two-sample difference-of-means test) and $H_0: \beta = 0$ (dummy variable regression) are equivalent when β refers to a difference between population means, and the standard errors $SE_{\overline{Y}_1 - \overline{Y}_0}$ and SE_b are the same. Consequently, the usual t test applied to b in a dummy variable regression will produce exactly the same t and P-values as a two-sample t test for the difference of means (see Chapter 13). Problems 51–53 illustrate dummy variable regression methods.

*51. Use the summary statistics regarding gender and MSAT scores in Table 13.6 (page 411) to answer the following. With these summaries you can obtain the answers *without actually conducting the regression.*

    a. Write out the regression equation we would obtain if we regressed MSAT (Y) on gender (X), with gender coded $0 = $ male, $1 = $ female.

    b. What t statistic would we obtain if we tested the null hypothesis H_0: $\beta = 0$?

*52. Chapter 11, Problem 22 (page 349) contains data on safety-related failures at 17 nuclear power plants. Nine are boiling-water reactors (BWR) and eight are pressurized-water reactors (PWR). Are there significant differences in the average number of failures at these two types of plant?

a. Regress standby failures (Y) on type (X), coding type as $0 =$ BWR, $1 =$ PWR. Write out the regression equation and explain what it tells us.

b. Apply a two-sided t test ($\alpha = .10$) to b. Is the difference between the two types significant?

*53. Repeat the steps of Problem 52 to determine if there are significant differences in the numbers of safety-related failures during operations at the two types of nuclear power plant.

NOTES

1. Time series analysis is often very technical, as are many of the relevant books. Among the most accessible introductions is that in McCleary et al. (1980). Business forecasting applications are covered in Nelson (1973). For a more thorough, current, but heavily mathematical treatment of time series methods see Shumway (1988), which includes a disk of microcomputer programs written in BASIC.

2. For example see Cook and Weisberg (1982).

3. *Curvilinear regression* refers to the results of linear regression applied to transformed variables. *Nonlinear regression,* not discussed in this book, refers to more difficult problems not solvable by linear regression methods.

4. The base-10 logarithm of x is the power to which you must raise 10 to get x (Chapter 6).

5. The curve in Figure 15.14 fits better than a straight line, but it does not fit perfectly. An even better fit might be obtained by using a different kind of curve, or by controlling for other variables (such as an oil-exporting economy) that confound the life expectancy–GNP relationship. Considering its simplicity, though, the curve in Figure 15.14 does a good job of summarizing the pattern of these data.

6. Contrast Esmeralda's density of .2 people/mi^2 and its fatality rate of 557.7 deaths/100,000 with Manhattan, New York. There the density is 64,000 people/mi^2, and the motor vehicle fatality rate is less than 1/200th of Esmeralda's: 2.5 deaths/100,000 (Baker, Whitfield, and O'Neill, 1987).

MULTIVARIATE ANALYSIS

PART **IV**

The possibility of third-variable explanations makes it hazardous to draw causal conclusions from bivariate analysis. Carefully designed experiments can help to rule out third-variable explanations, but often experiments are impractical. Researchers then turn instead to **multivariate analysis**, the study of more than two variables at once. Multivariate analysis allows us to study the effect of one variable on another, while **statistically controlling** for the effects of several additional variables.

Multiple regression is the most important and widely used multivariate technique. Its simplest form is a straightforward extension of bivariate regression techniques (Chapters 14 and 15). More complex variants of multiple regression can deal with problems as diverse as robust estimation, time series, measurement error, and any mix of categorical and measurement variables. This flexibility gives regression a central place among multivariate statistical methods.

Despite its difficult subject matter, Chapter 16 is generally less technical than Chapters 12–15. Few details of theory or calculation are presented, for that would require the vocabulary of *matrix algebra*, which is beyond the scope of a first course in applied statistics. Instead we shall look at how to understand multiple regression results. This brief introduction to a vast topic serves two purposes:

1. Current research in many fields leans heavily on multiple regression; this introduction may help you become a more educated consumer of such research reports.
2. For students who will go on to do original research or further study in statistics, Chapter 16 points the way ahead.

The previous 15 chapters laid the foundation for your study of multiple regression.

An Introduction to Multiple Regression*

Bivariate regression is an analytical tool of great power and flexibility. We met it as a technique for summarizing straight-line relationships between two measurement variables. But as we have seen, it readily extends to encompass curvilinear relationships or categorical X variables. This chapter presents a further extension, **multiple regression,** which allows for more than one X variable.

16.1   *THE MULTIPLE REGRESSION EQUATION*

Bivariate regression finds the linear equation that best fits a two-dimensional scatter of (X, Y) data. Such equations have the general form

$$\hat{Y} = a + b(X)$$

They define predicted values of the dependent variable ($\hat{Y}$) for any values of the independent variable (X). To discuss more than one X variable we must now modify our notation slightly, using the symbols X_1 (first X variable), X_2 (second X variable), X_3 (third X variable), and so forth. In this modified notation a bivariate regression equation, which has only one X variable, is written

$$\hat{Y} = a + b_1(X_1)$$

Regression analysis with one X variable (X_1):

$$\hat{Y} = a + b_1(X_1) \tag{16.1}$$

where

a (Y-intercept) is the predicted value of Y when X_1 equals 0; and
b_1 (regression coefficient on X_1) is the change in predicted Y per 1-unit increase in X_1.

The coefficient of determination, R^2, is the proportion of the variance of Y that is explained by X_1.

The simplest multiple regression involves two X variables, denoted X_1 and X_2. We find the linear equation that best fits the *three-dimensional* scatter of (X_1, X_2, Y) data. Such equations define predicted values of the dependent variable ($\hat{Y}$) for any *combination of values* of the independent variables (X_1 and X_2); see the box.

Regression analysis with two X variables (X_1 and X_2):

$$\hat{Y} = a + b_1(X_1) + b_2(X_2) \tag{16.2}$$

where

a (Y-intercept) is the predicted value of Y when both X_1 and X_2 equal 0;
b_1 (regression coefficient on X_1) is the change in predicted Y per 1-unit increase in X_1, if X_2 does not change; and
b_2 (regression coefficient on X_2) is the change in predicted Y per 1-unit increase in X_2, if X_1 does not change.

The coefficient of determination, R^2, is the proportion of the variance of Y that is explained by both X_1 and X_2 together.

The Y-intercept, regression coefficients, and coefficient of determination are now interpreted with reference to *two* independent variables instead of one.

The extension to regression with three X variables is equally straightforward: We seek the linear equation that best fits a *four-dimensional* scatter of (X_1, X_2, X_3, Y) data, as outlined in the box.

Regression analysis with three X variables (X_1, X_2, X_3):

$$\hat{Y} = a + b_1(X_1) + b_2(X_2) + b_3(X_3) \qquad\qquad [16.3]$$

where

a (*Y*-intercept) is the predicted value of Y when all X variables $(X_1, X_2,$ and $X_3)$ equal 0;

b_1 (regression coefficient on X_1) is the change in predicted Y per 1-unit increase in X_1, if all other X variables $(X_2$ and $X_3)$ do not change;

b_2 (regression coefficient on X_2) is the change in predicted Y per 1-unit increase in X_2, if all other X variables $(X_1$ and $X_3)$ do not change; and

b_3 (regression coefficient on X_3) is the change in predicted Y per 1-unit increase in X_3, if all other X variables $(X_1$ and $X_2)$ do not change.

The coefficient of determination, R^2, is the proportion of the variance of Y that is explained by all the X variables $(X_1, X_2,$ and $X_3)$ together.

Multiple regression analysis extends in this way to include any number of independent variables.

The geometric idea of **dimensions** is useful here:

Zero dimensions: A point
One dimension: A line
Two dimensions: A plane
Three dimensions: Ordinary space

Values on one measurement variable can be viewed geometrically as information in one dimension. Two measurement variables give information in two dimensions, and so on. Bivariate regression therefore amounts to summarizing two-dimensional information (a two-variable scatter plot) with a simpler, one-dimensional model (the regression line). Similarly, three-variable regression amounts to summarizing three-dimensional information (imagine a scatter of dots in three-dimensional space) with a simpler, two-dimensional model (the **regression plane**). Geometry in more than three dimensions is difficult to visualize but can be described mathematically. *Regression fits to K-dimensional data a simpler $(K - 1)$-dimensional model.*[1]

We judge fit in regression by the sum of squared residuals, RSS $= \Sigma e_i^2 = \Sigma(Y_i - \hat{Y}_i)^2$. The better the fit between data and regression equation (that is, between Y and $\hat{Y}$), the lower the RSS will be. Ordinary least squares (OLS) regression finds values for the intercept (a) and regression coefficients ($b_1, b_2, \ldots$) such that the sum of squared residuals is as low as possible. This least squares criterion applies to both bivariate and multiple regression.

PROBLEMS

*1. a. In the manner of Equations [16.1]–[16.3], write out a regression equation with five X variables.

b. Explain the meaning of a in your equation.

c. Explain the meaning of b_4 in your equation.

d. Explain the meaning of R^2 for your equation.

*2. If we regress income in dollars (Y) on years of work experience (X_1) and education (X_2), we will obtain an equation of the general form $\hat{Y} = a + b_1(X_1) + b_2(X_2)$. For this equation explain the meaning of the following.

a. a

b. b_1

c. b_2

d. R^2

16.2 INFERENCE AND MULTIPLE REGRESSION

The population counterpart of a regression equation with two X variables, [16.2], is

$$\mu_{Y|X_1, X_2} = \alpha + \beta_1(X_1) + \beta_2(X_2) \qquad [16.4]$$

where $\mu_{Y|X_1, X_2}$ represents the population conditional mean of Y, given X_1 *and* X_2. Values of a, b_1, and b_2 are calculated from sample data as estimates of the true population parameters α, β_1, and β_2. Equation [16.4] parallels the simpler bivariate version, Equation [15.1] in Chapter 15.

Hypothesis tests and confidence intervals for multiple regression resemble their bivariate counterparts. The t distribution is used with estimated standard errors to form intervals or tests regarding individual regression coefficients. Standard errors for multiple regression coefficients (SE_{b_k}) are usually estimated by computer.

To **test the null hypothesis** that an individual population regression coefficient is 0, H_0: $\beta_k = 0$, calculate

$$t = \frac{b_k - 0}{SE_{b_k}}$$

$$= \frac{b_k}{SE_{b_k}} \qquad [16.5]$$

where b_k is the sample regression coefficient on variable X_k, and SE_{b_k} is its estimated standard error. Compare this t statistic with a t distribution for $df_R = n - K$ degrees of freedom, where K is the number of coefficients.

> **Confidence intervals** for individual regression coefficients are found by
>
> $$b_k \pm t(SE_{b_k}) \qquad [16.6]$$
>
> where t is chosen from the t distribution with $df_R = n - K$ degrees of freedom.

The t statistic in Equation [16.5] tests hypotheses about individual regression coefficients. We can also use an F statistic to test hypotheses about *sets* of regression coefficients. The overall F statistic (Chapter 15), is used to test the null hypothesis that all β coefficients in the equation are 0.

> To test the null hypothesis that the population values of *all* β *(slope) coefficients in a regression equation* are 0, calculate
>
> $$F = \frac{ESS/df_E}{RSS/df_R} \qquad [16.7]$$
>
> Compare this F statistic with the theoretical F distribution with $df_E = K - 1$ and $df_R = n - K$ degrees of freedom.

F tests in multiple regression, Equation [16.7], take the same form as in bivariate regression (see Table 15.9, page 525).

Multivariate versions of the usual distributional assumptions are required with either t or F statistics. Multiple regression also faces some new problems besides those familiar from bivariate regression. We will later see that e vs. $\hat{Y}$ plots remain useful for checking a regression's validity.

PROBLEMS

Scientists have noticed a bivariate relationship between air pollution and death rates in American cities. Besides air pollution, many other factors are known to influence death rates. For example, death rates are higher in cities where the population is older, more nonwhite, denser, less educated, and poorer. Multiple regression can be used to test whether air pollution still has an effect on the death rate *after we control for effects* of these other predictors. A regression equation based on one sample of 98 U.S. metropolitan areas (Ozkaynak and Thurston, 1987) is

$$\hat{Y} = 69.6 + 57.2(X_1) + 2.8(X_2) + 2.3(X_3) + 33.8(X_4) - 9.4(X_5) + .2(X_6) + 5.4(X_7)$$
$$SE_b: \qquad\quad 4.8 \qquad 5.1 \qquad 0.7 \qquad 17.4 \qquad 2.5 \qquad 2.4 \qquad 1.5$$

Estimated standard errors for each b coefficient are given below the equation. Variable definitions are:

Y: Annual death rate per 100,000 population
X_1: Percentage of population 65 years or older
X_2: Median age of population
X_3: Percentage of population nonwhite
X_4: Base-10 logarithm of population density (people per square mile)
X_5: Percent of population with ≥ 4 years of college
X_6: Percent of population living below poverty level
X_7: Mean sulfate concentration in air ($\mu g/m^3$ air)

*3. Calculate t statistics and perform two-sided significance tests ($\alpha = .05$) for each of the seven slope coefficients. Does sulfate concentration (X_7) still have a significant effect, even after you control for the other six variables? What else remains a significant predictor of death rates?

*4. The sums of squares for this regression are ESS = 1,547,059 and RSS = 153,328. Find the overall F statistic, and test the hypothesis that all seven slope coefficients are 0 in the population.

16.3   *READING REGRESSION OUTPUT*

Chapter 8, Problem 19 (page 234) lists data from 118 college classes, collected to study whether higher teaching evaluations are given by classes receiving higher grades. Part-time instructors or graduate student teaching assistants taught 47 classes; regular faculty members taught the rest. Relationships between evaluations, grades, and class size differ somewhat for classes taught by these two kinds of instructor, so we will begin by analyzing them separately. (How we know that the relationships differ will be shown later on.)

Few modern researchers would attempt multiple regression without a computer; even bivariate regression is tedious when done by hand. Fortunately, computers make short work of such calculations. Results come back quickly, often in the form of output like that in Table 16.1. This table shows the regression of percentage of high teaching evaluations (Y) on percentage of A and B grades (X_1) for the 47 classes taught by part-time instructors and teaching assistants. At first glance Table 16.1 may look forbidding, but most of the contents should already be familiar.

The command that generated this output, "regress evals grades," is shown at the top of Table 16.1. The remainder of the table is the computer's response to that command, based on the raw data already in the computer's memory. Variables are identified by short names such as "evals" (Y, or percentage of high teaching evaluations) and "grades" (X_1, or percentage of A and B grades).

The bottom half of the output in Table 16.1 has four columns labeled Coefficient, Std. Error, t, and Prob > |t|. Their meanings follow.

TABLE 16.1   ***Regression of Percentage of High Evaluations on Percentage of High Grades for 47 Classes with Instructors and Assistants***

.regress evals grades

Source	SS	df	MS
Explained	2883.289	1	2883.289
Residual	26109.349	45	580.208
Total	28992.638	46	630.275

Number of obs = 47
F(1, 45)   = 4.97
Prob > F   = 0.031
R-square   = 0.0994
Adj R-square = 0.0794
Residual S.D. = 24.088

Variable	Coefficient	Std. Error	t	Prob > |t|
evals				
grades	.6690493	.3001275	2.229	0.031
_ cons	1.14822	19.17521	0.060	0.953

Source: Output adapted from the STATA statistical program.

Coefficient: This column gives the slopes and Y-intercept. Here the slope (right of the variable "grades") is b_1 = .6690493, and the Y-intercept (right of the label "_ cons"—meaning *constant*) is a = 1.14822. The regression equation is therefore

$$\hat{Y} = 1.14822 + .6690493(X_1)$$

or, rounding off,

$$\hat{Y} = 1.15 + .67(X_1) \qquad [16.8]$$

Std. Error: These are the estimated standard errors of the slope (SE_{b_1} = .3001275) and intercept (SE_a = 19.17521). They can be used for confidence intervals or hypothesis tests regarding b_1 or a.

t: This column represents the values in the Coefficient column divided by the values in the Std. Error column. The first, $t = b_1/SE_{b_1}$ = 2.229, is a test of the usual null hypothesis that the population regression slope is 0: H_0: β_1 = 0; see Equation [16.5]. The second, $t = a/SE_a$ = .060, is a test of the null hypothesis that the population intercept is 0: H_0: α = 0.

Prob > \|t\|: Read this as the "probability of a greater absolute value of *t*." These are the obtained *P*-values (two-sided) corresponding to the *t* statistics in the column to the left. If a one-sided test is desired, divide the obtained *P*-value by two.

The *P*-value for the grades variable is .031; since this is less than α = .05, we can reject the null hypothesis of no relationship. The grades–evaluations relationship is statistically significant.

Further information is given in the top half of the output. In the upper left are the three sums of squares, ESS = 2,883.289, RSS = 26,109.349, and TSS_Y = 28,992.638, together with their respective degrees of freedom. The upper right-hand side includes:

Number of obs: Sample size, $n = 47$.

F(1, 45):   Overall F statistic (see Equation [16.7]) with 1 and 45 (df_E and df_R) degrees of freedom.

Prob > F: "Probability of a greater F," the obtained P-value for the overall F test. Since this is a bivariate analysis, the P-value is the same as that obtained with a two-sided t test.

R-square: Coefficient of determination, R^2 = ESS/TSS_Y = .0994.

Adj R-square: R^2 adjusted for degrees of freedom; see Equation [16.9] in the next box.

Residual S.D.:   Standard deviation of the residuals, $s_e = \sqrt{RSS/df_R} = 24.088$.

Because adding X variables to an equation always increases R^2, in multiple regression the *adjusted* R^2 is generally preferred. The adjustment is based on the degrees of freedom.

Adjusted R^2 is defined as

$$R_a^2 = R^2 - \frac{df_E}{df_R}(1 - R^2) \qquad\qquad [16.9]$$

where R^2 is the unadjusted coefficient of determination, and df_E and df_R are the explained and residual degrees of freedom, respectively.

If there are many X variables relative to the sample size, then the ratio df_E/df_R will be large and the adjustment amounts to a substantial reduction in R^2. In larger samples or with fewer X variables, the adjustment makes less difference. In Table 16.1 we have

$$R_a^2 = .0994 - \frac{1}{45}(1 - .0994) = .0794$$

Thus, after adjustment for degrees of freedom, only 7.94% of the variance of teaching evaluations is explained by grades.

A regression output like Table 16.1 contains most of the regression statistics discussed in Chapters 14 and 15, compressed into a single table. With practice such tables become easy to read. In the next section we will look at similar output for a multiple regression.

PROBLEMS

* 5. Return to the sums of squares for the death rate–air pollution regression in Problem 4 to find these statistics.
     a. The coefficient of determination, R^2
     b. The adjusted coefficient of determination, R_a^2
     c. The standard deviation of the residuals, s_e

Most statistical programs produce regression output with essentially the same information as Table 16.1, but each program presents this information in a slightly different way. Your instructor can provide you with samples of *bivariate regression* output from a computer program used at your location. Use this output to answer Problems 6–12.

* 6. Write out the regression equation, and interpret the values of a and b_1.
* 7. Does X_1 have a significant effect on Y? Compare the P-values printed out for t and F tests with the probabilities that can be obtained by looking up the t and F statistics in Tables A.3 and A.5.
* 8. Show how the t statistic for b_1 is calculated from b_1 and SE_{b_1}.
* 9. Use the standard error of b_1 to construct 95% and 99% confidence intervals for β_1. Explain what these intervals mean.
* 10. Show how the F statistic can be calculated from ESS and RSS.
* 11. Show how the standard deviation of the residuals, s_e, can be calculated from RSS.
* 12. Show how R_a^2 can be calculated from R^2.

16.4 PREDICTING TEACHING EVALUATIONS FROM GRADES AND CLASS SIZE

Our bivariate regression (Table 16.1) shows that there is a significant relationship between teaching evaluations and grades. But could this apparent relationship be explained by some third variable? One obvious possibility is class size. Perhaps grades are higher in small classes, where instruction is more individual; moreover, if students like the smaller classes, their instructors may get better evaluations. In other words, class size might influence both grades and teaching evaluations. If this is true, the evaluations–grades relationship might be **spurious**: due entirely to the effects of an omitted third variable (size). Evaluations might be higher in classes with better grades simply because those classes also happen to be smaller.

Multiple regression allows us to evaluate such possibilities by including additional X variables in our analysis. Table 16.2 shows output for the regression of teaching evaluations (Y, "evals") on grades (X_1) *and* class size (X_2). The regression equation, rounded off, is

$$\hat{Y} = 11.56 + .73(X_1) - .30(X_2) \qquad [16.10]$$

The adjusted R^2 for this regression is .1609, so grades and class size together explain about 16% of the variance of teaching evaluations. The effects of both grades and class size are statistically significant.

The Y-intercept in Equation [16.10], $a = 11.56$, is the predicted percentage of high evaluations for a class with no A's and B's and no students:

$$\hat{Y} = 11.56 + .73(0) - .30(0) = 11.56$$

As often happens this Y-intercept represents a mathematical construct rather than a realistic prediction.

The slope coefficients have more sensible interpretations. The coefficient on grades, $b_1 = .73$, tells us that the predicted percentage of high evaluations goes up by .73 with every 1-point increase in the percentage of high grades, *if class size is held constant.* For example, suppose a class received $X_1 = 40\%$ A and B grades for an enrollment of $X_2 = 55$ students. Then the predicted percentage of high evaluations is

$$\hat{Y} = 11.56 + .73(40) - .30(55) = 24.26$$

TABLE 16.2   *Regression of Percentage of High Evaluations on Percentage of High Grades and Class Size*

. regress evals grades size

Source	SS	df	MS
Explained	5721.656	2	2860.828
Residual	23270.982	44	528.886
Total	28992.638	46	630.275

Number of obs	= 47
F(2, 44)	= 5.41
Prob > F	= 0.008
R-square	= 0.1973
Adj R-square	= 0.1609
Residual S.D.	= 22.998

Variable	Coefficient	Std. Error	t	Prob > |t|
evals				
grades	.7323049	.2878445	2.544	0.015
size	− .3031118	.1308428	− 2.317	0.025
__ cons	11.56337	18.85146	0.613	0.543

If we move up one percentage point to $X_1 = 41\%$ high grades, still with a class size of $X_2 = 55$, the predicted Y rises by $b_1 = .73$:

$$\hat{Y} = 11.56 + .73(41) - .30(55) = 24.99$$

The negative coefficient on class size, $b_2 = -.30$, tells us that the predicted percentage of high evaluations *decreases* by .30 with each one-student increase in class size, *if the percentage of high grades is held constant.* For instance, going from a class size of $X_2 = 55$ students to $X_2 = 56$, while X_1 stays at 41% A's and B's, the predicted Y will decrease by .30:

$$\hat{Y} = 11.56 + .73(41) - 30(56) = 24.69$$

Interpreting the regression coefficients in this manner does not imply that one variable really is somehow "held constant." It is actually impossible to go from 55 to 56 students without at least some change in the percentage of A's and B's. The interpretations above merely suggest how the coefficients can be individually understood. These coefficients have been calculated so as to control statistically for the effects of other variables in the equation. Each of the two slopes in Table 16.2 is calculated controlling for one other variable, so they are called **first-order partial** coefficients. A bivariate regression slope (like that in Table 16.1) is calculated controlling for no other variables, so it is a **zero-order partial.** If there were three X variables each slope would be a **second-order partial,** and so on.

Notice that the coefficient on grades (X_1) is actually higher in Equation [16.10], after we control for class size (X_2), than in the bivariate regression of Equation [16.8]. This shows that *the relationship between grades and evaluations is not spurious;* it is not "explained away" when class size is taken into account. In fact it slightly strengthens. Ruling out this possible alternative explanation bolsters the argument for a causal relationship between grades and evaluations. Of course we have not *proved* such a relationship—an infinite number of alternative explanations remain. Some of these other possibilities could also be evaluated by including them in our regression as third, fourth or fifth X variables.

PROBLEMS

*13. The regressions in Tables 16.1 and 16.2 are based on 47 of the 118 classes in the raw data. The other 71 classes were taught by full-time faculty members. Table 16.3 (page 577) shows the regression of teaching evaluations on grades and class size for these 71 classes. Write out the regression equation and interpret the values of a, b_1, b_2, and R_a^2.

*14. Discuss how the results for classes taught by full-time faculty (Table 16.3) compare with those for classes taught by part-time instructors and teaching assistants (Table 16.2), with respect to:

    a. Significance and magnitude of the effects of grades.

    b. Significance and magnitude of the effects of class size.

    c. Proportion of variance explained.

*15. Return to the death rate–air pollution regression equation in Problem 3. Interpret the numerical values of these coefficients for predicting the death rate (deaths per 100,000 people per year).

    a. b_1 = 57.2 (coefficient on percentage of population 65 years and older)

    b. b_7 = 5.4 (coefficient on mean sulfate concentration, $\mu g/m^3$ air)

16.5   *STANDARDIZED REGRESSION COEFFICIENTS ("BETA WEIGHTS")*

The regressions we have seen so far all used *unstandardized regression coefficients,* which express effects in terms of the natural units of the variables. *Standardized regression coefficients,* which express effects in terms of standard deviations, are popular in some fields.

The regression coefficient on any X variable X_k can be re-expressed as a **standardized regression coefficient,** b_k^*:

$$b_k^* = b_k \frac{s_k}{s_Y} \qquad [16.11]$$

where b_k is the unstandardized coefficient, s_k is the standard deviation of X_k, and s_Y is the standard deviation of Y.

Standardized regression coefficients are also called "beta weights," but this term could be confused with the population parameter β_k (*beta*)—which is actually estimated by b_k, not by the standardized coefficient b_k^*.

Equation [16.10] gave the unstandardized regression of teaching evaluations (Y) on grades (X_1) and class size (X_2):

$$\hat{Y} = 11.56 + .73(X_1) - .30(X_2)$$

The standard deviations for these three variables are s_Y = 25.11, s_1 = 11.83, and s_2 = 26.03. The standardized coefficient on X_1 is therefore

$$b_1^* = b_1 \frac{s_1}{s_Y} = .73\,\frac{11.83}{25.11} = .34$$

A 1-standard deviation increase in X_1 leads to a .34 standard deviation increase in predicted Y, if X_2 does not change. The standardized coefficient on X_2 is

$$b_2^* = b_2 \frac{s_2}{s_Y} = -.30\,\frac{26.03}{25.11} = -.31$$

A 1-standard deviation increase in X_2 leads to a .31 standard deviation *decrease* (because b_2^* is negative) in predicted Y, if X_1 does not change.

Standard scores (Chapter 8) are variables that were transformed so they are measured in standard deviations. Their mean is 0 and standard deviation is 1. Standardized regression coefficients are the coefficients we would obtain from a regression in which all variables were standard scores.

A **standardized regression equation** is written with standardized regression coefficients (b_k^*) and standard-score variables (Y^*, X_k^*):

$$\hat{Y}^* = b_1^* (X_1^*) + b_2^* (X_2^*) + \cdots + b_{K-1}^* (X_{K-1}^*) \qquad [16.12]$$

The standardized version of Equation [16.10] is

$$\hat{Y}^* = .34(X_1^*) - .31(X_2^*)$$

where Y^*, X_1^*, and X_2^* are evaluations, grades, and class size in standard-score form. Standardized regression equations always have a Y-intercept of 0. Coefficients of determination (R^2 or R_a^2) and hypothesis tests (t or F statistics) are the same for a standardized regression equation as for its unstandardized counterpart.

Standardized regression coefficients have these properties:

1. They resemble correlation coefficients (Chapter 14) in that their theoretical range is $-1 \le b^* \le +1$. The closer b^* is to -1 or $+1$, the stronger the relationship.

2. The coefficient on X_k, b_k^*, tells us *by how many standard deviations predicted Y changes, with each 1-standard deviation increase in X_k, if the other X variables do not change.*

Standardized coefficients are attractive because we can tell immediately how "strong" an effect is (how close to -1 or $+1$), without thinking about the variables' units. This ease of interpretation is deceptive, however. A practical understanding of relationships in the data is more likely to come from studying the unstandardized coefficients in their natural units.

PROBLEMS

*16. The following equalities hold for *bivariate regression only:*

$r^2 = R^2$ (The squared correlation coefficient equals the coefficient of determination.)

$r = b^*$ (The correlation coefficient equals the standardized regression coefficient.)

   a. Apply the first equality to the bivariate regression in Table 16.1 to find the correlation coefficient.

   b. Use Equation [16.11], given $s_Y = 25.11$ and $s_1 = 11.83$, to find the standardized regression coefficient from the unstandardized coefficient in Table 16.1.

   c. Are the correlation and bivariate standardized regression coefficients the same, as expected?

*17. Rewrite the regression equation for air pollution and death rates (Problem 3) in standardized form, using these standard deviations:

$$\hat{Y} = 69.6 + 57.2(X_1) + 2.8(X_2) + 2.3(X_3) + 33.8(X_4) - 9.4(X_5) + .2(X_6) + 5.4(X_7)$$

Std. dev.:   Std. dev.:

132.4   2.0   1.8   9.9   0.4   2.8   2.7   3.4

Refer to Problem 3 for variable definitions. Which variable is the "strongest" predictor of the death rate?

*18. Problem 15 had you interpret the unstandardized coefficients b_1 and b_7. Now write out the interpretation of the standardized coefficients b_1^* and b_7^*.

16.6   *INTERACTION EFFECTS*

When teaching evaluations (Y) are regressed on grades (X_1) and class size (X_2) among the 47 classes taught by part-time instructors, we obtain (Table 16.2)

$$\hat{Y} = 11.56 + .73(X_1) - .30(X_2) \tag{16.13}$$

The same regression using data on 71 classes taught by full-time faculty yields (Table 16.3)

$$\hat{Y} = 24.84 + .51(X_1) + .01(X_2) \tag{16.14}$$

Both grades and class size seem to have greater impact on the evaluations of part-time instructors. This is an example of an **interaction effect**—a situation where the *relationships* between variables change with the level of some other variable. The relationships between evaluations, grades, and class size change with instructor status.

Interaction effects can be analyzed with regression if we include special **interaction terms**—X variables created by multiplying other X variables. The simplest interaction terms are formed when one of the variables is a $\{0,1\}$ dichotomy or dummy variable. For example, let X_3 represent faculty status,

TABLE 16.3   **Regression of Percentage of High Evaluations on Percentage of High Grades and Class Size for Classes with Full-time Faculty**

. regress evals grades size

Source	SS	df	MS
Explained	4196.635	2	2098.317
Residual	31374.548	68	461.390
Total	35571.183	70	508.160

Number of obs $= 71$
$F(2, 68) = 4.55$
Prob $> F = 0.014$
R-square $= 0.1180$
Adj R-square $= 0.0920$
Residual S.D. $= 21.48$

Variable	Coefficient	Std. Error	t	Prob $> |t|$
evals				
grades	.5119427	.1700349	3.011	0.004
size	.0126773	.093712	0.135	0.893
__ cons	24.84383	11.11152	2.236	0.029

coded 0 for part-time instructors and 1 for full-time faculty. A term for the *interaction of grades with faculty status* is formed by multiplying status (X_3) by grades (X_1). A similar term for the interaction of class size with faculty status is formed by multiplying X_3 by X_2. These interaction terms may then be included among the X variables in a regression.

Including instructor status and both interaction effects in the teaching evaluations regression gives an equation of the form

$$\hat{Y} = a + b_1(X_1) + b_2(X_2) + b_3(X_3) + b_4(X_3)(X_1) + b_5(X_3)(X_2)$$

Table 16.4 (page 578) shows results of this regression run with all 118 cases from the original data. Rounded off, the equation is

$$\hat{Y} = 11.56 + .73(X_1) - .30(X_2) + 13.28(X_3)$$
$$- .22(X_3)(X_1) + .32(X_3)(X_2) \quad [16.15]$$

When fac $= 0$ (classes taught by part-time instructors) Equation [16.5] reduces to

$$\hat{Y} = 11.56 + .73(X_1) - .30(X_2) + 13.28(0) - .22(0)(X_1) + .32(0)(X_2)$$
$$= 11.56 + .73(X_1) - .30(X_2)$$

which is identical to Equation [16.13]. When fac $= 1$ Equation [16.15] reduces to

TABLE 16.4   *Regression of Percentage of High Evaluations on Percentage of High Grades, Class Size, and Faculty Status with Interaction Terms*

```
. regress evals grades size fac facgrade facsize
```

				Number of obs	=	118
Source	*SS*	*df*	*MS*	F(5, 112)	=	5.12
				Prob > F	=	0.0003
Explained	12481.249	5	2496.250	R-square	=	0.1859
Residual	54645.5307	112	487.907	Adj R-square	=	0.1496
Total	67126.7797	117	573.733	Residual S.D.	=	22.089

Variable	*Coefficient*	*Std. Error*	*t*	*Prob > |t|*
evals				
grades	.7323049	.2764682	2.649	0.009
size	−.3031118	.1256716	−2.412	0.017
fac	13.28046	21.41036	0.620	0.536
facgrade	−.2203622	.3271209	−0.674	0.502
facsize	.3157892	.1583666	1.994	0.049
___ cons	11.56337	18.10641	0.639	0.524

$$\hat{Y} = 11.56 + .73(X_1) - .30(X_2) + 13.28(1) - .22(1)(X_1) + .32(1)(X_2)$$
$$= 24.84 + .51(X_1) + .02(X_2)$$

which apart from rounding error is the same as Equation [16.14].

The regression with interaction terms implies the same two equations we obtain from separate regressions, but it also gives something more: tests of which differences between the two separate equations are statistically significant. In Table 16.4 we see that the interaction of class size with faculty status has a *P*-value, based on the usual *t* statistic, just below $\alpha = .05$ ($P = .049$). This indicates that *the effects of class size on teaching evaluations are significantly different* for faculty and for part-time instructors. In classes taught by part-time instructors, larger classes give lower teaching evaluations; see Equation [16.14]. In classes taught by faculty, class size has almost no effect on evaluations; see Equation [16.15]. By contrast, the effects of grades on teaching evaluations are *not* significantly different for the two kinds of instructors ($P = .502$).

Interaction terms created by multiplying a dummy variable by a measurement variable, as in this example, are called *slope dummy variables*.

> The regression coefficient on a **slope dummy variable** equals the *difference in slope* between the subgroups denoted by dummy variable categories 0 and 1. A *t* test of this coefficient tests whether the difference between slopes is statistically significant.

Multiplying two measurement variables creates another kind of interaction term.

Interaction terms often have undesirable statistical properties such as outliers, skewness, or high correlations with the component X variables (see next section). In general, treat regressions with interaction terms cautiously.

PROBLEM

Regression of College GPA on Age, VSAT, MSAT, Sex, and Interactions of Sex with VSAT and MSAT, for 9,380 College Students

`.regress gpa age vsat msat sex sexvsat sexmsat`

Source	SS	df	MS
Explained	531.591	6	88.598
Residual	2861.229	9373	.305
Total	3392.820	9379	.362

Number of obs	= 9380
F(6, 9373)	= 290.24
Prob > F	= 0.0001
R-square	= 0.1567
Adj R-square	= 0.1561
Residual S.D.	= .5525

Variable	Coefficient	Std. Error	t	Prob > |t|
gpa				
age	.0501299	.0021689	23.113	0.001
vsat	.0010801	.0001173	9.210	0.001
msat	.0012301	.0001073	11.467	0.001
sex	− .0400202	.0811013	− 0.493	0.622
sexvsat	.0004284	.0001609	2.663	0.008
sexmsat	.0000971	.0001493	0.650	0.516
__ cons	.2606103	.0753672	3.458	0.001

*19. The accompanying table shows the regression of college GPA on student age, VSAT, MSAT, sex, and interactions formed by multiplying a dummy variable for sex (0 = male, 1 = female) by VSAT and MSAT.

   a. Write out separate regression equations for men and women based on the output in the table.

b. Are either of the interaction effects significant? What other variables are significant predictors of college GPA?

c. Summarize what this regression tells us.

16.7   HOW MANY X VARIABLES?

There are two absolute limitations on how many X variables a regression may include:

1. Given a sample of size n, we can include no more than $n - 1$ independent variables.

2. Regression is impossible if any one of the X variables is a linear function of other X variables; that is, if the unadjusted R^2 from regressing any one X on all the other X's equals 1.

Before we run into these absolute limitations, we will encounter practical problems.

Until we reach the $n - 1$ limitation, we can always improve fit (decrease RSS and increase R^2) by adding more X variables to a regression. The resulting equation becomes more and more complex, however. A point of diminishing returns is soon reached, where adding more X variables brings only a small increase in R^2, outweighed by the difficulty of making sense of the growing equation.

Simple models that provide good summaries of complex data are said to be **parsimonious**. Parsimony is an important goal in science; if two models fit the data equally well, the simpler one is generally preferred. Parsimony in regression is reflected by the adjusted R^2 statistic, R_a^2. This statistic combines a measure of fit (R^2) with a measure of the difference in complexity between model and data: the residual degrees of freedom, $df_R = n - K$. Unlike R^2, R_a^2 will eventually begin to decrease if too many variables are added to the regression.

The second absolute limitation is that regression is impossible when there are perfect relationships among the X variables, a situation called **multicollinearity**. If X_1 and X_2 are perfectly correlated, we cannot calculate the effects of one while controlling for the other—they have no independent variation. In practice such perfect relationships often result from carelessly defined variables. A more common situation is strong, but not perfect, relationships among the X variables. This does not prevent analysis, but it does cause problems. The closer we get to true multicollinearity, the worse these problems are.

The difficulty is that if X_1 and X_2 are strongly correlated, it is hard to judge what the effect of a change in X_1 would be if X_2 "does not change." They have little independent variation, yet this small fraction must be used to estimate their separate effects. In such circumstances standard errors will

be high, warning of wild sample-to-sample fluctuations. Results from any
one sample thus cannot be trusted.

Multicollinearity can be tested by regressing each X variable on all of the
others. There are also danger signals, such as coefficient sign reversals and
suddenly high standard errors, that alert the experienced analyst. The simplest solution to the problem is to drop or combine the offending variables;
near-perfect interrelationships mean that most of their information is redundant anyway.

Multicollinearity is a potential problem in the teaching evaluations
regression of Table 16.4, because of the high correlations between interaction
terms like "facgrade" and its component parts "fac" and "grades." Also, two
of the coefficients in this regression are nonsignificant, meaning we cannot
reliably distinguish them from 0. We can reduce the multicollinearity problem and improve parsimony by dropping the two variables with nonsignificant effects (Table 16.5):

$$\hat{Y} = 21.24 + .57(X_1) - .30(X_2) + .32(X_3)(X_2) \qquad [16.16]$$

Dropping faculty status and the grades-by-status interaction brings only a
slight reduction in R^2 (from .1859 to .1826), while the adjusted R_a^2 actually
increases (from .1496 to .1611). Equation [16.16] fits the data about as well
as Equation [16.15], *and it is simpler.* Equation [16.16] is therefore more
parsimonious.

Simplifying a regression by dropping nonsignificant variables is called
backward elimination. It is best to drop variables carefully, one at a time, and

TABLE 16.5

*Regression of Percentage of High Evaluations on Percentage of High Grades,
Class Size, and Interaction of Faculty Status with Class Size*

. regress evals pab size facsize

Source	SS	df	MS			
				Number of obs	=	118
				F(3, 114)	=	8.49
Explained	12259.639	3	4086.546	Prob > F	=	0.0001
Residual	54867.141	114	481.291	R-square	=	0.1826
				Adj R-square	=	0.1611
Total	67126.780	117	573.7332	Residual S.D.	=	21.938

Variable	Coefficient	Std. Error	t	Prob > |t|
evals				
pab	.5746733	.1463444	3.927	0.001
size	−.2980006	.0937097	−3.180	0.002
facsize	.3153688	.0775545	4.066	0.001
___ cons	21.24491	9.572378	2.219	0.028

check the consequences of each elimination. Backward elimination is one of several regression strategies that provide objective rules about which X variables to keep in an analysis.

Even with such strategies, variable selection calls for careful judgment. The researcher must base these judgments on substantive as well as statistical knowledge.

PROBLEM

Regression Analysis with Four Cases and Three X Variables: Regression of Reindeer Cave Art on Reindeer Bones, Ibex Cave Art, and Ibex Bones[a]

.regress reinart reinbon ibexbon ibexart

Source	SS	df	MS
Explained	8.191	3	2.730
Residual	0.000	0	.
Total	8.191	3	2.730

Number of obs = 4
F(3, 0) = .
Prob > F = .
R-square = 1.0000
Adj R-square = .
Residual S.D. = 0.00

Variable	Coefficient	Std. Error	t	Prob > $|t|$
reinart				
reinbon	− 1.34609	0	.	.
ibexbon	.6006598	0	.	.
ibexart	− 4.279253	0	.	.
— cons	35.57547	0	.	.

[a]All prevalences in percentages.

*20. The accompanying table shows what happens when the number of X variables in a regression reaches its theoretical maximum, $n − 1$. The small data set on Spanish cave art in Problem 7 of Chapter 14 (page 473) is used for this example, with $n = 4$ cases and three independent variables. Since the regression equation is as complicated as the raw data, df_R equals 0. The computer was unable to calculate values for certain of the regression statistics.

   a. As measured by unadjusted R^2, how well does the regression equation predict actual Y values?

   b. Confirm your answer to part a by finding the predicted prevalences of reindeer art for each of the four regions in the original data.

   c. Why is it impossible to calculate t, F, or R_a^2 statistics? Answer with reference to the equations that define these statistics.

16.8 POLYNOMIAL REGRESSION

Chapter 15 showed how nonlinear transformations work for curvilinear regression. A variety of curves may be fit in this way, but they must be **monotonic**—go either up or down, but not both. **Nonmonotonic** curves such as U shapes or waves require different methods. One such method is *polynomial regression*, in which Y is regressed on more than one power of X—X, X^2, X^3, and so on. Second-order polynomial regression equations may describe curves with U or inverted-U shapes. That is, a second-order polynomial curve may *change direction once*, from increasing to decreasing or vice versa.

A **second-order polynomial regression** has the form

$$\hat{Y} = a + b_1(X) + b_2(X^2)$$

Figure 16.1 illustrates with survey data on income (Y) and age (X). The equation for this curve is

$$\hat{Y} = 3.431 + 1.389(X) - .017(X^2)$$

The curve indicates that predicted income rises with age until the early 40's; above this age income gradually declines.[2] The poorest people are the oldest, especially elderly widows. To perform this regression, given data on income and age:

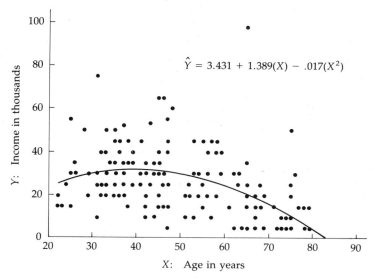

FIGURE 16.1   *Polynomial regression of income on age*

1. create a new variable equal to age squared; then
2. regress income on age and age squared.

Higher-order polynomials are required to fit curves that change directions more times. A third-order polynomial,

$$\hat{Y} = a + b_1(X) + b_2(X^2) + b_3(X^3)$$

can go twice from increasing to decreasing or vice versa. The coefficients are estimated by regressing Y on "three" X variables: X, X^2, and X^3. A fourth-order polynomial,

$$\hat{Y} = a + b_1(X) + b_2(X^2) + b_3(X^3) + b_4(X^4)$$

can change direction three times, and so on. Other X variables could be included along with polynomial terms, producing mixed equations such as

$$\hat{Y} = a + b_1(X_1) + b_2(X_1^2) + b_3(X_2) + b_4(X_3)$$

High-order polynomials may achieve a close fit to the data, but produce an uninterpretable sea-serpent curve. Like interaction effects, polynomial regression is prone to statistical problems. The various powers of X are often strongly correlated with each other, giving rise to multicollinearity. Moreover, transformations (X^2, X^3, . . .) tend to create positive skewness and high outliers. For these reasons the technique must be used carefully.

16.9   GRAPHS FOR MULTIPLE REGRESSION

Bivariate regression involves only two dimensions (X and Y), so the model and data are easily graphed in a two-dimensional (flat) scatter plot. Multiple regression involves K dimensions; K (number of coefficients) equals the total number of X and Y variables in the equation. When $K > 2$ it becomes harder to depict the model and data graphically. A variety of two-dimensional graphs are still useful in multivariate analysis, however.

The regression equation for predicting teaching evaluations from grades and class size (part-time instructors only) is

$$\hat{Y} = 11.56 + .73(X_1) - .30(X_2)$$

This equation describes a plane in three-dimensional space. The three-dimensional figure would be difficult to draw, but we can graph what the equation implies about the relationship between teaching evaluations (Y) and grades (X_1) *given any specific level of class size* (X_2).

For a class of $X_2 = 20$ students, the relationship between evaluations and grades is

$$\hat{Y}|(X = 20) = 11.56 + .73(X_1) - .30(20)$$
$$= 11.56 + .73(X_1) - 6$$
$$= 5.56 + .73(X_1)$$

When X_2 is held constant at 20, the relationship between Y and X_1 reduces to a straight line, the upper of the two in Figure 16.2. For a larger class of $X_2 = 100$ students,

$$. \hat{Y}|(X = 100) = 11.56 + .73(X_1) - .30(100)$$
$$= -18.44 + .73(X_1)$$

This equation is the lower line (Figure 16.2).

Graphs that show the relationship between Y and one X variable, holding other X variables constant, are called **conditional effect plots.** Figure 16.2 is a conditional effect plot for the relationship between Y and X_1, given two different levels of X_2. The two lines are parallel, reflecting the fact that the slope of this relationship is the same ($b_1 = .73$) regardless of the level of X_2. A similar plot in Figure 16.3 (page 586) shows the relationship between evaluations and class size (X_2), for classes with $X_1 = 30\%$ A and B grades and classes with $X_1 = 90\%$ A's and B's. If X_1 is held constant at any value between 30 and 90, we would see a line between the two in Figure 16.3.

Figures 16.2 and 16.3 depict relationships for the subsample of classes ($n = 47$) taught by part-time instructors and teaching assistants. Figures 16.4 and 16.5 show the implications of the full sample ($n = 118$) regression in Table 16.5:

$$\hat{Y} = 21.24 + .57(X_1) - .30(X_2) + .32(X_3)(X_2)$$

The slope of the relationship between evaluations and grades is the same for both kinds of instructor (Figure 16.4). To show these relationships between Y and X_1 as lines, X_2 (class size) is held constant at its mean, $\overline{X}_2 = 51.7$.

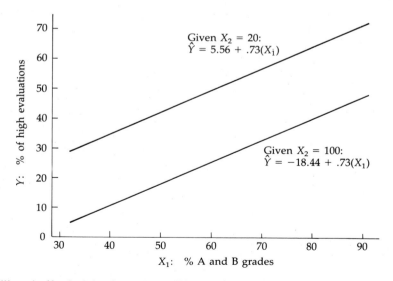

FIGURE 16.2   *Conditional effect plots given size = 20 and 100*

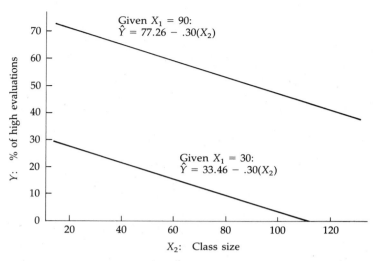

FIGURE 16.3   *Conditional effect plots given grades = 30 and 90*

The relationship between evaluations and class size is negative for part-time instructors and near 0 for full-time faculty (Figure 16.5). To show these relationships between Y and X_2 as lines, X_1 (percentage of high grades) is held constant at its mean, $\overline{X}_1 = 56.9$.

Conditional effect plots can be constructed in this manner from regression equations with any number of X variables. We simply trace the relationship between Y and one X variable, with all the other X variables assigned some arbitrary set of values. We can then assign different arbitrary values and see how this changes the line. With straightforward linear equations the resulting graphs are unsurprising (Figures 16.2 and 16.3), but conditional

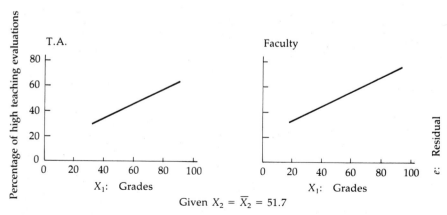

Conditional effect plots of relationship between teaching evaluations

FIGURE 16.4   *and grades, $X_3 = 0$ or 1*

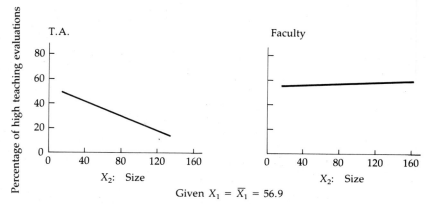

Given $X_1 = \bar{X}_1 = 56.9$

FIGURE 16.5   *Conditional effect plots of relationship between teaching evaluations and class size, $X_3 = 0$ or 1*

effect plots become more valuable with interaction effects (Figure 16.5) and curvilinear relationships.

In Chapter 15 we discussed several statistical problems that cause trouble in regression analysis (see Tables 15.10 and 15.11). These problems can be even more troublesome in multiple regression. Influential cases in particular are a common but easily overlooked source of trouble.

Special graphs have been invented to diagnose problems in multiple regression, but they are beyond the scope of this book. The simple e vs. $\hat{Y}$ plot introduced in Chapter 15 is also quite useful. Figure 16.6 shows an e vs.

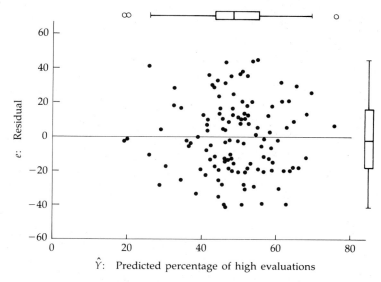

FIGURE 16.6   *e vs. $\hat{Y}$ plot for regression of Table 16.5*

$\hat{Y}$ plot for the teaching evaluations regression of Table 16.5. This plot is reassuring about the regression's validity: There is no evidence of nonlinearity, outliers, nonnormality, or heteroscedasticity (compare with Figures 15.7 and 15.8).

The next section looks at graphs for another regression example, where the results are less straightforward.

PROBLEM

*21. Show how the equations for the two lines shown in Figure 16.3 can be obtained from the regression equation, using X_1 values of 30 and 90.

16.10 *BIRTH RATE, GNP, AND CHILD MORTALITY*

Table 16.6 contains data on crude birth rates, per capita gross national products (GNP), and child mortality rates for 109 countries in 1985. Birth rates are known to be higher in poor (low GNP) countries, where population growth places increasing demands on already thin resources. One reason for the high birth rates in poor areas is the fact that many children die before adulthood. If parents want grown children who will support them in their old age, they must bear many and hope some survive. This argument suggests that birth rates will be higher in countries with high child mortality rates.

When birth rates (Y) from Table 16.6 are regressed on per capita GNP (X_1) we obtain

$$\hat{Y} = 38.92 - .00192(X_1) \qquad\qquad [16.17]$$

SE_b: .00023
t: -8.35
P-value: $P < .001$
$R_a^2 = .39$

The significant ($P < .001$) negative regression coefficient confirms that birth rates are higher in poor countries. When we control for child mortality rates, however, the coefficient on GNP drops noticeably:

$$\hat{Y} = 20.57 - .00049(X_1) + .176(X_2) \qquad\qquad [16.18]$$

SE_b: .00018   .013
t: -2.72 13.54
P-value: $P < .01$ $P < .001$
$R_a^2 = .77$

Although it is still significant, the coefficient on GNP (X_1) is only about one fourth as large after we control for child mortality rate (X_2). Child mortality

TABLE 16.6   *1985 Birth Rate, per Capita GNP, and Child Mortality Rate in 109 Nations*

Country	Births per 1,000 pop.	Dollars of GNP per person	Deaths per 1,000 children < 5
	Y	X_1	X_2
1. Ethiopia	46	110	206
2. Bangladesh	40	150	141
3. Burkina Faso	49	150	173
4. Mali	48	150	217
5. Bhutan	43	160	153
6. Mozambique	45	160	145
7. Nepal	43	160	153
8. Malawi	54	170	191
9. Zaire	45	170	122
10. Burma	30	190	66
11. Burundi	47	230	141
12. Togo	49	230	109
13. Madagascar	47	240	130
14. Niger	51	250	168
15. Benin	49	260	134
16. Ctr. African R.	42	260	164
17. India	33	270	100
18. Rwanda	52	280	153
19. Somalia	49	280	185
20. Kenya	54	290	107
21. Tanzania	50	290	132
22. Sudan	45	300	130
23. China	18	310	37
24. Haiti	35	310	145
25. Guinea	50	320	187
26. Sierra Leone	48	350	218
27. Senegal	46	370	164
28. Ghana	46	380	105
29. Pakistan	44	380	131
30. Sri Lanka	25	380	38
31. Zambia	49	390	99
32. Mauritania	45	420	157
33. Bolivia	42	470	137
34. Lesotho	41	470	120
35. Liberia	49	470	150
36. Indonesia	32	530	108
37. Yemen P.D.R.	46	530	175
38. Yemen Arab R.	48	550	188
39. Morocco	36	560	100
40. Philippines	33	580	52

(continued)

TABLE 16.6   *1985 Birth Rate, per Capita GNP, and Child Mortality Rate (Continued)*

Country	Births per 1,000 pop.	Dollars of GNP per person	Deaths per 1,000 children < 5
	Y	X_1	X_2
41. Egypt	36	610	104
42. Ivory Coast	45	660	120
43. Papua N.G.	37	680	75
44. Zimbabwe	47	680	84
45. Honduras	42	720	83
46. Nicaragua	43	770	75
47. Dominican R.	32	790	76
48. Nigeria	50	800	130
49. Thailand	26	800	46
50. Cameroon	47	810	99
51. El Salvador	38	820	70
52. Botswana	46	840	82
53. Paraguay	35	860	45
54. Jamaica	25	940	21
55. Peru	33	1,010	105
56. Turkey	30	1,080	93
57. Mauritius	20	1,090	26
58. Congo	45	1,110	84
59. Ecuador	35	1,160	72
60. Tunisia	32	1,190	86
61. Guatemala	40	1,250	70
62. Costa Rica	29	1,300	19
63. Colombia	27	1,320	51
64. Chile	22	1,430	23
65. Jordan	39	1,560	52
66. Syria	44	1,570	58
67. Brazil	29	1,640	72
68. Uruguay	19	1,650	30
69. Hungary	12	1,950	21
70. Portugal	14	1,970	20
71. Malaysia	30	2,000	30
72. South Africa	37	2,010	85
73. Poland	19	2,050	20
74. Yugoslavia	16	2,070	29
75. Mexico	33	2,080	53
76. Panama	26	2,100	26
77. Argentina	23	2,130	35
78. South Korea	21	2,150	29
79. Algeria	41	2,550	89
80. Venezuela	31	3,080	39

(*continued*)

TABLE 16.6   *1985 Birth Rate, per Capita GNP, and Child Mortality Rate (Continued)*

Country	Births per 1,000 pop.	Dollars of GNP per person	Deaths per 1,000 children < 5
	Y	X_1	X_2
81. Greece	13	3,550	17
82. Israel	23	4,990	14
83. Trinidad–Tobago	25	6,020	23
84. Hong Kong	14	6,230	9
85. Oman	44	6,730	126
86. Singapore	17	7,420	9
87. Libya	45	7,170	100
88. Saudi Arabia	42	8,850	65
89. Kuwait	34	14,480	23
90. United Arab Em.	30	19,270	36
91. Spain	13	4,290	10
92. Ireland	19	4,850	10
93. Italy	10	6,520	12
94. New Zealand	16	7,010	11
95. Belgium	12	8,280	11
96. United Kingdom	13	8,460	9
97. Austria	12	9,120	11
98. Netherlands	12	9,290	8
99. France	14	9,540	8
100. Australia	15	10,830	9
101. Finland	13	10,890	6
102. West Germany	10	10,940	10
103. Denmark	10	11,200	7
104. Japan	13	11,300	6
105. Sweden	11	11,890	6
106. Canada	15	13,680	8
107. Norway	12	14,370	8
108. Switzerland	11	16,370	8
109. U.S.A.	16	16,690	11

Source: Data from the World Bank (1987).

rate is a significant predictor of the birth rate even controlling for GNP, and the two X variables together explain about 77% of the variance of birth rate.[3]

Figure 16.7 (page 592) is an e vs. $\hat{Y}$ plot for this regression. The plot hints at trouble: There is a suggestion of inverted-U curvilinearity in the scatter of residuals (compare with the upper-right plot in Figure 15.8). The regression of Equation [16.18] may thus be invalid, since it assumes linearity.

Figure 16.8 is a **scatter plot matrix** containing scatter plots for each possible pairing of the variables in Equation [16.18]: Y vs. X_1, Y vs. X_2, and X_2 vs.

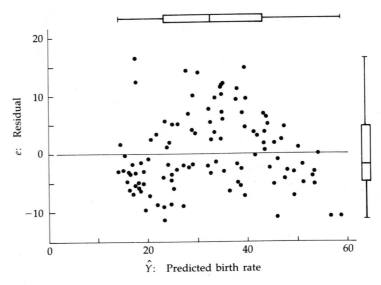

FIGURE 16.7   *e vs. Ŷ plot for regression of birth rate on GNP and child mortality*

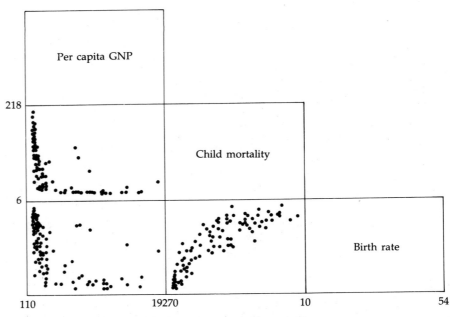

FIGURE 16.8   *Scatter plot matrix of GNP, birth rate, and child mortality*

X_1. Such plots give a snapshot of the numerous bivariate interrelationships in multivariate data. Here we can see that all the bivariate relationships are definitely curvilinear. They become more linear if we work instead with the logarithm of per capita GNP, $\log(X_1)$, and the square root of child mortality rate, $\sqrt{X_2}$, as shown by the scatter plot matrix of Figure 16.9. These transformations also make the positively skewed distributions of GNP and child mortality more symmetrical.

When birth rate (Y) is regressed on the logarithm of per capita GNP and the square root of child mortality rate we obtain

$$\hat{Y} = .829 + 1.078[\log(X_1)] + 3.550\sqrt{X_2} \qquad [16.19]$$

SE_b:	1.582	.266
t:	0.68	13.34
P-value:	$P > .20$	$P < .001$

$R_a^2 = .84$

This curvilinear equation fits better than the linear equation: $R_a^2 = .84$ vs. .77. The coefficient on $\log(X_1)$ is nonsignificant and weakly positive; if we dropped this variable from the equation, the proportion of variance explained would stay almost the same.

The curvilinearity of Equation [16.19] allows a better fit to the data, but also makes interpretation trickier. The coefficient on $\log(X_1)$, $b_1 = 1.078$, tells

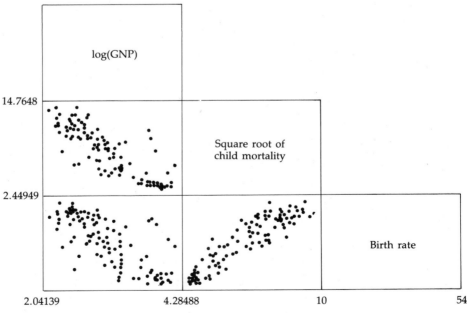

FIGURE 16.9   *Scatter plot matrix with transformed variables*

us that the predicted birth rate increases by 1.078 births per 1,000 population, if the logarithm of per capita GNP goes up by 1 and the square root of child mortality rate does not change. This literal reading is obviously unsatisfactory. As with bivariate regression, to help understand the implications of curvilinear equations, we turn to graphs.

The median child mortality rate among the 109 countries in Table 16.6 is 71 deaths per 1,000 children under age 5. The median provides a reasonable arbitrary value at which to hold this variable constant, while we look at the conditional effects of GNP. With child mortality held constant at $X_2 = 71$, the relationship between birth rate and GNP is

$$\hat{Y} = .829 + 1.078[\log(X_1)] + 3.550\sqrt{71}$$
$$= .829 + 1.078[\log(X_1)] + 29.913$$
$$= 30.742 + 1.078[\log(X_1)] \tag{16.20}$$

A conditional effect plot based on Equation [16.20] appears in Figure 16.10. The feebleness of this relationship is indicated by the near-horizontal curve; predicted birth rate rises by very little across the whole range of GNP, *if child mortality is held constant.*

We need a second conditional effect plot to show the relationship between birth rate and child mortality. If we hold GNP constant at its median, $1,010 per person,

$$\hat{Y} = .829 + 1.078[\log(1,010)] + 3.55\sqrt{X_2}$$
$$= .829 + 3.239 + 3.55\sqrt{X_2}$$
$$= 4.068 + 3.55\sqrt{X_2} \tag{16.21}$$

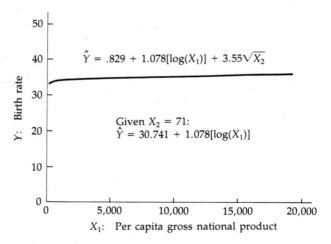

FIGURE 16.10   *Conditional effect plot given mortality = 71*

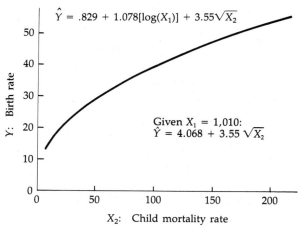

$$\hat{Y} = .829 + 1.078[\log(X_1)] + 3.55\sqrt{X_2}$$

Given $X_1 = 1,010$:
$\hat{Y} = 4.068 + 3.55\sqrt{X_2}$

Y: Birth rate

X_2: Child mortality rate

FIGURE 16.11   *Conditional effect plot given GNP = 1,010*

This curve is graphed in Figure 16.11. Predicted birth rate rises steeply as we move from low to medium-low child mortality rates, holding GNP constant. It levels off slightly at higher mortality rates.

What if we held the controlled variable constant at some value other than its median? The curves in Figures 16.10 and 16.11 would start higher or lower but otherwise look the same. That is, different values of the other variables will yield an infinite number of parallel curves, just as with linear regression we get parallel lines (Figures 16.2 and 16.3). *This is not true if the Y variable has been transformed,* however. If Y has been subjected to a nonlinear transformation the conditional effect curves may look different at different values of the other X variables. This complicates interpretation, but conditional plots can still help you think out the details.

> When the dependent variable in a multiple regression is a nonlinear transformation of the Y variable that is of substantive interest, *all the X variables' effects are implicitly interactions.* That is, the effect of each X variable changes with the values of the other X variables.

PROBLEMS

*22. The independent variables in the death rates–air pollution regression of Problem 3 have the following means:

$\overline{X}_1 = 10.5$
$\overline{X}_2 = 29.8$

$$\overline{X}_3 = 16.1$$
$$\overline{X}_4 = 2.6$$
$$\overline{X}_5 = 9.8$$
$$\overline{X}_6 = 10.8$$
$$\overline{X}_7 = 11.1$$

Draw a conditional plot showing the effect of sulfate concentrations (X_7) over the range 0–20 on the death rate, if all other X variables are held constant at their means. Use a Y-axis scale from 760 to 900 deaths per 100,000 people per year in your graph. Describe what you see.

*23. The regression in Problem 3 includes a curvilinear effect of population density: $X_4 = \log_{10}$(population density). Graph the conditional effect of population density on death rate, holding the other X variables constant at their means. Again, use a Y-axis scale from 760 to 900. To approximate the regression curve, connect points for population densities of 10, 100, 1,000, and 10,000 people per square mile. Describe what you see.

16.11 *DUMMY VARIABLE REGRESSION*

Problems 51–53 in Chapter 15 introduced dummy variable regression, by which categorical X variables can be included in a regression. Since dummy variables take on only two values (coded as 0 and 1), bivariate regression cannot handle categorical variables with more than two categories. With multiple regression, enough dummy variables can be created to represent any number of categories.

> A categorical variable with K categories can be represented in a regression by $K - 1$ **dummy variables.**

For example, to represent the categorical variable religion, with the four categories {Catholic, Protestant, Jewish, other}, we need three dummy variables: Catholic {0 = no, 1 = yes}, Protestant {0 = no, 1 = yes}, and Jewish {0 = no, 1 = yes}. A fourth dummy variable for "others" is not needed because we know that any cases for which Catholic = 0, Protestant = 0, and Jewish = 0 must fall in the "other" category. Indeed, if we created a fourth dummy variable for "other" and tried to include all four in a regression, we would find the regression impossible due to multicollinearity.

Table 16.7 contains data on a drinking scale and year in college for 19 college students. In Chapter 13 (Tables 13.9–13.12) we analyzed these data by ANOVA and saw that the relationship between drinking behavior and year in college fell just short of statistical significance. Year in college was earlier coded as a four-category categorical variable, but Table 16.7 shows how this

TABLE 16.7   *Data on Student Drinking by Year in College (from Table 13.9)*

Student	Year	Drinking scale[a] Y	Freshman dummy variable X_1	Sophomore dummy variable X_2	Junior dummy variable X_3
1	Freshman	16	1	0	0
2	Freshman	19	1	0	0
3	Freshman	25	1	0	0
4	Freshman	17	1	0	0
5	Sophomore	33	0	1	0
6	Sophomore	22	0	1	0
7	Sophomore	32	0	1	0
8	Sophomore	18	0	1	0
9	Sophomore	22	0	1	0
10	Sophomore	14	0	1	0
11	Junior	17	0	0	1
12	Junior	22	0	0	1
13	Junior	14	0	0	1
14	Junior	10	0	0	1
15	Junior	14	0	0	1
16	Senior	8	0	0	0
17	Senior	18	0	0	0
18	Senior	18	0	0	0
19	Senior	9	0	0	0

[a]Ranges from 4 (abstinence) to 33 (heavy and frequent drinking).

translates into three dummy variables: Students with 0 values on the dummy variables for "freshman," "sophomore," and "junior" must be seniors.

At the top of Table 16.8 (page 598) is a dummy variable regression analysis of these data. The regression equation is

$$\hat{Y} = 13.25 + 6(X_1) + 10.25(X_2) + 2.15(X_3) \qquad [16.22]$$

For a freshman, this equation reduces to

$$\hat{Y} = 13.25 + 6(1) + 10.25(0) + 2.15(0) = 19.25$$

which is the mean drinking scale for freshmen (see Table 13.9, page 421). For sophomores, the equation reduces to

$$\hat{Y} = 13.25 + 6(0) + 10.25(1) + 2.15(0) = 23.5$$

which is the mean for sophomores. Equation [16.22] will likewise produce the means for juniors (15.4) and for seniors (13.25).

Coefficients in dummy variable regressions have these interpretations:

1. The Y-intercept, a, is the mean of Y for the left-out category.

2. Each regression coefficient b_k is the difference between the mean of Y for category k, and the mean for the left-out category.

The sums of squares, F statistic, and P-value for the overall F test will be identical for ANOVA and the corresponding dummy variable regression. This can be seen by comparing the top (dummy variable regression) and bottom (ANOVA) of Table 16.8.

Displaying regression and ANOVA versions of the same analysis in Table 16.8 demonstrates that regression conveys more information. Either analysis gives the same overall F test, but regression also supplies individual coefficients reflecting the mean differences and t tests for which of those differences are significant. We see that although the overall null hypothesis is not

TABLE 16.8   *Drinking Scores and Year in College Analyzed by Dummy Variable Regression and by Analysis of Variance*

. regress drink fresh soph junior

Source	SS	df	MS		
				Number of obs	= 19
				F(3, 15)	= 3.06
				Prob > F	= 0.0605
Explained	309.905	3	103.302	R-square	= 0.3797
Residual	506.20	15	33.747	Adj R-square	= 0.2557
Total	816.105	18	45.339	Residual S.D.	= 5.8092

Variable	Coefficient	Std. Error	t	Prob > |t|
drink				
fresh	6	4.107716	1.461	0.165
soph	10.25	3.749815	2.733	0.015
junior	2.15	3.896922	0.552	0.589
_ cons	13.25	2.904594	4.562	0.001

. anova drink year

Source	SS	df	MS	F	Prob > F
Between groups	309.905	3	103.302	3.06	0.0605
Within groups	506.20	15	33.747		
Total	816.105	18	45.339		

rejected ($P = .0605$), we can reject the null hypothesis of no difference between the means for sophomores and seniors ($P = .015$).

PROBLEMS

Regression equations often include a mix of dummy and measurement variables. In such equations the dummy variables' coefficients should not be viewed as differences in means, but as differences in intercepts—hence the term **intercept dummy variable.**

For a sample of $n = 45$ professional basketball players, the number of rebounds caught in one season (Y) is regressed on the player's height in inches (X_1), average minutes played (X_2), and a dummy variable (X_3) coded 1 if the player's position is guard and 0 if the player is a center or forward:

$$\hat{Y} = -1,498.4 + .14(X_1) + 20.9(X_2) - 134.5(X_3)$$
$$SE_b: \qquad\qquad .02 \qquad 7.4 \qquad 61.9$$

$R_a^2 = .75$. Problems 24–26 refer to this regression.

*24. Interpret the values of a, b_1, b_2, b_3, and R_a^2 in this equation.

*25. Use t tests to determine whether height, minutes, and the guard position have significant partial effects on season rebounds.

*26. Write out two separate versions of this regression equation: one for guards, and one for nonguards. How do these equations demonstrate that dummy variable coefficients represent differences in intercept?

SUMMARY

Regression equations estimated from sample data have the general form

$$\hat{Y} = a + b_1(X_1) + b_2(X_2) + b_3(X_3) + \cdots + b_{K-1}(X_{K-1})$$

where K, the total number of a and b_k coefficients in the equation, must be no greater than the sample size: $K \leq n$. Ordinary least squares (OLS) regression obtains values for a (Y-intercept) and b_k (regression coefficients) such that the residual sum of squares, $\Sigma(Y_i - \hat{Y}_i)^2$, is as low as possible. Bivariate regression is simply OLS with $K = 2$ (a and one b):

$$\hat{Y} = a + b_1(X_1)$$

The arithmetic mean is an OLS regression with $K = 1$ (a only):

$$\hat{Y} = a$$

In this sense the arithmetic mean, bivariate regression, and multiple regression are just increasingly complex expressions of the same statistical idea.

Focusing on the regression coefficients, b_k, we can examine the effects of one X variable on Y while *statistically controlling for* the effects of all the other X variables that are in the regression equation. The t distribution is used for hypothesis tests and confidence intervals regarding such partial effects.

The coefficient of determination, R^2, summarizes how well all the X variables together predict Y. Generally, the more X variables in the equation the better predictions will be, but we seek a parsimonious model that achieves the best balance of predictive power and simplicity. The adjusted R^2 statistic, R_a^2, is preferred because it supports this balancing act. The F distribution is used to test hypotheses about combinations of coefficients, such as the overall null hypothesis that all of the slopes are 0.

Interaction terms, nonlinear transformations, polynomials, and dummy variables extend the possible applications of multiple regression. Interaction terms let us test the hypothesis that a relationship changes with the values of other X variables. Nonlinear transformations remain indispensable in taming ill-behaved distributions. Transformations and polynomial regression also provide simple ways to test hypotheses about curvilinear relationships. Any difference-of-means test or ANOVA can be reformulated as a regression, and we can work with mixtures of categorical and measurement X variables. These extensions to regression bring diverse analytical problems within the scope of a single technique.

We have barely scratched the surface of multiple regression in this chapter. A more advanced course, recommended for anyone considering a research career, would cover the basics of regression in greater depth. Beyond ordinary least squares, regression extends still further to encompass such real-world challenges as categorical Y variables, robust estimation, time series analysis, variables measured with error, and causal models with multiple Y variables. Each extension opens up new avenues for research.

PROBLEMS

*27. In Table 2.4 (page 32) we saw a measure of "environmental voting," based on an assessment of how pro-environment were the votes of U.S. congressional delegations. Some of the variance of this measure is explained by state demographic characteristics:

$$\hat{Y} = -292.1 + 3.54(X_1) + .0018(X_2) + 18.3(X_3) - 2.34(X_4)$$
$$SE_b: \qquad\qquad 1.44 \qquad .0014 \qquad 13.1 \qquad .59$$

$R_a^2 = .47$. Variable definitions are as follows:

Y: Environmental voting index (percentage)

X_1: Median age (years)

X_2: Per capita income (dollars)

X_3: Median education (years)

Y_4: Percent growth in population, 1980 to 1983

a. Which variables are significant predictors of environmental voting?

b. Interpret the numerical values of the significant coefficients identified in part a.

c. Interpret the numerical value of R_a^2.

d. What environmental voting index value would you predict for a state where the median age is 30 years, the per capita income is $12,000, the median education is 12.5 years, and the population grew by 4% over 1980–1983?

*28. In another 50-state analysis, researchers were surprised to discover a substantial correlation between rape rates and circulation rates for sex-oriented men's magazines (Baron and Straus, 1987). Numerous variables were included in regression analyses to see if the rape–sex magazine relationship might be spurious. Results of one such regression based on data for 1979 are

$$\hat{Y} = -3.37 + .0033(X_1) - .46(X_2) + .098(X_3) + 1.5(X_4) + .04(X_5)$$
$$SE_b: \qquad\qquad .0008 \qquad .37 \qquad 1.071 \qquad\quad .3 \qquad\quad .01$$

$R_a^2 = .78$. Variable definitions are:

Y: Rapes per 100,000 population

X_1: Copies of *Playboy* magazine per 100,000 adults

X_2: Percentage of population below poverty

X_3: Males as a percentage of population aged 15–24

X_4: Homicides per 100,000 population

X_5: Assaults per 100,000 population

a. Which coefficients are statistically significant at $\alpha = .01$?

b. Interpret the values of the significant coefficients.

c. Interpret the value of R_a^2.

The table at the top of page 602 contains computer output from the regression analysis of data on unemployment rates for graduates of 44 British universities. A complete data listing is in Chapter 5, Problem 15 (page 143). CAT is included as a dummy variable, with values of 1 for CAT schools and 0 for others. Use the results in the table for Problems 29–33.

Regression of Graduate Unemployment Rates on Mean Entrance Examination Score, Male/Female Sex Ratio, Academic Staff Cost per Student, and CAT[a] Status for 44 British Universities

.regress unemp exam sexrat cost CAT

Source	SS	df	MS
Explained	535.849	4	133.962
Residual	486.208	39	12.467
Total	1022.057	43	23.769

Number of obs	= 44
F(4, 39)	= 10.75
Prob > F	= 0.001
R-square	= 0.5243
Adj R-square	= 0.4755
Residual S.D.	= 3.531

Variable	Coefficient	Std. Error	t	Prob > |t|
unemp				
exam	$-$.8105914	.3714778	-2.182	0.035
sexrat	-3.729617	1.228865	-3.035	0.004
cost	$-$.4059358	.2651708	-1.531	0.134
CAT	$-$.1711953	2.49281	-0.069	0.946
cons	68.34019	25.43677	2.687	0.011

[a]CAT is a dummy variable coded 1 for ex-Colleges of Advanced Technology, 0 otherwise.
Source: Taylor (1984).

*29. a. Write out the regression equation.

  b. Interpret the numerical values of any coefficients that are statistically significant at $\alpha = .05$.

  c. Interpret the value of R_a^2.

  d. Construct a 95% confidence interval for the coefficient on X_1.

*30. Write out equations for predicting unemployment rates for graduates of CAT (CAT $= 1$) and other (CAT $= 0$) universities.

*31. Draw a conditional plot for the effect of the male/female sex ratio, holding average exam scores and staff costs at their respective medians ($Md_1 = 9.2$; $Md_3 = 98.85$), for a non-CAT school. How would the appearance of this graph change if we held average exam scores instead at their low extreme, $X_1 = 6.3$? At their high extreme, $X_1 = 13.6$?

*32. Figure 16.12 is an e vs. $\hat{Y}$ plot for the graduate unemployment regression. Describe any possible problems you see in this plot.

*33. A scatter plot matrix for the graduate unemployment data is shown in Figure 16.13.

  a. Explain the strange appearance of scatter plots involving the variable CAT.

  b. Do you see any possible statistical problems in this plot? How might they relate to your observations about the e vs. $\hat{Y}$ plot in Problem 32?

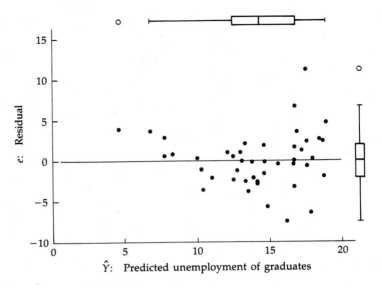

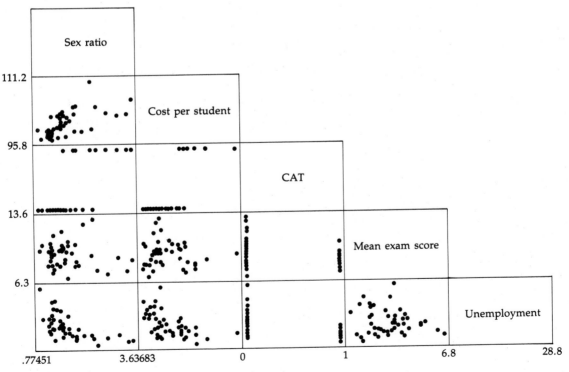

FIGURE 16.12   e vs. Ŷ plot for unemployment regression

FIGURE 16.13   Scatter plot matrix of British university data

*34. The coefficient b_1 in Equation [16.2] is a first-order partial, meaning it represents the effect of X_1 on Y controlling for one other X variable. What orders of partial are the b coefficients in the following regressions?

 a. Death rates in American cities, Problem 3

 b. Teaching evaluations, Equation [16.8]

 c. Teaching evaluations, Equation [16.10]

 d. Reindeer cave art, Problem 20

 e. Birth rates, Equation [16.18]

 f. Environmental voting, Problem 27

NOTES

1. The letter K here has the same meaning assigned in Chapter 15: the number of coefficients (a's and b's). K also equals the total number of X and Y variables in the analysis—hence the number of dimensions involved. Furthermore, we will see that this is equivalent to its meaning in Chapter 13: the number of categories in ANOVA.

   Since a point has *zero dimensions,* when we calculate a variable's mean we are summarizing one-dimensional information (the variable's distribution) with a simpler zero-dimensional model (the mean).

   Describing the mean in this way emphasizes its connection with regression—the arithmetic mean is actually the simplest kind of regression, with *zero* independent variables: $\hat{Y} = a$.

   The "best" value of a, which results in the lowest possible sum of squared errors, $\Sigma e_i^2 = \Sigma(Y_i - \hat{Y}_i)^2$, is $a = \overline{Y}$. Thus ordinary least squares regression, applied to a problem with zero X variables, reduces to finding the mean of Y.

2. Given a second-order polynomial,

$$\hat{Y} = a + b_1(X) + b_2(X^2)$$

the predicted Y values will have one minimum or maximum point, $\hat{Y}_m$. This point occurs at X value

$$X_m = \frac{-b_1}{2b_2}$$

For the equation $\hat{Y} = 3.431 + 1.389(X_1) - .017(X_2)$,

$$X_m = \frac{-1.389}{2(-.017)} = 40.85$$

$\hat{Y}_m$ is then found by substitution:

$$\hat{Y} = 3.431 + 1.389(40.85) - .017(40.85^2) = 31.8$$

Predicted income in this sample reaches its highest value of 31.8 thousand dollars (the high point of the curve in Figure 16.1, $\hat{Y}_m = 31.8$) when age equals 40.85 years ($X_m = 40.85$).

3. Child mortality undoubtedly influences the birth rate. The birth rate, in turn, may influence child mortality: short intervals between births worsen survival chances. The true causal relationship of these variables is therefore more complicated than this simple analysis depicts.

APPENDICES

Answers to In-Chapter Problems

CHAPTER 1

3. Frequency distribution of expected grades:

Grade	f	$\hat{\pi}$
D	1	.0345
C	6	.2069
B	16	.5517
A	6	.2069
Totals	29	1.0000

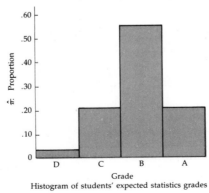

Histogram of students' expected statistics grades

The majority expect a B grade. Equal numbers expect C or A; one a D; and none an F.

5. *Sample:*   the collection of students actually interviewed or surveyed; *population:*   the larger universe of students from which the sample was selected.

6. Frequency distribution of GPAs:

True limits	Interval	f	$\hat{\pi}$	$\hat{p}$
1.795–1.995	1.80–1.99	1	.036	3.6%
1.995–2.195	2.00–2.19	1	.036	3.6
2.195–2.395	2.20–2.39	3	.107	10.7
2.395–2.595	2.40–2.59	7	.250	25.0
2.595–2.795	2.60–2.79	5	.179	17.9
2.795–2.995	2.80–2.99	2	.071	7.1
2.995–3.195	3.00–3.19	3	.107	10.7
3.195–3.395	3.20–3.39	4	.143	14.3
3.395–3.595	3.40–3.59	2	.071	7.1
Totals		28	1.000	100.0%

7.

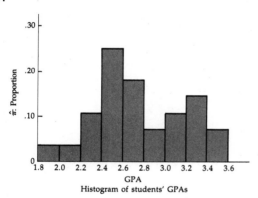

GPA
Histogram of students' GPAs

8. Frequency distribution of ratings of violent TV shows:

True limits	Interval	f	$\hat{\pi}$	$\hat{p}$
69.95–79.95	70.0–79.9	2	.182	18.2%
79.95–89.95	80.0–89.9	1	.091	9.1
89.95–99.95	90.0–99.9	2	.182	18.2
99.95–109.95	100.0–109.9	3	.273	27.3
109.95–119.95	110.0–119.9	2	.182	18.2
119.95–129.95	120.0–129.9	1	.091	9.1
Totals		11	1.000	100.0%

9.

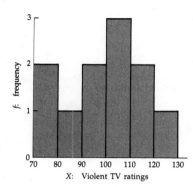

X: Violent TV ratings

11. Five of the 7 students missing VSAT values are older, nontraditional students. The average age of the 23 students with a VSAT score is 19.6 years; the average age of the 7 students who did not report one is 26 years. Since they graduated from high school years ago, the older students may be less likely to remember their SAT scores. They may also have been admitted in a special "continuing education" status that does not require the SAT.

CHAPTER 2

1. a. Passengers in smoking section:

Class limits	f	$\hat{p}$
0–9	1	5%
10–19	7	35
20–29	11	55
30–39	1	5
Totals	20	100%

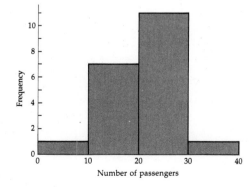

Number of passengers

b. Number of cigarettes smoked:

Class limits	f	$\hat{p}$
0–19	6	30%
20–39	9	45
40–59	2	10
60–79	0	0
80–99	2	10
100–119	0	0
120–139	1	5
Totals	20	100%

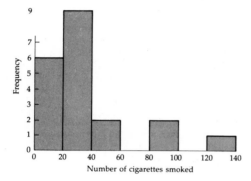

2.

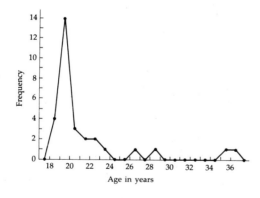 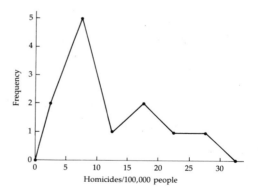

3.

Class interval	Frequency	Cumulative frequency	Cumulative proportion	Cumulative percentage
0–1,999	14	14	.583	58.3
2,000–3,999	1	15	.625	62.5
4,000–5,999	3	18	.750	75.0
6,000–7,999	0	18	.750	75.0
8,000–9,999	1	19	.792	79.2
10,000–11,999	2	21	.875	87.5
12,000–13,999	1	22	.917	91.7
14,000–15,999	1	23	.958	95.8
16,000–17,999	0	23	.958	95.8
18,000–19,999	0	23	.958	95.8
20,000–21,999	0	23	.958	95.8
22,000–23,999	1	24	1.000	100.0

The ogive rises steeply, then flattens, indicating that the distribution is positively skewed. Level regions in the ogive indicate that there are gaps in the upper tail of the frequency distribution.

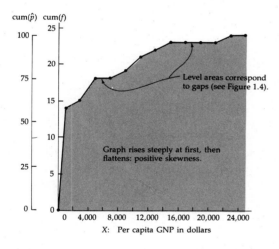

4. Airline baggage complaints per 1,000 passengers:

```
 2 | 9
 3 | 9
 4 | 01          Stems digits 1's, leaves .1's.
 5 | 5           7 | 1 means 7.1
 6 | 05
 7 | 12344
 8 |
 9 |
10 | 36
```

The distribution is negatively skewed, with a gap and two outliers (Northwest and United) at the high end.

5. Percentage of flights arriving on time:

```
6 | 0                Stems digits 10's, leaves 1's.
7 | 4567799          8 | 0 means 80%
8 | 033456
```

All values are crowded onto three stems, so we do not see much detail in the shape.

6. Consumer complaints per 100,000 passengers:

```
either   1 | 69        or        0 | 1122223459
         2 | 2556                1 | 1279
         3 | 4             stems digits 10's,
         4 | 0             leaves 1's.
         5 | 1
         6 |
         7 |
         8 |
         9 | 5
        10 |
        11 | 8
        12 | 4
        13 |
        14 |
        15 |
        16 |
        17 | 2
        18 |
        19 | 1
```

stems digits 1's,
leaves .1's.

The left-hand version has too many stems, so the data are too spread out. The right-hand version has the opposite problem of too few stems and not enough detail.

7. Percentage of flights arriving on time:

```
6* | 0
6. |                Stems digits 10's,
7* | 4              leaves 1's.
7. | 567799
8* | 0334           8* | 0 means 80%
8. | 56
```

We can now see a slight gap between Pacific Southwest, only 60.3% of whose flights arrived on time, and the rest of the distribution.

8. Airline consumer complaints:

```
0* | 11222234       Stems digits 10's,
0. | 59             leaves 1's.
1* | 12
1. | 79             1. | 7 means 17 complaints
```

We now have a more readable display, without too few or too many stems. The distribution is positively skewed: Most airlines had relatively few complaints, but some had many.

9. Nicotine concentration on airliners:

0*	00000000000011111
0t	233
0f	4455
0s	7
0.	
1*	1

Stems digits 100's,
leaves 10's.

1* | 1 means 110 μg/m³

The stem-and-leaf display has the same shape as the histogram in Figure 2.1, but rotated to an upright position.

10. Errors in self-reported VSAT scores:

−1*	43100
−0.	7
−0*	321100
0*	001
0.	6

Stems digits 100's,
leaves 10's.

0. | 6 means true score
is 60 points higher than
self-reported score

The display shows two peaks and a central gap. One peak is for large negative errors of 100 points or more, exaggerating the actual score; the other peak is for small or no reporting errors. The small errors were also more likely to be exaggerations. This distribution shows a pattern common in self-report data: People tend to give responses that present themselves in a more favorable light.

11. The three whale species were hunted to near extinction in order of their economic value: first blue whales, then fin, then sei.

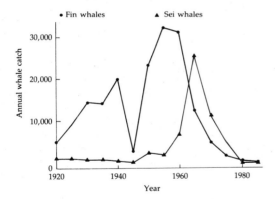

12. The dropoff in tent camping can be explained only partly by a switch to R.V.s. R.V. camping peaked in the late 1970's and is now declining.

    The popularity of tent camping in the early 1960's may reflect the huge generation of "baby boom" children, who were then at an age for family camping. A few years later they were old enough to backpack by themselves, hence the mid-1970's peak in back country camping. As the baby boomers aged further, back country camping too declined.

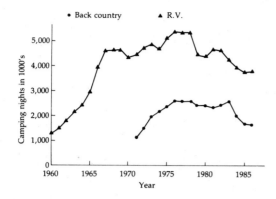

13. a.

KINDERGARTEN ENROLLMENTS

Year	Raw data	Smooth	Rough
1980	82	82	0
1981	77	82	−5
1982	99	84	15
1983	84	93	−9
1984	93	93	0
1985	122	114	8
1986	114	121	−7
1987	121	121	0

b.

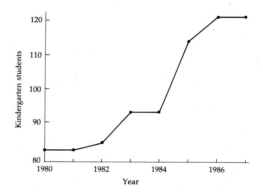

c. The smoothed plot shows a climb from enrollments in the low 80's to the low 120's.

14.

GRADE 12 ENROLLMENTS

Year	Raw data	Smooth	Rough
1980	150	150	0
1981	147	144	3
1982	132	134.5	−2.5
1983	127	127.25	−0.25
1984	123	129.5	−6.5
1985	145	139	6
1986	143	142.25	0.75
1987	138	138	0

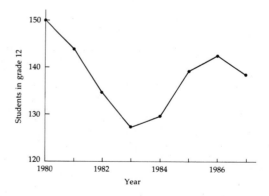

Enrollments dropped to a low point in 1983, returned to near their earlier levels, then dropped again in 1987.

15. Many possible answers. Obvious choices include the distributions of student age (Figure 1.1) and nicotine concentration in airliners (Figure 2.1).

CHAPTER 3

1. a. Strongly unimodal at 19
   b. Unimodal in 5–9.9 class
   c. Weakly multimodal at 2–3.9, 8–9.9, and 10–11.9 classes
   d. Weakly unimodal at 50–59 class
2. a. Mode = motor vehicle accident
   b. Mode = 3,000–3,999 class (thousands)
   c. Modes = 2 and 3
   d. Mode = 0–999 class
3. *High*: $n = 7$, so median is $(7 + 1)/2 = $ 4th case in ordered list; $Md = 17.4$. *Medium*: $n = 4$, so median is $(4 + 1)/2 = 2.5$th case, or the average of the 2nd and 3rd cases in ordered list; $Md = 11.7$. *Low*: $n = 2$, so median is $(2 + 1)/2 = 1.5$th case, or the average of 1st and 2nd cases in ordered list; $Md = 8.35$.
   These results suggest that lead concentrations in tree leaves increase as traffic density increases.
4. a. At position $(n + 1)/2 = 3$: $199.3 million
   b. At position $(n + 1)/2 = 3.5$ (halfway between Pomona and Delaware: $215.75 million. Adding Harvard raises the median by $26.45 million.
5. a. Median at depth 7.5: $Md = 883$
   b. Median at depth 5.5: $Md = 948$
   c. The median score for high expenditure communities is 65 points higher than that for low expenditure communities.
6. a. Md depth: $(n + 1)/2 = (61 + 1)/2 = 31$; Q depths: $(tmd + 1)/2 = (31 + 1)/2 = 16$
   b. Md depth: $(n + 1)/2 = (250 + 1)/2 = 125.5$; Q depths: $(tmd + 1)/2 = (125 + 1)/2 = 63$
   c. Md depth: $(n + 1)/2 = (4 + 1)/2 = 2.5$; Q depths: $(tmd + 1)/2 = (2 + 1)/2 = 1.5$
7. Median depth: $(28 + 1)/2 = 14.5$; Quartile depth: $(14 + 1)/2 = 7.5$; First quartile: $Q_1 = 99.81$; Third quartile: $Q_3 = 115.33$. About one-fourth of the skiers finished in less than 99.81 sec., and about one-fourth took more than 115.33 sec. The middle half took between 99.81 and 115.33 sec.
8. Median depth: $(14 + 1)/2 = 7.5$; Quartile depth: $(7 + 1)/2 = 4$; First quartile: $Q_1 = 2.5$; Third quartile: $Q_3 = 11.8$.
9. Low expenditure: $\overline{X} = 894.1$; High expenditure: $\overline{X} = 922.4$. The difference in mean SAT scores is only 28.3 points, compared with the 65-point difference in medians. The high expenditure communities still appear to have higher test scores, but not nearly as much as the medians suggest.
10. The mean for high expenditure communities is pulled down by Boston's very low value (762).
12. a. Mild positive skewness; the mean is pulled up by a few students with high GPA.
   b. Approximately symmetrical distribution.
   c. Positively skewed; mean age is pulled up by a few older students.
   d. Mild negative skewness; the mean is pulled down by a few students expecting low grades.
   e. Positively skewed; the mean is pulled up by a few heavy students.
13. a. $Md < \overline{X}$
   b. $Md < \overline{X}$
   c. $Md \approx \overline{X}$
14. a. Either could be used. The mean is preferable with mild skewness.
   b. The mean and median are nearly the same; the mean is preferable for its statistical properties.
   c. The median (19) better describes the typical age.
   d. Either could be used. The mean is preferable with mild skewness.
   e. The median is a better summary; the mean is pulled up by one heavy student.

15.

X interval	X midpoint	f	fX
0	0	2,041	0
1–10	5.5	645	3,547.5
11–20	15.5	466	7,223
21–25	23	206	4,738
26–30	28	84	2,352
31–40	35.5	222	7,881
41–50	45.5	214	9,737
51–99	75	389	29,175
100	100	185	18,500

$$n = 4,452 \qquad\qquad \Sigma X_i = 83,153.5$$

$$\overline{X} = \frac{\Sigma X_i}{n} = \frac{83,153.5}{4,452} = 18.68$$

16.

$$Md = X_L + w\left(\frac{n/2 - \mathrm{cum}(f)}{f}\right)$$

$$= 0.5 + 10\left(\frac{4,452/2 - 2,041}{645}\right) = 3.37$$

The median, $Md = 3.37$, is much lower than the estimated mean, $\overline{X} = 18.68$. This difference reflects the severe positive skewness of the distribution. The mean is pulled up by the numerous residency programs with high percentages of FMG, even though nearly half the programs have none.

17.

$$Md = X_L + w\left(\frac{n/2 - \mathrm{cum}(f)}{f}\right)$$

$$= 9,999.5 + 5,000\left(\frac{10,131,000/2 - 3,896,000}{1,635,000}\right)$$

$$= 13,576$$

The median income in female headed households is $Md = \$13,576$, less than half as much as the median income for married couple households ($\$29,954$). The proportion of millionaires in either type of household should have little effect on the medians, but it could affect a less resistant statistic such as the mean.

18. a. Unweighted mean: $\overline{X} = 59.5$ years

   b. Weighted mean:

Country	Life expectancy X	Population f	fX
Mauritania	44	1,561,000	68,684,000
France	74	54,432,000	4,027,968,000
Austria	72	7,574,000	545,328,000
Zaire	48	30,336,000	1,456,128,000
Sums:		$n = 93,903,000$	$\Sigma X_i = 6,098,108,000$

$$\overline{X} = \frac{\Sigma X_i}{n} = 64.94$$

   c. The weighted mean (64.94) exceeds the unweighted mean (59.5). In two countries life expectancies are high, and in two low. The unweighted mean counts each of the four countries equally. The weighted mean takes into account that about two-thirds of the 93 million people in these four countries live in Austria or France, where life expectancies are high.

CHAPTER 4

1. a. Low extreme: 3 thousand hectares/year; High extreme: 1,480 thousand hectares/year
   b. Median, at depth $(9 + 1)/2 = 5$: $Md = 190$; Quartiles, at depth $(5 + 1)/2 = 3$: $Q_1 = 87$, $Q_3 = 340$
   c. Range: $1,480 - 3 = 1,477$
       IQR : $Q_3 - Q_1 = 340 - 87 = 253$
2. The IQR is in the same units as the raw data, thousands of hectares per year. Deforestation rates in the middle 50% of these countries span a range of about 253 thousand hectares per year.
3. a. High expenditure communities: $Md = 948$, $Q_1 = 860$, $Q_3 = 987$, IQR $= 127$
   b. Low expenditure communities: $Md = 883$, $Q_1 = 836$, $Q_3 = 956$, IQR $= 120$
   c. The similar IQRs (127 and 120) tell us that there is about the same amount of variation or spread among the middle 50% of the communities in both categories.
4. and 5. $\overline{X} = 368.7$

X_i	$X_i - \overline{X}$	$(X_i - \overline{X})^2$
3	-365.7	133,736
3	-365.7	133,736
87	-281.7	79,355
125	-243.7	59,390
190	-178.7	31,934
270	-98.7	9,742
340	-28.7	824
820	451.3	203,672
1,480	1,111.3	1,234,988

$$\text{TSS}_X = 1,887,377$$

The Brazilian deforestation rate was 1,111.3 thousand hectares per year higher than the South American average. The Paraguayan rate was 178.7 thousand hectares per year below the average. Such deviation scores are measured in thousands of hectares per year (from the mean). The total sum of squares, TSS_X, is measured in (thousands of hectares per year) squared.

6. The variance is $s_X^2 = \text{TSS}_X/(n - 1) = 235,922$, measured in (thousands of hectares per year) squared. The standard deviation is the square root of the variance: $s_X = \sqrt{s_X^2} = \sqrt{235,922} = 485.7$, which is in the variable's natural units, 1,000's of hectares per year.
7. Mean $= 14.1$; standard deviation $= 12.1$.
8. Mean $= 21.5$; standard deviation $= 16.7$.
   a. The boys tend to play rough-and-tumble more often than the girls (mean $= 21.5$ episodes per boy vs. 14.1 per girl).
   b. There is more variation in the number of episodes among boys than among girls (16.7 vs. 12.1).
9. Sample: those subscribers actually surveyed. Population: all 10,000 subscribers. Sample statistics: the mean $(\overline{X})$ and standard deviation (s_X) of incomes reported by the subscribers surveyed. Once the survey is done, these statistics will be known. Population parameters: the mean (μ) and standard deviation (σ_X) of incomes of all 10,000 subscribers. These are unknown, but they may be estimated using the sample statistics.
10. a. Mean $= 922.4$, $s_X = 85.8$
   b. Mean $= 894.1$, $s_X = 63.7$
   c. Average SAT scores tend to be higher in the high expenditure communities (922.4 vs. 894.1), which are also more variable, however (85.8 vs. 63.7).
11. Normal: $\overline{X} = .158$, $s_X = .111$; impaired: $\overline{X} = .381$, $s_X = .150$
   a. The impaired subjects used gestures during a larger proportion of their conversation time (.381 vs. .158).
   b. There was more variation among the impaired subjects than among normal ones (.150 vs. .111).
12. a. PSD $= 94.1$, so $s_X <$ PSD. The distribution tails are comparatively light. Fewer communities have exceptionally high or low average SAT scores than would be expected in a normal distribution.
   b. PSD $= 88.9$, so $s_X <$ PSD. Again, the distribution tails are comparatively light, with few outlying cases.
   c. PSD $= 592.6$, so $s_X >$ PSD. The distribution of water savings has heavier than normal tails: More households made large positive and negative savings than we would expect if the distribution were normal.

13. The mean is nearly twice the median, indicating severe positive skewness. A comparison of PSD and s_x would be misleading in so skewed a distribution, but we already know it cannot resemble a normal distribution. Positive skewness indicates that most countries have comparatively low or moderate rates of deforestation, but there are much higher rates in a few countries—namely, Colombia and Brazil. The mean is pulled up so much by Colombia and Brazil that it exceeds the values for all seven other countries.

14. Girls: $Md = 3$, IQR = 3. Boys: $Md = 2.5$, IQR = 3. Based on a comparison of medians, the girls tended to participate in more aggressive episodes. Boys and girls were equally variable, as the identical IQRs show.

15. Girls: $\overline{X} = 3.4$, $s_x = 3.1$. Boys: $\overline{X} = 4.1$, $s_x = 5.3$. Based on a comparison of means, the boys tended to participate in more episodes and had considerably more variable behavior. These conclusions seem to contradict those of Problem 14.

16. Girls   Boys

Girls		Boys
100	0*	001111
332	0t	2233
44	0f	444
	0s	7
98	0.	
	1*	
	1t	3
	1f	
	1s	
	1.	
	2*	0

Stems digits 10's, leaves digits 1's.
2* | 0 means 20 episodes

Both distributions are positively skewed. Most children of either gender participated in only 0–3 episodes of aggressive fighting or chasing, but a few boys and girls were much more aggressive. The two highly aggressive boys pulled up the boys' mean and standard deviation and account for most of the apparent differences between boys and girls.

CHAPTER 5

1. a. At depth 7: $Md = 135.6$; at depth 4: $Q_1 = 45.3$, $Q_3 = 213.1$; IQR = $213.1 - 45.3 = 167.8$
   b. Low inner fence: $Q_1 - 1.5(\text{IQR}) = -206.4$; low adjacent: 1.7; high inner fence: $Q_3 + 1.5(\text{IQR}) = 464.8$; high adjacent: 387.8
   c. Low outer fence: $Q_1 - 3(\text{IQR}) = -458.1$; high outer fence: $Q_3 + 3(\text{IQR}) = 716.5$; no mild outliers
   d. 1979 is a high severe outlier, because 723.5 lies outside the high outer fence (716.5).

2.

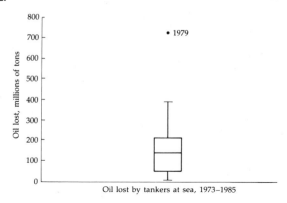

Oil lost by tankers at sea, 1973–1985

The distribution shows some positive skewness: the upper tail is longer and there is one high severe outlier—1979, when 723.5 million metric tons of oil were lost.

3. a. The boys have a higher median number of episodes than the girls do ($Md = 19$ episodes for boys vs. 9.5 for girls).

   b. The boys are more variable in the number of episodes they participate in: IQR = 25.5 for boys and 15 for girls.

   c. Both distributions are positively skewed. Most children participated in a moderate number of episodes, but a few had many.

4. a. The neurologically impaired subjects tend to use gestures during a higher proportion of their conversation ($Md = .39$ for impaired subjects, .13 for normal).

   b. The normal subjects show somewhat less variation (IQR = .205 for impaired, .13 for normal).

5. $Q_1 = 80.75$; $Md = 109.9$; $Q_3 = 132.35$; IQR = 51.6; Low adjacent value = 51.3; High adjacent value = 164.3.

   The box plots show that the 8088 computers are much slower and more variable in speed than are the 80286 or 80386 computers. The 8088 distribution is light tailed but roughly symmetrical.

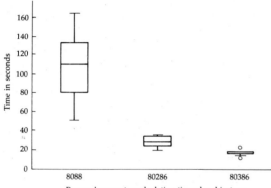

Personal computer calculation times by chip type

6.

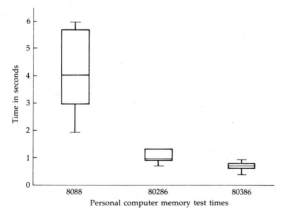

Personal computer memory test times

8088:

$$Q_1 = 2.95 \quad IQR = 2.75$$
$$Md = 4.00 \quad \text{Low adjacent value} = 1.90$$
$$Q_3 = 5.70 \quad \text{High adjacent value} = 6.00$$

80826:

$$Q_1 = .900 \quad IQR = 0.40$$
$$Md = .965 \quad \text{Low adjacent value} = .71$$
$$Q_3 = 1.300 \quad \text{High adjacent value} = 1.32$$

80386:

$$Q_1 = .61 \quad IQR = 0.16$$
$$Md = .69 \quad \text{Low adjacent value} = .40$$
$$Q_3 = .77 \quad \text{High adjacent value} = .94$$

The box plots again show the 8088 computers to be slower and more variable in speed than are the 80286 or 80386 computers. There are no outliers.

7. The median sales increase substantially over these three years, going from $527.7 million in 1983 to $913.65 million in 1985. The exception to this trend is the state of Washington, where gross sales actually declined. As a result the low extremes in the three box plots go down 1983–1985, at the same time the high extremes are rising. Overall the amount of variation is increasing (IQRs go from 252.4 in 1983 to 554.2 in 1985). By 1985 the distribution is negatively skewed, indicating that most of the lotteries had high sales, and only a few were low.

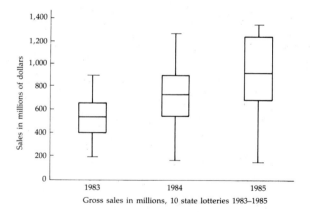

Gross sales in millions, 10 state lotteries 1983–1985

8. Nicotine concentrations in the smoking sections (IQR = 37.4) are far more variable than those in boundary (IQR = 14.35) or nonsmoking (IQR = 3.6) seats. In all three sections the distributions are positively skewed, because most flights had low levels of nicotine but a few were much higher.

On at least one flight the concentration in the nonsmoking section was as high as the smoking section median. On a number of flights the boundary section concentration was well above the smoking section median.

These data suggest that:

1. You are least likely to encounter high nicotine concentrations if you sit in the nonsmoking section; avoid boundary seats.

2. Even in the nonsmoking sections, nicotine concentrations are sometimes as high as a typical smoking section.

3. In smoking sections nicotine concentrations are sometimes many times higher than usual.

9. Center: The blue-collar median ($Md = 31.9$) is much higher than the white-collar median ($Md = 6.5$). Spread: Death rates are also far more spread out or variable among blue-collar (IQR = 22.1) than among white-collar jobs (2.7). Symmetry: Both distributions are positively skewed, with a drawn-out upper tail of high-risk occupations. Outliers: Three blue-collar occupations are high outliers; one of these, logger, is a severe outlier. Likewise three white-collar occupations are outliers, one of them severe (pilot). The death rate for pilots is high even compared with most blue-collar jobs, but the other white-collar outliers (sales manager and office worker) are well below the blue-collar median.

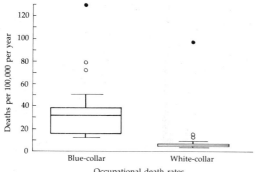

Occupational death rates

11. Median life expectancy is highest in Europe, followed by the Americas and Asia. Africa's is strikingly low ($Md = 48$). Europe is also the least variable continent, with only a 3-year IQR: European countries all have high life expectancies. The single low outlier in the European distribution is Turkey, which is geographically remote from Europe proper. Asia is the most varied continent, with life expectancies ranging from the highest (Japan's 77) to the lowest (Cambodia's 37) in the world. Life expectancies in the Americas also vary widely, but not as much as in Asia. Both Asia and the Americas have negatively skewed distributions, meaning that most countries have relatively high life expectancies but a few are much lower.

Life expectancies are very low in all of the African countries. Only one outlier is as high as the world median: Mauritius, a tiny island nation in the Indian Ocean. Like Turkey, Mauritius is geographically apart from the continent it has been grouped with. The fact that they are both geographical and statistical outliers suggests that Turkey and Mauritius do not really belong with the continents they were assigned to in this analysis.

CHAPTER 6

1. Neither country is a severe outlier, so they do not greatly affect their respective means. We usually do not think of Turkey as part of Europe, however, and Mauritius is an isolated island a thousand miles east of the African coast. It could be argued that neither outlier really belongs with its continental grouping in this analysis. To compare Africa with Europe, as those continents are usually understood, it would be better to set Turkey and Mauritius aside. This would result in (1) lowering the mean life expectancy among African countries, and (2) raising the mean life expectancy among European countries. Standard deviations for both continents would also decline.

2. There is no reason to delete the heavy student. Although no others in the 8-student sample approach his weight, many certainly would in a larger sample. A 255-pound student is simply part of the distribution's

upper tail, so outlier deletion is inappropriate here. It would lead to a less realistic comparison of marathoners and students.

3. a.   1.69897
  b.   2.99388
  c.   0.66276
  d. -3
  e.   0
  f. -0.04576
  g.   0.04139
  h. Error (impossible)
  i. Error (impossible)

4. Because of rounding and the limits of machine precision, most of the values will not come out exactly as expected. They should all be close, though.

5. *Male weight (kg)* Log_{10}(*Male weight*)

0	011344677899		$-0*$	0
1	1		$0*$	00
2	0		$0.$	566888999
3			$1*$	03
4			$1.$	8
5				
6	9			

Stems digits 10's.   Stems digits 1's.
6 | 9 means 69 kg   1. | 8 means 1.8 log(kg)

The distribution of weights in kilograms is extremely skewed, with one severe outlier, *Pongo pygmaeus*. The logarithmic distribution is much better behaved—nearly symmetrical, with no outliers. Taking logarithms has brought dramatic statistical improvement.

6. *Female weight (kg)* Log_{10}(*Female weight*)

0	0012234466679		$-0*$	11
1	0		$0*$	0444
2			$0.$	6688889
3	7		$1*$	0
4			$1.$	5

Stems digits 10's.   Stems digits 1's.
3 | 7 means 37 kg   1. | 5 means 1.5 log(kg)

As in problem 5, taking logs made the distribution more symmetrical. The raw-data distribution is positively skewed, with *Pongo pygmaeus* again appearing as a high severe outlier. The logarithmic distribution is less skewed, with no outliers.

7. *Population density*   *Square root(Pop. density)*

0*	011		0*	1
0t	22222333		0t	
0f	45		0f	4444555555
0s			0s	67
0.	89		0.	99
1*			1*	1
1t	2			

Stems digits 10,000's.
1t | 2 means
12,000 people   Stems digits 100's.
per square mile.   1* | 1 means 110.

The square root distribution still appears to be positively skewed, but less so than the raw data. The transformation was perhaps not strong enough. The next stronger transformation in the ladder of powers is the logarithm.

8. In order of increasing population ≥ 45:

City	Pop. ≥ 45	Neg. reciprocal(Pop. ≥ 45)	
1 Irving	24,134	− .0000414 or	− 4.14E-05
2 Pasadena	24,865	− .0000402	− 4.02E-05
3 Berkeley	25,934	− .0000386	− 3.86E-05
4 Chesapeake	29,477	− .0000339	− 3.39E-05
5 Aurora	31,468	− .0000318	− 3.18E-05
6 Stamford	37,480	− .0000267	− 2.67E-05
7 Lubbock	40,543	− .0000247	− 2.47E-05
8 Springfield	41,694	− .0000240	− 2.40E-05
9 Winston–Salem	43,206	− .0000231	− 2.31E-05
10 Arlington	49,101	− .0000204	− 2.04E-05
11 Tacoma	51,334	− .0000195	− 1.95E-05
12 Warren	52,473	− .0000191	− 1.91E-05
13 Albuquerque	89,102	− .0000112	− 1.12E-05
14 Atlanta	121,676	− .0000082	− 8.22E-06
15 Buffalo	128,752	− .0000078	− 7.77E-06
16 Philadelphia	599,987	− .0000017	− 1.67E-06

9.

Population ≥ 45			Negative reciprocal (Pop. ≥ 45)	
0	2222334444558		− 4	10
1	22		− 3	831
2			− 2	64430
3			− 1	991
4			− 0	871
5	9			

Stems digits 100,000's.
5 | 9 means 590,000 people.

Stems digits .00001's.
− 1 | 9 means − .000019.

The negative reciprocal transformation has made an extremely skewed distribution almost symmetrical.

10. The mean and median of the negative reciprocal values are nearly the same: $\overline{X}^* = -.0000233$ and $Md^* = -.0000236$. In contrast, the raw-data mean is more than twice the median: $\overline{X} = 86,952$ and $Md = 42,450$. The skewed raw-data distribution has no clear center, unlike the more symmetrical transformed distribution.

11. a. 5.75 kg
  b. 4.07 kg
  c. 4.17 kg
  d. 3,698 people/square mile
  e. 3,101 people/square mile
  f. 42,990 people
  g. 42,436 people

12. a.

X	$X^* = \sqrt{X}$
100	10
200	14.1421
300	17.3205

Mean of $X^* = 13.8209$, median of $X^* = 14.1421$. The mean is less than the median, so the distribution is negatively skewed.

   b. Squared mean square root = 13.8209^2 = 191; squared median square root = 14.1421^2 = 200.
   c. Taking square roots changed the relative distances between values. The inverse transformed mean will generally not be the same as the original mean.
   d. Taking square roots had no effect on the relative positions of the values. If there are an odd number of cases, with one exactly in the middle, the inverse transformed median will be exactly the same as the original median. With an even number of cases the median is really a mean of the two middle values, so the transformation shifts it as it does the mean.

CHAPTER 7

1. Possible examples:
   a. Drawing an ace, and drawing anything but an ace
   b. Drawing an ace, and drawing a jack
   c. Drawing an ace, and drawing a diamond
   d. Drawing a face card, and drawing a jack
2. **a.** $P(Q)$ = 4/52 = .077
   b. $P(not\ Q)$ = 1 − $P(Q)$ = 1 − .077 = .923
   c. $P(H)$ = 13/52 = .25
   d. Since Q and H are independent events: $P(Q\ and\ H)$ = $P(Q)P(H)$ = (4/52) (13/52) = 1/52 = .019
   e. Since Q and H are not mutually exclusive: $P(Q\ or\ H)$ = $P(Q)$ + $P(H)$ − $P(Q\ and\ H)$ = 4/52 + 13/52 − 1/52 = 16/52 = .308
3. Possible examples:
   a. X = card drawn is both a queen and a jack
   b. Y = card drawn is either a heart or not a heart
4. Let $P(1)$ represent the probability that pump 1 fails, $P(2)$ the probability that pump 2 fails, etc. Under independence,
   $P(1\ and\ 2\ and\ 3\ and\ 4)$ = $P(4)P(3)P(2)P(1)$ = (.05)(.05)(.05)(.05) = .00000625, or one four-pump failure in 160,000 emergencies.
5. $P(1\ and\ 2\ and\ 3\ and\ 4)$ = $P(4)P(3|4)P(2|3\ and\ 4)P(1|2\ and\ 3\ and\ 4)$.
   In the worse case, the conditional probabilities are all equal to 1: if pump 4 fails, 3 must fail, and so on. Then the probability of a four-pump failure would be $P(1\ and\ 2\ and\ 3\ and\ 4)$ = $P(4)(1)(1)(1)$ = .05, or one four-pump failure in 20 emergencies.
6. Pessimistic reasoning leads to a probability estimate 8,000 times higher than that obtained under the optimistic assumption of independence.
7. **a.**

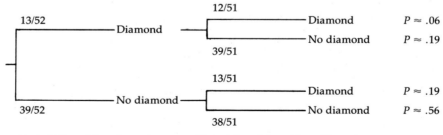

   Probability of drawing at least one diamond ≈ .06 + .19 + .19 = .44

b.

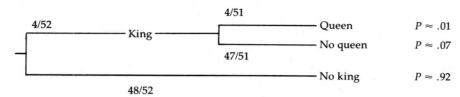

Probability of drawing first a king, then a queen ≈ .01

c.

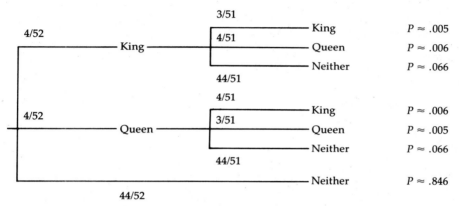

Probability of drawing one queen and one king ≈ .006 + .006 = .012

8. a.

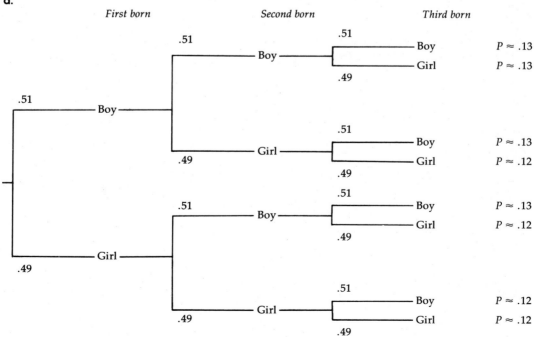

b. There are six possible ways to end up with at least one boy and one girl.

c. P{three boys} ≈ .13; P{three girls} ≈ .12. Probability of three girls or three boys (mutually exclusive events) can be found by adding: .13 + .12 = .25.

9. a. $P(E) = .9; P(L) = .447$

  b. $P(E) = .9; P(S) = .1. P(E \text{ and } L) = P(E)P(L) = (.9)(.447) = .4023$ if L and E are independent. $P(S \text{ and } L) = P(S)P(L) = (.1)(.447) = .0447$ if L and E are independent.

  c. Students training to teach secondary-school social studies probably would have better geographical knowledge than those training for elementary education; their studies and interests presumably include more geography, history, etc.

  d. $P(L|E) < P(L); P(L|S) > P(L)$

10. a. $n = 282. P(E)n = (.9)(282) \approx 254; P(L)n = (.447)(282) \approx 126$

  b. $P(L)P(E)n = (.447)(.9)(282) \approx 113.5$

  c. Number expected under independence = 113.5. Number actually observed = 99. Fewer elementary education majors were able to locate England on a map than we would have expected, if major and ability were independent. This suggests that they are *not* independent: Ability to locate England is dependent on major and is lower for elementary education majors.

11. a. $P(X \geq 2.5) = 1 - .420 = .580$

  b. $P(X \geq 3.5) = 1 - .936 = .064$

  c. $P(2.5 \leq X < 3.5) = .936 - .420 = .516$

  d. $P(0.5 \leq X < 2.5) = .420 - .004 = .416$

  e. $P(X < 0.5 \text{ or } X \geq 3.5) = .004 + .064 = .068$

12. a. $P(X \geq 10,000) = .799$

  b. $P(X \geq 50,000) = .147$

  c. $P(5,000 \leq X < 20,000) = .348$

  d. $P(X < 5,000 \text{ or } X \geq 50,000) = .224$

  e. $P(X < 10,000 \text{ or } 20,000 \leq X < 35,000 \text{ or } X \geq 50,000) = .618$

13. $N = 88,458,000$

  a. $P(X \geq 50,000)N = 13,003,326$

  b. $P(35,000 \leq X < 50,000)N = 13,976,364$

14. a. $1 - .9974 = .0026$

  b. $P\{[X < (\mu - 2\sigma)] \text{ or } [X > (\mu + 2\sigma)]\} = 1 - .9544 = .0456$, so $P[X < (\mu - 2\sigma)]$ must be half this: $.0456/2 = .0228$

  c. $P[(\mu - 1\sigma) < X < (\mu + 1\sigma)] = .6826$, so $P[X < (\mu - 1\sigma)] = (1 - .6826)/2 = .1587$. $P[(\mu - 2\sigma) < X < (\mu + 2\sigma)] = .9544$, so $P(X > 2\sigma) = (1 - .9544)/2 = .0228$. Since they are mutually exclusive events, $P\{[X < (\mu - 1\sigma)] \text{ or } [X > (\mu + 2\sigma)]\} = .1587 + .0228 = .1815$

  d. 0 (mutually exclusive events)

15. a. $z = 0$

  b. $z = 1$

  c. $z = -2$

16. The mean $(\overline{X} = 2.58)$ is slightly lower than the median $(Md = 2.61)$, indicating mild negative skewness. The standard deviation $(s_X = .62)$ is slightly larger than the pseudo-standard deviation (PSD = IQR/1.35 = .58), so the distribution is a bit heavy-tailed.

17. a. $\mu + 2\sigma_X = 2.58 + 2(.62) = 3.82$

  b. A 1.34 GPA is $2.58 - 1.34 = 1.24$ points, or 2 standard deviations, below the mean. The probability of values more than 2 standard deviations *below* the mean is $P[X < (\mu - 2\sigma)] = .0228$. This is the same distance below the mean (2σ) as the answer in part a is *above* the mean.

CHAPTER 8

1. a. Let (33333) represent the event all five digits are 3's. $P(33333) = P(3)P(3)P(3)P(3)P(3) = (.1)(.1)(.1)(.1)(.1) = .00001$. We can use the multiplication rule, since these events are independent.

  b. $P(24679) = P(2)P(4)P(6)P(7)P(9) = (.1)(.1)(.1)(.1)(.1) = .00001$, which is the same answer as part a. *Any* five-digit sequence has this probability.

c. The probability that any one digit is not 0 is $P(not\ 0) = 9/10 = .9$. The probability that the event *not 0* occurs five times in a row is therefore: $P(not\ 0)P(not\ 0)P(not\ 0)P(not\ 0)P(not\ 0) = (.9)(.9)(.9)(.9)(.9) = .59$.

d. The probability that any one digit is odd is $5/10 = .5$. The probability that all five are odd is therefore: $P(\text{all five are odd}) = P(odd)P(odd)P(odd)P(odd)P(odd) = (.5)(.5)(.5)(.5)(.5) = .03125$.

Notice that parts a–d can be calculated fastest as exponents: $P(\text{all five are odd}) = P(odd)^5$, etc.

5. Interpret the digit 1 as "one dot showing," 2 as "two dots showing," etc. We could simply read through the table, ignoring digits of 0 or 7–9, to simulate successive rolls of the die. To simulate rolling a pair of dice, read successive two-digit combinations, again ignoring 0's and 7–9's.

6. Pairs from Table 8.1: 52, 75, **24**, 36, 66, 65, 77, 31, 88, 43, **24** (ignore), **23**, **22**, 65, 88, **27**, 59, 68, 92, 45, 44, **13**, 35, **03**, ...

Student	Gender	GPA
24	Female	2.20
23	Female	1.96
22	Male	2.50
27	Female	2.80
13	Male	3.00
03	Female	

7. Pairs from Table 8.1: 52, 75, **24**, 36, 66, 65, 77, 31, 88, 43, **24**, **23**, **22**, 65, 88, **27**, 59, 68, 92, 45, 44, **13**, ...

Student	Gender	GPA
24	Female	2.20
24 (again)	Female	2.20
23	Female	1.96
22	Male	2.50
27	Female	2.80
13	Male	3.00

8.

School	Endowment	Random Number
Washington University	972.5	14608
University of Texas, Austin	244.6	23380
Duke University	338.7	26928
Brown University	315.4	33778
Yale University	1,750.7	42895
Cornell University	711.7	43186
Princeton Theological Seminary	284.2	44487
Tulane University	182.1	65809
St. Louis University	120.5	67743
Dartmouth College	520.6	77412
Georgetown University	174.0	90568
Rockefeller University	475.7	94363
Wellesley College	265.0	96725
Emory University	731.8	98078

a. The first three schools are selected for an $n = 3$ sample: Washington, Texas, and Duke.

b. Brown, Yale, and Cornell.

c. For an approximately 70% random sample, we could choose each random number less than 70000: St. Louis University and the eight schools preceding it in the sorted table. The sample then consists of 9 schools, so it is actually about $100(9/14) = 64\%$ of the population.

10.

Number	State	Teen births
08	Delaware	16.7%
24	Mississippi	23.2
24	Mississippi	23.2 (again)
26	Montana	12.4
42	Tennessee	19.9
46	Virginia	15.5

$\overline{X} = 18.48; s_X = 4.37$

11. a. $\overline{X} = 18.48$, the mean of the percentage of teenage births in our 6-state random sample.
   b. $s_X = 4.37$, the standard deviation of the percentage of teenage births in our 6-state random sample.
   c. $n = 6$, the number of cases (states) in the sample.
   d. μ is the population mean: the mean percentage of teenage births in the population consisting of all 50 states.
   e. σ_X is the population standard deviation: the standard deviation of the percentage of teenage births in the population consisting of all 50 states.
   f. $N = 50$, the number of cases (states) in the population.

12. a. Second random sample:

Number	State	Teen births
08	Delaware	16.7%
13	Illinois	15.7
19	Maine	15.3
21	Massachusetts	10.7
42	Tennessee	19.9
46	Virginia	15.5

$\overline{X}_2 = 15.63; s_2 = 2.96$

   b. Third random sample:

Number	State	Teen births
03	Arizona	16.5%
05	California	13.9
12	Idaho	13.1
14	Indiana	17.3
14	Indiana	17.3 (again)
25	Missouri	16.9

$\overline{X}_3 = 15.83; s_3 = 1.85$

   c. Means in the three samples range from 15.63 to 18.48. Standard deviations range from 1.85 to 4.37. Although we might guess that the true population parameters will be somewhere in these ranges, we cannot be very specific.

13. a. $\overline{\overline{X}} = 16.65$
   b. $s_{\overline{X}} = 1.59$
   c. $\overline{s}_X = 3.06$
   d. $s_{s_X} = 1.26$

14. a. The mean of the three sample means, $\overline{\overline{X}} = 16.65$, is not equal to the true population mean, $\mu = 15.31$. This discrepancy should not surprise us because the Central Limit Theorem states only that $\overline{X}$ will approach μ over an infinite number of large samples. Here we have only three small samples.
   b. $\sigma_{\overline{X}} = \sigma_X/\sqrt{n} = 3.49/\sqrt{6} = 1.42$
   c. If we collect many large samples, the standard deviation of the sample means should approach the standard error of the mean: $s_{\overline{X}} \rightarrow \sigma_{\overline{X}}$.

15. a. The larger the standard deviation, the greater the standard error will be, because $\sigma_{\bar{x}} = \sigma_x/\sqrt{n}$. Thus sample-to-sample variability increases with the population variability of X.

   b. The larger the sample size n, the smaller the standard error will be. There is less sample-to-sample variation with large samples than with small ones.

16. a. $SE_{\bar{x}} = s_x/\sqrt{n} = 4.37/\sqrt{6} = 1.78$

   b. $SE_{\bar{x}} = s_x/\sqrt{n} = 2.96/\sqrt{6} = 1.21$

   c. $SE_{\bar{x}} = s_x/\sqrt{n} = 1.85/\sqrt{6} = .76$

   d. The estimated standard errors reflect random sample-to-sample variations in standard deviations. These are, after all, small samples. The Central Limit Theorem describes only what happens with *large* samples.

17. a. The estimate is unconvincing because it is based on only five cases. In small samples, the standard error will be relatively large, since $\sigma_{\bar{x}} = \sigma_x/\sqrt{n}$. A large standard error implies a lot of sample-to-sample variability; the mean in any one sample could be far from the true population value.

   b. A more convincing estimate would result from a larger random sample. The larger the sample, the lower the standard error, implying little sample-to-sample variability. The mean from any one sample is then more likely to be near the true population value.

CHAPTER 9

1. a. As suggested graphically in Figure 4.1, we can check for approximate normality in two steps:

   1. Compare the median with the mean. If they are approximately equal, the distribution can be considered to be *symmetrical*.

   2. If a mean–median comparison shows that the distribution is symmetrical, next compare the standard deviation with the pseudo-standard deviation (PSD). If they are about equal, the distribution can be considered more or less *normal*. (There is no point in making this second comparison if the distribution is not even symmetrical.)

   b. The mean (231.6) and median (233) are close, so the distribution is roughly symmetrical. The PSD = IQR/1.35 = 13.7, which is substantially less than the standard deviation (19.2). Therefore this distribution is *heavy-tailed* relative to the normal distribution.

   c. The mean (225.5) and median (225) are nearly identical. The PSD = IQR/1.35 = 21.5, which is close to the standard deviation (20.9), so this distribution has approximately normal tails.

2. The comparisons in Problem 1, parts b and c, and the visual evidence in Figures 9.1 and 9.2, show that the $n = 1,000$ distribution is much more nearly normal than the $n = 36$ distribution. We also see that the sample mean for the 1,000 sums (225.5) is much closer to the population mean (225). In both respects, the $n = 1,000$ sample more resembles the theoretical population distribution than the smaller ($n = 36$) sample does.

3. a. A typical stem-and-leaf display of 200 random digits might be:

```
                   0 | 00000000000000000000000000000
stems digits 1's,  1 | 00000000000000000000
leaves digits .1's.2 | 000000000000000000
                   3 | 00000000000000000000000
9 | 0 means 9.     4 | 00000000000000000000000
                   5 | 00000000000000000000
                   6 | 00000000000000000
                   7 | 000000000000000000
                   8 | 0000000000000000
                   9 | 0000000000000000
```

The expected theoretical distribution is rectangular. By chance, the observed distribution of this sample of 200 random digits is slightly different from a uniform distribution.

b. The distribution for 40 five-digit sums of the digits shown in the answer to part a is:

1*	0134
1.	555667778999
2*	01111222222333444
2.	6666
3*	00
3.	5

stems digits 10's,
leaves digits 1's.

3. | 5 means 35.

The distribution of sums should have a distinct central peak and lower-frequency tails, quite unlike the roughly rectangular distribution of the individual random digits. The distribution of sums is not really normal, but it is much closer to normal than is the original distribution of digits.

In the example shown here, notice that the slight positive skewness evident in the distribution of random digits (part a) is also reflected in the distribution of sums. Sums of *many* small random influences will be normally distributed, but we created these sums by adding up only five digits each. This is too few to completely overcome the shape of the original distribution.

c. The mean of random digits should theoretically be $\mu = 4.5$. For sums of five random digits, it should be $5(4.5) = 22.5$.

5. a.

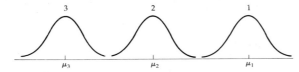

b.

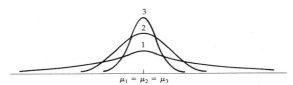

c.

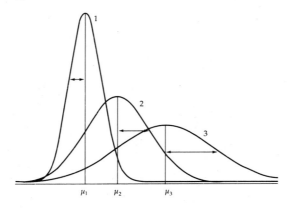

d.

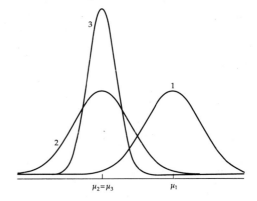

6. We can guess that the mean test scores will be highest in the advanced class, lowest in the remedial class, and intermediate in the general class. Since the advanced class contains no below-average students, it is more homogeneous and will have a lower standard deviation than the general class. For similar reasons the remedial class will also have a lower standard deviation. Since a normal distribution is symmetrical, its top half (advanced students) and bottom half (remedial students) should be equally spread out. In summary: $\mu_a > \mu_g > \mu_r$, $\sigma_a = \sigma_r < \sigma_g$

7. **a.** $z = 4$, or 4 standard deviations above the mean
   b. $z = -3$, or 3 standard deviations below the mean
   c. $z = 2.5$, or 2.5 standard deviations above the mean
   d. $z = -.507$, or .507 standard deviation below the mean
   e. $z = 1$, or 1 standard deviation above the mean
   f. $z = -2$, or 2 standard deviations below the mean

8. **a.** $x = 725$
   b. $x = 425$
   c. $x = 21.734$
   d. $x = 2,075$
   e. $x = .1$
   f. $x = -15$

9.
Confidence interval	Two-sided tests	One-sided tests	Critical value
.82	.18	.09	1.34

10.
	Confidence interval	Two-sided tests	One-sided tests	Critical value
a.	.34	.66	.33	.44
b.	.94	.06	.03	1.88
c.	.39	.61	.305	.51
d.	.97	.03	.015	2.17

11. **a.** To include the middle 50%, or .50 confidence interval, requires $z = .674$. Thus the middle 50% is contained within $0 \pm .674$. The width of the IQR is $2(.674) = 1.348$.
   b. Since IQR $= 1.348\sigma_X$, it follows that $\sigma_X = $ IQR$/1.348$.
   c. In a normal distribution, the IQR equals 1.348 standard deviations; therefore the standard deviation approximately equals the PSD. If the standard deviation and PSD are not equal, this is evidence that the distribution is not normal.

12. Nicaragua's per capita 1982 GNP is .610 standard deviation below the 24-nation mean.

13. **a.** $\bar{X} = 943.67$; $s_x = 217.96$; $Z = (X - 943.67)/217.96$

Year	Flow (Raw)	Flow (Standard Score)
1960	845	-.45
1961	1,109	.76
1962	1,027	.38
1963	751	-.88
1964	1,066	.56
1965	833	-.51
1966	441	-2.31
1967	1,059	.53
1968	1,167	1.02
1969	1,139	.90
1970	1,117	.80
1971	770	-.80

b. 1966: $X = 441$, $X^* = -2.31$. In 1966 White River flow was 2.31 standard deviations below the mean for these 12 years.

1968: $X = 1,167$, $X^* = 1.02$. In 1968 White River flow was 1.02 standard deviations above the mean for these 12 years.

c. $\overline{X}^* = 0.0$, $s_{X^*} = 1.0$. Apart from small rounding errors, the mean and standard deviation of X^* equal 0 and 1, as they theoretically should.

d. $\overline{X}^* = (\overline{X} - \overline{X})/s_X = (943.67 - 943.67)/217.96 = 0$

14. a. $Md = 1,043$, $\overline{X} = 943.67$. The distribution is negatively skewed; the mean is pulled down by the particularly low flow of 1966.

b. $Md^* = .455$, $\overline{X}^* = 0$. This distribution too is negatively skewed; the mean is pulled down by the low flow of 1966.

c. Raw data mean–median difference: $Md - \overline{X} = 1,043 - 943.67 = 99.33$ cu ft/sec.

Standard scores mean–median difference: $Md^* - \overline{X}^* = .455 - 0 = .455$ standard deviation.

Since one standard deviation is 217.96 cu ft/sec, a difference of .455 standard deviation is $.455(217.96)$ = 99.17 cu ft/sec. Apart from rounding error, this is the same as the mean–median difference in the raw data (part a). Both distributions show identical negative skewness.

15. a. $\overline{X} = 12.97$, $s_X = 2.42$, $n = 155$, and $\mathrm{SE}_{\overline{X}} = .19$. For 90% c.i. (from Table A.2): $z = 1.645$.

$$\overline{X} \pm z\mathrm{SE}_{\overline{X}} = 12.97 \pm 1.645(.19) = 12.97 \pm .31$$

We are 90% confident that $12.66 \leq \mu \leq 13.28$.

b. For 99% c.i. (from Table A.2): $z = 2.576$.

$$\overline{X} \pm z\mathrm{SE}_{\overline{X}} = 12.97 \pm 2.576(.19) = 12.97 \pm .49$$

We are 99% confident that $12.48 \leq \mu \leq 13.46$.

c. The 90% confidence interval is the narrowest of the three. Over the long run such intervals will be accurate (contain μ) only 90% of the time, however. The 99% confidence interval is the widest interval. Over the theoretical long run, 99% confidence intervals are more likely to be accurate, that is, to contain the true parameter μ.

16. a. $\overline{X} = 1.4$ µg/g, $s_X = 0.4$ µg/g, and $n = 44$.

$$\mathrm{SE}_{\overline{X}} = s_X/\sqrt{n} = 0.4/\sqrt{44} = .06$$

b. For 95% c.i. (from Table A.2): $z = 1.96$.

$$\overline{X} \pm z\mathrm{SE}_{\overline{X}} = 1.4 \pm 1.96(.06) = 1.4 \pm .118$$

We are 95% confident that $1.282 \leq \mu \leq 1.518$.

c. For 99.9% c.i. (from Table A.2): $z = 3.29$.

$$1.4 \pm 3.29(.06) = 1.4 \pm .197$$

We are 99.9% confident that $1.203 \leq \mu \leq 1.597$.

d. The 99.9% interval is wider than the 95% interval. To be more confident, we would use the wider interval.

17. a. $\overline{X} = 39.1$ hr/wk, $s_X = 15.7$ hr/wk, and $n = 110$. For 95% c.i. (from Table A.2): $z = 1.96$.

$$\overline{X} \pm z\mathrm{SE}_{\overline{X}} = \overline{X} \pm z(s_X/\sqrt{n}) = 39.1 \pm 1.96(15.7/\sqrt{110})$$
$$= 39.1 \pm 2.93$$

We are 95% confident that $36.17 \leq \mu \leq 42.03$.

Informal interpretation: Based on our analysis of these 110 schools, we are 95% confident that the mean usage rate for microcomputers in the population consisting of all Chicago-area elementary schools is between 36.17 and 42.03 hours per week.

Formal interpretation: If we took many random samples of $n = 110$ schools each from this population, and calculated confidence intervals in this manner for each one, 95% of those confidence intervals should include the true population mean μ.

b. $\overline{X} = 58\%$, $s_x = 22.6\%$, and $n = 110$. For 99% c.i. (from Table A.2): $z = 2.576$.

$$\overline{X} \pm zSE_{\overline{x}} = \overline{X} \pm z(s_x/\sqrt{n})$$
$$= 58 \pm 2.576(22.6/\sqrt{110}) = 58 \pm 5.55$$

We are 99% confident that $52.45 \leq \mu \leq 63.55$.

Informal interpretation: Based on our analysis of these 110 schools, we are 99% confident that the mean percentage of students regularly using computers is between 52.45% and 63.55%, in the population of all Chicago-area elementary schools.

Formal interpretation: If many 110-school samples were chosen randomly from the population of all Chicago-area elementary schools, and confidence intervals were calculated in this manner from each sample, about 99% of those confidence intervals should include the true population mean μ.

18. $\hat{\pi} = f/n = 32/232 = .138$ and $n = 232$. For 95% c.i. (from Table A.2): $z = 1.96$.

$$\hat{\pi} \pm z(SE_{\hat{\pi}}) = \hat{\pi} \pm z\sqrt{\hat{\pi}(1 - \hat{\pi})/n}$$
$$= .138 \pm 1.96\sqrt{.138(1 - .138)/232} = .138 \pm .044$$

We are 95% confident that $.094 \leq \pi \leq .182$. That is, based on this sample we are 95% confident that among the population of all adult local residents in this community, the true proportion who would drink the water (π) is between .094 and .182.

19. $\hat{\pi} = .63$ and $n = 1{,}021$. For a 95% c.i. (from Table A.2): $z = 1.96$.

$$\hat{\pi} \pm z(SE_{\hat{\pi}}) = \hat{\pi} \pm z\sqrt{\hat{\pi}(1 - \hat{\pi})/n}$$
$$= .63 \pm 1.96\sqrt{.63(1 - .63)/1{,}021} = .63 \pm .0296$$

We are 95% confident that $.6004 \leq \pi \leq .6596$. That is, based on this sample we are 95% confident that among the population of all American adults, between 60.04% and 65.96% believed the Vietnam War to be a mistake.

20. a. $\hat{\pi} = .63$ and $n = 50$.

$$\hat{\pi} \pm z(SE_{\hat{\pi}}) = \hat{\pi} \pm z\sqrt{\hat{\pi}(1 - \hat{\pi})/n}$$
$$= .63 \pm 1.96\sqrt{.63(1 - .63)/50} = .63 \pm .1338$$

The width of the interval would be $\pm .1338$, or .2676.

b. $\hat{\pi} = .63$ and $n = 50{,}000$.

$$\hat{\pi} \pm z(SE_{\hat{\pi}}) = \hat{\pi} \pm z\sqrt{\hat{\pi}(1 - \hat{\pi})/n}$$
$$= .63 \pm 1.96\sqrt{.63(1 - .63)/50{,}000} = .63 \pm .0042$$

The width of the interval would be $\pm .0042$, or .0084.

c. The interval is much narrower based on 50,000 cases than based on 1,021 or 50 cases. The larger the sample, the more precisely we can estimate the population proportion π.

21. a. $\hat{\pi} = .38$ and $n = 4{,}955$.

$$\hat{\pi} \pm z(SE_{\hat{\pi}}) = \hat{\pi} \pm z\sqrt{\hat{\pi}(1 - \hat{\pi})/n}$$
$$= .38 \pm 3.29\sqrt{.38(1 - .38)/4{,}955} = .38 \pm .0227$$

We are 99.9% confident that $.3573 \leq \pi \leq .4027$. That is, if this were a random sample we could be 99.9% confident that in the population of all adult male homosexuals, between 35.73% and 40.27% were infected with HIV.

b. Since this sample was not randomly selected, the confidence interval calculations provide little guidance about the true population percentage of HIV infection. It is possible that the men volunteering for the study were particularly worried about AIDS, either because of health problems or because they suspected exposure to AIDS. The prevalence of AIDS may also be higher in large cities than in other communities. For these and other reasons, the rate of HIV infection among the homosexual population nationwide might be well outside of the confidence interval calculated in part a.

22. a. We are 95% confident that $.6004 \leq \pi \leq .6596$, where π is the proportion thinking Vietnam was a mistake. We can therefore reject the hypothesis that a majority do not think so ($\pi \leq .50$) with at least 95% confidence.

b. 95% confidence interval: $53.78 \leq \mu \leq 62.22$, so we should not reject the hypothesis that $\mu = 60$.
c. 95% confidence interval: $1.28 \leq \mu \leq 1.52$, so we may confidently reject the hypothesis that $\mu = 1.0$.
23. a. 99% confidence interval: $.122 \leq \pi \leq .168$. Based on this sample we are 99% confident that the proportion of contaminated shellfish being sold in this state was between .122 and .168.
    b. Our interval leads us to reject this hypothesis with at least 95% confidence, because the interval does not contain the hypothesized value $\pi = .10$.

CHAPTER 10

1. a. $Z = (.86 - .76)/\sqrt{.76 (1 - .76)/497} = 5.22$. Obtained P-value is well below .05 (from Table A.2, $P < .000001$); reject H_0—significantly different.
    b. $Z = (.45 - .25)/\sqrt{.25 (1 - .25)/497} = 10.30$. $P < .000001$; reject H_0—significantly different.
    c. $Z = (.79 - .61)/\sqrt{.61 (1 - .61)/497} = 8.23$. $P < .000001$; reject H_0—significantly different.
    d. $Z = (.23 - .27)/\sqrt{.27 (1 - .27)/497} = -2.01$. $P < .05$; reject H_0—significantly different.
These tests suggest that compared with students nationwide, River City students are more likely to consume beer or wine, marijuana, liquor, or cocaine occasionally. They are less likely to use illegal stimulants.
2. a. H_0: $\pi \geq .10$, H_1: $\pi < .10$, where π represents the proportion of defective parachutes in the population of all parachutes we manufacture. Hypothesis H_1 specifies direction, so a one-sided test is needed.
    b. H_0: $\pi \leq .50$, H_1: $\pi > .50$, where π represents the proportion of women in the population of all U.S. college students. H_1 calls for a one-sided test.
    c. H_0: $\mu \leq 18.1$, H_1: $\mu > 18.1$, where μ represents the mean test score for the population of all students who receive coaching. H_1 calls for a one-sided test.
    d. H_0: $\pi = .60$, H_1: $\pi \neq .60$, where π represents the proportion of all small businesses that do not pay taxes. Since H_1 does not specify direction, a two-sided test is needed.
    e. H_0: $\mu \leq 64.8$, H_1: $\mu > 64.8$, where μ represents the mean July temperature in Seattle this year. Calls for a one-sided test.
    f. H_0: $\mu \geq 9$, H_1: $\mu < 9$, where μ represents the mean length of recovery time for patients with this new surgery. Calls for a one-sided test.
3. The null and alternative hypotheses are: H_0: $\pi \geq .10$, H_1: $\pi < .10$, where π represents the proportion of defective parachutes in the population of all parachutes we manufacture. We next need to collect a random sample of parachutes, perhaps by assigning each parachute a number and using a random number table or device to select certain ones for testing. Once the sample of parachutes is chosen, we test for defects and calculate the *sample* proportion defective, $\hat{\pi}$. A P-value is obtained to tell us "How likely are we to obtain this many defective parachutes, if the sample came randomly from a population where H_0: $\pi \geq .10$ is true?" If the P-value is low enough we can reject H_0 and believe H_1, that the proportion of defects among all our parachutes (not just the sample) is below 10% (H_1: $\pi < .10$). If the P-value is not low enough, then we cannot reject the hypothesis that the proportion is 10% or more (H_0: $\pi \geq .10$).
    A Type I error is possible only if we decide to reject H_0. If we fail to reject H_0, only a Type II error is possible.
4. *Type I error*: Reject H_0 when H_0 is true. This means we conclude that fewer than 10% of our parachutes are defective (H_1: $\pi < .10$) when in fact 10% or more are defective (H_0: $\pi \geq .10$). Such a Type I error might lead us to be complacent and to continue producing parachutes with a relatively high defect rate—which could lead to disaster.
    Type II error: Do not reject H_0 when H_1 is true. This means we conclude that 10% or more of our parachutes are defective (H_0: $\pi \geq .10$), when in fact the defect rate is lower than 10% (H_1: $\pi < .10$). We might then take costly steps to try to reduce the proportion of defects. Our mistake would be one of overcaution, which would likely have less serious consequences than the overoptimism of a Type I error.
5. The higher the α (other things being equal), the higher the probability of making a disastrous Type I error—believing that our parachutes are safer than they are. Therefore we are better off choosing a lower α, such as $\alpha = .01$.
6. b. *Type I error*: concluding that at least half of U.S. college students are women, when less than half really are.
    Type II error: concluding that less than half of U.S. college students are women, when actually half or more are.

c. *Type I error:* concluding that coached students score higher than the national average, when in fact they do not.
Type II error: concluding that coached students score no higher than the national average, when in fact they do.
d. *Type I error:* concluding that the claim is inaccurate, when it actually is true.
Type II error: concluding that the claim is accurate, when it actually is false.
e. *Type I error:* concluding that this year's mean July temperature is higher than 64.8°F, when in fact it is not.
Type II error: concluding that this year's mean July temperature is not higher than 64.8°F, when it actually is.
f. *Type I error:* concluding that the average recovery time for patients with this new type of surgery is less than nine weeks, when it is actually nine weeks or more.
Type II error: concluding that the average recovery time for patients with this new type of surgery is nine weeks or more, when it is actually less than nine weeks.

7. a. Only $n = 14$ cases are required for $\pi = .4$, so $n = 18$ is more than enough for $\pi = .45$.
  b. For $\pi = .1$, $n = 81$ is required, so $n = 73$ will not be enough for $\pi = .08$.
  c. Hard to tell from Table 10.5; may not be enough.
  d. For $\pi = .3$, only $n = 21$ cases are needed.
  e. Hard to tell, but $n = 1,500$ may not be enough.

8. a. min $\{n\pi/(1 - \pi), n(1 - \pi)/\pi\} = 14.7$; large enough n
  b. min $\{n\pi/(1 - \pi), n(1 - \pi)/\pi\} = 6.3$; not large enough n
  c. min $\{n\pi/(1 - \pi), n(1 - \pi)/\pi\} = 6.2$; not large enough n
  d. min $\{n\pi/(1 - \pi), n(1 - \pi)/\pi\} = 404$; large enough n
  e. min $\{n\pi/(1 - \pi), n(1 - \pi)/\pi\} = 3.76$; not large enough n

9. a. .057
  b. .033
  c. .025
  d. .021

10. a. The larger the sample size (n), the smaller the standard error.
  b. Standard errors indicate the amount of sample-to-sample variation in a statistic. Low standard errors make it more likely that a statistic from any one sample (e.g., $\hat{\pi}$) will be close to the corresponding population parameter (e.g., π).
  c. As sample size enlarges, each additional 100 cases causes a smaller decrease in the standard error. This suggests "diminishing returns" from sample size. Increasing sample size in an already large sample may not much improve our ability to estimate population parameters.

11. $n = 30$, $\hat{\pi} = .267$, $H_0: \pi = .5$, $H_1: \pi \neq .5$

$$Z = (.267 - .5)/\sqrt{.5(1 - .5)/30} = -2.55$$

$.01 < P < .02$; do not reject H_0. The sample evidence (26.7% of these 30 students are male) is not quite strong enough to reject the null hypothesis that these students came randomly from a population that is 50% male. The Z statistic is *almost* large enough for significance at $\alpha = .01$, and it would easily be significant at the less stringent $\alpha = .05$ level.

12. a. $n = 200$, $\hat{\pi} = .6$, $H_0: \pi = .53$, $H_1: \pi \neq .53$

$$Z = (.6 - .53)/\sqrt{.53(1 - .53)/200} = 1.98$$

$P < .05$; reject H_0. The evidence from this sample of $n = 200$ children is strong enough to reject the hypothesis that the percentage from Korea is unchanged; we conclude that in the population of all 1985 adoptions, it *has* changed from 1980.
  b. $n = 200$, $\hat{\pi} = .07$, $H_0: \pi = .13$, $H_1: \pi \neq .13$

$$Z = (.07 - .13)/\sqrt{.13(1 - .13)/200} = -2.52$$

$P < .02$; reject H_0. We can also reject the hypothesis that the percentage of Colombian adoptions is unchanged, based on this sample. Specifically, it appears that the percentage of the overseas adoptions coming from Korea has increased, while the percentage from Colombia has decreased.

13. a. min $\{n\pi_0/(1 - \pi_0), n(1 - \pi_0)/\pi_0\} = 177.4$; large enough n
   b. min $\{n\pi_0/(1 - \pi_0), n(1 - \pi_0)/\pi_0\} = 29.9$; large enough n
14. a. 95% confidence interval: $\overline{D} \pm 1.96(SE_{\overline{D}})$
   b. Hypothesis test: $Z = (\overline{D} - \delta_0)/SE_{\overline{D}}$ (reject if $Z > 2.576$)
15. $H_0: \mu = 0.0, H_1: \mu \neq 0.0, \overline{X} = 1.5, n = 60, s_X = 3.0$

$$Z = (\overline{X} - \mu_0)/SE_{\overline{X}} = (1.5 - 0.0)/(3.0/\sqrt{60}) = 3.87$$

$P < .0005$, so we must reject H_0 at the $\alpha = .01$ level. We conclude that the mean improvement in cognitive test scores is significantly different from zero. That is, there was significant improvement following Logo experience.
16. $Z = 3.87$ implies a distance of 3.87 estimated standard errors between the sample mean (1.5) and the null hypothesis population mean ($\mu_0 = 0.0$).
17. a. $Z = 1.92, .05 < P < .10$. We cannot reject H_0 at $\alpha = .01$; the mean improvement is not significantly different from 0.
18. $\hat{\pi} = .37, n = 178, H_0: \pi \geq .49, H_1: \pi < .49$

$$Z = (\hat{\pi} - \pi_0)/SE_{\hat{\pi}} = (.37 - .49)/\sqrt{.49(1 - .49)/178} = -3.20$$

One-sided $P < .0025$, reject H_0. The proportion aged 20–39 was significantly lower among pick-up participants than it was in the city population as a whole.
19. $\hat{\pi} = .06, n = 178, H_0: \pi \geq .33, H_1: \pi < .33$

$$Z = (\hat{\pi} - \pi_0)/SE_{\hat{\pi}} = (.06 - .33)/\sqrt{.33(1 - .33)/178} = -7.66$$

One-sided $P < .0000005$, reject H_0. The proportion of apartment dwellers was significantly lower among pick-up participants than in the city-wide population.
20. $\hat{\pi} = .8, n = 30, H_0: \pi \leq .55, H_1: \pi > .55$ (One-sided test)

$$Z = (\hat{\pi} - \pi_0)/\sqrt{\pi_0(1 - \pi_0)/n} = (.8 - .55)/\sqrt{.55(1 - .55)/30} = 2.75$$

$.0025 < P < .005$, reject H_0. The proportion of black players at skill positions is significantly higher than the proportion of black players in the NFL as a whole.
21. a. $\hat{\pi} = .54, n = 600, N = 1,500,000, Z = 2.576$ for 99% confidence interval. Finite population correction:

$$\sqrt{(1,500,000 - 600)/(1,500,000 - 1)} = .9998$$

$$SE_{\hat{\pi}} = \sqrt{\hat{\pi}(1 - \hat{\pi})/600} = 0.0203$$

We are 99% confident that $.488 \leq \pi \leq .592$. Since this interval includes proportions of .5 or less, we cannot be 99% confident that Smith is supported by a majority, or $\pi > .50$.
   b. $\hat{\pi} = .54, n = 600, N = 1,000, Z = 2.576$ for 99% confidence interval. Finite population correction:

$$\sqrt{(1,000 - 600)/(1,000 - 1)} = .6323$$

$$SE_{\hat{\pi}} = \sqrt{\hat{\pi}(1 - \hat{\pi})/600} = 0.0203$$

$$.6323 (SE_{\hat{\pi}}) = 0.0128$$

We are 99% confident that $.507 \leq \pi \leq .573$. Since this interval is entirely above .5, we can be 99% confident that Smith *is* supported by a majority, or $\pi > .50$.
   c. The correction makes standard errors smaller, hence makes confidence intervals narrower, when the sample is a large fraction of the population. It makes sense that we can estimate the population parameter π with greater precision if our sample includes more than half of the population, as in part b.
22. a. $\overline{X} = 8.6, s_x = 19.1, n = 36, N = 168$

$$SE_{\overline{X}}^* = (19.1/\sqrt{36}) \sqrt{(168 - 36)/(168 - 1)} = 2.83$$

$$H_0: \mu \leq 0, H_1: \mu > 0 \text{ (one-sided test)}$$

$$Z = (\overline{X} - \mu_0)/SE_{\overline{X}}^* = (8.6 - 0)/2.83 = 3.04$$

.001 < P < .005, reject H_0. Based on the evidence from this sample of 36 cities, we can reject the hypothesis that the mean growth rate in all 168 cities was 0 or less.

b. $\overline{X}$ = 10.4, s_X = 9.2, n = 36, N = 168, Z = 1.96 for 95% confidence interval

$$SE_{\overline{X}}^* = (9.2/\sqrt{36}) \sqrt{(168 - 36)/(168 - 1)} = 1.363$$

$$\overline{X} \pm Z(SE_{\overline{X}}^*) = 10.4 \pm 1.96(1.363)$$

We are 99% confident that $7.73 \le \mu \le 13.07$; the mean homicide rate in the 168 cities is between 7.73 and 13.07 victims per 100,000 population.

CHAPTER 11

1. a. df = 24, t = 2.064
 b. df = 5, t = 6.869
 c. df = 13, t = 2.160
 d. df = 11, t = 2.718
2. a. P < .05, reject H_0
 b. .05 < P < .10, do not reject H_0
 c. P < .001, reject H_0
 d. .10 < P < .20, do not reject H_0
 e. P < .001, reject H_0
 f. P > .25, do not reject H_0
3. A normal distribution is a theoretical continuous probability distribution, defined by Equation [9.1]. When graphed it has a symmetrical bell-shaped appearance.
 a. skewness, outliers, multiple peaks or gaps
 b. skewness (mean–median); light or heavy tails (s_X–PSD)
 c. skewness, outliers
4. a. "Fewer than three in one million": P < .000003
 b. A severe outlier (Chapter 5) is a case beyond the outer fences, which lie at $Q_1 - 3(IQR)$ and $Q_3 + 3(IQR)$. Table A.2 shows that standard normal distribution quartiles, which enclose the middle .50 of the area, are $\pm.674$. The IQR must be approximately 2(.674) = 1.348. This makes the outer fence locations

$$Q_1 - 3(IQR) = -.674 - 3(1.348) = -4.718$$
$$Q_3 + 3(IQR) = .674 + 3(1.348) = 4.718$$

Values beyond the outer fences therefore lie more than ± 4.718 standard deviations from the mean. We can see from Table A.1 that the probability of values more than 4.7 standard deviations above the mean is .5 − .4999987 = .0000013, so the probability of values more than ± 4.7 standard deviations from the mean is 2(.0000013) = .0000026. The outer fences are slightly beyond this, so in a normal distribution the probability of values beyond the outer fences must be P < .0000026, which rounded off is "fewer than three in one million."

5. a. $\overline{X} \pm t(SE_{\overline{X}}) = 34 \pm 2.060(9/\sqrt{26})$; $30.4 \le \mu \le 37.6$
 b. $\overline{X} \pm t(SE_{\overline{X}}) = 0.3 \pm 2.567(5/\sqrt{18})$; $-2.73 \le \mu \le 3.33$
 c. $\overline{X} \pm t(SE_{\overline{X}}) = 237 \pm 1.699(100/\sqrt{30})$; $206 \le \mu \le 268$
6. $\overline{X} \pm t(SE_{\overline{X}}) = \overline{X} \pm 2.262(s_X/\sqrt{n})$
              $= 498 \pm 2.262(392/\sqrt{10}) = 498 \pm 280$

The interval is $218 \le \mu \le 778$. That is, we are 95% confident that in 1979 the mean DDT concentration in lower sediments of lakes along Bear Creek was between 218 and 778 μg/kg. Note that our confidence that the population mean μ lies in this interval rests on viewing these ten sites as a random sample of all possible sites along the creek.

7. The large-sample formula used, based on the normal approximation, does not actually produce 95% confidence intervals when it is applied to smaller samples. In n = 5 samples the normal approximation leads us to construct intervals that are much too narrow. A similar but less extreme problem affects the n = 25 intervals. The intervals based on $n = 125$ samples seem to be valid, in that they contain π about 95% of the time.

8. Equation [10.5] states that a sample is large enough for the normal approximation to be applied if

$$\min\{n\pi/(1 - \pi), n(1 - \pi)/\pi\} \geq 9.$$

Given $\pi = .171$,

     for $n = 5$ $\min\{5(.171)/(1 - .171), 5(1 - .171)/.171\} = 1.03$
     for $n = 25$ $\min\{25(.171)/(1 - .171), 25(1 - .171)/.171\} = 5.16$
     for $n = 125$ $\min\{125(.171)/(1 - .171), 125(1 - .171)/.171\} = 25.78$

By this rule only the $n = 125$ samples are large enough to justify the normal approximation. This is consistent with the fact that only the $n = 125$ confidence intervals in Problem 7 actually contained π about 95% of the time.

9. $\overline{X} = 2.07$, $s_X = 2.17$, $n = 19$, $SE_{\overline{X}} = s_X/\sqrt{n} = .498$
The sample mean is now *greater than* $\mu_0 = 1.59$, so a test here cannot possibly lead to rejection of H_0: $\mu \geq 1.59$.
    a. Hospital 14 was responsible for the low mean operating margin. Deleting this outlier reveals that the remaining 19 "underfunded" hospitals' mean operating margins are above the state average.
    b. The outlier was also inflating the standard deviation, which is only half as large once hospital 14 is deleted.
    c. Since the standard error is estimated from the standard deviation, it too is smaller when the outlier is deleted.
    d. The overall conclusion of this test, do not reject H_0, remains the same with or without hospital 14. The "underfunded" hospitals do not have significantly lower operating margins.

10. $\overline{X} = 238.7$, $s_X = 714.4$, $n = 19$, H_0: $\mu = 115.8$; H_1: $\mu \neq 115.8$

$$t = (\overline{X} - \mu_0)/(s_X/\sqrt{n}) = (238.7 - 115.8)/(714.4/\sqrt{19}) = .75$$

With df $= n - 1 = 18$, $.20 < P < .50$; do not reject H_0. The mean operating margin gain (loss) among "underfunded" hospitals is not significantly different from the state average.

11. Stem-and-leaf display of operating margin gain (loss):

```
−1*  | 0
−0.  |
−0*  | 0000
 0*  | 000011122334
 0.  | 6
 1*  |                    Stems digits 1,000's, leaves digits 100's.
 1.  |
 2*  |
 2.  | 8                  2. | 8 means operating margin gain of 2,800%
```

The display contains two severe outliers, one high and one low. Since such outliers are uncommon in normal distributions, their presence here casts doubt on the normality assumption. The conclusion reached in Problem 10 is thus also in doubt.

12. $\overline{X}^* = 2.623$, $s_X = .765$, $n = 21$, $SE_{\overline{X}} = .170$

$$\overline{X} \pm t(SE_{\overline{X}}) = 2.623 \pm 2.086(1.70)$$

Thus, $2.268 \leq \mu^* \leq 2.978$ or $185.35 \leq \mu_G \leq 950.60$.

13. Binomial probabilities given $n = 4$ and $\pi = .3$:

f	$P(f)$
0	.2401
1	.4116
2	.2646
3	.0756
4	.0081

$P(f \geq 3) = P(3) + P(4)$
$= .0756 + .0081$
$= .0837$

14. Binomial probabilities given $n = 6$ and $\pi = .067$:

f	$P(f)$
0	.6596
1	.2842
2	.0510
3	.0049
4	.0003
5	<.0001
6	<.0001

$P(f = 0) = .6596$

There is about a 66% chance that those hired would, "by chance," include no women. Since the probability is high (above $\alpha = .05$) we should not reject H_0.

15.
$$P(f \geq 3) = 1 - P(f < 3)$$
$$= 1 - [P(0) + P(1) + P(2)]$$
$$= 1 - (.6596 + .2842 + .0510) = .0052$$

The obtained P-value, $P = .0052$, is well below $\alpha = .05$, so we should reject H_0.

18. Poisson probabilities given $\mu = .51$:

x	$P(X = x)$
0	0.6005
1	0.3063
2	0.0781

$$P(X \geq 3) = 1 - P(X < 3)$$
$$= 1 - (.6005 + .3063 + .0781) = .0151$$

CHAPTER 12

1.

Spouse abuse		Intact	Remarried	Recon.	All
None		743	92	78	913
	Row	81%	10%	9%	
	Column	95%	91%	88%	
	Total	77%	9%	8%	
Any		36	9	11	56
	Row	64%	16%	20%	
	Column	5%	9%	12%	
	Total	4%	1%	1%	
All		779	101	89	969

Family type heading spans Intact, Remarried, Recon.

2. a. Sixteen percent of the adults reporting some spouse abuse lived in remarried families.
  b. Ninety-five percent of the adults in intact families reported no spouse abuse.
  c. Seventy-seven percent of the adults surveyed lived in intact families and reported no spouse abuse.
3. It seems possible that family type influences spouse abuse, so family type is the independent variable and we should use column percentages.
   Spouse abuse was reported by only 5% of the adults in intact families, whereas 9% of those in remarried families and 12% of those in reconstituted families reported spouse abuse. The probability of spouse abuse seems to be lowest in intact families and highest in reconstituted families.

4.

Adolescent drinking	Drinking of mothers			
	Abstainer	Moderate	Heavy	All
Abstainer	7	20	7	34
E	5.38	19.83	8.78	
Moderate	7	45	13	65
E	10.29	37.92	16.79	
Heavy	5	5	11	21
E	3.33	12.25	5.42	
Totals	19	70	31	120

a. Abstainer/abstainer: $O = 7$, $E = 5.38$. There are slightly more abstaining adolescents, whose mothers also abstained, than we would expect if mother's and child's drinking habits were independent.

b. Heavy/moderate: $O = 5$, $E = 12.25$. If mother's and child's drinking habits were independent, we would expect to see more heavy-drinking adolescents with moderate-drinking mothers than are actually observed in the table.

c. Heavy/heavy: $O = 11$, $E = 5.42$. More of the children of heavy-drinking mothers become heavy drinkers themselves than we would expect if mother's and child's drinking were independent.

5. The largest discrepancies occur among adolescents with moderate-drinking mothers. More of them become moderate drinkers, and fewer of them become heavy drinkers, than we would expect if mother's drinking were unrelated to adolescent's drinking.

6. a. df = 1; $.10 < P < .25$; do not reject H_0
   b. df = 6; $P < .001$; reject H_0
   c. df = 16; $.05 < P < .10$; do not reject H_0
   d. df = 5; $P > .50$; do not reject H_0
   e. df = 16; $.25 < P < .50$; do not reject H_0
   f. df = 10; $.01 < P < .025$; reject H_0

7.

Spouse abuse	Family type			
	Intact	Remarried	Reconst.	All
None	743	92	78	913
E	733.98	95.16	83.86	
$(O - E)^2/E$	0.11	0.11	0.41	
Any	36	9	11	56
E	45.02	5.84	5.14	
$(O - E)^2/E$	1.81	1.71	6.67	
Totals	779	101	89	969

$X^2 = \Sigma(O - E)^2/E = .11 + .11 + .41 + 1.81 + 1.71 + 6.67 = 10.81$
df $= (K_r - 1)(K_c - 1) = (2 - 1)(3 - 1) = 2$
$P < .005$; reject H_0

The strongest evidence (largest $(O - E)^2/E$ value) against the independence hypothesis is in the any abuse/reconstituted cell. There are more than twice as many abuse reports here as we would expect if abuse and family type were independent.

There is a significant relationship between spouse abuse and family type. As described using column percentages in Problem 3, spouse abuse is least likely to be reported in intact families, and most likely to be reported in reconstituted families.

8. a. df = $(K_r - 1)(K_c - 1) = 2$

	Injury group		
Skiing ability	Injured	Uninjured	All
Beginner	20	80 − 20 = 60	80
Intermediate	9	93 − 9 = 84	93
Advanced/expert	31 − 29 = 22	183 − 144 = 39	41
All	31	183	214

b. df = $(K_r - 1)(K_c - 1) = 2$

	Family type			
Spouse abuse	Intact	Remarried	Reconstituted	All
None	743	92	913 − 835 = 78	913
Any	779 − 743 = 36	101 − 92 = 9	89 − 78 = 11	56
All	779	101	89	969

c. df = 4

Adolescent drinking	Drinking of mother			
	Abstainer	Moderate	Heavy	All
Abstainer	7	20	34 − 27 = 7	34
Moderate	7	45	65 − 52 = 13	65
Heavy	19 − 14 = 5	70 − 65 = 5	31 − 20 = 11	21
All	19	70	31	120

9. $X^2 = 26.4$, df = 1, $P < .001$. We can reject the null hypothesis that vaccine type and getting typhoid fever are independent. There is a significant difference in the effectiveness of the two vaccines.

10. The column variable, type of vaccine, is the independent variable here. We should therefore use column percentages:

	Vi	Pneumo
Typhoid fever	0.4%	1.7%
No typhoid	99.6%	98.3%

The Vi vaccine was evidently more effective than the Pneumo vaccine. Only 0.4% (4 in 1,000) of the subjects vaccinated with Vi contracted typhoid fever, whereas 1.7% (17 in 1,000) of those vaccinated with Pneumo did so.

11. $X^2 = 20.48$, df = $K − 1 = 2$, $P < .001$. We can reject the null hypothesis; the racial distribution in these ads is significantly different from that for the U.S. population. More specifically, there are more whites and fewer nonwhites (blacks and others) in the ads than we would expect if these ads were representative of the U.S. population.

12. $X^2 = 1.05$, df = $K − 2 = 4$, $P > .5$. The observed distribution of V−1 hits is not significantly different from a Poisson distribution. The discrepancies between the observed and expected distributions of bomb hits are small enough that they could easily be due to chance ($P > .5$).

13. Combining drinking categories is a bad idea. The pattern in this table is not as simple as with the skiing data. It can be studied most easily in the column percentages shown in the next table.

Adolescent drinking	Drinking of mothers			
	Abstainer	Moderate	Heavy	All
Abstainer	7	20	7	34
	37%	29%	23%	
Moderate	7	45	13	65
	37%	64%	42%	
Heavy	5	5	11	21
	26%	7%	35%	
Totals	19	70	31	120

The lowest rate of heavy drinking is among adolescents whose mothers drank moderately (7%); the percentage of heavy-drinking adolescents with abstinent mothers is actually higher (26%), closer to the percentage for heavy-drinking adolescents whose mothers drank heavily (35%). If we combined categories we would lose this information and weaken the contrast between adolescents whose mothers were moderate and heavy drinkers.

Since only one out of nine cells (11%) has an expected frequency below five—and not very far below five ($E = 3.33$)—it seems reasonable to go ahead and perform a standard chi-square test. Such a test produces a X^2 value of 14.94, with df = 4, $P < .005$: There is a significant relationship between drinking habits of mothers and their adolescent children.

14.

Self-reported water savings	Actual water savings		
	None	Some	All
None: O^* (E)	9.5 (8.7)	12.5 (13.3)	22
$(O^* - E)^2/E$	.07	.05	
Column %	77%	60%	
Some: O^* (E)	3.5 (4.3)	7.5 (6.7)	11
$(O^* - E)^2/E$	.15	.10	
Column %	23%	40%	
All	13	20	33

$X^2 = 0.37$, df = 1, $P > .5$

The relationship between actual and self-reported water savings is not statistically significant. We cannot reject the hypothesis that these data came from a population in which there is no relationship.

There is a weak relationship in the sample: 40% of those who actually saved water also reported so, whereas only 23% of those who actually didn't save water said they had. Nearly half the sample (15/33) was mistaken about whether or not they actually conserved water.

15.

Self-reported water savings	Actual water savings		
	None	Some	All
None: O (E)	1,000 (866.7)	1,200 (1,333.3)	2,200
$(O - E)^2/E$	20.51	13.33	
Column %	77%	60%	
Some: O^* (E)	300 (433.3)	800 (666.7)	1,100
$(O^* - E)^2/E$	41.03	26.67	
Column %	23%	40%	
All	1,300	2,000	3,300

$X^2 = 101.54$, df = 1, $P < .005$

We can now reject the hypothesis of independence; the relationship between self-reported and actual water savings *is* statistically significant. With a much larger sample, we can be confident that the sample relationship results from a relationship in the population, rather than from random chance. The strength of this relationship, as described by the column percentages, is unchanged.

CHAPTER 13

2.

1. North America	2. Europe
$n_1 = 7$	$n_2 = 11$
$\overline{Y}_1 = .41$	$\overline{Y}_2 = .01$
$s_1 = .38$	$s_2 = .69$

At the North American sites in this sample, the average sulfate concentration declined by .41 mg/l. The European sites had almost no overall improvement: The average reduction was only .01 mg/l. The average reduction of 0.17 mg/l seen across all 18 sites in Table 13.2 results from combining the North American sites (substantial reduction) with the European sites (no reduction).

3. $H_0: \mu_2 - \mu_1 = 0$, where μ_1 is the mean sulfate concentration in the population consisting of all North American sites; and μ_2 is the mean sulfate concentration in the population consisting of all European sites.

The null hypothesis states that there is no difference between these two *population* means. There is a substantial difference between the two *sample* means. The mean sulfate concentration at European sites ($\overline{Y}_2 = 1.53$) is more than twice as high as the mean concentration at North American sites ($\overline{Y}_1 = .59$). Thus the sample evidence appears not to support H_0.

4. a. Figure 5.8: Distributions are both positively skewed, which will pull up both means. The boys' distribution shows more variation than the girls'.

   b. Figure 5.10: The speeds of 80286 computers are more varied than those of 80386 computers. It is difficult to judge skewness.

   c. Figure 5.11: All three SAT score distributions are approximately normal, with similar variability.

   d. Figure 5.12: We see no support for either the normality or the constant variance assumptions. All three nicotine distributions are positively skewed, so the means (especially in the smoking seats) are pulled up. The variation in nicotine concentration is greatest in the smoking section seats.

   e. Problem 15: There are great differences in variation across the four groups of countries. European countries have almost uniformly high life expectancies, whereas Asia includes both the highest and the lowest life expectancies in these data.

5. $H_0: \mu_2 - \mu_1 = 0$, $H_1: \mu_2 - \mu_1 \neq 0$ (two-sided test)

$s_p = .53$ $SE_{\overline{Y}_2 - \overline{Y}_1} = .20$ df = 32

$t = (\overline{Y}_2 - \overline{Y}_1)/SE_{\overline{Y}_2 - \overline{Y}_1} = (2.89 - 2.84)/.20 = .25$

$P > .50$, do not reject H_0. The difference between men's and women's mean GPA is not statistically significant.

6. Mean grams of cocaine by psychotic symptoms:

1. Psychotic	2. Nonpsychotic
$n_1 = 66$	$n_2 = 25$
$\overline{Y}_1 = 16.9$	$\overline{Y}_2 = 4.6$
$s_1 = 31.9$	$s_2 = 7.5$

$H_0: \mu_2 - \mu_1 \geq 0$, $H_1: \mu_2 - \mu_1 < 0$ (one-sided test)

$s_p = 27.54$ $SE_{\overline{Y}_2 - \overline{Y}_1} = 6.47$ df = 89

$t = (\overline{Y}_2 - \overline{Y}_1)/SE_{\overline{Y}_2 - \overline{Y}_1} = (4.6 - 16.9)/6.47 = -1.90$

One-sided $P < .05$, reject H_0: Patients with psychotic symptoms did use significantly more cocaine.

7. Mean grams of cocaine by violent behavior:

1. Violent	2. Nonviolent
$n_1 = 29$	$n_2 = 62$
$\overline{Y}_1 = 22.4$	$\overline{Y}_2 = 9.8$
$s_1 = 46.6$	$s_2 = 11.3$

$H_0: \mu_2 - \mu_1 = 0$, $H_1: \mu_2 - \mu_1 \neq 0$ (two-sided test)
$s_p = 27.76$ $SE_{\overline{Y}_2 - \overline{Y}_1} = 6.25$ df $= 89$
$t = (\overline{Y}_2 - \overline{Y}_1)/SE_{\overline{Y}_2 - \overline{Y}_1} = (9.8 - 22.4)/6.25 = -2.02$
One-sided $P < .05$, reject H_0. Violent patients used significantly more cocaine than nonviolent patients.

8. a. Positive skewness implies that most users consume relatively smaller amounts of cocaine, but a few consume much larger amounts. This seems likely to be the case in the general population of cocaine users, so we can expect the population distribution of grams consumed to be positively skewed just as this sample distribution is.

 b. Sample standard deviations of grams consumed by psychotic or violent patients are much larger than the standard deviations for nonpsychotic or nonviolent patients. This suggests that the second distributional assumption (constant population variance) is also untrue.

9. df $= 32$, $t = 2.041$, $SE_{\overline{Y}_2 - \overline{Y}_1} = .20$
$(\overline{Y}_2 - \overline{Y}_1) \pm t(SE_{\overline{Y}_2 - \overline{Y}_1}) = (2.89 - 2.84) \pm 2.041(.20)$
$-.36 \leq (\mu_2 - \mu_1) \leq .46$

 We are 95% confident that in the population of students this sample came from, the mean difference (female–male) in cumulative GPA is between $-.36$ and $+.46$ point. Since this interval contains 0, we cannot reject the null hypothesis that $\mu_2 - \mu_1$ equals 0.

10. df $= 116$, $t = 1.98$, $s_p = 23.57$, $SE_{\overline{Y}_2 - \overline{Y}_1} = 4.43$
$(\overline{Y}_2 - \overline{Y}_1) \pm t (SE_{\overline{Y}_2 - \overline{Y}_1}) = (52.7 - 43.2) \pm 1.98(4.43)$
$0.73 \leq (\mu_2 - \mu_1) \leq 18.27$

 We are 95% confident that in the population of college courses this sample came from, the mean percentage of high evaluations received by full-time faculty and part-time instructors differs by between .73 and 18.27 percentage points (full-time higher). Since this interval (barely) does not contain 0, we may reject the null hypothesis of no difference between the two categories of instructors with respect to mean percentage of high evaluations received.

11. df $= 116$, $t = 1.98$, $s_p = 13.95$, $SE_{\overline{Y}_2 - \overline{Y}_1} = 2.62$
$(\overline{Y}_2 - \overline{Y}_1) \pm t (SE_{\overline{Y}_2 - \overline{Y}_1}) = (53.0 - 62.8) \pm 1.98(2.62)$
$-14.99 \leq (\mu_2 - \mu_1) \leq -4.61$

 We are 95% confident that in the population of college courses this sample came from, the mean percentage of A and B grades awarded by full-time faculty and part-time instructors differs by between 14.99 and 4.61 percentage points (full-time lower). Since this interval does not contain 0, we may reject the null hypothesis of no difference between the two categories of instructor with respect to mean percentage of high grades awarded.

12. Given $D = $ Posttest $-$ Pretest:
$\overline{D} = 2.00$, $s_D = 1.53$, $n = 24$ $H_0: \delta \leq 0$
df $= 23$ $H_1: \delta > 0$ (one-sided)
$t = (\overline{D} - \delta_0)/SE_{\overline{D}}$
  $= (2.00 - 0)/(1.53/\sqrt{24}) = 6.40$
$P < .0005$; reject H_0. Students' use of evidence in writing significantly improved following this course.

13.

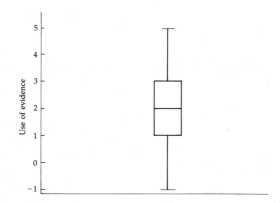

The box plot shows a very symmetrical, well-behaved distribution, giving us no reason to question the normality assumption.

14. Given D = True − Self-reported GPA:

$\bar{D} = -.235$, $s_D = .480$, $n = 34$ H_0: $\delta \geq 0$

df = 33 (approx. 30) H_1: $\delta < 0$ (one-sided)

$t = (\bar{D} - \delta_0)/\text{SE}_{\bar{D}}$

   $= (-.235 - 0)/(.480/\sqrt{34}) = -2.85$

$P < .01$; reject H_0. Students' true GPAs are significantly lower than their self-reports.

15. a. $\text{df}_B = K - 1 = 4 - 1 = 3$, $\text{df}_W = n_. - K = 34 - 4 = 30$; $P < .001$; reject H_0

     b. $\text{df}_B = K - 1 = 3 - 1 = 2$, $\text{df}_W = n_. - K = 45 - 3 = 42$ (approx. 40); $.05 < P < .10$; do not reject H_0

     c. $\text{df}_B = 1$, $\text{df}_B = 11$ (approx. 10); $.001 < P < .01$; do not reject H_0

     d. $\text{df}_B = 10$, $\text{df}_W = 113$ (approx. 120); $P < .001$; reject H_0

     e. $\text{df}_B = 2$, $\text{df}_W = 7$; $P > .25$; do not reject H_0

     f. $\text{df}_B = 4$, $\text{df}_W = 20$, $.01 < P < .05$; reject H_0

16. BSS = 9,873.5, $\text{df}_B = 2$

WSS = 1,992.5, $\text{df}_W = 9$

$F = (\text{BSS}/\text{df}_B)/(\text{WSS}/\text{df}_W) = (9{,}873.5/2)/(1{,}992.5/9) = 22.3$

$P < .001$; reject H_0. The mean frequency of injury claims is not the same for all three car sizes.

17. BSS = 375.56, $\text{df}_B = 2$

WSS = 48,761.51, $\text{df}_W = 11{,}197$ (approx. ∞)

$F = (\text{BSS}/\text{df}_B)/(\text{WSS}/\text{df}_W) = 43.12$

$P < .001$; reject H_0. The differences between school types are statistically significant.

18. $s_p = \sqrt{\text{WSS}/(n_. - K)} = \sqrt{46{,}609/(11{,}200 - 3)} = 2.04$

1. Public	2. Catholic	3. Private Non-Catholic
$\text{SE}_{\bar{Y}_1} = s_p/\sqrt{n_1}$	$\text{SE}_{\bar{Y}_2} = s_p/\sqrt{n_2}$	$\text{SE}_{\bar{Y}_3} = s_p/\sqrt{n_3}$
$= .020$	$= .065$	$= .144$

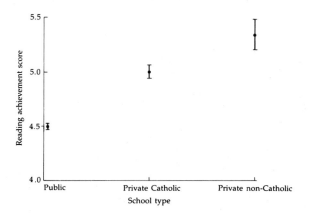

The group means in the figure should not be connected because there is no natural ordering among the categories; we could just as well have listed Catholic or private non-Catholic schools first.

19. Error-bar widths reflect the size of the standard errors, $SE_{\overline{Y}_k} = s_p/\sqrt{n_k}$. The pooled standard deviation, s_p, is the same for each group, so the larger the n_k (group size), the smaller the standard error. Since there are many more public school students (10,000 vs. 200), the standard error for their mean is relatively small. A smaller standard error reflects the fact that we expect less sample-to-sample variation, and our sample mean is more likely to be close to the true population mean, for a sample of 10,000 than for 200.

20. $\overline{Y}. = 103$; $s_p = \sqrt{WSS/(n. - K)} = \sqrt{1{,}992.5/(12 - 3)} = 14.88$

1. Small	2. Midsize	3. Large
$SE_{\overline{Y}_1} = s_p/\sqrt{n_1}$	$SE_{\overline{Y}_2} = s_p/\sqrt{n_2}$	$SE_{\overline{Y}_3} = s_p/\sqrt{n_3}$
$= 7.44$	$= 7.44$	$= 7.44$

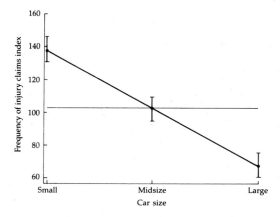

The group means in the figure should be connected because there is an obvious ordering to the categories small, midsize, and large.

22. $s_p = \sqrt{WSS/(n. - K)} = \sqrt{875.35/(60 - 3)} = 10.77$

1. Very Poor	*2. Poor*	*3. Normal*
$SE_{\bar{Y}_1} = s_p/\sqrt{n_1}$	$SE_{\bar{Y}_2} = s_p/\sqrt{n_2}$	$SE_{\bar{Y}_3} = s_p/\sqrt{n_3}$
$= 3.25$	$= 2.47$	$= 1.97$

The standard error is highest for boys with very poor reading ability, because this group has the fewest cases ($n_1 = 11$). In smaller samples we expect more sample-to-sample variation; the large standard error reflects greater uncertainty about the location of the true population mean μ_1.

23. a.

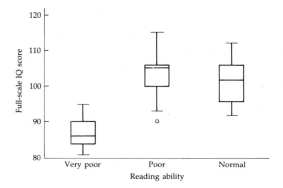

The box plots in the figure appear reasonably well behaved, with similar spread and no severe skewness.

b.

1. Very Poor	*2. Poor*	*3. Normal*	*All*
$n_1 = 11$	$n_2 = 19$	$n_3 = 30$	$n. = 60$
$\bar{Y}_1 = 86.91$	$\bar{Y}_2 = 103.47$	$\bar{Y}_3 = 101.57$	$\bar{Y}. = 99.48$
$s_1 = 4.30$	$s_2 = 6.23$	$s_3 = 6.07$	$s. = 8.36$

c. $BSS = 2,171.58$ $df_B = 2$
$WSS = 1,952.03$ $df_W = 57$ (approx. 60)
$F = 31.71$ $P < .001$; reject H_0

The relationship between reading ability group and full-scale IQ score is statistically significant.

d.

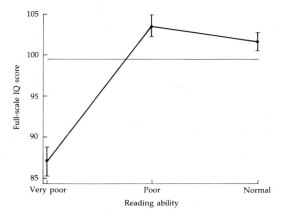

$s_p = 5.85$; $SE_{\bar{Y}_1} = 1.76$, $SE_{\bar{Y}_2} = 1.34$, $SE_{\bar{Y}_3} = 1.07$

The highest mean IQ is among poor readers. Normal readers are slightly lower, and very poor readers are substantially lower. The larger standard error for the mean IQ of very poor readers indicates that we can be less precise about the corresponding population mean for this group.

24. a. Sulfate concentraton ranks (two-sample):

1. North American sites	Rank	2. European sites	Rank
Edson, Canada	4	Ejde, Denmark	13
Kelowna, Canada	1	Sodankyla, Finland	3
Mould Bay, Canada	5.5	Monte Cimone, Italy	7.5
Sable Island, Canada	2	Trapani, Italy	9
Wynyard, Canada	7.5	Kise, Norway	12
Caribou, Maine, U.S.	5.5	Suwalki, Poland	16
Huron, S. Dakota, U.S.	10	Velen, Sweden	11
		Lazaropole, Yugoslavia	17
		Kurgan, U.S.S.R.	18
		Turukhansk, U.S.S.R	14
		Svratouch, Czech.	15

b.

North America	Europe
$n_1 = 7$	$n_2 = 11$
$\bar{R}_1 = 5.07$	$\bar{R}_2 = 12.32$
$s_1 = 3.10$	$s_2 = 4.48$

c. $s_p = 4.02$ $SE_{\bar{R}_2 - \bar{R}_1} = 1.94$ $df = 16$
$t = (\bar{R}_2 - \bar{R}_1)/SE_{\bar{R}_2 - \bar{R}_1} = (12.32 - 5.07)/1.94 = 3.73$
$.001 < P < .002$; reject H_0 (two-sided)
Sulfate concentrations in European precipitation are significantly different.

25. a. Sulfate concentration ranks (paired-difference):

Site	Sulfate 1975-1978 R_{78}	Sulfate 1979-1982 R_{82}	Reduction in Sulfate $D_R = R_{78} - R_{82}$
Edson, Canada	12	6	6
Kelowna, Canada	3	1	2
Mould Bay, Canada	20	8	12
Sable Island, Canada	26.5	2	24.5
Wynyard, Canada	14	10.5	3.5
Caribou, Maine, U.S.	20	8	12
Huron, S. Dakota, U.S.	17	16	1
Ejde, Denmark	29	23.5	5.5
Sodankyla, Finland	8	4	4
Monte Cimone, Italy	20	10.5	9.5
Trapani, Italy	15	13	2
Kise, Norway	26.5	22	4.5
Suwalki, Poland	35	31	4
Velen, Sweden	23.5	18	5.5
Lazaropole, Yugoslavia	34	32	2
Kurgan, U.S.S.R.	26.5	36	-9.5
Turukhansk, U.S.S.R.	5	26.5	-21.5
Svratouch, Czech.	33	30	3

b. $\overline{D}_R = 3.89$, $s_D = 9.21$, $n = 18$, $df = 17$
c. $t = (\overline{D}_R)/(s_D/\sqrt{n}) = 3.89/(9.21/\sqrt{18}) = 1.79$
$.025 < P < .05$, reject H_0 (one-sided)
The decrease in sulfate concentration is statistically significant.

CHAPTER 14

1. It appears that joint damage problems were more common on later flights; the problem was getting worse, despite efforts to fix it.

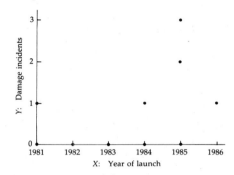

2. The taller players tend to pull in more rebounds.

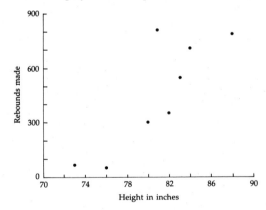

4. a. Slope: $b = 5.6$
For every inch of increase in height, average or predicted weight increases by 5.6 pounds.
b. Intercept: $a = -238$
A college student 0 inches tall is predicted to weigh -238 pounds. This makes no sense because a height of 0 inches lies far outside of the normal range of this variable—no college student is 0 inches tall.
5. a. Slope: $b = .99$
For every additional smoking section passenger, predicted nicotine concentration increases by .99 $\mu g/m^3$ of air.

b. Intercept: $a = 9.07$
Even if there are no passengers in the smoking section, the nicotine concentration is predicted to be 9.07 $\mu g/m^3$.

c.

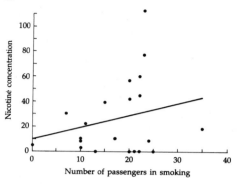

6. a. Slope: $b = .62$. For every 1-point increase in the reindeer bone prevalence index, the predicted percentage of reindeer art increases by .62 percentage point. Intercept: $a = -10.8$. A region where no reindeer bones were found would be predicted to have -10.8% reindeer art—an impossible figure.

b.

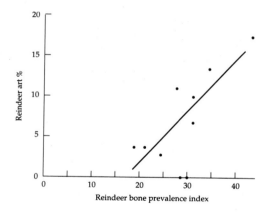

7. Mean reindeer bone prevalence: $\overline{X} = 2.91$
Mean reindeer art percentage: $\overline{Y} = 1.69$

Region	$X - \overline{X}$	$(X - \overline{X})^2$	$Y - \overline{Y}$	$(Y - \overline{Y})^2$	$(X - \overline{X})(Y - \overline{Y})$
East	4.35	18.92	1.92	3.69	8.35
Central	.16	.03	−1.02	1.04	−.16
West	−2.60	6.76	.78	.61	−2.03
Far West	−1.90	3.61	−1.69	2.86	3.21
Sums:	≈0	29.32	≈0	8.20	9.37
		$TSS_X = 29.32$		$TSS_Y = 8.20$	$SCP = 9.37$

a. Slope: $b = SCP/TSS_X = 9.37/29.32 = .32$
b. Intercept: $a = \bar{Y} - b\bar{X} = 1.69 - .32(2.91) = .76$
c. Correlation: $r = SCP/(\sqrt{TSS_X}\sqrt{TSS_Y}) = 9.37/(\sqrt{29.32}\sqrt{8.2}) = .60$
d. $\hat{Y} = .76 + .32(X)$
For every 1-point increase in reindeer bone prevalence, reindeer cave art increased by .76 percentage point. A region in which no reindeer bones were found would be predicted to have less than 1% ($a = .76\%$) reindeer art.

8. $\bar{X} = 21.69$ $\bar{Y} = 10.05$ $TSS_X = 88.25$ $TSS_Y = 6.85$ $SCP = 13.77$
a. Slope: $b = SCP/TSS_X = 13.77/88.25 = .16$
b. Intercept: $a = \bar{Y} - b\bar{X} = 10.05 - .16(21.69) = 6.58$
c. Correlation: $r = SCP/(\sqrt{TSS_X}\sqrt{TSS_Y}) = 13.77/(\sqrt{88.25}\sqrt{6.85}) = .56$
d. Equation: $\hat{Y} = 6.58 + .16(X)$
For every 1-point increase in ibex bone prevalence, ibex cave art increased by .16 percentage point. A region in which no ibex bones were found would be predicted to have 6.58% ibex art.

9. a. If $X_i = 0$, $\hat{Y}_i = 3.04 - .064(0) = 3.04$
b. If $X_i = 10$, $\hat{Y}_i = 3.04 - .064(10) = 2.4$

10. a. Regression of reindeer art on reindeer bones:

Region	X	Y	$\hat{Y} = .76 + .32(X)$	$e = Y - \hat{Y}$
East	7.26	3.61	3.08	.53
Central	3.07	.67	1.74	−1.07
West	.31	2.47	.86	1.61
F. West	1.01	0	1.08	−1.08

b. The largest residual is in the West. Few reindeer bones were found there ($X_3 = .31$), so we would predict little reindeer art ($\hat{Y}_3 = .86$). Actually this region had the second highest percentage of reindeer art ($Y_3 = 2.47$), so our prediction is too low by 1.61 percentage points ($e_3 = +1.61$).

c.

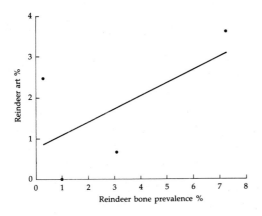

11. $ESS = \Sigma(Y_i - \bar{Y})^2$
$= (3.08 - 1.69)^2 + (1.74 - 1.69)^2 + (.86 - 1.69)^2 + (1.08 - 1.69)^2$
$= 3.00$
$RSS = \Sigma e_i^2$
$= (.53)^2 + (-1.07)^2 + (1.61)^2 + (-1.08)^2$
$= 5.18$

12. Regression of ibex art on ibex bones:

Region	X	Y	$\hat{Y} = 6.66 + .16(X)$	$e = Y - \hat{Y}$
East	23.11	8.43	10.36	−1.93
Central	13.67	9.11	8.85	.26
West	24.73	11.11	10.62	.49
F. West	25.25	11.54	10.70	.84

The largest residual is in the East. Many ibex bones were found there ($X_1 = 23.11$), so we would predict much ibex art ($\hat{Y}_1 = 10.36$). Actually this region had the lowest percentage of ibex art ($Y_1 = 8.43$), so our prediction is too high by 1.93 percentage points ($e_1 = -1.93$).

13. a. $R^2 = ESS/TSS_Y = 4.302/13.652 = .315$; 31.5% of the variation in number of booster rocket field joint damage incidents on these 23 shuttle flights is explained by the joint temperature at launch.

b. $R^2 = ESS/TSS_Y = 50,026/92,104 = .54$; 54% of the variation in students' weights is explained by their heights.

c. $R^2 = ESS/TSS_Y = 865.5/17,867.6 = .0484$; only 4.8% of the variation in nicotine concentration on these flights is explained by the number of passengers in the smoking section.

14. a. $b = SCP/TSS_X = 8,366.75/152.875 = 54.729$; for every 1-inch increase in height, the predicted number of rebounds increases by 54.729.

b. $a = \bar{Y} - b\bar{X} = 450.75 - 54.729(80.875) = -3,975.46$; a basketball player 0 inches tall is predicted to get $-3,975.46$ rebounds. Obviously this is unrealistic, an extreme example of prediction outside the range of X values in the data.

c. $R^2 = ESS/TSS_Y = 457,907/645,209.5 = .71$; 71% of the variation in rebounds is explained by height.

d. $1 - R^2 = 1 - .71 = .29$; 29% of the variation in rebounds is not explained by height.

15. $\hat{Y} = -3,975.46 + 54.729(X)$.

Player	Height	Rebounds	Predicted	Residual
Willis	84	704	621.78	82.22
Bird	81	805	457.59	347.41
Walton	83	544	567.05	−23.05
Leavell	73	67	19.76	47.24
Lee	82	351	512.32	−161.32
Brewer	76	53	183.95	−130.95
Sampson	88	781	840.69	−59.69
Reid	80	301	402.86	−101.86

The largest residual belongs to Larry Bird. He is "only" 6'9" tall, so would be predicted to pull in $\hat{Y}_2 = 457.59$ rebounds this season. He actually pulled in $Y_2 = 805$ rebounds, making our prediction $e_2 = 347.41$ rebounds too low. Bird is a much better rebounder than would be expected on the basis of his height.

16. Some b and RSS values from these data are:

b	RSS
30.000	280,816
40.000	220,493
50.000	190,746
54.000	187,408
54.729	**187,327**
55.000	187,338
60.000	191,573
70.000	222,976
80.000	284,953

As these examples show, RSS is lowest when b equals the least squares value, $b = 54.729$. Any other b value leads to a higher RSS; the farther it is from the least squares b, the higher the RSS.

17. a. There is a weak positive correlation between SAT and GPA; students with higher SAT scores tend to have higher GPAs. Only about 3% of the variance in GPA ($R^2 = .029$) is explained by SAT.

  b. The correlation between weight and GPA is very weak and negative; heavier students tend to have lower GPAs. Weight explains only 1% ($R^2 = .01$) of the variance in GPA.

  c. There is a moderate positive correlation between serum chlorpropamide level and serum insulin level among these chlorpropamide-overdose patients: Higher insulin levels tend to occur with higher chlorpropamide levels. Chlorpropamide level explains about 20% ($R^2 = .2025$) of the variance of serum insulin level.

  d. The correlation between number of safety-related standby and operation failures is positive and moderate to strong. Plants with more standby failures tend also to have more operation failures. Number of standby failures explains about 46% ($R^2 = .4624$) of the variance of operation failures.

  e. There is a weak to moderate negative correlation between attention/concentration ability and spatial ability. Boys with higher attention/concentration ability tended to have lower spatial ability. Attention/concentration ability explains about 12% ($R^2 = .1156$) of the variance of spatial ability.

18. $r = SCP/(\sqrt{TSS_X}\sqrt{TSS_Y}) = 8,366.75/(\sqrt{152.875}\sqrt{645,209.5}) = .84$; $R^2 = r^2 = .84^2 = .71$; or, $R^2 = ESS/TSS_Y = 457,907/645,209.5 = .71$

19. The correlation should be negative. The higher the percentage of gross sales a lottery pays out in prizes, the lower the percentage it can return to the state.

20. a. $\hat{Y} = -222.55 + 34.09(X)$

  For every 1-bed increase, the predicted number of patient admissions increases by 34.09. A hospital with 0 beds is predicted to admit -222.55 patients.

  b. $R^2 = .97$

  Number of beds explains 97% of the variance in number of patient admissions.

21. Predicted hospital admissions:

Hospital	Beds (X)	Predicted Admissions $\hat{Y} = -222.55 + 34.09(X)$
7. Fairlawn	104	3,322.81
9. Anna Jaques	156	5,095.49
10. Milton	161	5,265.94
13. Mercy	311	10,379.44

Complete data with missing values replaced:

Hospital	Admissions	Doctors accused
1. Ludlow	2,041	10
2. Brookline	2,126	20
3. Amesbury	2,364	14
4. N.S. Children's	2,398	11
5. Southwood	2,425	12
6. J. B. Thomas	2,857	16
7. Fairlawn	3,322.81	15
8. Sancta Maria	3,916	18
9. Anna Jaques	5,095.49	22
10. Milton	5,265.94	24
11. Cambridge	5,527	27
12. Choate Symmes	8,560	43
13. Mercy	10,379.44	49
14. Newton Wellesley	12,129	64

22. $\hat{Y} = .93 + .005(X)$ $R^2 = .96$

The missing value replacement caused little change in the regression results.

23. The largest negative residual is now that of Anna Jaques hospital. We estimate Anna Jaques admitted 5,095.49 patients, which leads to the prediction that about 26 physicians would be sued, disciplined, or charged. (The exact predicted value is very sensitive to how we round off the slope coefficient b). This gives a large negative residual of about -4, indicating that at Anna Jaques hospital about 4 fewer doctors were accused than we would predict based on the hospital's size.

24. $b = SCP/TSS_X = 5,110.49/.5629 = 9,078.86$
$a = \overline{Y} - b\overline{X} = 19,991.5 - 9,078.86(12.47) = -93,221.9$
$\hat{Y} = -93,221.9 + 9,078.86(X)$

25. $r = SCP/(\sqrt{TSS_X}\sqrt{TSS_Y}) = 5,110.49/(\sqrt{.5629}\sqrt{197,774,971.6}) = .484$
$R^2 = r^2 = .484^2 = .235$
There is a moderate positive relationship between both variables. States with higher median education levels tend also to have higher average teacher salaries. Median education levels explain 23.5% of the variation in average teacher salaries.

26. In Vermont, the average teacher salary is relatively low: $Y_{20} = 16,299$. Based on Vermont's above-average median education level ($X_{20} = 12.6$), we would predict a much higher average salary, $\hat{Y}_{20} = 21,171.7$. Vermont thus has the largest negative residual among these 21 states, $e_{20} = Y_{20} - \hat{Y}_{20} = -4,872.7$.

In New York, the average teacher salary is relatively high: $Y_{13} = 25,000$. Based on a middling education level ($X_{13} = 12.5$) we would predict lower teacher salaries in New York: $\hat{Y}_{13} = 20,263.85$. New York thus has the largest positive residual, $e_{13} = 4,736.15$.

27. The equation suggests that for every 1-gallon increase in per capita alcohol sales, predicted marriage rates rise by 27.4 marriages per 1,000 population. Alcohol sales explain 81% of the variance in marriage rates.

28.

a. In 12 states there is little relationship. Nevada is a severe high outlier on both marriage rate and alcohol sales, and it obviously exerts great influence on the regression line—essentially "creating" a marriage–alcohol relationship.

b. The most sensible corrective action would be to delete Nevada from the analysis, since its high marriage and alcohol rates largely reflect the behavior of tourists.

c. If Nevada is deleted, we could expect both the slope and the coefficient of determination to become much closer to 0. In fact this does happen: Based on $n = 12$ states, the slope becomes $b = .34$ instead of 27.4, and the coefficient of determination is $R^2 = .01$ instead of .81.

29. $\hat{Y} = -1,255,621 + 639.445(X)$
$Y_{2000} = -1,255,621 + 639.445(2000) = 23,269$
This prediction will be accurate only if present trends continue. In reaction to scientific warnings, chlorofluorocarbon release may be curtailed over the next decade, so that the actual cumulative release will be below the trend line prediction. Then again, warnings might not be enough, and chlorofluorocarbon release might actually rise.

30.

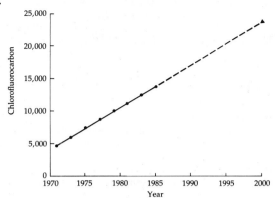

The line fits the scatter of dots almost perfectly, so r and R^2 will be very close to 1.0. In fact, they are $r = .9993$ and $R^2 = .9987$.

CHAPTER 15

1. a. $s_e = \sqrt{RSS/(n - K)} = \sqrt{381.18/8} = 6.903$
   b. $df_R = n - K = 10 - 2 = 8$
   c. $SE_b = s_e/\sqrt{TSS_X} = 6.903/\sqrt{38.5} = 1.113$
   d. $SE_a = s_e\sqrt{(1/n) + (\bar{X}^2/TSS_X)} = 6.903\sqrt{(1/10) + (1.5^2/38.5)} = 2.748$
   e. $SE_{\hat{\mu}Y|X_0} = s_e\sqrt{(1/n) + (X_0 - \bar{X})^2/TSS_X} = 6.903\sqrt{(1/10) + (5 - 1.5)^2/38.5} = 4.464$
   f. $SE_{\hat{Y}_0} = s_e\sqrt{(1/n) + (X_0 - \bar{X})^2/TSS_X + 1} = 6.903\sqrt{(1/10) + (5 - 1.5)^2/38.5 + 1} = 8.221$

2. If the $s_e = 7$, the points would be more scattered, with little trend for a regression line to fit. But if $s_e = 1$, the points would be less scattered around the regression line; the fit between line and data would be much closer.

3. Each case added to the sample increases the total sum of squares, TSS_X, and therefore *decreases* the estimated standard error of b, SE_b. A lower standard error means less sample-to-sample variation in b. The b value obtained from any one sample is therefore more likely to be near β, if it is based on a large sample rather than a small one.

4. a. $\hat{Y} = 398.8 - .00565(X)$
   b. $RSS = 4,598.5$
   c. $s_e = 39.15$
   d. $SE_b = .034$

5. $H_0: \beta \leq 0;$ $H_1: \beta > 0$ $df_R = n - 2 = 87$
   $t = (b - \beta_0)/SE_b = (.0025 - 0)/.0007 = 3.57$
   $P < .0005;$ reject H_0
   There is a significant positive relationship between Mathematics majors' GPAs and VSAT scores.

6. $H_0: \beta \leq 0;$ $H_1: \beta > 0$ $df_R = n - 2 = 434$
   $t = (b - \beta_0)/SE_b = (.001 - 0)/.0003 = 3.33$
   $P < .0005;$ reject H_0
   There is a significant positive relationship between English majors' GPAs and MSAT scores.

7. $H_0: \beta = 0;$ $H_1: \beta \neq 0$ $df_R = n - 2 = 3$
   $t = (b - \beta_0)/SE_b = (-.00565 - 0)/.034 = -.166$
   $P > .50;$ do not reject H_0
   The relationship between speed and price is not statistically significant.

8. $H_0: \beta \leq .0005;$ $H_1: \beta > .0005$ $df_R = n - 2 = 434$
   $t = (b - \beta_0)/SE_b = (.001 - .0005)/.0003 = 1.67$
   $.025 < P < .05;$ do not reject H_0 at $\alpha = .01$

Mean English major GPA does not rise by significantly more than .0005 point for every one additional MSAT point.

9. a. $df_R = 31$, $t = 1.697$

$b \pm t(SE_b) = 37.992 \pm 1.697(13.987) = 37.992 \pm 23.736 \rightarrow 14.256 \leq \beta \leq 61.728$

We are 90% confident that in the population of all city households, mean summer 1980 water use rose between 14.256 and 61.728 ft³ per $1,000 increase in income.

b. $df_R = 31$, $t = 2.75$

$b \pm t(SE_b) = 37.992 \pm 2.75(13.987) = 37.992 \pm 38.464 \rightarrow -0.472 \leq \beta \leq 76.456$

We are 99% confident that in the population of all city households, mean summer 1980 water use changed between -0.472 and $+76.456$ ft³ per $1,000 increase in income.

The 90% confidence interval is the narrowest, followed by the 95% interval. To be 99% confident, we need a relatively wide interval.

10. $df_R = 494$, $t = 1.96$

$b \pm t(SE_b) = 45.61 \pm 1.96(5.68) = 45.61 \pm 11.13 \rightarrow 34.48 \leq \beta \leq 56.74$

The $n = 496$ interval is much narrower than the $n = 33$ interval. We can be more precise in estimating population parameters from a large sample than we can from a small sample.

11. a. $df_R = 10$, $t = 1.372$

$-1.0326 \pm 1.372(.365) \rightarrow -1.533 \leq \beta \leq -.532$

b. $t = 2.228$

$-1.0326 \pm 2.228(.365) \rightarrow -1.846 \leq \beta \leq -.219$

c. $t = 4.537$

$-1.0326 \pm 4.537(.365) \rightarrow -2.689 \leq \beta \leq .623$

12. $\hat{Y} = -381.66 + .7285(X)$; $s_e = 212.58$ $SE_b = .0965$

$b \pm t(SE_b) = .7285 \pm 2.447(.0965) \rightarrow .4924 \leq \beta \leq .9646$

We are 95% confident that in the population of 1986–1987 season NBA players, the mean number of season points rose between .4924 and .9646 with each additional minute played.

13. a. Since the 95% confidence interval does not include zero, we can reject H_0: $\beta = 0$ at the $\alpha = .05$ level (two-tailed). The relationship between points and minutes is statistically significant.

b. $r = .95$, $R^2 = .90$

This is a very strong positive relationship. Minutes played explain over 90% of the variance of season points scored.

14. a. $\hat{Y} = 1,665.97 + 37.992(X) = 1,665.97 + 37.992(19) = 2,387.82$ ft³

b. $-762.22 \leq Y_0 \leq 5,501.86$

Negative household water use is impossible, however, so it makes sense to give 0 as the lower bound of this interval.

c. The same as in part a: 2,387.82

d. $1,810.23 \leq \mu_{Y|X_0} \leq 2,965.41$

15. The interval will be narrowest at $X_0 = \bar{X} = 26.82$. At this point the interval is $2,152.24 \leq \mu_{Y|X_0} \leq 3,217.6$, and the width is 1,065.36.

16. a. $\hat{Y} = 1.3 + .005(X) = 1.3 + .005(4,000) = 21.5$

b. $df_R = 8$, $t = 3.355$, $s_e = 3.4108$, $SE_{\hat{Y}_0} = 3.5803$

(Your exact results may differ slightly, depending on rounding off and whether you used summaries from Table 14.9 or calculated directly.)

$\hat{Y}_0 \pm t(SE_{\hat{Y}_0}) = 21.5 \pm 3.355(3.5803) \rightarrow 9.31 \leq Y_0 \leq 33.34$

We are 99% confident that an individual hospital with 4,000 patient admissions would have between 9.31 and 33.34 physicians sued, disciplined, or charged.

17. $F = (ESS/df_E)/(RSS/df_R) = (7,757,530/1)/(38,411,561/31) = 6.26$

$P < .05$; reject H_0

The relationship between income and water savings is statistically significant.

18. $F = (ESS/df_E)/(RSS/df_R) = (1,255,761/1)/(44,913,330/31) = .87$

$P > .25$; do not reject H_0

The relationship between education and water savings is not statistically significant.

19. a. $R^2 = \text{ESS/TSS}_Y = 7{,}757{,}530/46{,}169{,}091 = .168$

16.8% of the variance in water savings is explained by income. The square root of R^2 is .41. Since the slope has a negative sign, the correlation must also be negative: $r = -.41$. This relationship is of weak to moderate strength.

b. $R^2 = \text{ESS/TSS}_Y = 1{,}255{,}761/46{,}169{,}091 = .027$

Only 2.7% of the variance in water savings is explained by education. The square root of R^2 is .165. The correlation takes the same sign as the slope: $r = -.165$, indicating a weak negative relationship.

21. There is a weak suggestion of a U-shaped pattern in Figure 15.6. This suggests that a curve might fit the data better than a straight line (Figure 15.1) does.

22. The residuals distribution contains a severe high outlier—the Dordogne River region.

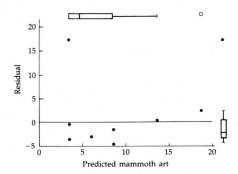

23. $\hat{Y} = -2.37 + .87(X)$

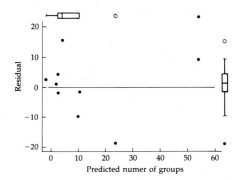

The residuals fan out left to right (heteroscedasticity). There are also two high outliers, one of them severe, in the $\hat{Y}$ distribution.

24.

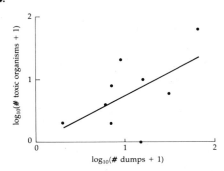

The outliers are pulled in and the heteroscedasticity eliminated in the log–log plot.

25.

0	0113579
1	
2	0
3	
4	
5	
6	3

Stem–and–leaf display of number of anti–toxic waste groups.

Stems digits 10's, leaves digits 1's.
6 | 3 means 63 groups.

The distribution is positively skewed, so a transformation such as $\log(Y)$ is appropriate. A Y^2 transformation would be appropriate if the distribution were *negatively* skewed.

The logarithm of 0 is undefined. Adding 1 to each value raises the lowest value to 1; the logarithm of 1 is 0.

26. $\log(\hat{Y} + 1) = .004 + .74\,[\log(X + 1)]$

Intercept: In a state with no toxic waste dump sites, we predict that the logarithm of the number of organizations (plus one) will be .004. *Slope:* For every factor-of-ten (or 1-logarithm) increase in the number of toxic waste sites (plus one), the log of the number of organizations (plus one) will increase by .74.

27. $s_e = .486$, $SE_b = .39$, $t = 1.90$, $P < .05$, reject H_0.

There is a significant positive relationship.

28.

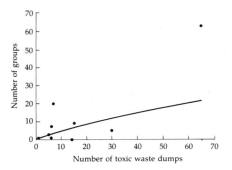

State	Dumps: X	Log(X + 1)	Groups: Y	Log($\hat{Y}$ + 1)	$\hat{Y}$
Colorado	15	1.2041	9	.8965	6.8791
Delaware	14	1.1761	0	.8757	6.5110
Hawaii	6	.8451	1	.6303	3.2687
New York	65	1.8195	63	1.3528	21.5295
Oklahoma	5	.7782	3	.5807	2.8077
S. Dakota	1	.3010	1	.2269	.6863
Tennessee	8	.9542	20	.7112	4.1430
W. Virginia	6	.8451	7	.6303	3.2687
Wisconsin	30	1.4914	5	1.1094	11.8658

29.

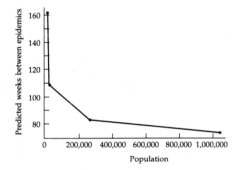

Population: X	Mean Weeks Between Epidemics: $-(\hat{Y}^{-1}) = .00245 - .00268[log_{10}(X)]$
1,700	161.09
22,000	108.84
269,000	82.63
1,046,000	73.09

Predicted weeks between epidemics drops off steeply as population goes from 1,700 to 22,000, then less steeply as population continues to increase.

30. a. Several explanations are possible. Wealthier people may tend to care more about green lawns and shiny cars; they are more likely to have swimming pools and labor–saving, water–using appliances; perhaps they even shower more often.
  b. Possibly households with higher incomes are households where more people live; more people use more water.
31. a. Poor people may find life more aggravating and crime more attractive, either of which could lead to violence.
  b. Both median incomes and homicide rates could reflect differences among the cities with respect to culture, demography, unemployment, or other variables.

CHAPTER 16

1. a. $\hat{Y} = a + b_1X_1 + b_2X_2 + b_3X_3 + b_4X_4 + b_5X_5$
  b. a is the predicted value of Y when $X_1 = X_2 = X_3 = X_4 = X_5 = 0$.

c. b_4 is the change in predicted Y per 1-unit increase in X_4, if X_1, X_2, X_3, and X_5 remain constant.

d. R^2 gives the proportion of the variance of Y explained by X_1, X_2, X_3, X_4, and X_5 together.

2. a. a: The predicted income for a worker with no experience and no education.

b. b_1: The change in predicted income for each one-year increase in work experience, if education stays the same.

c. b_2: The change in predicted income for each one-year increase in education, if work experience stays the same.

d. R^2: The proportion of the variance of income explained by work experience and education.

3. $\hat{Y} = 69.6 + 57.2(X_1) + 2.8(X_2) + 2.3(X_3) + 33.8(X_4) - 9.4(X_5) + .2(X_6) + 5.4(X_7)$

SE_b:	4.8	5.1	0.7	17.4	2.5	2.4	1.5
t:	11.9	.5	3.3	1.9	−3.8	.1	3.6
$P < .05$?	Yes	No	Yes	No	Yes	No	Yes

Sulfate concentration, percentage with ≥4 years of college, percentage nonwhite, and percentage ≥65 each has a significant partial effect, after the other six variables in this equation are controlled for.

4. $F = (ESS/df_E)/(RSS/df_R) = (1,547,059/7)/(153,328/90) = 129.7$

$P < .001$; we can reject the hypothesis that all seven regression coefficients are 0 in the population.

5. a. $R^2 = ESS/TSS_Y = 1,547,059/(1,547,059 + 153,328) = .91$

b. $R_a^2 = R^2 - (df_E/df_R) (1 - R^2) = .91 - (7/90) (1 - .91) = .903$

c. $s_e = \sqrt{RSS/df_R} = \sqrt{153,328/90} = 41.3$

13. $\hat{Y} = 24.84 + .51(X_1) + .01(X_2)$

$a = 24.84$: In a class with no A's and B's and no students, we predict 24.84% high evaluations.

$b_1 = .51$: The predicted percentage of high evaluations increases by .51 point with each 1-point increase in the percentage of A's and B's, if class size stays the same.

$b_2 = .01$: The predicted percentage of high evaluations increases by .01 point with each one-student increase in class size, if the percentage of A's and B's stays the same.

$R_a^2 = .09$: Adjusted for degrees of freedom, grades and class size explain about 9% of the variance in teaching evaluations.

14. a. Grades have a statistically significant ($P < .05$) positive effect on teaching evaluations for both types of teacher. This effect is weaker in classes taught by full-time faculty, however ($b_1 = .51$ vs. $b_1 = .73$).

b. Class size has significant negative effect ($b_2 = -.30$) on teaching evaluations for part-time instructors and teaching assistants. For classes taught by full-time faculty the effect is weak, positive ($b_2 = .01$), and not significant.

c. Grades and class size explain about 16% of the variance of teaching evaluations in classes taught by part-time instructors and teaching assistants, as compared with only 9% in classes taught by full-time faculty.

15. a. The predicted death rate increases by 57.2 deaths per 100,000 people per year if there is a 1-point increase in the percentage of population ≥65 years, while the median age, percentage nonwhite, population density, percentage of college graduates, and mean sulfate concentration stay the same.

b. The predicted death rate increases by 5.4 deaths per 100,000 people per year if there is a 1-microgram per cubic meter increase in the mean sulfate concentration, while the percentage ≥65 years, median age, percentage nonwhite, population density, and percentage of college graduates stay the same.

16. a. $r = \sqrt{R^2} = \sqrt{.0994} = .315$ (since b_1 is positive, the correlation must also be positive).

b. $b_1^* = b_1 (s_1/s_Y) = .6690493(11.83/25.11) = .315$

c. As expected, $b_1^* = r$.

17. $\hat{Y}^* = .86(X_1^*) + .04(X_2^*) + .17(X_3^*) + .1(X_4^*) - .2(X_5^*) + .004(X_6^*) + .14(X_7^*)$

The percentage of the population ≥65 years has by far the strongest effect on the death rate: $b_1^* = .86$.

18. $b_1^* = .86$: A 1-standard deviation increase in the percentage ≥65 leads to a .86 standard deviation increase in the predicted mortality rate, if the other six X variables do not change.

$b_7^* = .14$: A 1-standard deviation increase in the mean sulfate concentration leads to a .14 standard deviation increase in the predicted mortality rate, if the other six X variables do not change.

19. a. The overall equation is

$$\hat{Y} = .2606103 + .0501299(\text{age}) + .0010801(\text{VSAT}) + .0012301(\text{MSAT}) - .0400202(\text{sex}) + .0004284(\text{sex})(\text{VSAT})$$
$$+ .0000971(\text{sex})(\text{MSAT})$$

For men (sex = 0) this reduces to

$$\hat{Y} = .2606103 + .0501299(\text{age}) + .0010801(\text{VSAT}) + .0012301(\text{MSAT}) - .0400202(0) + .0004284(0)(\text{VSAT})$$
$$+ .0000971(0)(\text{MSAT})$$
$$= .2606103 + .0501299(\text{age}) + .0010801(\text{VSAT}) + .0012301(\text{MSAT})$$

For women (sex = 1) this reduces to

$$\hat{Y} = .2606103 + .0501299(\text{age}) + .0010801(\text{VSAT}) + .0012301(\text{MSAT}) - .0400202(1) + .0004284(1)(\text{VSAT})$$
$$+ .0000971(1)(\text{MSAT})$$
$$= (.2606103 - .0400202) + .0501299(\text{age}) + (.0010801 + .0004284)(\text{VSAT})$$
$$+ (.0012301 + .0000971)(\text{MSAT})$$
$$= .2205901 + .0501299(\text{age}) + .0015085(\text{VSAT}) + .0013272(\text{MSAT})$$

b. The interaction between sex and verbal SAT (sexVSAT) is significant at $P < .01$. Other significant predictors are age, verbal SAT, and math SAT.

c. Verbal and math SATs are statistically significant predictors of college GPA. Older students also have significantly higher grades. The predictive effect of verbal SAT is significantly stronger for women than for men. All of these variables together explain less than 16% of the variance in GPA, however.

20. a. $R^2 = 1$, so 100% of the variance in reindeer art prevalence is predicted by reindeer bone prevalence, ibex bone prevalence, and ibex art prevalence.

b. East region: $Y_1 = 3.61$

$$\hat{Y}_1 = 35.57547 - 1.34609(7.26) + .6006598(23.11) - 4.279253(8.43) = 3.61$$

Central region: $Y_2 = .67$

$$\hat{Y}_2 = 35.57547 - 1.34609(3.07) + .6006598(13.67) - 4.279253(9.11) = .67$$

West region: $Y_3 = 2.47$

$$\hat{Y}_3 = 35.57547 - 1.34609(.31) + .6006598(24.73) - 4.279253(11.11) = 2.47$$

Far west region: $Y_4 = 0$

$$\hat{Y}_4 = 35.57547 - 1.34609(1.01) + .6006598(25.25) - 4.279253(11.54) = 0$$

The predicted values equal the actual values of Y for each region.

c. With one a and three b coefficients, $K = 4$. The residual degrees of freedom are therefore $df_R = n - K = 4 - 4 = 0$. Also (see part b) the residual sum of squares is 0. As a result the program calculates the residual standard deviation, $s_e = \sqrt{\text{RSS}/df_R}$ as 0, which in turn leads to zero values for all the coefficient standard errors SE_{b_k}.

The t statistic, $t = b_k / SE_{b_k}$, therefore requires division by 0, an impossible operation. The same happens with the F statistic, $F = (\text{ESS}/df_E)/(\text{RSS}/df_R)$, and the adjusted R^2, $R_a^2 = R^2 - (df_E/df_R)(1 - R^2)$.

21. If $X_1 = 30$: $\hat{Y} = 11.56 + .73(30) - .30(X_2)$
                          $= 33.46 - .30(X_2)$ (lower line)

    If $X_1 = 90$: $\hat{Y} = 11.56 + .73(90) - .30(X_2)$
                          $= 77.26 - .30(X_2)$ (upper line)

22.

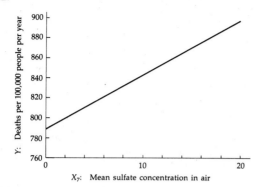

Predicted mortality rate increases from about 789 to 897 deaths per 100,000 people per year, as the sulfate concentration increases from 0 to 20, with all other variables held at their means.

23.

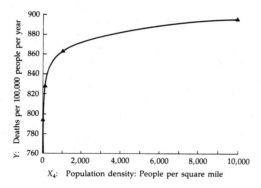

The predicted mortality rate climbs from about 794 to 896 deaths per 100,000 people per year as population density goes from 10 to 10,000 people per square mile, with other variables held constant at their means. The effect of population density is most dramatic going from 10–1,000 people per square mile; the curve levels off at higher densities.

24. $a = -1,498.4$: A player who is 0 inches tall, played no minutes, and is a center or forward is predicted to pull in $-1,498.4$ rebounds in a season.

$b_1 = .14$: The predicted number of rebounds rises .14 with each additional inch of height, if minutes and position do not change.

$b_2 = 20.9$: The predicted number of rebounds rises by 20.9 for each additional minute of average playing time, if height and position do not change.

$b_3 = -134.5$: The predicted number of rebounds is 134.5 lower for a guard than for a center or forward with the same height and average playing time.

$R_a^2 = .75$: Height, average minutes played, and position (guard/other) together explain about 75% of the variance of number of rebounds.

25. $\hat{Y} = -1,498.4 + .14(X_1) + 20.9(X_2) - 134.5(X_3)$

SE_b:	.02	7.4	61.9
t:	7.00	2.82	-2.17
$P < .05$?	Yes	Yes	Yes

All three variables have significant partial effects.

26.   For guards $(X_3 = 1)$:

$\hat{Y} = -1,498.4 + .14(X_1) + 20.9(X_2) - 134.5(1)$

$\quad = -1,632.9 + .14(X_1) + 20.9(X_2)$

For nonguards $(X_3 = 0)$:

$\hat{Y} = -1,498.4 + .14(X_1) + 20.9(X_2) - 134.5(0)$

$\hat{Y} = -1,498.4 + .14(X_1) + 20.9(X_2)$

The only difference between these two equations is that the guards' equation has a lower intercept.

Statistical Tables

TABLE A.1 *Probabilities for the Standard Normal Distribution*

Probability $P(0 < Z < z)$

z	.00	.01	.02	.03	.04	.05	.06	.07	.08	.09
0.0	.0000	.0040	.0080	.0120	.0160	.0199	.0239	.0279	.0319	.0359
0.1	.0398	.0438	.0478	.0517	.0557	.0596	.0636	.0675	.0714	.0753
0.2	.0793	.0832	.0871	.0910	.0948	.0987	.1026	.1064	.1103	.1141
0.3	.1179	.1217	.1255	.1293	.1331	.1368	.1406	.1443	.1480	.1517
0.4	.1554	.1591	.1628	.1664	.1700	.1736	.1772	.1808	.1844	.1879
0.5	.1915	.1950	.1985	.2019	.2054	.2088	.2123	.2157	.2190	.2224
0.6	.2257	.2291	.2324	.2357	.2389	.2422	.2454	.2486	.2517	.2549
0.7	.2580	.2611	.2642	.2673	.2704	.2734	.2764	.2794	.2823	.2852
0.8	.2881	.2910	.2939	.2967	.2995	.3023	.3051	.3078	.3106	.3133
0.9	.3159	.3186	.3212	.3238	.3264	.3289	.3315	.3340	.3365	.3389
1.0	.3413	.3438	.3461	.3485	.3508	.3531	.3554	.3577	.3599	.3621
1.1	.3643	.3665	.3686	.3708	.3729	.3749	.3770	.3790	.3810	.3830
1.2	.3849	.3869	.3888	.3907	.3925	.3944	.3962	.3980	.3997	.4015
1.3	.4032	.4049	.4066	.4082	.4099	.4115	.4131	.4147	.4162	.4177
1.4	.4192	.4207	.4222	.4236	.4251	.4265	.4279	.4292	.4306	.4319
1.5	.4332	.4345	.4357	.4370	.4382	.4394	.4406	.4418	.4429	.4441
1.6	.4452	.4463	.4474	.4484	.4495	.4505	.4515	.4525	.4535	.4545
1.7	.4554	.4564	.4573	.4582	.4591	.4599	.4608	.4616	.4625	.4633
1.8	.4641	.4649	.4656	.4664	.4671	.4678	.4686	.4693	.4699	.4706
1.9	.4713	.4719	.4726	.4732	.4738	.4744	.4750	.4756	.4761	.4767
2.0	.4772	.4778	.4783	.4788	.4793	.4798	.4803	.4808	.4812	.4817
2.1	.4821	.4826	.4830	.4834	.4838	.4842	.4846	.4850	.4854	.4857
2.2	.4861	.4864	.4868	.4871	.4875	.4878	.4881	.4884	.4887	.4890
2.3	.4893	.4896	.4898	.4901	.4904	.4906	.4909	.4911	.4913	.4916
2.4	.4918	.4920	.4922	.4925	.4927	.4929	.4931	.4932	.4934	.4936
2.5	.4938	.4940	.4941	.4943	.4945	.4946	.4948	.4949	.4951	.4952
2.6	.4953	.4955	.4956	.4957	.4959	.4960	.4961	.4962	.4963	.4964
2.7	.4965	.4966	.4967	.4968	.4969	.4970	.4971	.4972	.4973	.4974
2.8	.4974	.4975	.4976	.4977	.4977	.4978	.4979	.4979	.4980	.4981
2.9	.4981	.4982	.4982	.4983	.4984	.4984	.4985	.4985	.4986	.4986
3.0	.4987	.4987	.4987	.4988	.4988	.4989	.4989	.4989	.4990	.4990
3.1	.4990	.4991	.4991	.4991	.4992	.4992	.4992	.4992	.4993	.4993
3.2	.4993	.4993	.4994	.4994	.4994	.4994	.4994	.4995	.4995	.4995
3.3	.4995	.4995	.4995	.4996	.4996	.4996	.4996	.4996	.4996	.4997
3.4	.4997	.4997	.4997	.4997	.4997	.4997	.4997	.4997	.4997	.4998

z	P	z	P	z	P
3.5	.49977	4.0	.499968	4.5	.4999960
3.6	.49984	4.1	.499979	4.6	.4999979
3.7	.49989	4.2	.499987	4.7	.4999987
3.8	.499928	4.3	.4999915	4.8	.4999992
3.9	.499952	4.4	.4999946	4.9	.4999995

TABLE A.2   *Critical Values for the Standard Normal Distribution*

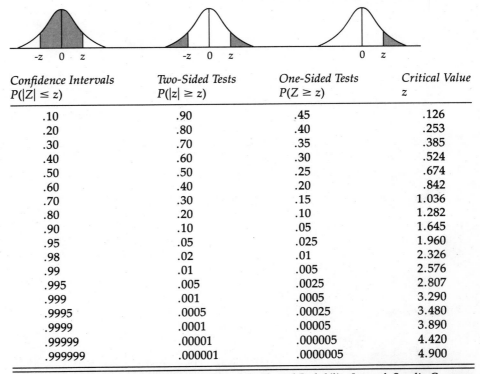

Confidence Intervals $P(\lvert Z \rvert \le z)$	Two-Sided Tests $P(\lvert z \rvert \ge z)$	One-Sided Tests $P(Z \ge z)$	Critical Value z
.10	.90	.45	.126
.20	.80	.40	.253
.30	.70	.35	.385
.40	.60	.30	.524
.50	.50	.25	.674
.60	.40	.20	.842
.70	.30	.15	1.036
.80	.20	.10	1.282
.90	.10	.05	1.645
.95	.05	.025	1.960
.98	.02	.01	2.326
.99	.01	.005	2.576
.995	.005	.0025	2.807
.999	.001	.0005	3.290
.9995	.0005	.00025	3.480
.9999	.0001	.00005	3.890
.99999	.00001	.000005	4.420
.999999	.000001	.0000005	4.900

Source: D. B. Owen and D. T. Monk, *Tables of the Normal Probability Integral,* Sandia Corporation Technical Memo 64-57-51 (March 1957).

TABLE A.3   *Critical Values for Student's t Distribution*

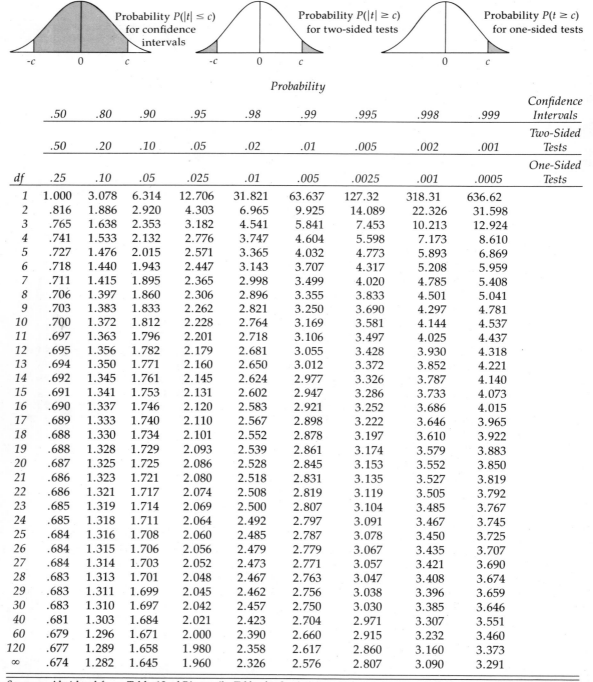

| | Probability P(\|t\| ≤ c) for confidence intervals | | Probability P(\|t\| ≥ c) for two-sided tests | | Probability P(t ≥ c) for one-sided tests |

					Probability					
	.50	.80	.90	.95	.98	.99	.995	.998	.999	*Confidence Intervals*
	.50	.20	.10	.05	.02	.01	.005	.002	.001	*Two-Sided Tests*
df	.25	.10	.05	.025	.01	.005	.0025	.001	.0005	*One-Sided Tests*
1	1.000	3.078	6.314	12.706	31.821	63.637	127.32	318.31	636.62	
2	.816	1.886	2.920	4.303	6.965	9.925	14.089	22.326	31.598	
3	.765	1.638	2.353	3.182	4.541	5.841	7.453	10.213	12.924	
4	.741	1.533	2.132	2.776	3.747	4.604	5.598	7.173	8.610	
5	.727	1.476	2.015	2.571	3.365	4.032	4.773	5.893	6.869	
6	.718	1.440	1.943	2.447	3.143	3.707	4.317	5.208	5.959	
7	.711	1.415	1.895	2.365	2.998	3.499	4.020	4.785	5.408	
8	.706	1.397	1.860	2.306	2.896	3.355	3.833	4.501	5.041	
9	.703	1.383	1.833	2.262	2.821	3.250	3.690	4.297	4.781	
10	.700	1.372	1.812	2.228	2.764	3.169	3.581	4.144	4.537	
11	.697	1.363	1.796	2.201	2.718	3.106	3.497	4.025	4.437	
12	.695	1.356	1.782	2.179	2.681	3.055	3.428	3.930	4.318	
13	.694	1.350	1.771	2.160	2.650	3.012	3.372	3.852	4.221	
14	.692	1.345	1.761	2.145	2.624	2.977	3.326	3.787	4.140	
15	.691	1.341	1.753	2.131	2.602	2.947	3.286	3.733	4.073	
16	.690	1.337	1.746	2.120	2.583	2.921	3.252	3.686	4.015	
17	.689	1.333	1.740	2.110	2.567	2.898	3.222	3.646	3.965	
18	.688	1.330	1.734	2.101	2.552	2.878	3.197	3.610	3.922	
19	.688	1.328	1.729	2.093	2.539	2.861	3.174	3.579	3.883	
20	.687	1.325	1.725	2.086	2.528	2.845	3.153	3.552	3.850	
21	.686	1.323	1.721	2.080	2.518	2.831	3.135	3.527	3.819	
22	.686	1.321	1.717	2.074	2.508	2.819	3.119	3.505	3.792	
23	.685	1.319	1.714	2.069	2.500	2.807	3.104	3.485	3.767	
24	.685	1.318	1.711	2.064	2.492	2.797	3.091	3.467	3.745	
25	.684	1.316	1.708	2.060	2.485	2.787	3.078	3.450	3.725	
26	.684	1.315	1.706	2.056	2.479	2.779	3.067	3.435	3.707	
27	.684	1.314	1.703	2.052	2.473	2.771	3.057	3.421	3.690	
28	.683	1.313	1.701	2.048	2.467	2.763	3.047	3.408	3.674	
29	.683	1.311	1.699	2.045	2.462	2.756	3.038	3.396	3.659	
30	.683	1.310	1.697	2.042	2.457	2.750	3.030	3.385	3.646	
40	.681	1.303	1.684	2.021	2.423	2.704	2.971	3.307	3.551	
60	.679	1.296	1.671	2.000	2.390	2.660	2.915	3.232	3.460	
120	.677	1.289	1.658	1.980	2.358	2.617	2.860	3.160	3.373	
∞	.674	1.282	1.645	1.960	2.326	2.576	2.807	3.090	3.291	

Source: Abridged from Table 12 of *Biometrika Tables for Statisticians*, Vol. 1, edited by E. S. Pearson and H. O. Hartley (London: Cambridge University Press, 1962).

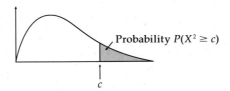

Probability $P(X^2 \geq c)$

Probability

df	.500	.250	.100	.050	.025	.010	.005	.001
1	.455	1.323	2.706	3.841	5.024	6.635	7.879	10.83
2	1.386	2.773	4.605	5.991	7.378	9.210	10.60	13.82
3	2.366	4.108	6.251	7.815	9.348	11.34	12.84	16.27
4	3.357	5.385	7.779	9.488	11.14	13.28	14.86	18.47
5	4.351	6.626	9.236	11.07	12.83	15.09	16.75	20.52
6	5.348	7.841	10.64	12.59	14.45	16.81	18.55	22.46
7	6.346	9.037	12.02	14.07	16.01	18.48	20.28	24.32
8	7.344	10.22	13.36	15.51	17.53	20.09	21.96	26.12
9	8.343	11.39	14.68	16.92	19.02	21.67	23.59	27.88
10	9.342	12.55	15.99	18.31	20.48	23.21	25.19	29.59
11	10.34	13.70	17.28	19.68	21.92	24.72	26.76	31.26
12	11.34	14.85	18.55	21.03	23.34	26.22	28.30	32.91
13	12.34	15.98	19.81	22.36	24.74	27.79	29.82	34.53
14	13.34	17.12	21.06	23.68	26.12	29.14	31.32	36.12
15	14.34	18.25	22.31	25.00	27.49	30.58	32.80	37.70
16	15.34	19.37	23.54	26.30	28.85	32.00	34.27	39.25
17	16.34	20.49	24.77	27.59	30.19	33.41	35.72	40.79
18	17.34	21.60	25.99	28.87	31.53	34.81	37.16	42.31
19	18.34	22.72	27.20	30.14	32.85	36.19	38.58	43.82
20	19.34	23.83	28.41	31.41	34.17	37.57	40.00	45.32
21	20.34	24.93	29.62	33.67	35.48	38.93	41.40	46.80
22	21.34	26.04	30.81	33.92	36.78	40.29	42.80	48.27
23	22.34	27.14	32.01	35.17	38.08	41.64	44.18	49.73
24	23.34	28.24	33.20	36.42	39.36	42.98	45.56	51.18
25	24.34	29.34	34.38	37.65	40.65	44.31	46.93	52.62
26	25.34	30.43	35.56	38.89	41.92	45.64	48.29	54.05
27	26.34	31.53	36.74	40.11	43.19	46.96	49.64	55.48
28	27.34	32.62	37.92	41.34	44.46	48.28	50.99	56.89
29	28.34	33.71	39.09	42.56	45.72	49.59	52.34	58.30
30	29.34	34.80	40.26	43.77	46.98	50.89	53.67	59.70
40	39.34	45.62	51.81	55.76	59.34	63.69	66.77	73.40
50	49.33	56.33	63.17	67.50	71.42	76.15	79.49	86.66
60	59.33	66.98	74.40	79.08	83.30	88.38	91.95	99.61
70	69.33	77.58	85.53	90.53	95.02	100.4	104.2	112.3
80	79.33	88.13	96.58	101.9	106.6	112.3	116.3	124.8
90	89.33	98.65	107.6	113.1	118.1	124.1	128.3	137.2
100	99.33	109.1	118.5	124.3	129.6	135.8	140.2	149.4

Source: Abridged from Table 8 of *Biometrika Tables for Statisticians*, Vol. 1, edited by E. S. Pearson and H. O. Hartley (London: Cambridge University Press, 1962).

TABLE A.5   *Critical Values for the F Distribution*

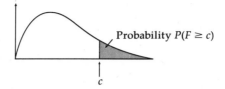

Probability $P(F \geq c)$

c

					Numerator Degrees of Freedom, df_1							
	P	1	2	3	4	5	6	8	10	20	40	∞
1	.25	5.83	7.50	8.21	8.58	8.82	8.98	9.19	9.32	9.58	9.71	9.85
	.10	39.86	49.50	53.59	55.83	57.24	58.20	59.44	60.19	61.74	62.53	63.33
	.05	161.4	199.5	215.7	224.6	230.2	234.0	238.9	241.9	248.0	251.1	254.3
2	.25	2.57	3.00	3.15	3.23	3.28	3.31	3.35	3.38	3.43	3.45	3.48
	.10	8.53	9.00	9.16	9.24	9.29	9.33	9.37	9.39	9.44	9.47	9.49
	.05	18.51	19.00	19.16	19.25	19.30	19.33	19.37	19.40	19.45	19.47	19.50
	.01	98.50	99.00	99.17	99.25	99.30	99.33	99.37	99.40	99.45	99.47	99.50
	.001	998.5	999.0	999.2	999.2	999.3	999.3	999.4	999.4	999.4	999.5	999.5
3	.25	2.02	2.28	2.36	2.39	2.41	2.42	2.44	2.44	2.46	2.47	2.47
	.10	5.54	5.46	5.39	5.34	5.31	5.28	5.25	5.23	5.18	5.16	5.13
	.05	10.13	9.55	9.28	9.12	9.01	8.94	8.85	8.79	8.66	8.59	8.53
	.01	34.12	30.82	29.46	28.71	28.24	27.91	27.49	27.23	26.69	26.41	26.13
	.001	167.0	148.5	141.1	137.1	134.6	132.8	130.6	129.2	126.4	125.0	123.5
4	.25	1.81	2.00	2.05	2.06	2.07	2.08	2.08	2.08	2.08	2.08	2.08
	.10	4.54	4.32	4.19	4.11	4.05	4.01	3.95	3.92	3.84	3.80	3.76
	.05	7.71	6.94	6.59	6.39	6.26	6.16	6.04	5.96	5.80	5.72	5.63
	.01	21.20	18.00	16.69	15.98	15.52	15.21	14.80	14.55	14.02	13.75	13.46
	.001	74.14	61.25	56.18	53.44	51.71	50.53	49.00	48.05	46.10	45.09	44.05
5	.25	1.69	1.85	1.88	1.89	1.89	1.89	1.89	1.89	1.88	1.88	1.87
	.10	4.06	3.78	3.62	3.52	3.45	3.40	3.34	3.30	3.21	3.16	3.10
	.05	6.61	5.79	5.41	5.19	5.05	4.95	4.82	4.74	4.56	4.46	4.36
	.01	16.26	13.27	12.06	11.39	10.97	10.67	10.29	10.05	9.55	9.29	9.02
	.001	47.18	37.12	33.20	31.09	29.75	28.84	27.64	26.92	25.39	24.60	23.79
6	.25	1.62	1.76	1.78	1.79	1.79	1.78	1.78	1.77	1.76	1.75	1.74
	.10	3.78	3.46	3.29	3.18	3.11	3.05	2.98	2.94	2.84	2.78	2.72
	.05	5.99	5.14	4.76	4.53	4.39	4.28	4.15	4.06	3.87	3.77	3.67
	.01	13.75	10.92	9.78	9.15	8.75	8.47	8.10	7.87	7.40	7.14	6.88
	.001	35.51	27.00	23.70	21.92	20.81	20.03	19.03	18.41	17.12	16.44	15.75
7	.25	1.57	1.70	1.72	1.72	1.71	1.71	1.70	1.69	1.67	1.66	1.65
	.10	3.59	3.26	3.07	2.96	2.88	2.83	2.75	2.70	2.59	2.54	2.47
	.05	5.59	4.74	4.35	4.12	3.97	3.87	3.73	3.64	3.44	3.34	3.23
	.01	12.25	9.55	8.45	7.85	7.46	7.19	6.84	6.62	6.16	5.91	5.65
	.001	29.25	21.69	18.77	17.19	16.21	15.52	14.63	14.08	12.93	12.33	11.70
8	.25	1.54	1.66	1.67	1.66	1.66	1.65	1.64	1.63	1.61	1.59	1.58
	.10	3.46	3.11	2.92	2.81	2.73	2.67	2.59	2.54	2.42	2.36	2.29
	.05	5.32	4.46	4.07	3.84	3.69	3.58	3.44	3.35	3.15	3.04	2.93
	.01	11.26	8.65	7.59	7.01	6.63	6.37	6.03	5.81	5.36	5.12	4.86
	.001	25.42	18.49	15.83	14.39	13.49	12.86	12.04	11.54	10.48	9.92	9.33

Denominator Degrees of Freedom, df_2

Numerator Degrees of Freedom, df_1

	P	1	2	3	4	5	6	8	10	20	40	∞
9	.25	1.51	1.62	1.63	1.63	1.62	1.61	1.60	1.59	1.56	1.54	1.53
	.10	3.36	3.01	2.81	2.69	2.61	2.55	2.47	2.42	2.30	2.23	2.16
	.05	5.12	4.26	3.86	3.63	3.48	3.37	3.23	3.14	2.94	2.83	2.71
	.01	10.56	8.02	6.99	6.42	6.06	5.80	5.47	5.26	4.81	4.57	4.31
	.001	22.86	16.39	13.90	12.56	11.71	11.13	10.37	9.89	8.90	8.37	7.81
10	.25	1.49	1.60	1.60	1.59	1.59	1.58	1.56	1.55	1.52	1.51	1.48
	.10	3.28	2.92	2.73	2.61	2.52	2.46	2.38	2.32	2.20	2.13	2.06
	.05	4.96	4.10	3.71	3.48	3.33	3.22	3.07	2.98	2.77	2.66	2.54
	.01	10.04	7.56	6.55	5.99	5.64	5.39	5.06	4.85	4.41	4.17	3.91
	.001	21.04	14.91	12.55	11.28	10.48	9.92	9.20	8.75	7.80	7.30	6.76
12	.25	1.46	1.56	1.56	1.55	1.54	1.53	1.51	1.50	1.47	1.45	1.42
	.10	3.18	2.81	2.61	2.48	2.39	2.33	2.24	2.19	2.06	1.99	1.90
	.05	4.75	3.89	3.49	3.26	3.11	3.00	2.85	2.75	2.54	2.43	2.30
	.01	9.33	6.93	5.95	5.41	5.06	4.82	4.50	4.30	3.86	3.62	3.36
	.001	18.64	12.97	10.80	9.63	8.89	8.38	7.71	7.29	6.40	5.93	5.42
14	.25	1.44	1.53	1.53	1.52	1.51	1.50	1.48	1.46	1.43	1.41	1.38
	.10	3.10	2.73	2.52	2.39	2.31	2.24	2.15	2.10	1.96	1.89	1.80
	.05	4.60	3.74	3.34	3.11	2.96	2.85	2.70	2.60	2.39	2.27	2.13
	.01	8.86	5.51	5.56	5.04	4.69	4.46	4.14	3.94	3.51	3.27	3.00
	.001	17.14	11.78	9.73	8.62	7.92	7.43	6.80	6.40	5.56	5.10	4.60
16	.25	1.42	1.51	1.51	1.50	1.48	1.48	1.46	1.45	1.40	1.37	1.34
	.10	3.05	2.67	2.46	2.33	2.24	2.18	2.09	2.03	1.89	1.81	1.72
	.05	4.49	3.63	3.24	3.01	2.85	2.74	2.59	2.49	2.28	2.15	2.01
	.01	8.53	6.23	5.29	4.77	4.44	4.20	3.89	3.69	3.26	3.02	2.75
	.001	16.12	10.97	9.00	7.94	7.27	6.81	6.19	5.81	4.99	4.54	4.06
18	.25	1.41	1.50	1.49	1.48	1.46	1.45	1.43	1.42	1.38	1.35	1.32
	.10	3.01	2.62	2.42	2.29	2.20	2.13	2.04	1.98	1.84	1.75	1.66
	.05	4.41	3.55	3.16	2.93	2.77	2.66	2.51	2.41	2.19	2.06	1.92
	.01	8.29	6.01	5.09	4.58	4.25	4.01	3.71	3.51	3.08	2.84	2.57
	.001	15.38	10.39	8.49	7.46	6.81	6.35	5.76	5.39	4.59	4.15	3.67
20	.25	1.40	1.49	1.48	1.46	1.45	1.44	1.42	1.40	1.36	1.33	1.29
	.10	2.97	2.59	2.38	2.25	2.16	2.09	2.00	1.94	1.79	1.71	1.61
	.05	4.35	3.49	3.10	2.87	2.71	2.60	2.45	2.35	2.12	1.99	1.84
	.01	8.10	5.85	4.94	4.43	4.10	3.87	3.56	3.37	2.94	2.69	2.42
	.001	14.82	9.95	8.10	7.10	6.46	6.02	5.44	5.08	4.29	3.86	3.38

Denominator Degrees of Freedom, df_2

(continued)

TABLE A.5 **(continued)**

	P	\multicolumn{11}{c}{Numerator Degrees of Freedom, df_1}
		1	2	3	4	5	6	8	10	20	40	∞
30	.25	1.38	1.45	1.44	1.42	1.41	1.39	1.37	1.35	1.30	1.27	1.23
	.10	2.88	2.49	2.28	2.14	2.05	1.98	1.88	1.82	1.67	1.57	1.46
	.05	4.17	3.32	2.92	2.69	2.53	2.42	2.27	2.16	1.93	1.79	1.62
	.01	7.56	5.39	4.51	4.02	3.70	3.47	3.17	2.98	2.55	2.30	2.01
	.001	13.29	8.77	7.05	6.12	5.53	5.12	4.58	4.24	3.49	3.07	2.59
40	.25	1.36	1.44	1.42	1.40	1.39	1.37	1.35	1.33	1.28	1.24	1.19
	.10	2.84	2.44	2.23	2.09	2.00	1.93	1.83	1.76	1.61	1.51	1.38
	.05	4.08	3.23	2.84	2.61	2.45	2.34	2.18	2.08	1.84	1.69	1.51
	.01	7.31	5.18	4.31	3.83	3.51	3.29	2.99	2.80	2.37	2.11	1.80
	.001	12.61	8.25	6.60	5.70	5.13	4.73	4.21	3.87	3.15	2.73	2.23
60	.25	1.35	1.42	1.41	1.38	1.37	1.35	1.32	1.30	1.25	1.21	1.15
	.10	2.79	2.39	2.18	2.04	1.95	1.87	1.77	1.71	1.54	1.44	1.29
	.05	4.00	3.15	2.76	2.53	2.37	2.25	2.10	1.99	1.75	1.59	1.39
	.01	7.08	4.98	4.13	3.65	3.34	3.12	2.82	2.63	2.20	1.94	1.60
	.001	11.97	7.76	6.17	5.31	4.76	4.37	3.87	3.54	2.83	2.41	1.89
120	.25	1.34	1.40	1.39	1.37	1.35	1.33	1.30	1.28	1.22	1.18	1.10
	.10	2.75	2.35	2.13	1.99	1.90	1.82	1.72	1.65	1.48	1.37	1.19
	.05	3.92	3.07	2.68	2.45	2.29	2.17	2.02	1.91	1.66	1.50	1.25
	.01	6.85	4.79	3.95	3.48	3.17	2.96	2.66	2.47	2.03	1.76	1.38
	.001	11.38	7.32	5.79	4.95	4.42	4.04	3.55	3.24	2.53	2.11	1.54
∞	.25	1.32	1.39	1.37	1.35	1.33	1.31	1.28	1.25	1.19	1.14	1.00
	.10	2.71	2.30	2.08	1.94	1.85	1.77	1.67	1.60	1.42	1.30	1.00
	.05	3.84	3.00	2.60	2.37	2.21	2.10	1.94	1.83	1.57	1.39	1.00
	.01	6.64	4.61	3.78	3.32	3.02	2.80	2.51	2.32	1.88	1.59	1.00
	.001	10.83	6.91	5.42	4.62	4.10	3.74	3.27	2.96	2.27	1.84	1.00

Denominator Degrees of Freedom, df_2

Source: Abridged from Table 18 of *Biometrika Tables for Statisticians,* Vol. 1, edited by E. S. Pearson and H. O. Hartley (London: Cambridge University Press, 1962).

References

ACHARYA, ISWAR L., et al. (1987). "Prevention of typhoid fever in Nepal with the Vi capsular polysaccharide of *Salmonella typhi.*" *New England Journal of Medicine* 317(18): 1101–1104.

ALEXANDER, JEFFREY A., and JOAN R. BLOOM (1987). "Collective bargaining in hospitals: An organizational and environmental analysis." *Journal of Health and Social Behavior* 28(1): 60–71.

BAKER, SUSAN P., R. A. WHITFIELD, and BRIAN O'NEILL (1987). "Geographic variations in mortality from motor vehicle crashes." *New England Journal of Medicine* 316(22): 1384–1387.

BARNES, GRACE M., MICHAEL P. FARRELL, and ALLEN CAIRNS (1986). "Parental socialization factors and adolescent drinking behaviors." *Journal of Marriage and the Family*, 48 (February): 27–36.

BARON, LARRY, and MURRAY A. STRAUS (1984). "Sexual stratification, pornography, and rape in the United States." In *Pornography and Sexual Aggression*, ed. Neil Malamuth and Edward Donnerstein, 185–209. San Francisco: Academic Press.

BARRY, THOMAS E., MARY C. GILLY, and LINDLEY E. DORAN (1985). "Advertising to women with different career orientations." *Journal of Advertising Research* 25(2): 26–35.

BARTLETT, MAURICE S. (1978). "Epidemics." In *Statistics: A Guide to the Unknown*, ed. Judith Tanur, Frederick Mosteller, William Kruskal, Richard Link, Richard Pieters, Gerald Rising, and E. L. Lehmann, 86–96. San Francisco: Holden–Day.

BEYER, W. NELSON, W. JAMES FLEMING, and DOUGLAS SWINEFORD (1987). "Changes in litter near an aluminum reduction plant." *Journal of Environmental Quality* 16(3): 246–250.

BHATTACHARYYA, GOURI K., and RICHARD A. JOHNSON (1977). *Statistical Concepts and Methods*. New York: Wiley.

BORDIERI, JAMES E., MAURICE L. SOLODKY, and KATHLEEN A. MIKOS (1985). "Physical attractiveness and nurses' perceptions of pediatric patients." *Nursing Research* 34(1): 24–26.

BROSS, IRWIN D., and NEAL S. BROSS (1987). "Do atomic veterans have excess cancer? New results correcting for the healthy soldier bias." *American Journal of Epidemiology* 126(6): 1042–1050.

BROWER, KIRK J., FREDERICK C. BLOW, and THOMAS P. BERESFORD (1988). "Forms of cocaine and psychiatric symptoms." *The Lancet* 1(8575/6): 50.

BROWN, L. R., W. U. CHANDLER, C. FLAVIN, C. POLLOCK, S. POSTEL, L. STARKE, and E. C. WOLF (1986). *State of the World 1986*. New York: W. W. Norton.

BURT, BARBARA J., and MAX NEIMAN (1985). "Urban legislative response to regulatory energy policy." *Policy Studies Review* 5(1): 81–88.

CASTRO, FELIPE G., EBRAHIM MADDAHIAN, MICHAEL D. NEWCOMB, and P. M. BENTLER (1987). "A multivariate model of the determinants of cigarette smoking among adolescents." *Journal of Health and Social Behavior* 28(3): 273–289.

CHAMBERS, JOHN M., WILLIAM S. CLEVELAND, BEAT KLEINER, and PAUL A. TUKEY (1983). *Graphical Methods for Data Analysis*. Belmont, Calif.: Wadsworth.

CHMIEL, JOAN S., et al. (1987). "Factors associated with prevalent immunodeficiency virus (HIV) infection in the multicenter AIDS cohort study." *American Journal of Epidemiology* 126(4): 568–576.

CLARKE, R. D. (1946). "An application of the Poisson distribution." *Journal of the Institute of Actuaries* 22: 48.

CLEVELAND, WILLIAM S. (1985). *The Elements of Graphing Data*. Monterey, Calif.: Wadsworth.

COHEN, LAWRENCE E., and KENNETH C. LAND (1987). "Age structure and crime: Symmetry versus asymmetry and the projection of crime rates through the 1990s." *American Sociological Review* 52 (April): 170–183.

COLEMAN, JAMES, THOMAS HOFFER, and SALLY KILGORE (1982). "Cognitive outcomes in public and private schools." *Sociology of Education* 55 (April/July): 65–76.

COOK, R. DENNIS, and SANFORD WEISBERG (1982). "Criticism and influence analysis in regression." In *Sociological Methodology 1982*, ed. Samuel Leinhardt, 313–361. San Francisco: Jossey–Bass.

COOPER, C. M., F. E. DENDY, J. R. McHENRY, and J. C. RITCHIE (1987). "Residual pesticide concentrations in Bear Creek, Mississippi, 1976 to 1979." *Journal of Environmental Quality* 16(1): 69–72.

COUNCIL ON ENVIRONMENTAL QUALITY (1987). *Environmental Quality 1985.* Washington, D.C.: Council on Environmental Quality.

CROWLEY, ANNE E., and SYLVIA I. ETZEL (1986). "Graduate medical education." *Journal of the American Medical Association* 256(12): 1585–1594.

DEMETRULIAS, D. M., and N. R. ROSENTHAL (1985). "Discrimination against females and minorities in microcomputer advertising." *Computers and the Social Sciences* 1(2): 91–95.

EVERITT, B. S. (1977). *The Analysis of Contingency Tables.* London: Chapman and Hall.

FATOKI, O. S. (1987). "Colorimetric determination of lead in tree leaves as indicators of atmospheric pollution." *Environment International* 13: 369–373.

FEYEREISEN, PIERRE (1982). "Temporal distribution of co-verbal hand gestures." *Ethology and Sociobiology* 3: 1–9.

FRANK, ELLEN, LINDA L. CARPENTER, and DAVID J. KUPFER (1988). "Sex differences in recurrent depression: Are there any that are significant?" *American Journal of Psychiatry* 145(1): 41–45.

GAULIN, STEVEN J. C., and LEE DOUGLAS SAILER (1985). "Are females the ecological sex?" *American Anthropologist* 87: 111–119.

GOFMAN, JOHN W. (1981). *Radiation and Human Health.* San Francisco: Sierra Club Books.

GOODMAN, LEO A., JAMES A. DAVIS, and JAY MAGIDSON (1978). *Analyzing Qualitative/Categorical Data: Log-Linear Models and Latent-Structure Analysis.* Cambridge, Mass.: Abt.

HAMILTON, ALICITA V. (1968). *A Comparative Study of the Sex–Role Concepts of Disadvantaged and Advantaged Preschool Children* (M.A. thesis). Denver: University of Denver.

HAMILTON, LAWRENCE C. (1980). "Grades, class size, and faculty status predict teaching evaluations." *Teaching Sociology* 8(1): 47–62.

HAMILTON, LAWRENCE C. (1981). "Self-reports of academic performance: Response errors are not well behaved." *Sociological Methods and Research* 10(2): 165–185.

HAMILTON, LAWRENCE C. (1982). "Response to water conservation campaigns: An exploratory look." *Evaluation Review* 6(5): 673–688.

HAMILTON, LAWRENCE C. (1983). "Saving water: A causal model of household conservation." *Sociological Perspectives* 26(4): 355–374.

HAMILTON, LAWRENCE C. (1985). "Concern about toxic wastes: Three demographic predictors." *Sociological Perspectives* 28(4): 463–486.

HERMAN, WAYNE L., JR., MICHAEL HAWKINS, and CHARLES BERRYMAN (1985). "World place name location skills of elementary pre-service teachers." *Journal of Educational Research* 79(1): 33–35.

HOAGLIN, DAVID C., FREDERICK MOSTELLER, and JOHN W. TUKEY, eds. (1983). *Understanding Robust and Exploratory Data Analysis.* New York: Wiley.

HOAGLIN, DAVID C., FREDERICK MOSTELLER, and JOHN W. TUKEY, eds. (1985). *Exploring Data Tables, Trends, and Shapes.* New York: Wiley.

HOUGHTON, A., E. W. MUNSTER, and M. V. VIOLA (1978). "Increased incidence of malignant melanoma after peaks of sunspot activity." *The Lancet*, April 8, 759–760.

HUFF, DARRELL (1954). *How to Lie with Statistics.* New York: W. W. Norton.

The Information Please Almanac (1988). Boston: Houghton Mifflin.

JACOBSEN, GRACE, JOAN E. THIELE, JOAN H. McCUNE, and LARRY D. FAR-RELL (1985). "Handwashing: Ring-wearing and number of microorganisms." *Nursing Research* 34(3): 186–188.

JOHANSON, DAVID C., and HAROLD T. PHEENY (1988). "A new look at the loss of consciousness experience within the U.S. naval forces." *Aviation, Space, and Environmental Medicine* (January): 6–9.

KALMUSS, DEBRA, and JUDITH A. SELTZER (1986). "Continuity of marital behavior in remarriage: The case of spouse abuse." *Journal of Marriage and the Family* 48: 113–120.

KATON, W., M. L. HALL, J. RUSSO, L. CORMIER, M. HOLLIFIELD, P. P. VITAL-IANO, and B. D. BEITMAN (1988). "Chest pain: Relationship of psychiatric illness to coronary arteriographic results." *The American Journal of Medicine* 84(1): 1–9.

KEANE, ANNE, JOSEPH DUCETTE, and DIANE C. ADLER (1985). "Stress in ICU and non-ICU nurses." *Nursing Research*, 34(4): 231–236.

KLONOFF, DAVID C. (1988). "Association of hyperinsulinemia with chlorpropamide toxicity." *The American Journal of Medicine* 84(January): 33–38.

KOHL, RANDALL L., JERRY L. HOMICK, NITZA CINTRON, and DICK S. CALKINS (1987). "Lack of effects of astemizole on vestibular ocular reflex, motion sickness, and cognitive performance in man." *Aviation, Space, and Environmental Medicine* (December): 1171–1174.

KOOPMANS, LAMBERT H. (1987). *Introduction to Contemporary Statistical Methods.* 2d ed. Boston: Duxbury.

KORT, FRED (1986). "Considerations for a biological basis of civil rights and liberties." *Journal of Social Biological Structures* 9: 37–51.

LAMY, PHILIP, and JACK LEVIN (1985). "Punk and middle-class values: A content analysis." *Youth and Society* 17(2): 157–170.

LARSEN, RICHARD J., and MORRIS L. MARX (1985). *An Introduction to Probability and Its Applications.* Englewood Cliffs, N.J.: Prentice–Hall.

LEAGUE OF CONSERVATION VOTERS (1985). *How Congress Voted on Energy and the Environment.* Washington, D.C.

LEVINE, ADELINE GORDON (1982). *Love Canal: Science, Politics, and People.* Lexington, Mass.: Lexington Books.

LINSKY, ARNOLD, JOHN COLBY, and MURRAY STRAUS (1985). "Social stress, normative constraints, and alcohol problems in American states." Paper presented at the annual meeting of the Society for the Study of Social Problems, Washington D.C., August 24.

McCLEARY, RICHARD, RICHARD A. HAY, JR., ERROL E. MEIDINGER, and DAVID McDOWALL (1980). *Applied Time Series Analysis for the Social Sciences.* Beverly Hills: Sage.

McCLURG, PATRICIA A., and CHRISTINE CHAILLE (1987). "Computer games: Environments for developing spatial cognition?" *Journal of Educational Computing Research* 3(1): 95–111.

McGEE, GLENN W. (1987). "Social context variables affecting the implementation of microcomputers." *Journal of Educational Computing Research* 3(2): 189–206.

MODIANOS, DOAN T., ROBERT C. SCOTT, and LARRY W. CORNWELL (1987). "Testing intrinsic random-number generators." *Byte: The Small Systems Journal* 12(1): 175–178.

MOSTELLER, FREDERICK, and JOHN W. TUKEY (1977). *Data Analysis and Regression.* Reading, Mass.: Addison–Wesley.

MURRAY, M. L., D. B. CHAMBERS, R. A. KNAPP, and S. KAPLAN (1987). "Estimation of long-term risk from Canadian uranium mill tailings." *Risk Analysis* 7(3): 287–295.

MUSSALO–RAUHAMAA, H., S. S. SALMELA, A. LEPPÄNEN, and H. PYYSALO (1986). "Cigarettes as a source of some trace and heavy metals and pesticides in man." *Archives of Environmental Health* 41(1): 49–55.

NASH, JAMES, and LAWRENCE SCHWARTZ (1987). "Computers and the writing process." *Collegiate Microcomputer* 5(1): 45–48.

NELSON, CHARLES R. (1973). *Applied Time Series Analysis for Managerial Forecasting.* San Francisco: Holden–Day.

OLDAKER, GUY B., III, and FRED C. CONRAD, JR. (1987). "Estimation of effect of environmental tobacco smoke on air quality within passenger cabins of commercial aircraft." *Environmental Science and Technology* 21(10): 994–998.

OZKAYNAK, HALUK, and GEORGE D. THURSTON (1987). "Associations between 1980 U.S. mortality rates and alternative measures of airborne particle concentration." *Risk Analysis* 7(4): 449–461.

PALM, RISA (1986). "Racial and ethnic influences in real estate practices." *The Social Science Journal* 23(1): 45–53.

PERROW, CHARLES (1984). *Normal Accidents: Living with High-Risk Technologies.* New York: Basic Books.

PIERSON, WILLIAM R., WANDA W. BRACHACZEK, ROBERT A. GORSE, JR., STEVEN M. JAPAR, and JOSEPH M. NORBECK (1987). "Acid rain and atmospheric chemistry at Allegheny Mountain." *Environmental Science and Technology* 21(7): 679–690.

PRESIDENTIAL COMMISSION ON THE SPACE SHUTTLE *CHALLENGER* ACCIDENT (1986). *Report.* Washington D.C.

REDDY, MICHAEL M., TIMOTHY D. LIEBERMANN, JAMES C. JELINSKI, and NEL CANE (1985). "Variation in pH during summer storms near the continental divide in central Colorado, U.S.A." *Arctic and Alpine Research* 17(1): 79–88.

REID, DON, and CHRIS FALKENSTEIN (1983). *Rock Climbs of Tuolumne Meadows.* Denver: Chockstone Press.

REILLY, THOMAS F., LARRY J. WHEELER, and LEONARD E. ETLINGER (1985). "Intelligence versus academic achievement: A comparison of juvenile delinquents and special education classifications." *Criminal Justice and Behavior* 12(2): 193–208.

RICE, PATRICIA C., and ANN L. PATERSON (1985). "Cave art and bones: Exploring the interrelationships." *American Anthropologist* 87: 94–100.

RICE, PATRICIA C., and ANN L. PATERSON (1986). "Validating the cave art–archeofaunal relationship in Cantabrian Spain." *American Anthropologist* 88: 658–667.

SHUMWAY, ROBERT (1988). *Applied Statistical Time Series Analysis.* Englewood Cliffs, N.J.: Prentice–Hall.

SIVARD, RUTH LEGER (1985). *World Military and Social Expenditures 1985.* Washington, D.C.: World Priorities.

SMITH, PETER K., and KATHRYN LEWIS (1985). "Rough-and-tumble play, fighting, and chasing in nursery school children." *Ethology and Sociobiology* 6: 175–181.

SPIKA, JOHN S., FRANCOIS DABIS, NANCY HARGRETT–BEAN, JOACHIM SALCEDO, SERGE VEILLARD, and PAUL A. BLAKE (1987). "Shigellosis at a Caribbean resort." *American Journal of Epidemiology* 126(6): 1173–1181.

STOTO, MICHAEL A., and JOHN D. EMERSON (1983). "Power transforms for data analysis." In *Sociological Methodology 1983–84*, ed. Samuel Leinhardt, 126–168. San Francisco: Jossey–Bass.

TAYLOR, JIM (1984). "The unemployment of university graduates." *Research in Education* 31(May): 11–24.

TREACY, PAUL (1985). *WISC-R Profiles of Truly Reading-Disabled Boys* (Ph.D. dissertation). Kensington University, California.

TUFTE, EDWARD R. (1983). *The Visual Display of Quantitative Information*. Cheshire, Conn.: Graphics Press.

TUKEY, JOHN W. (1977). *Exploratory Data Analysis*. Reading, Mass.: Addison–Wesley.

UNGERHOLM, S., and J. GUSTAVSSON (1985). "Skiing safety in children: A prospective study of downhill skiing injuries and their relation to the skier and his equipment." *International Journal of Sports Medicine* 6(6): 353–358.

UNITED STATES ATOMIC ENERGY COMMISSION (1974). *Reactor Safety Study*. Washington, D.C.: U.S. Department of Energy.

UNITED STATES BUREAU OF THE CENSUS (1983). *Statistical Abstract of the United States*. Washington, D.C.: U.S. Department of Commerce.

UNITED STATES BUREAU OF THE CENSUS (1986). *Statistical Abstract of the United States*. Washington, D.C.: U.S. Department of Commerce.

UNITED STATES BUREAU OF THE CENSUS (1987). *Statistical Abstract of the United States*. Washington, D.C.: U.S. Department of Commerce.

UNITED STATES NATIONAL PARK SERVICE (1986). *1986 National Park Statistical Abstract*. Denver: U.S. Department of Interior.

UPTON, GRAHAM J. G. (1978). *The Analysis of Cross-Tabulated Data*. New York: Wiley.

VELLEMAN, PAUL F. (1982). "Applied nonlinear smoothing." In *Sociological Methodology 1982*, ed. Samuel Leinhardt, 141–178. San Francisco: Jossey–Bass.

VELLEMAN, PAUL F., and DAVID C. HOAGLIN (1981). *Applications, Basics, and Computing of Exploratory Data Analysis*. Boston: Duxbury.

WAINER, HOWARD (1986). "Five pitfalls encountered while trying to compare states on their SAT scores." *Journal of Educational Measurement* 23(1): 69–81.

WALSH, ANTHONY (1985). "The role of the probation officer in the sentencing process." *Criminal Justice and Behavior* 12(3): 289–303.

WEISMAN, CAROL S., and MARTHA A. TEITELBAUM (1987). "The work–family role system and physician productivity." *Journal of Health and Social Behavior* 28(3): 247–257.

WORLD BANK (1987). *World Development Report 1987*. New York: Oxford University Press.

WRIGHT, JEFF R., MARK H. HOUCK, JAMES T. DIAMOND, and DEAN RANDALL (1986). "Drought contingency planning." *Civil Engineering Systems* 3 (December): 210–215.

Index

SYMBOL	MEANING
$P(A \text{ or } B)$	Probability that either A occurs, or B occurs, or both do
$P(A \text{ and } B)$	Probability that both A and B occur
$P(A\|B)$	Probability of A given B
$P(\text{not } A)$	Probability that A does not occur
PSD	Pseudo-standard deviation
q	"Power" used in ladder of powers transformations
Q_1, Q_3	Sample first and third quartiles
r	*In crosstabulation:* Row marginal *In regression*: Sample correlation coefficient
R^2	Coefficient of determination
R_a^2	Adjusted R^2; coefficient of determination adjusted for degrees of freedom
RSS	Residual sum of squares
s	Sample standard deviation
$s_X, s_Y, \ldots$	Sample standard deviation of variables X, Y, etc.
$s_1, s_2, \ldots$	Sample standard deviations in samples 1, 2, etc.
s_p	Pooled standard deviation
s_e	Sample standard deviation of regression residuals
σ	(*sigma*) Population standard deviation
$\sigma_X, \sigma_Y, \ldots$	Population standard deviation of variables X, Y, etc.
$\sigma_1, \sigma_2, \ldots$	Population standard deviation in populations 1, 2, etc.
SCP	Sum of cross products
$SE_{\overline{X}}$	Estimated standard error of mean $\overline{X}$
$\sigma_{\overline{X}}$	Standard error of mean $\overline{X}$
$SE_{\overline{Y}_2 - \overline{Y}_1}$	Estimated standard error of difference of means $\overline{Y}_2 - \overline{Y}_1$
$SE_{\hat{\pi}}$	Estimated standard error of proportion $\hat{\pi}$
SE_a	Estimated standard error of regression intercept a
SE_b	Estimated standard error of regression coefficient b
$SE_{\hat{Y}_0}$	Estimated standard error of individual predicted Y, $\hat{Y}_0$